Vita Mathematica

Volume 18

Edited by
Martin Mattmüller

More information about this series at http://www.springer.com/series/4834

Hugo Steinhaus

Mathematician for All Seasons

Recollections and Notes, Vol. 1 (1887–1945)

Translated by Abe Shenitzer

Edited by Robert G. Burns, Irena Szymaniec
and Aleksander Weron

Author
Hugo Steinhaus (1887–1972)

Translator
Abe Shenitzer
Brookline, MA, USA

Editors
Robert G. Burns
York University
Dept. Mathematics & Statistics
Toronto, ON, Canada

Irena Szymaniec
Wrocław, Poland

Aleksander Weron
The Hugo Steinhaus Center
Wrocław University of Technology
Wrocław, Poland

Vita Mathematica
ISBN 978-3-319-21983-7   ISBN 978-3-319-21984-4   (eBook)
DOI 10.1007/978-3-319-21984-4

Library of Congress Control Number: 2015954183

Springer Cham Heidelberg New York Dordrecht London
© Springer International Publishing Switzerland 2015
This work is subject to copyright. All rights are reserved by the Publisher, whether the whole or part of the material is concerned, specifically the rights of translation, reprinting, reuse of illustrations, recitation, broadcasting, reproduction on microfilms or in any other physical way, and transmission or information storage and retrieval, electronic adaptation, computer software, or by similar or dissimilar methodology now known or hereafter developed.
The use of general descriptive names, registered names, trademarks, service marks, etc. in this publication does not imply, even in the absence of a specific statement, that such names are exempt from the relevant protective laws and regulations and therefore free for general use.
The publisher, the authors and the editors are safe to assume that the advice and information in this book are believed to be true and accurate at the date of publication. Neither the publisher nor the authors or the editors give a warranty, express or implied, with respect to the material contained herein or for any errors or omissions that may have been made.

Cover credit: Photo of Hugo Steinhaus. Courtesy of Hugo Steinhaus Center Archive, Wrocław University of Technology

Printed on acid-free paper

Springer International Publishing AG Switzerland is part of Springer Science+Business Media (www.birkhauser-science.com)

Foreword to the First Polish Edition (1992)

You hold in your hands a record of the memories of Hugo Steinhaus, eminent mathematician, a founder of the Polish School of Mathematics, first-rate lecturer and writer, and one of the most formidable minds I have encountered. His steadfast gaze, wry sense of humor (winning him enemies as well as admirers), and penetrating critical and skeptical take on the world and the people in it, combined in an impression of brilliance when I, for the first time, conversed with him. I know that many others, including some of the most eminent of our day, also experienced a feeling of bedazzlement in his presence.

In my first conversation with professor Steinhaus, he attempted to explain to me, someone who never went beyond high school mathematics, what that discipline is and what his own contribution to it was. He told me then—I took notes for later perusal—the following, more or less. It is often thought that mathematics is the science of numbers; this is in fact what Courant and Robbins claim in their celebrated book *What is Mathematics?*. However, this is not correct: higher mathematics does indeed include the study of number relations but a welter of non-numerical concepts besides. The essence of mathematics is the deepest abstraction, the purest logical thought, with the mind's activity mediated by pen and paper. And there is no resorting to the senses of hearing, sight, or touch beyond this in the exercise of pure ratiocination.

Moreover, of any given piece of mathematics it can never be assumed that it will turn out to be "useful". Yet many mathematical discoveries have turned out to have amazingly effective applications—indeed, the modern world would be nothing like what it is without mathematics. For instance, there would be no rockets flying to other planets, no applications of atomic energy, no steel bridges, no Bureaux of Statistics, international communications, number-based games, radio, radar, precision bombardment, public opinion surveys, or regulation of processes of production. However, despite all this, mathematics is not at its heart an applied science: whole branches of mathematics continue to develop without there being any thought given to their applicability, or the likelihood of applications. Consider, for instance, "primes", the whole numbers not factorable as products of two smaller whole numbers. It has long been known that there are infinitely many such numbers, and among them there are "twins", such as 3 and 5, 5 and 7, 11 and 13, 17 and 19, and so on. It is probable that there are infinitely many such twins, but no one has as yet managed to prove this, despite a great many attempts, all without the slightest potential practical application in view.

The late Zygmunt Janiszewski, a brilliant mathematician, wrote: "I do mathematics in order to see how far one can get by means of pure reason."

The number of problems thought up by mathematicians but still waiting to be solved is unlimited. And among those for which solutions are found, only a few will find practical application. But it is mathematical abstraction that attracts the best minds—those capable of the purest kind of human mental activity, namely abstract thought. Hugo Steinhaus was of the opinion that the progress of mathematics is like a great march forward of humanity. But while the great mass of mankind has reached no further than the level of the cave-dweller, and a few have attained the level of the best of the middle ages, and even fewer the level of the eighteenth century, the question arises as to how many have reached the present level. He also said: "There is a continuing need to lead new generations along the thorny path which has no shortcuts. The Ancients said there is no royal road in mathematics. But the vanguard is leaving the great mass of pilgrims further and further behind, the procession is ever more strung out, and the leaders are finding themselves alone far out ahead."

* * *

However, Hugo Steinhaus's recollections are to be read not so much in order to learn any mathematics—although one can glean from them interesting facts about what mathematicians have achieved. The main reasons for reading them are as follows: First, he led an interesting life, active and varied—although this is not to say that it was an easy one since the epithet "interesting" as used of life in our part of the world has often enough been a euphemism for experiences one would not wish on anyone. Second, his great sense of humor allows him to describe his experiences in unexpected ways. Third, his vast acquaintance—people fascinated him—included many interesting, important, and highly idiosyncratic individuals. And fourth and last, he always said what he thought, even though this sometimes brought trouble on him. Since he had no definite intention of publishing his writings, it follows that he was even franker in them. This truth-telling in response to difficult questions, this reluctance to smooth edges, not shrinking from assertions that may hurt some and induce in others uneasy feelings of moral discomfiture: this is perhaps the main virtue of these notes.

The following were the chief character traits of the author of these notes: a sharp mind, a robust sense of humor, a goodly portion of shrewdness, and unusual acuteness of vision. For him, there was no spouting of slogans, popular myths, or propaganda, or resorting to comfortable beliefs. He frequently expressed himself bluntly, even violently, on many of the questions of his time—for instance, questions concerning interwar politics as it related to education (even though he, as a former Polish Legionnaire, might have been a beneficiary of them), general political problems, totalitarianism in its hitlerian and communist manifestations, and issues of anti-Semitism and Polish-Jewish relations. He said many things people did not like back then, and things they don't like today.

I believe that especially today, when our reality is so different from that of Steinhaus's time, it is well worthwhile to acquaint oneself with his spirit of contrariness and his sense of paradox, since these are ways of thinking that are today even more useful than in past times.

* * *

Steinhaus believed deeply in the potential for greatness and even perfection of the well-trained human mind. He often referred to the so-called "Ulam Principle" (named for the famous Polish mathematician Stanisław Ulam, who settled in the USA) according to which "the mathematician will do it better", meaning that if two people are given a task to carry out with which neither of them is familiar, and one of them is a mathematician, then that one will do it better. For Steinhaus, this principle extended to practically every area of life and especially to those associated with questions related to economics.

A particular oft-reiterated claim of his was that people who make decisions pertaining to large facets of public life—politics, the economy, etc.—should understand, in order to avoid mistakes and resultant damage, that there are things they don't understand but which others do. But of course such understanding is difficult to attain and remains rare.

Hugo Steinhaus represented what was best in that splendid flowering of the Polish intelligentsia of the first half of the twentieth century, without which our nation could never have survived to emerge reborn. This constituted a great impetus for good, triumphing over tanks, guns, and the secret police combined—a truly Polish strike force.

Kazimierz Dziewanowski[1]

---

[1] Kazimierz Dziewanowski (1930–1998), Polish writer, journalist, and diplomat. Polish ambassador to Washington 1990–1993.

Introduction to the English Edition

There are two well-known romantic anecdotes concerning Hugo Steinhaus. Following a period of military service in the early part of World War I, he was given a desk job in Kraków. In the summer of 1916, he went on a "random walk" from his Kraków residence at 9 Karmelicka Street to Planty Park, where he overheard the words "Lebesgue integral" spoken by one of two young men seated on a park bench—none other than the self-taught lovers of mathematics Stefan Banach and Otto Nikodým. Later Steinhaus would create, with Banach, the famous Lwów school of mathematics, one of the two prominent Polish mathematics schools—the other was in Warsaw—flourishing in Poland between the wars. According to the second anecdote, in the 1930s Steinhaus, Banach, and others used to frequent the "Scottish Café" in Lwów, where they would engage in animated mathematical discussions, using the marble tabletops to write on.[1] At some point, Banach's wife Łucja gave them a thick exercise book, and the "The Scottish Book"[2] was born, in final form a collection of mathematical problems contributed by mathematicians since become legendary, with prizes for solutions noted, and including some solutions. It was destined to have a tremendous influence on world mathematics.

In addition to "discovering" Banach and collaborating with him, Steinhaus pioneered the foundations of probability theory, anticipating Kolmogorov, and of game theory, anticipating von Neumann. He is also well known for his work on

---

[1] The mathematical activity connected with "The Scottish Café" has inspired a cycle of poems by Susan H. Case, published by Slapering Hol Press, 2002. From the review by Charles Martin: "This series of poems is loosely based upon the experiences of the mathematicians of The Scottish Café, who lived and worked in Lvov [Lwów], Poland, now [in] Ukraine. There is no theme more important for poetry to address in our time, when that life is imperiled by barbarisms from within and without. By recalling with celebratory joy the vigor, the messiness, the courage of that life as it was once lived in a terrible time by the patrons of The Scottish Café in Lvov, these poems do us a great service."

[2] Available in English as: R. Daniel Mauldin (ed.), *The Scottish Book*, Birkhäuser Boston, Boston, MA, 1981. Steinhaus contributed ten problems to *The Scottish Book*, including the last, dated May 31, 1941, just days before the Nazis occupied Lwów.

trigonometric series and his result concerning the problem of "fair division", a forerunner of the "ham sandwich theorem". These are just a few among the many notable contributions he made to a wide variety of areas of mathematics.[3] He was the "father" of several outstanding mathematicians, including, in addition to Banach, the well-known mathematicians Kac, Orlicz, and Schauder, to name but three of those he supervised. He published extensively on both pure and applied topics. He was an inspired inventor. His popularization, entitled in English *Mathematical Snapshots*, is still in print. There is also an English translation of his *One Hundred Problems in Elementary Mathematics* published by Dover.

However, although his reminiscences and diary entries contain much of direct mathematical interest or interest for the history of mathematics, and the mathematical theme recurs throughout, they are of much wider interest. Steinhaus was a man of high culture: he was well versed in science, read widely in philosophy and literature, knew Latin, German, French, and English, was a great stickler for linguistic accuracy—a disciple of Karl Kraus in this—and reveled in the vital cosmopolitan culture of Lwów, where he was professor and dean between the wars. Being also of penetrating intelligence, unusual clarity of understanding, acerbic wit, given to outspokenness, and a Polish Jew, he was well equipped to pass comment on the period he lived through (1887–1972).

Thus, we have here a historical document of unusual general appeal reporting on "interesting times" in an "interesting" part of the world—the inside story, recounted unemotionally, with flair and sometimes scathing humor, and featuring a cast of thousands. First, the halcyon pre-Great War days are chronicled: a rather idyllic, if not privileged, childhood centered on his hometown Jasło in the region of southern Poland known as Galicia, then part of the Austro-Hungarian Empire, a first-class education at the regional Gymnasium, and a brief period as a student at the University of Lwów before going off to Göttingen to do his Ph.D. under Hilbert. (Here, in addition to a fascinating description of that university town and its student culture, we get interesting sketches of many of the mathematical and scientific luminaries of those days.) Next we have a description of his role in the early part of World War I as a member of a gun-crew, trundling their artillery piece about the eastern theater of the war. This is followed by an elaboration of the interwar years—a period of Polish independence following well over a hundred years of foreign domination—which witnessed the above-mentioned blossoming of Polish mathematics of which he, at the University of Lwów, was a central figure, but also an intensifying nationalism and anti-Semitism.

There then ensue the horrors of the two occupations. Just prior to the Soviet invasion we are given a chilling account of the chaotic situation at the Hungarian border whither many Poles—especially representatives of the Polish government—flee seeking refuge in Hungary. The indecision as to what the best course of action might be in appalling circumstances and the reigning sense of helplessness in the face of impending disaster are conveyed in vivid concrete terms without recourse

---

[3] See: Hugo Steinhaus, *Selected Papers*, PWN, Warsaw 1985.

Introduction to the English Edition xi

to emotional props. After assessing the situation insofar as that were possible, the Steinhauses decide to return to Lwów, where they are greeted by the sight of Red Army soldiers already in the streets. This first, Soviet, occupation, from September 1939 to June 1941, is characterized by summary arrests and mass deportations, hallmarks of Stalinist repression, hidden behind a thin veneer of normalcy. The second occupation, this time by the Nazis, lasting from June 1941 to early 1945, is marked by a more blatant, racially motivated brutality. Following a terrifying period of evading arrest by moving from one friend's residence to another, the Steinhauses manage to find a provisional hiding place in the countryside. (This makes for especially gripping, though harrowing, reading.)

At the end of the war, following on the westward flight of the German army (and their spiteful razing of his beloved Jasło), the Steinhauses are able to emerge from their second hiding place. But then Poland is translated westwards by some hundreds of kilometers, so that Lwów becomes L'viv, a Ukrainian city, and in the west, Breslau on the Oder (completely destroyed by the war), formerly German, becomes Wrocław,[4] capital of Lower Silesia, later to become a great industrial and agricultural region of Poland. It is to this ruined city that Steinhaus eventually goes to assist in re-establishing the university and polytechnic. He helps to realize the goal of reconstituting in Wrocław what had been lost in Lwów by founding a mathematics school in Wrocław, this time of applied mathematics, and renewing the tradition of "The Scottish Book" with "The New Scottish Book".[5]

There now follows, in the form of diary entries, a long semi-tirade, laced with irony and interspersed with assessments of local and international developments, concerning the frustrations of living in a communist vassal state where distorted ideology trumps basic common sense—a Poland subjugated to and exploited by the Soviet behemoth. (Thus we have here a sort of potted history of postwar Europe and America as viewed from inside Poland.)

In the words of his former student Mark Kac, "[Hugo Steinhaus] was one of the architects of the school of mathematics which flowered miraculously in Poland between the two wars and it was he who, perhaps more than any other individual, helped to raise Polish mathematics from the ashes to which it had been reduced by the Second World War to the position of new strength and respect which it now occupies. He was a man of great culture and in the best sense of the word a product of Western Civilization."

The overall impression of Steinhaus's *Recollections and Notes* is of the compelling record of a man of intelligence and steadfast intellectual honesty, good sense and natural dignity pursuing a life of integrity and demanding scientific and

---

[4]See: N. Davies and R. Moorhouse, *Microcosm. Portrait of a Central European City*, Jonathan Cape, 2002.

[5]In fact, he was the chief organizer and first dean of the Faculty of Mathematics, Physics, and Chemistry, when, at this initial stage, the university and polytechnic in Wrocław were not yet separate institutions.

intellectual enquiry in the face of encroaching calamity and chaos brought about chiefly by human ignorance and evil.

In Wrocław, Steinhaus remains a well-known and very popular figure. In 1990, a Hugo Steinhaus Center was established, affiliated with the Wrocław Polytechnic. A "Café and Restaurant Steinhaus" was opened in 2012, and in 2013 his bust was put on display in the Wrocław Pantheon, located in the famous Wrocław City Hall.

* * *

Publication History and Acknowledgments

When Steinhaus's diary ends in 1968, he is 81 years old, and the USSR seems to be a fixture of the world's political scene. That is the year of the "Prague Spring" and widespread Polish student protests, and their brutal suppression, in the first case by Soviet tanks and in the second by police batons. Although some early portions of Steinhaus's *Recollections* were published in the Polish magazine *Znak* in 1970, full publication was at that time out of the question for reasons which a perusal of the later pages of the diary makes clear. The first complete Polish edition was brought out by the London firm Aneks in 1992, while second and third editions were published by the publishing house "Atut" in 2002 and 2010, under the auspices of the Hugo Steinhaus Center. A German translation was published in 2010.[6]

The present English translation by Abe Shenitzer was edited first by Robert G. Burns, who also added footnotes considered necessary for an Anglophone reader, and chapter headings to facilitate cross-referencing among the footnotes. Since a great many inaccuracies had inevitably crept in, it was judged essential that a Polish expert edit the English text a second time, a task fulfilled to the letter by Irena Szymaniec, who also corrected and rationalized the footnotes. Aleksander Weron, the Director of the Hugo Steinhaus Center, oversaw the whole process, providing encouragement and final authority and expertise.

We wish to thank all others who helped with the editorial process, in particular Edwin Beschler, Aleksander Garlicki, Ina Mette, Martin Muldoon, Patrick O'Keefe, Jim Tattersall, and Wojbor A. Woyczyński. Special thanks are due to Martin Mattmüller for many corrections and improvements, to Dorothy Mazlum for her great rapport in connection with the production process, and to Carolyn King, cartographer in the Geography Department of York University, for her superlative work making five of the maps.

We wish the reader of these *Recollections and Notes* much pleasure from them.

Robert G. Burns
Abe Shenitzer
Irena Szymaniec
Aleksander Weron

---

[6]Hugo Steinhaus, *Erinnerungen und Aufzeichnungen*, Neisse-Verlag, 2010.

Editors' Note on Polish Feminine Endings of Personal Names
In the original work the Polish feminine endings *-owa*, indicating a woman's married name, and *-ówna*, indicating her maiden name, are frequently used. These have been preserved in the present translation, including the index. Thus the index entry *Steinhausówna (Kottowa), Lidia (Lidka), the author's daughter* refers to a female whose maiden name is Steinhaus, married name Kott, and first name Lidia, of which Lidka is an affectionate or diminutive version. The use of these endings was not uniform in the original, nor is it in the present translation. Thus, e.g., we have Mrs. Kossak instead of Kossakowa, and occasionally there occur hybrid forms where a married woman uses her maiden name.

Contents

Part I

1. Jasło 3
2. The Gymnasium 21
3. In the Capital Lwów 43
4. Göttingen 51
5. The Return Home 91
6. The Life of a Private Scholar 105

Part II

7. In the University Town Lwów 129
8. The First Occupation 229

Interlude: Flashes of Memory 279

9. The Second Occupation 287
10. Homeless Wandering 295
11. Osiczyna 315

Interlude: Flashes of Memory 331

12. Stróże 335
13. Diary Entries 385

A Flash of Memory 390

Index of Names 467

Part I

Part 1

Chapter 1
Jasło

The town[1] was situated at the junction of three rivers, in a shallow basin surrounded by low hills. The nearest hill, lying to the north—"above the quarry"—was only a half-hour's walk away. If you made the effort you were rewarded with a view of white houses, the blue Carpathian Mountains, the silver Wisłoka,[2] and green meadows. Within a reasonable hike to the west lay the village of Podzamcze[3] and a roadside inn near the ruins of the castle of the Firlejs.[4] Some of the neighboring inns had quaint names—such as "Bumpcatcher," "The Little Mill," "The Candlestick," and "Spree".[5] The road leaving the Wisłoka at Podzamcze, leading to Kołaczyce and Tarnów, was the most important of the region—which doubtless explains why the original castle was built there. I too thought it important since it was the route taken every week by the wagon laden with tobacco drawn by three great draught horses arriving from my Father's[6] main warehouse in Tarnów. Tobacco and tobacco products were then a state monopoly, and my Father held a lease permitting him to trade in these wares. The wagon, covered in tent canvas, would halt in front of our coach house. The horses were given feed and the driver Lewandowski would sit by the door of the coach house and eat his dinner—a loaf of bread and a jug of buttermilk. The horses, the imposing bulk of the wagon, and the size of this meal altogether made the arrival of the wagon a big event for us children. Lewandowski was strong enough to protect the wagon and its load by himself over

---

[1] Jasło is a town in southeastern Poland, then part of Austria-Hungary, where the author was born in 1887.

[2] A tributary of the Vistula.

[3] Literally "by the castle".

[4] A powerful sixteenth century Polish family who built on the ruins of what had been a medieval fortress.

[5] In Polish, *Łapiguz, Młynek, Lichtarz,* and *Pohulanka.*

[6] The capitalization of "Father" and "Mother" when referring to the author's parents has been taken over from the original.

© Springer International Publishing Switzerland 2015
H. Steinhaus, *Mathematician for All Seasons*, Vita Mathematica 18,
DOI 10.1007/978-3-319-21984-4_1

Fig. 1.1 Partitioned Poland. The map shows the partition from 1871 (the date of the unification of Germany) to 1918. However the total partition of Poland between Austria-Hungary, Russia and Prussia dates from 1795, with certain changes in the regions ruled by these powers instituted following the Napoleonic wars. Thus, for example, Warsaw came under Russian domination only in 1815 (Map courtesy of Carolyn King, Department of Geography, York University, Toronto.)

the seven mile route. Once, at night, when a thief tried to help himself to the tobacco, Lewandowski chained him to the wagon and made him keep up with the horses for a mile. Sometimes he was permitted to arrange the tobacco on the shelves himself: the square boxes labelled "Dames" separately from the long brown boxes labelled "Drama," and from the small circular packets of "Herzegowina". The grains of feed scattered by the horses attracted pigeons until at any one time there were more than fifty of them—and their number did not decline since we children refused to eat pigeons. For us they were holy birds, like the ibis for the Egyptians.

In addition to the inner courtyard there was an outer yard between our house and the offices of the district authorities. We called this yard "where the horses go".

1 Jasło

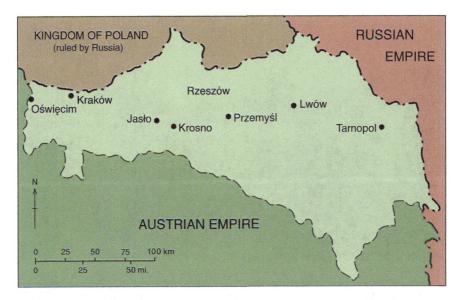

Fig. 1.2 The Kingdom of Galicia and Lodomeria, the south-eastern portion of Poland ruled by Austria-Hungary from 1795 to 1918. Note the town of Jasło, scene of Steinhaus's youth, and also the city Lwów, provincial capital (Map courtesy of Carolyn King, Department of Geography, York University, Toronto.)

The courtyard was enclosed by warehouses, coach houses, the stable, and the house of the local haberdasher Menasse. Through the two ground floor windows of our house you looked into the "hall," so-called, formerly part of the grocery established by my grandfather Józef. There was a billiard table in this hall, and festive dinners organized by the local golden youth or by local notables used to be held there. The house was probably a monastery originally, and became the property of an official named Praxmayer only after the closure of the monasteries during the reign of Emperor Joseph II.[7] It was around that time—that is, towards the end of the eighteenth century—that the house was appraised at 4000 florins. On the ground floor the walls were a meter thick, and the ceilings of both the store and the smaller rooms were vaulted. Our family acquired the house in 1853. The insurance policy on the house, the first issued by the Kraków firm *Florianka*, was dated from that year. Adjoining the house there was a large garden shielded from the yard by the coach house and from the neighbors by high walls, ideally isolated since it was also at a level one storey above the street, nicknamed Targowica. It was in that garden that I learned to ride a grey pony, milder and more patient than other steeds.

The most important part of the house was the store. It was advertised as a "grocery and storehouse of wines, alcoholic distillations, and liqueurs". The young men who did the selling were usually to be found between the garden and the warehouse, washing bottles with water and sand or corking them by means of a

---

[7] Holy Roman Emperor 1765–1790, son of the Empress Maria Theresa.

Fig. 1.3 Two views of the old Jasło market place (Courtesy of the HSC Archive, Wrocław University of Technology.)

special machine once they had been filled with wine. They used to roll the empty barrels standing up on them, and climb up on the roofs to pelt the salesmen of our neighbor Tytus Brąglewicz with chestnuts. Because of its low ceiling, the store was cool in summer even when its doors were open to the market square, shimmering in the dazzling sunshine. On the shelves stood hundreds of bottles, deep blue, cherry red, white, some of clay or stone. On the counter there might lie a ham held with tongs, and the show window was decorated with an oversized champagne bottle. In those days, although bottles were sometimes used as potential weapons to threaten people with, there was no thought then of filling them with phosgene

Fig. 1.4 Ewelina Steinhausowa, the author's Mother (Courtesy of Tomasz Duski.)

or trinitrotoluene.... I never felt any interest in using alcohol or tobacco, possibly because they were abundantly to hand. When I was ten years old, my Father assured me that if ever I should feel so inclined I might have as many cigars and as much wine as I wanted.

For me the center of attraction was the stable. Usually two horses could be found there, but since they were frequently sold and new ones bought, there would sometimes be two pair. I remember some of them. Once there was a pair of beautiful grey ponies, another time a chestnut and a dun pair, and at yet another time two black, nimble, but vicious ponies. The larger breeds worked in the fields, but were also used for transporting bricks, stones, and timber. This had to do with the fact that when I was still not of school age my Father left the running of the store to the eldest of the salesmen and became involved in the building trade. One of his first buildings

Fig. 1.5 Bogusław Steinhaus, the author's Father (Courtesy of Tomasz Duski.)

Fig. 1.6 Hugo, Felicja, Olga, and Irena (Courtesy of Tomasz Duski.)

was the post office. When it came time to lay the cornerstone, Father organized a dedication ceremony to which he invited friends and acquaintances and some of the more distinguished citizens of Jasło. Breakfast was served in the shed where tables were set up laden with copious quantities of wine and vodka. As a seven-year-old circulating among the guests, my eyes registered more readily than the grownups' more elevated ones the liqueur glasses left untouched or not fully drained. I quaffed the lot. At home Mother noticed that I was running a fever, easily diagnosed by the

Fig. 1.7 Felicja, Irena, Olga, and Hugo (Courtesy of Tomasz Duski.)

doctor Józef Kadyj as caused by "acute alcohol poisoning". This episode taught me rather early that alcohol is a treacherous friend.

The ponies were needed for the buggy, in which we would drive in summer to the Wisłoka. We would depart in the morning, sometimes taking lunch with us, and return in the afternoon. We would set up camp in a willow brake, usually on an island that could be reached by fording the river in the buggy. At least this was the procedure when my Mother and sisters were with us; in their absence it was more "masculine": we would walk to the river with Father, and, after a swimming lesson in what was sometimes muddy and cold water, we used to visit "The Little Mill" for beer, whole meal bread and dried fish. The latter was a specialty of the inn: sundried whiting stretched on wooden hoops, with dry, somewhat salty, pale yellow flesh. The local boys always swam naked. They used to smear their bodies with mud from the river's edge, transforming themselves into "blacks", then wash the mud off in the river and dry themselves with sand—much as ink used to be dried using a powder in those days—or just lie down on the gravel to dry off. No one ever used a towel. Experience taught us that clothes put on after swimming in the river did not get wet—on the contrary, the body got dry and warm. It seems that man is adapted to life near water, since bathing in a river affords a kind of satisfaction not obtainable from immersion in bath or swimming pool. Perhaps it is the effect of the mud and sand, or the roots, blades of grass, and leaves that it picks up that lend a fast flowing stream the power to make the eyes of those who bathe in it sparkle.

My first riding horse was acquired from the brewery in Trzcinica, although at that time horses were usually bought in Żmigród, whither horse traders would bring them from Hungary via Dukla. This pony, a grey, was mild-mannered and could walk up and down stairs; this test was carried out at my behest once when I was ill in bed. However, I was really too young to have a pony, and was not allowed a saddle but had to make do with a saddlecloth—so of course without stirrups. The successor to the grey one easily bucked me off by lowering his head and kicking up his hind legs. Sometimes he would then jump over me as I lay on the ground. (These ponies were little peasant horses. In view of the manner in which they negotiated potholes and hollows, I think the horse traders were right when they claimed they were closely related to the tarpan.[8])

When we went riding, a popular destination was the brickyard behind the railway. We often had to wait at the crossing because wagons were being shunted: a locomotive would position the wagons on a sloping portion of branching track, a railroad worker would jump down and crawl under each wagon to unhitch it, and then dash from between the wheels, wave a red flag, and shout in singsong fashion to the pointsman on the tower: "the first, the second,..., the sixth,...." The wagons rolled by themselves in succession as the points changed noisily. Finally, the proud locomotive would come, and you could see the rails bend and squeeze the ties[9] into the ground. A penny placed on the rail was transformed into a warm little smooth elliptical plate. The points mystified me. Once my Father had drawn my attention to the flanges on the wheels, I felt puzzled as to how it was possible for a wagon to move onto a branching track. So he sketched a set of points for me. Then I tried to draw them for myself—and ended up making at least a hundred such attempts. Beyond the railroad the *Heizraum*[10] was visible, and sometimes one might see there a locomotive being slowly rotated on a turntable. And beyond that were the open fields, where forget-me-nots grew in the ditches and dragonflies flew over the brick pits. Next to a little spring one could find fragrant herbs whose leaves provided homes to cockchafers[11] whose carapace glistened with a metallic iridescence, and against the sky evanescent larks—whose song is harder to imitate than the nightingale's trill.

---

[8] A prehistoric type of wild horse ranging over what is now Europe.

[9] Or sleepers.

[10] A Galician railroad term of German origin, for the Polish *remiza*, meaning "engine house".

[11] A kind of beetle.

I was allowed to participate in making bricks, which were moulded by hand out of clay on tables. The finished product was smooth and imprinted with the initials B. S.—for Bogusław Steinhaus, the owner of the brickyard and my Father. The brickyard was of circular type, and had on the upper floor a number of iron kiln doors. When you opened one of these and threw in a shovelful of slag, there burst forth a shower of sparks, and then one could see below a blazing, pink-gold brick.

That was our shortest trip. Much longer was the one to Iwonicz, the site of the first summer holiday I remember. We stayed in a house called *Ustronie*.[12] The forest was so close that in the morning the sunlight filtering through the leaves formed shifting patches on the floor, and the birds' twittering seemed to come from within the room itself. "Iwon's son"[13] made strange impressions on me. The scalding hot water from the Bełkotka[14] would have delighted Chesterton![15] And Wincenty Pol,[16] one of whose poems was etched on a marble table above this little spring, would not have believed that methane drawn from shafts in the Lower Carpathians such as those of Iwonicz would in the future be used as fuel in the factories in Radom.[17]

A second wonder was the fish pond in the woods. One threw bread rolls into the water and watched the huge carp compete for every last crumb. The scent of the pines and the aroma of the bilberries made this land where the rocks oozed oil feel even more exotic. Salamanders and snakes abounded. On the entertainment side, there was a shooting gallery with backdrop a painting of a room; if your bullet hit the clock in that room, then it struck the hour and the painted baby was heard to cry. One would come across young people from Congress Poland[18] dressed in Russian Gymnasium uniforms, white with shiny buttons, and with the three-headed Tsar's eagle reproduced on hat and belt. The most famous visitor to the spa at that time was Lucyna Ćwierczakiewiczowa[19] from Warsaw, by then an old woman confined to a wheelchair. She was always surrounded by a gaggle of gentlemen who seemed to be having a wonderful time—she was presumably demonstrating her wit. Perhaps the guests owed the excellent quality of the food in the restaurant to her presence. However that may be, when I visited Iwonicz a great many years later the toasted fish fillets were as tasty as they had been in that distant past. And the belts on the

---

[12] Meaning "retreat".

[13] That is, Iwonicz.

[14] The name of the hot spring, meaning "babbler".

[15] Gilbert Keith Chesterton (1874–1936), English Catholic writer; popular then in Poland in view of his friendliness towards Poles.

[16] A nineteenth century Polish poet and geographer.

[17] A city in central Poland.

[18] Informal name for the part of Poland assigned as protectorate to Russia by the 1815 Congress of Vienna, but eventually absorbed into the Russian empire. The other two of the three regions into which the Congress of Vienna partitioned Poland went to Austria-Hungary and Prussia. This was essentially a reinstatement of the partition of Poland imposed just prior to the Napoleonic wars.

[19] Polish writer, journalist, and author of the first book on Polish cooking. Lived from 1829 till 1901.

1 Jasło

pumps raising the oil from deep in the shafts sang the same refrain as then: "Is there? There isn't! Is there? There isn't!...".

From the hillock just behind the hot baths, one could view another impressive automaton of the locale. I would stare mesmerized by it. This was the orchestra of the Auber brothers from Tarnów, which played twice a day in the pavilion behind the promenade. One might help the conductor by conducting along with him with a stick such as that used for propelling a hoop. Around that time my Mother arranged for one of the Aubers to teach me the violin. These lessons were a failure then as later, except for developing my ear for music slightly.

There was a fountain in front of the pavilion with adjustable outlet. Thus sometimes the stream of water spouted straight up, sometimes it simply spilled out, and at still other times four arcs of water cleverly turned a little water mill and when the sun shone a rainbow was seen. At the edges of the pool one could catch musk beetles, dark green creatures with shiny metallic carapaces.

The villa *Ustronie* belonged to the Ostaszewski family, who lived on the ground floor. Once, after an elderly gentleman approached and spoke French to me, my Father informed me that he was Count Załuski, the Austrian consul in Egypt. So when I began collecting postage stamps I wrote to him in Cairo asking him to send me some Egyptian stamps. He replied very politely to me—then eight years old—and enclosed a few stamps with the letter.

The Załuski family were dominant in Iwonicz, and in the high season arranged parties for the younger of the spa's guests. I remember attending one such party with about a hundred children. Every child was given a honey cake, just two of which had almonds baked in them. The boy and girl who by chance got these cakes became king and queen of the festival. On this occasion it so happened that I got a cake with an almond in it, as a result of which I was placed along with my partner on a throne and crowned, and the other children marched past in pairs bowing to us. However, as well as partaking of the grandeur of royalty, I was also a party to its problematic side. Being then so young, I was not used to the more adult clothing I'd been dressed in, and had not put on suspenders. Hence for the period of my reign I had to avoid sudden movement lest I make a very unkingly spectacle of myself. While this immovability on my part may have made an external impression of regal gravity, internally I was filled with dark misgivings. I was greatly relieved when the entertainment came to an end.

* * *

When I was seven a political event occurred which made a lasting and deep impression on me. A priest by the name of Stojałowski began to agitate for certain measures in the *powiat*.[20] The Austrian authorities, the governor, and the *starosta*[21] did their best to nullify his activity, but the priest managed nonetheless to incite the local peasants to carry out assaults on Jewish property, especially inns. The attacks erupted suddenly. Father came home one evening and told us that peasants

---

[20] Administrative district, or county.

[21] A county official with certain administrative responsibilities.

had set fire to the inn across the river, and, having positioned themselves on the bridge over the Jasiołka river, were threatening to enter the town. A few citizens of the town, including my Father, organized a guard, and, armed with guns, began to patrol Targowica Street. The peasant band retreated when they heard the steady footsteps of the patrol. At dawn a military unit arrived from Rzeszów.[22] It was raining, and at six in the morning I saw soldiers line up in two rows on the street. Their commanding officer was given orders by the *starosta*, and they marched off. Martial law was proclaimed and military courts set up. The commotion died down, but just in case, a company of the 57th Infantry Regiment was quartered in town throughout that summer. An actual army detachment, including an officer on a horse, stacks of arms, tents—all this added up to a splendid sight and left a deep impression on me.

Years earlier, before the railroad line had been extended through the town, there had been a garrison stationed in Jasło. Father used to tell me stories of those times. One such story concerned a captain who always distributed his monthly pay among boxes labelled "for rent", "for beer", "for cigars", "for the shoemaker", and so on. When towards the end of the month there was no money left in the cigar box, he would stop smoking till the next payday, for, as he said, it is beneath the dignity of a captain in the Austrian imperial army to be in debt to a shoemaker. Although the town was no longer garrisoned, occasionally a regiment of cavalry would pass through: Uhlans[23] in stout blue jackets and red trousers, dragoons in antique helmets, or hussars with braided sashes fastened to their left arm—it was as beautiful a sight as in the lithographed color pictures I had seen of the former Polish army. Small wonder then that, when I heard in the distance the *Generalmarsch* sounded on a trumpet, there was no keeping me at home. (I had ample free time since I was privately tutored and did not attend school. Incidentally, it is safe to say that such an education takes half the time of a standard school education.)

I was sometimes asked by grownups visiting my parents such questions as "What do you want to be?", "Whom do you like more, Father or Mother?", "How old are you?"—and so on. I would always answer without reflecting. When I was asked how I managed to answer so unhesitatingly, I replied immediately: "I know what you are likely to ask me and have an answer prepared beforehand."

* * *

I enjoyed visits to my Mother's family in Tarnów for two reasons: one could see soldiers every day because both foot soldiers and uhlans were stationed there, and there was a sweet shop whose owner had the promising name Delekta. His marzipan cakes were sprinkled with chocolate and his chocolate cakes were half dark chocolate and half cream. Grandfather pampered me to such an extent that on

---

[22] A city in south-east Poland.

[23] Polish light cavalry armed with lances, sabres, and pistols. In the Turkic Tatar language, "Uhlan" means a brave warrior, among other things. The first Uhlan regiments were created in the early eighteenth century, and the tradition continued right up to the beginning of World War II.

one occasion I came down with a high fever from eating too many nuts. The doctor prescribed some awful mixture, and I begged my aunt not to force me to take it. Of course, she relented—and next morning the pharmacy let us know that someone had made a mistake and the medicine I was supposed to take was in fact worse than my illness.

When I turned nine, I was enrolled in grade four at the county school. The school was situated in the "old town hall," which had once been a monastery. The school rooms were on the upper floor, the ground floor being occupied by a jail, or town lock-up. The traditions at this school were rather strange. In particular, beatings played an important role both as a means to a better upbringing and as a method of payment among the pupils. Thus by way of payment for a Swiss stamp, say, obtained from a fellow student, you might give him the right to hit your outstretched palm three times with a ruler. The teacher, on the other hand, used a cane. The teacher of religion was Count Wiśniewski, an elegant-looking priest with a pale face and black eyes, punctual about entering the classroom on time. If any Jewish pupil[24] were still in the room when he came in, he would lay him out on the front bench and flog him with the cane. Later it turned out that he was not only a sadist but also a pederast, and by way of penance he was ordered to join the Carmelites.

Once my fellow pupils pointed out to me what looked like a mentally disturbed person visible through the bars of the lock-up. They soon discovered what most infuriated him, and taunted him mercilessly, making him shake the bars, turn pale, and froth at the mouth. Our town boasted a few harmless such creatures: "Gypsum Charlie", "Peach", and "Casper-pow", who would always be taking aim at us with a stick as gun and firing by means of the vocal "Pow!". And there were a dozen or so beggars who had the traditional right of making the rounds of the town once a week.

* * *

From the first floor window of our house one could always see something of interest. A butcher might be seen taking a calf to the slaughterhouse with one hand on its neck and the other on its crupper, or swineherds herding a few dozen pigs to the railroad. The latter would run along on either side of the herd using staves to keep them together. Or one might see a dogcatcher throw a noose with great skill over the head of a stray, causing disquiet among the people and whining and squealing among the dogs in the market place.

Except for Fridays the market place was always well swept and dry. On Fridays, which were market days, there was always plenty to see from very early in the morning. Shoemakers from Kołaczyce set up horizontal poles from which they hung hundreds of pairs of knee boots with tops of thick yellow leather. Where the crowd was densest an organ grinder would be playing *Daisy, Daisy* or the *Danube Waltz*. White mice ran on a wheel that caused little cards inscribed with predictions to be extracted from a basket. Sometimes a bagpipe player turned up. He would squeeze the bellows of the bagpipe under an arm, beat a drum with his free hand, and jerk

---

[24]Naturally, Jewish pupils were exempt from Christian instruction.

a string connected to a bell on his head. There was often a specialist in diamond grindstones for sharpening scythes laying it on thick to the gawkers gathered round him. Peasants came in carts from all directions. By ten o'clock the market place was so crowded that anyone wanting to leave had to somehow negotiate six or more carts so jammed together that it was barely possible for their shafts to be dislodged. The horses would in the meantime munch on the hay in neighboring carts while their owners went to the stores to buy horseshoes, belts, axles, chains, knives, and vodka. Some of the merchants bought grain and sold flour. The main store with a liquor licence was packed. A crowd, like an informal outdoor club, would gather in front of the tobacco shop on the ground floor of our house, convenient for its proximity to the grocery and the offices of the *starosta* just a few steps away. The carts would start leaving towards evening. At sunset pigeons circled the market place and helped themselves to the seed inadvertently scattered about. Occasionally rain would change the picture. It was entertaining to watch the vendors struggling to erect sail-like canopies over their stalls, and people running as fast as they could to avoid a soaking. A downpour would transform the market place into a yellow sea pockmarked by the large drops, and the dark wind-driven clouds would obscure the view of the houses.

* * *

To see a real flood one had to go north to the Jasiołka river. I would sometimes go there with Father when he had business to transact with the Credit Society, whose offices occupied the ground floor of a large house in which my great grandfather had lived at one time. However, the only thing of interest to me there were the stamps for imprinting seals which I was allowed to play with. Opposite this house there was a small apartment house, on the ground floor of which lived my Aunt Frydman, a sister of my grandfather. Her home was full of trinkets, miniatures, little rugs, and there was even a miniature garden whose tiny beds had stone borders. But all was clean and neat. This appealed to me very much though it did strike me as a little strange. My aunt had three children. One son, Marceli, a journalist, was the editor of the semi-official Viennese daily *Fremdenblatt*. He was later ennobled, choosing for his crest the image of an old lime tree behind the apartment house. Her second son was a doctor in Gorlice,[25] and her daughter Ludmiła lived alone in Florence, where she made a living by translating Polish novels (including those of Sienkiewicz[26]). Some 20 years later I happened to meet her in Venice and was struck by her beauty. Marceli's grandson was until recently director of the Vienna Opera; he had inherited his musical gift from his father Richard.

If there was nothing of entertainment value going on, one had to stay home and read. My reading matter was then entirely haphazard. There were the weekly "The

---

[25] A town in south-eastern Poland.

[26] Henryk Sienkiewicz (1846–1916), Polish journalist and novelist. Author of *Quo vadis*. Nobel laureate for literature in 1905.

Children's Friend" and "Family Evenings",[27] neither of which interested me. The bound annual "The World"[28] was more to my liking. It contained short stories and novellas which I read with delight beginning in the middle and understanding less than every other word. Failing these, one could turn to a German encyclopedia we had and look at the tables and pictures with captions such as "Dog races," "Horse races," "Diamonds," "Birds," "Flags," "Cross section of a coal mine," and so on. As a last resort one might unscrew a lens from a pair of binoculars and using it to focus sunlight, and burn black monograms on cork or on the window frame. If this too failed, one had to stand at the window and simply wait for something extraordinary to happen. And indeed extraordinary things did happen. A peasant wedding group might be seen making its way to church. The men rode horses whose manes were braided with ribbons and the bridesmaids sat in two rows on a wagon, with the bride, adorned with flowers, standing between. A second wagon carried the older folk and members of a band tenaciously playing a melody both cacophonous and strange, as if of Tartar origin. Or sometimes a Gypsy appeared leading a bear by a chain through its nose; the bear danced on its hind legs and onlookers threw its master pennies wrapped in paper. In the winter came carollers carrying a crib containing a likeness of the infant Christ, or one might see a *turoń*[29] with a long muzzle pass by, or Jews dressed for *Purim*.[30] Sometimes I would see a Czech couple, the man playing a violin and the woman a harp, going from house to house playing *Where Is My Home?*

Such were the more routine episodes. There were others on a higher level, so to speak, such as the "running footman".[31] He would be dressed somewhat like a Byronic Pied Piper of Hamelin, in tights and a long-sleeved vest with silver bells instead of buttons, and a pointed hat. He had a riding whip to chase away dogs and munched on a lemon. He would run around the market place leading a gaggle of beggars amidst the barking of dogs. The public was delighted. The running footman was testimony to the fact that the Middle Ages was still with us.

The arrival of the travelling circus was another noteworthy event, obviously of a higher sort. The troupe entered the market place triumphantly and circled it accompanied by the full circus orchestra. The ringmaster, in a red dress-coat, rode in front of the procession on a black stallion, with Madame Ringmaster riding by his side in a black frock coat with a little top hat on her head and a riding whip in her hand. They were followed by the prima ballerina on a grey mare, "stupid Augustus"[32] on a piebald horse with a little monkey on his back, and a pony pulling

---

[27] In Polish, *Przyjaciel Dzieci* and *Wieczory Rodzinne*.

[28] In Polish, *Świat*.

[29] A Christmas mummer dressed up as a barnyard animal.

[30] A festival to celebrate the deliverance of the Jewish people living in the ancient Persian empire from a plot to annihilate them.

[31] Such as used to run beside the carriages of the nobility, or run ahead to prepare for their lord's arrival.

[32] In German *dummer August*, a clown figure in the circus dating as far back as Roman times.

a big drum. The rear guard consisted of an Arab on a camel, animal trainers, piebald great danes, and the rest of the live circus inventory. Or a troupe of acrobats might come to town. These would erect a pole guyed with ropes in the middle of the market place, stretch a rope from the top of the pole to the attic of the nearest tenement, and then *à la Blondin* walk the tightrope there and back.

If nothing so special was on the agenda, one had to make do with a wedding or a funeral. Processions of various kinds passed by frequently. The most sublime of these was the Corpus Christi procession, which circled the market place from altar to altar amidst showers of flowers and the pealing of little bells. The market place would on such occasions be half full of worshippers and onlookers. For the Easter morning service nine short mortars were set up in the market place, and the town police, functioning as gunners, produced deafening salvos. To celebrate the emperor's birthday, uniformed officials wearing three-cornered hats and bearing swords could be seen passing through the market place on their way to church. Gendarmes in gala uniforms sporting gold braids wore hats embroidered with dark green crests. Once, when an errant pig got in the way of the *cortège*, a certain tax official chased the "enemy" off with his sword. This was reported to the *starosta* and led to disciplinary measures being taken against the official.

Sometimes the Almighty arranged an unexpected contretemps. Once a little cart laden with a hundred or so siphons of soda water turned over while rounding the corner from Kościuszko Street into the market place. The siphons all exploded, producing a crashing and spattering perhaps reminiscent of the bomb outrage against Grand Prince Sergiusz[33] which I knew of from the illustrated papers. (This prompts me to mention by the way that a light white wine mixed half-and-half with soda water tastes better than champagne.)

* * *

One morning two black monstrosities appeared in the market place. These were steam boilers in the process of being moved from the train station to the electrical station under construction on the bank of the Jasiołka. They were too heavy to be transported by horse and cart—and in those days there were of course no tractors or bulldozers or trucks to do the job. They had to be rolled along on wooden cylinders rotated using crowbars; the cylinders freed behind were successively moved to the front to take the weight of the boilers as they moved along. This took at least a week. At last, one summer evening Father took me to the market place to see the electric lights. It was an incredible sight: the light was steady and warm, flowing from golden wires looped in the interior of what looked to me like glass pears. These were Edison bulbs with filaments of carbonized bamboo, and the electric current was then—and for a considerable time afterwards—direct, so that the light did not tire the eye.

---

[33]This may be Grand Duke Sergei Alexandrovich (1857–1905), a son of Tsar Alexander III of Russia, assassinated in 1905 by means of a bomb thrown into his carriage. However, the author would have been in his late teens at that time.

The construction of a through railroad line, completed in 1882, when my Father married my Mother, marks the American period in the history of our town. When she first arrived from Tarnów in a horse-drawn carriage, my Mother saw a sleepy little village without what one could really call streets, and without pavements. The only activity observable back then had been that associated with the construction of the district court and the railroad. The "pioneering" atmosphere of our town's development extended even beyond my childhood on account of the discovery of oil in the vicinity. Oil deposits had been found a dozen or so years before I was born, in Rogi, Bóbrka, and Wietrzno.[34] Wells were drilled by the Canadian MacGarvey,[35] perhaps the first oil wells on the European continent. Then the deposit at Potok was discovered and the famous "crazy" well drilled, and oil fever grew—as described by Sewer[36] in one of his novels. Some of the extravagant personages that emerged at this time must have served as models for the characters of Nowaczyński's[37] *Szambelan Sar*. I got to know one of these people, a Mr. Klobassa, who, just one mile from Jasło, drilled an oil well so productive that his annual income was estimated at 400,000 gulden, equivalent at that time to about 160,000 gold dollars. He had a palace built with a splendid park, and maintained horses and carriages in London, Paris, and Vienna. He once had the whole cast of a Parisian operetta brought to Vienna. He had at one time gone to Rome to court the daughter of Count Ledóchowski, a niece of Cardinal Rampolli. A dogshow of purebreds was then in progress in Rome, and the day after the countess had been heard to praise some pointer or greyhound, Mr. Klobassa's valet brought her the dog in question with a dogcollar set with precious stones worth over 10,000 guldens. Of course, my memories do not go back that far, but I do remember his palace in Skołyszyn and the park, and also the trotters that he used to race the train to Jasło. He was so much on the lookout for good business opportunities, that in 1898, when the Spanish-American war over the Antilles threatened, he bought 40,000 crowns' worth of Havana cigars for fear of missing out. After losing his fortune, he married a peasant woman and settled in a village, claiming that this was the happiest time of his life. He was an excellent cook, and one of his old lady friends sent him olive oil from Italy expressly so that he could prepare roast hare his own special way.

* * *

Soon industry began to develop. Although the oil refinery in Ułaszowice, one of the first to be considered by representatives of Standard Oil, failed, others soon appeared. This had a significant effect on the socio-economic structure of the district, since the peasants were no longer restricted to just one type of work, that is, to agriculture. A new sort of worker came into being, namely the peasant who

---

[34] Small localities in south-eastern Poland.

[35] William Henry MacGarvey (1843–1914), petroleum pioneer and industrialist.

[36] Sewer for "Severus" (Roman Emperor 193–211 A.D.), pseudonym of Ignacy Maciejowski (1835–1901), novelist of the Polish peasantry.

[37] Adolf Nowaczyński (or Neuwert-Nowaczyński) (1876–1944), Polish satirical novelist and playwright.

supplemented his agricultural earnings by working in a brickyard or refinery, or on the railroad or in construction. Since he had a few morgs[38] of land and a shack, he was not afraid of being homeless when a job came to an end. And he was no longer so concerned when the price of grain fell since he now had the chance of finding work in a factory. Experience shows that this type of person, what you might call an "economic biped", is more likely to acquire an education and have a progressive outlook on life than either a simple villager or an urban proletarian. A large proportion of these people were oil workers. Many emigrated from this part of Poland to Borysław and Stanisławów, or even further afield to the Caucasus, Trinidad, and the Dutch East Indies.[39] Thus it is that miners' jargon in Sumatra and Borneo includes, in addition to Dutch and Malay, Polish words. Of course, of greatest significance in this respect was emigration to America. The dollars that emigrants brought back with them ultimately enriched the *powiat*, and as a result the large estates broke up of their own accord. The pulse of our town, no longer sleepy or provincial, continued to beat steadily, its citizenry stimulated by ever new sights.

---

[38] A morg (German *Morgen*) is a little over half a hectare.
[39] Now Indonesia.

Chapter 2
The Gymnasium

My freedom, the freedom of a child, came to an end when I was enrolled in the regional Gymnasium. The principal of the Gymnasium at the time was Klemens Sienkiewicz, a Ruthenian,[1] a grey-haired jovial gentleman with a liking for Pilzner beer, and reminding one more of a country squire than a teacher. One of the oldest teachers was Władysław Węgrzyński, who was a veteran of the uprising of 1863,[2] and, as an historian, had written a monograph on the castle Golesz at Podzamcze. Of the rest I recall Kawecki, who wrote a drama about Kalina[3] and the schools; Jerzy Żuławski,[4] author of the play *Eros i Psyche* ("Eros and Psyche"), among other works; the poet Wiśniewski, son-in-law of the principal; Jedlicz, who later became a translator and the artistic director of the Lwów theater; Womela, who was quoted by Irzykowski[5] in *Pałuba*; the poet Eminowicz....

I was nine years old. In Grade One I began learning Latin and German, among other things. Learning was treated as a serious business in the school. History caused me considerable difficulties because I could not learn facts by heart and couldn't think of a better way of mastering the subject. At least a third of my fellow students were sons of peasants—another indication of the prosperity of our *powiat*—and the rest were sons of low-ranking railroad workers, postmen, leaseholders, craftsmen, and city folk from Kołaczyce, Dembowiec, Pilzno, or Krosno. Some were sons of Jewish merchants, some were Ruthenian, and there were sons of orthodox priests.

---

[1] The Ruthenians are an East Slavic people from the region of the northern Carpathians bordering on present-day western Ukraine, southern Poland, Slovakia, and Romania.

[2] The "January Uprising", a great Polish national rebellion (1863–1865) against the Russian overlord.

[3] Antoni Kalina (1846–1906), Polish slavicist, ethnologist, and philologist.

[4] Jerzy Żuławski (1874–1915), Polish writer, playwright, and poet; a pioneer of Polish science fiction.

[5] Karol Irzykowski (1873–1944), Polish writer, literary critic, and film theoretician. His novel *Pałuba* ("The Hag") appeared in 1903.

Fig. 2.1 The Jasło Gymnasium, from which Steinhaus graduated on May 5, 1905 (Courtesy of the HSC Archive, Wrocław University of Technology.)

Only a small proportion came from the "intelligentsia", or white-collar class. Some of the teachers—especially the younger ones—saw fit to abuse their position by showing their scorn for students from poorer homes. For such as these I lost all respect.

Of course, my lifestyle changed; I now had to wear a uniform conforming to a decree of the Imperial Council for State Schools specifying that the tunic had to be navy blue, the trousers grey, and the gymnasist's soul pure and chaste. The reality was otherwise: only a tiny fraction of the students observed this dress code. In

2 The Gymnasium

particular, the grey pants were not popular and most wore pants of the same material as the tunic. I personally disliked the school cap which had the form of a peaked elliptical drum with a stiff crown, and soon changed it for one with a soft crown similar to a French officer's *képi*. There was a change at home too: I was transferred from the children's room to the office, and my former bed, next to my younger sisters', was given to a nursemaid. Thus I no longer had a proper bed but slept on a hard sofa, with deer antlers, two shotguns, and a parlor rifle[6] hanging on the wall next to it. One morning on waking I noticed a crease in the sheet which had caused me a little discomfort. It turned out that the evening before a bayonet had fallen from its perch on the wall and the servant girl who made the bed in the dark had covered it with the sheet—and then I had slept on it without really feeling anything much.

The office was on the western side of the house, and of a summer morning I would see through the window the intense reflection of the sunlight off the cream-colored wall of the *starostwo*[7] and the cool blue of the sky. The scent of lilacs wafted in from the garden, one could hear a murmur coming from the market place and the patter of children's footsteps as they rushed to school. The Angelus bell rang, cool air came in through the open window, and one felt like drinking the summer day, or immersing oneself in the glorious colors and joyfulness of the morning.

* * *

When I was about 12 I fell ill with whooping cough. I became very thin. My parents held the opinion that I should eat properly, but I detested milk. However, I was lucky because I actually liked the taste of cod liver oil—like an Eskimo perhaps—so the situation was not very serious. To recuperate I was sent with my aunt to Kołobrzeg.[8] On the way we passed through Wrocław[9] and Kraków. I knew Kraków quite well since I had once stayed there with Father in the hotel *Pod Różą*[10] high above the street. I remember how horse-drawn streetcars would come along the street with their bells ringing to warn pedestrians. In Wrocław I was impressed only by the asphalted streets, the bicycles and tricycles, and the zoological gardens. On arriving in Kołobrzeg, my aunt and I went looking for a place for us to stay. She stopped in front of a hotel on *Bahnstrasse* and got absorbed in detailed negotiations of some sort with the concierge. After standing patiently behind her for some time I decided that, now that I was so close to the sea, I simply had to see it without further delay. So I left my aunt and walked on till I reached a river with a stone rampart. The river was crammed with barges, and I even saw a steamer there. Reasoning that the river must flow to the sea, I followed it to the end of the rampart and jumped down onto the sandy bank. There still being no sea visible, I walked further, until, rounding

---

[6] A low-powered rifle formerly used for indoor target shooting.

[7] The offices of the *starosta*.

[8] A Polish town on the Baltic coast dating from the Middle Ages, for several centuries prior to 1945 under Prussian or German rule. Its name literally means "by the shore".

[9] Then Breslau.

[10] "By the Rose"

a sand dune, I saw an incredible sight: row upon row of the white crests of waves moving towards me and crashing down into green depths. I listened to the unending melody of the waves' roar and tasted the salty wind.... I stood there alone, gaping, until I finally remembered to close my mouth, which had remained wide open in wonder for several minutes. Meanwhile, my aunt was looking for me in a state of panic; she had notified the police and given my description to all passers-by. At last she found me, and was so relieved to see me that she forgave me my escapade.

I was fascinated by the sea. It is in fact not true that there are no high and low tides in the Baltic Sea, for I definitely observed these at Kołobrzeg. At that time there was a garrison stationed there. Officers would often be seen in large numbers at the tables in the restaurant in the Gardens—especially when a military orchestra played there. At their request the orchestra would play a musical version of a battle, including reveilles, marches, alarums, and other military signals, until finally there came a horrid cacophony of trumpets, drums, and timpani, reinforced by gunshots fired by soldiers hidden in the park's bushes. A more Prussian kind of music would be hard to imagine.

On returning home I came down with a leg disease: the doctors were at a loss to explain the large red blotches on my calf. I was made to stay in bed for close to three months, and was bored to tears. I must have read Kramsztyk's[11] *Wieczory czwartkowe* ("Thursday Evenings"), a most interesting book, at least ten times. This is a collection of splendid expositions of physics for children. My mother hit on the idea of giving me a copy of Mark Twain's *The Adventures of Tom Sawyer* as well as a German translation of some of his tales. I knew no German apart from a few words I had picked up in Kołobrzeg, and had no German dictionary. Initially I would ask my parents for the meanings of some of the words, but soon made less and less use of their help. I ended up learning German without being conscious of the process. Although German was a required language at the Gymnasium, few of my fellow students from the earlier classes could read German with any ease. I conclude that the accepted methods of learning foreign languages are inferior to the best one, namely, the reading of books in the language, even if one initially understands next to nothing.

* * *

The manoeuvres of the Austrian Imperial Army that took place around this time were especially memorable. Preparations began, I think, at least a month beforehand. The "supposition", so-called, was that northern divisions, which were to play the role of the invading Russian army, would move towards the Dukla Pass under the command of General Waldstätten. The defenders of the pass were to be led by General Galgótzy, who was to move from Koszyce against the aggressor. It was planned that these two forces meet in Jasło, and that is why the emperor Franz Joseph I had his headquarters there. For the then *starosta* Count Michałowski

---

[11] Stanisław Kramsztyk (1841–1906), Polish scholar of Jewish origin, physicist and mathematician, popularizer of science.

this was anything but an easy test. Every citizen of Jasło who owned a presentable home had to make rooms available to military notables. For instance, Prince Georg of Bavaria[12] was billeted in the house of the *starosta*, the minister of war Baron Kriegshammer in my uncle's house, and an army inspector by the name of Reinlender in our house, and so on. The emperor himself was put up in the offices of the *powiat*, and ate his main meals in the company of some hundred members of his retinue in a pavilion specially built in the park; the kitchen, commandeered for the occasion, was in a house that later became the restaurant managed by Mr. Dunaj, and was connected to the pavilion by means of a gallery. The emperor's horses, the famous grey Arabs from Lipica, had arrived two weeks earlier; they were stabled in Jaszczwa, and over this two-week period they hauled carriages every day to Jasło and back to familiarize them with the route and prevent them bolting. Emperor Franz Joseph did not use an automobile. He had brought with him an Irish mare, a priceless "hunter", which could negotiate every slope without stumbling. A special train was kept at all times under steam in case the emperor needed to travel to neighboring stations closer to the battlefield. This train was driven by the director of railroads himself, but as luck would have it, on one occasion on a slope near Tarnowiec the wheels of the train began nonetheless to slip—despite the fact that railroad dignitaries had in person sprinkled sand on the rails beforehand.

The monarch arrived at noon on a beautiful summer's day. Before the offices of the *powiat* stood delegations from the *Sejm*,[13] the local squires and clergy, gendarmes in field uniforms, a citizens' guard, uniformed police, as well as representatives of the secret police, peasants in traditional Kraków costume, and a large crowd of onlookers. The regional governor and marshal were also present, of course.... Now the emperor/king[14] was about to arrive. The *starosta* Count Michałowski was standing in his carriage with his bare handsome grey head turned in the direction of the anticipated arrival of the emperor, serving as a sort of sentinel. Suddenly there was a shriek right next to me: thieves were taking advantage of the crush to lift the skirt of a village girl—in the august presence of the highest officials of the imperial and royal sovereign's court, and a mere hundred feet from his Highness and the supreme representatives of Galicia and Lodomeria.[15] The thieves managed to get away. (There is a witty remark about Władysław Michałowski in Boy-Żeleński's[16] marvellous book *Znaszli ten kraj?* ("Knowest Thou Not This Land?").) I remember the sequence of Jasło *starostas* of my time: Gabryszewski,

---

[12]Possibly the Prince Georg of Bavaria who lived from 1880 to 1943.

[13]The Polish parliament, which still functioned to some extent in partitioned Poland.

[14]Emperor of Austria and king of Hungary.

[15]The Kingdom of Galicia and Lodomeria, a region overlapping present-day southern Poland and western Ukraine made up that part of divided Poland ruled by Austria from 1772 to 1918.

[16]Tadeusz Boy-Żeleński (1874–1941), Polish gynecologist, writer, poet, critic, and translator of over 100 French literary classics into Polish. *Enfant terrible* of the Polish literary scene of the first half of the twentieth century. Murdered by the Nazis in July 1941 in the "Massacre of Lwów Professors".

Michałowski, Prince Sapieha, and then Rawski, Łęczyński, Leszczyński, Antoni Zoll, and Maroszányi.

But to come back to the manoeuvres. One morning Father told a worker to harness a horse to a buggy and took me and my friend Wacek Pasterczyk to watch the manoeuvres. After travelling for about an hour in the direction of Krosno, we suddenly came across a squadron of Uhlans led by a lieutenant. As it happened Father knew the lieutenant, a squire from the Jasło area called Bobrowski. He had been leading his squadron through the woods around Warzyce, Sieklówka, Birówka, and other localities for the last 24 hours, and complained that he hadn't eaten anything since the previous morning. On such expeditions Father used to take along a box containing six little bottles of wine, liqueur, vodka, medicinal alcohol, *eau-de-cologne*, and cognac. After curing the lieutenant's thirst, and, with bread and sausage, his hunger, and giving him a cigar to cap the meal off with, Father was much surprised when his patient suddenly uttered the word *Hofequipage*, set spurs to his horse, and galloped off. And indeed a carriage with gilded wheel-spokes was just then to be seen approaching us from the direction of Jaszczwa. However, when it reached us we could see it had no passengers. By this time Bobrowski's squadron was a few hundred yards away. Suddenly a second squadron of riders appeared wearing white headbands, whereupon the first one took off into the bush with the second chasing it. The squadrons were so compact and nimble that they looked from a distance like a single rider hard on the heels of another. A little later we overtook a whole division of foot soldiers marching towards Krosno; the navy blue column stretched to the horizon. Then suddenly this long blue serpent trembled when from the left-hand side, out of the forest, came the thunder of cannon fire. It was clear that the attack by camouflaged artillery had taken the marching column completely by surprise. This was, naturally, scandalous. Adjutants galloped hither and thither, Archduke Leopold Salvator appeared on a foaming mount uttering fervent curses aimed at some guilty colonel, and the whole column moved into the ditch to the right of the road and marched on. A while later it was drawn up into square battalions and set off towards Tarnowiec, with us bringing up the rear. On a hill near Wrocanka we came across a battery of fifty cannon firing at an invisible enemy. We took shelter in a barn, but the noise was so tremendous that it was difficult to endure. Wacek ran out and was confronted by an extended line formation. He was hit in the buttock by a blank cartridge made of wood. Then lines of white-clad soldiers appeared on the hill opposite—Waldstätten's infantry. At this the mediators monitoring the manoeuvres called a halt to the cannon fire, and a commission began to estimate the number of overrun trenches.

* * *

Every Gymnasium had its "majówka", or May picnic, and ours was no exception. Its name notwithstanding, this outing usually took place on a June holiday. We marched four abreast under the watchful eyes of the teachers. Behind this youthful regiment came a supply column of one or two wagons—depending on the number of classes participating—laden with barrels of beer—vodka was not allowed—and if music was planned for, a trumpet and drums, which were to be played alternately by our drummer and trumpeter. The outing would usually be to Podzamcze. We bivouacked on the grass, ate bread and sausage, and drank beer. We sang songs in chorus, and pestered the teachers with offers of beer in order to get them drunk. On the way back in the evening, we advertised our return by means of the trumpet and drums as well as fireworks, and as we arrived in town lit flares. Sometimes rain spoiled the outing; it was those occasions that led me to the conclusion that a solitary walk through the woods is incomparably more pleasant under a summer shower than in fine weather.

I mentioned earlier that around the year 1900 my parents had hired a nanny to look after us children. The first two attempts at finding such help were not very successful, but the third applicant, a Frenchwoman, adjusted splendidly. This was Eugenia[17] Heurtier, from Lyon. She liked us very much, was cheerful, and knew her orthography and grammar thoroughly. When Mother could not handle me, Eugenia would bribe me with a morsel of fried bread, a tidbit that I have relished ever since. When she left after a dozen or so years, she used to write to us. After World War I ended we received a letter from her on fine paper imprinted with a vignette of a *château*. It turned out that during the war she had met a military doctor, a very rich man from Chile or Argentina, and they had married. He died a short time later leaving her a beautiful *château* and estate in France. At that time the concept of planned propaganda was perhaps unknown, but in any case the French Ministry of Foreign Affairs would have had difficulty finding a propagandist capable of convincing us to like France as effectively as did this gay, unaffected, and cheerful girl with a quality very rare in our country—a woman's sense of humour (so different from a man's, which usually resembles exploding firecrackers).

* * *

It struck one more and more that radical changes were occurring—and not only in the way people dressed. Around this time the watchword heard in all of Galicia was the development of national industry. There appeared *The League for the Support of National Industry*, founded by Prince Lubomirski, owner of a sugar refinery in Przeworsk competing with the Morawy refineries, especially those in Chropin. Our teachers supported this movement, with the result that we, the students, were always supposed to specify when buying exercise books, pencils, and ink that we wanted "national" ones. The old established merchants and shopkeepers were on the whole not enthusiastic about this trend, although there were those who supported it. Thus Mr. Apfel put a sign in the shop window of his little store in the market place advertising "nationally-made" wares. This attracted so many customers that all his family members were kept busy serving them. Certain wholesalers attached labels to their goods with the picture of a little bee and the words "Produced nationally". However, since no trademark rights had been obtained for these labels, no one could really be sure they were buying goods made in Poland.

Another fad at that time was abstinence. One active society of anti-alcoholics called itself *Eleuteria*.[18] More demanding of its initiates was the brotherhood

---

[17] Formally, Eugénie.

[18] A word of Greek origin signifying freedom.

Eleusis,[19] founded by Wincenty Lutosławski, a philosopher who taught at Jagiellonian University[20] in Kraków, but had to resign his post because of his eccentric pronouncements. He and his fellow anti-alcoholics were, however, well-meaning. The rules of their society had a mystical base related to ideas of Towiański.[21] This fashion appealed to those gymnasists who understood an abstinent to be a jaded convert who had already experienced life's pleasures, that is, who was both experienced and mature, such a pose being very attractive to us 15-year-olds. However, only a small minority of Gymnasium students fell for this type of posing, since the majority viewed beer and vodka as God's gifts to mankind, to be scorned only by fools. Most smoked cigarettes whenever they were available, and there were those sporting mustaches who flirted *comme il faut* with the maids and other girls in the boarding houses. The gymnasists who were sons of peasants or low-level officials took lodgings in boarding houses, many of which were run by widows of craftsmen or clerks. There were also suburban single-storey dwellings with small gardens, many not so very different from peasant huts, with rooms to let to students. In many cases—especially when the lodger came from a peasant family—the rent was paid with farm products: cornmeal, flour, cheese, and bacon.

Most such rooms were shared by several boys. This meant, of course, that the conditions were not conducive to study. On the other hand, they facilitated the pursuit of certain pleasures: chatting, card-playing, smoking, and drinking. The obligatory kind of speech practised by the students among themselves was the peasant argot, but colloquial speech with a Masurian[22] accent rather than the local folk dialect. If you didn't want to risk being told that you were "putting it on", then you had to "croak like them"—and that's also why sons of members of the intelligentsia, and even Jews, spoke that way. The atmosphere of the contemporary school was perfectly described by Zegadłowicz[23] in his novel *Nightmares*. (This is the Zegadłowicz after whom the town of Wadowice named one of its streets, but then, concluding that he didn't after all write to their liking, had his name officially removed.[24])

* * *

We were soon seized by another craze: literature. The end of the nineteenth century, which came to be called *fin-de-siècle* everywhere, was in Poland the age of *Young Poland*. The movement started in Kraków, or perhaps even further away, but in any case could not but reach us. We began to read Wyspiański, Tetmajer, and Przybyszewski.[25] Some members of our parents' generation said that such reading

---

[19] Originally a village near Athens known for the Eleusinian mysteries.

[20] One of only two Polish-language universities in Galicia, the other being Lwów University.

[21] Andrzej Towiański (1799–1878), Polish religious philosopher and mystic.

[22] Masuria is an area in northeastern Poland famous for its many lakes.

[23] Emil Zegadłowicz (1888–1941), Polish poet, novelist, art scholar, and translator.

[24] However, in 1968 a monument to Emil Zegadłowicz was unveiled in Wadowice.

[25] Polish playwrights, poets, or novelists of the time. In particular, Stanisław Przybyszewski (1868–1927), influenced in his early twenties by Nietzsche and satanism, was a poet of the decadent naturalistic school and associated with the symbolist movement.

matter was "inappropriate for the young," with the result that the young immersed themselves in it all the more. A "secret library" was established in the following way. A man was found who was willing to rent a room in his own name, books were acquired on credit from booksellers to stock it, and soon a lending library of some 500 titles began to function. The credit for this enterprise goes to certain students in the class two years ahead of mine, in the first place to Emil Żychiewicz, who was later to become head of the Atlas Library in Lwów. There was no monitoring of our reading matter after school hours. We read essentially at random, or rather we subconsciously yielded to the wave flowing to us from the West via Kraków. Thus above all we read northern writers such as Ibsen, Bjørnson, and Heijermans without having any idea of their caliber. We also had translations of Zola and Maupassant, and of Hauptmann and Sudermann—and no one seemed to realize that in fact Sudermann was not at all superior to our own Rodziewiczówna[26].... None of us knew about Strindberg or German writers such as Karl Kraus,[27] nor had we heard of Flaubert or Stendhal—perhaps because the older generation had neglected them. There was no Boy's Library[28] yet, so none of us had heard of *The Physiology of Marriage* by Balzac, but we regarded *Love and Marriage* by the naive advocate of women's rights Ellen Key[29] as the model of progressive philosophy. It was due to Przybyszewski that Scandinavian writers became fashionable in Poland. He lived in Germany for considerable periods, and was regarded there as a literary revitalizer of German style. (The cultivation of Scandinavian writers in Europe was initiated by the European Jewish intelligentsia, possibly because the works of the Scandinavians first became known in Berlin and were then translated into German and praised in German and Viennese newspapers.) No one had heard of Dickens then, let alone Meredith. Somehow Oscar Wilde got through to us—though rather late—as a representative English author. On the other hand, our own authors, such as Żeromski, Reymont, and Tetmajer, did not need to struggle to win our hearts— we were overcome at once. For several years from the time I was given a thick notebook as a gift, I recorded faithfully the titles of the books I was reading, with accompanying dates, short excerpts, and critical evaluations. Suddenly my choice of reading matter took a very different turn; I got hold of some of the writings of Nietzsche, first in translations by Łącek, Berent, and others, and then in the original German. Here I was exposed to a completely different scale of values. Nietzsche convinced me that the values of antiquity had been preserved by nations

---

[26] Maria Rodziewiczówna (1863–1944), Polish writer of the interwar years. Her works often idealized rural life and the peasantry.

[27] Austrian writer and journalist, satirist, aphorist, essayist, playwright, and poet. Regarded as one of the foremost German-language satirists of the twentieth century. One of his aphorisms: "Children play soldiers—that makes sense. But why do soldiers play children?" Attacked hypocrisy, psychoanalysis, corruption in the Habsburg empire, the nationalism of the pan-German movement, *laissez-faire* economic policies, and many other subjects. Lived from 1874 till 1936.

[28] A reference to Boy-Żeleński's translations of French literary works—see above.

[29] Ellen Karolina Sofia Key (1849–1926), Swedish feminist writer.

deriving from Rome, and that the works of northern European writers represented the continuing revolt of the German vanguard against the clarity and orderliness of Roman culture. He managed to pry apart the discordant pair Schiller-Goethe. Unfortunately, Goethe was not on the Gymnasium curriculum. I wandered about in my reading for a long time before coming across French literature, and of Russian literature I knew only of Gorky and Leo Tolstoy; Pushkin[30] and Chekhov were unknown to me. Traveling theatrical troupes put on Gorky's *The Petty Bourgeois*, and the false revolutionary pathos of the characters was very much to our liking. I think that if it were not for our Imperial/Royal Gymnasium, we would never have read Mickiewicz's *Sir Thaddeus*.[31] We failed to understand Mickiewicz;[32] we considered him as being accorded official recognition and exaggerated importance for patriotic reasons. Even more difficult for us to appreciate was the wild beauty of Malczewski's[33] *Maria*. Although we liked Sienkiewicz, we regarded him as a reactionary and were ashamed that we read him—after all, the older generation sang his praises, and the local judge Bochniewicz, a robust drunkard, had changed his name to Bochniewicz-Zagłoba.[34] And of course every Gymnasium boasted a dozen or so imitators of Kmicic and Skrzetuski.[35] I myself was deputized to assume the role of Ketling[36] throughout the territory of the Jasło Gymnasium. There is perhaps no phenomenon more paradoxical than a poet's fame: in every Polish city there is a Mickiewicz Street, but every inhabitant of a street so named would doubtless agree deep down with the lady who reacted to my rapture with the words "*Sir Thaddeus* beautiful? Certainly not! But very instructive!" And then she quoted "Warsaw, our city, how many are thy charms!" as an example of poetry. So I had to revise my scale of literary values yet once more. Evaluation of literary quality is a difficult and complex affair.

* * *

As I have already mentioned, I had sufficient patience and talent to acquire an understanding of music only on a very mediocre level. It should not be assumed that this was due to provincialism; some of my fellow citizens had considerable musical culture—for instance, Teodor, the son of the principal of our Gymnasium

---

[30] Aleksander Sergeevich Pushkin (1799–1837), great Russian Romantic poet. Regarded as the founder of modern Russian literature.

[31] *Pan Tadeusz*, an epic tale in verse, considered by many the national epic of Poland, and set as compulsory reading in Polish schools to this day.

[32] Adam Mickiewicz (1798–1855), Polish Romantic poet. Ranked with Pushkin as among the greatest poets writing in Slavic languages.

[33] Antoni Malczewski (1793–1826), Polish Romantic poet, known for his only work *Maria*, a narrative poem of dire pessimism.

[34] A Falstaffian character in Sienkiewicz's famous trilogy of novels.

[35] Other characters in Sienkiewicz's trilogy.

[36] Hassling-Ketling of Elgin, another fictional character from Sienkiewicz's trilogy—an expatriate Scot who comes to Poland to serve in the army of the King of Poland.

Sienkiewicz, who had settled down in Jasło as a lawyer. This was a gentleman who refused to yield to the provincialism of the locality, and determined on becoming a "European"—a word having a special meaning for him. Thus, in his view a dress coat and top hat were a must at official appearances. However, since a "European" cannot possibly be seen walking in the street in broad daylight and in formal dress, he would sometimes have to take a cab to cross the street. When he wished to bathe in the river, he would always ride in a droshka,[37] taking with him in a handbag fresh underclothes, a bottle of cologne, face powder, a bath wrap, and a blanket. The natives—that is, the non-Europeans—considered this elaborate river bath an absurdity, but, being a man of unwavering principles, Mr. Teodor was unperturbed. He was of a high musical ability, and decided to do something one would have thought impossible in Jasło, to wit, arrange a performance of one of Beethoven's concertos for piano and orchestra. This work called for at least a hundred musicians, so he tracked down in the town, its vicinity, and even more remote places all the local fiddlers, bass players, cellists, oboists, trumpeters, and percussionists, and over a period of three months held practice sessions for individual players and small groups, and, finally, full rehearsals of the whole orchestra. The concert was held in the village of Sokół, the only affordable "European" locale. It may be that the impossible was accomplished on this occasion by virtue of the special character of our town, one with too many components to analyse. However, the most important of these was undoubtedly oil.

The oil dealers were people of a special kind. (They are described in Sewer's novel *Nafta*.) I mentioned two of these earlier, namely Klobassa and MacGarvey; another such was Teodor's father-in-law Mr. Macher, the Belgian consul, whose father, a school inspector, is mentioned in Sewer's novel. Following on these, others soon arrived: Perkins, MacIntosh, Trzecieski, Sroczyński, and so on. I got to know many of them by sight and from descriptions such as those in Neuwert-Nowaczyński's satire *Szambelan Sar*. I recall a visit to the entrepreneur Ochmann, a supplier of drilling pipes in Krosno. Father let a friend and me, then ten years old, accompany him there in the buggy. Mr. Ochmann received us in a disused hangar that served him as office, bedroom, and warehouse. When we arrived at about ten in the morning, he gave us a breakfast of bread and butter, enough caviar to fill a liter jug, and bottles of white wine. We ate the caviar with a big spoon, and during the feast our host would occasionally jump up and run over to a crank-handle that he used to raise a heavy weight to the ceiling, which powered a gramophone as it slowly fell.

When Father gave me a bicycle as a present, I could go on trips without him. My bicycle and my cousin Dyk's[38] were acquired directly from the factory in Stryj, and arrived in large wooden cages wrapped in paper—a nice surprise for us! Sometime in the distant past my Father had been one of the first in the *powiat* to ride a bicycle. His had been a penny-farthing, with a large front wheel and a small back one, and,

---

[37] A kind of two- or four-wheeled carriage.

[38] Józef Schoenborn.

ridden downhill, it was not unusual for the rider to fall on his face accompanied by the saddle and the little rear wheel—much as when a horse lowers his head and bucks with his hind legs. Little wonder that the modern bicycle was considered a great achievement of contemporary technology when it first appeared.

At this point I must say more about my cousin. His father, who had worked as an engineer in an oil refinery in Libusza, near Gorlice, died young, and his widow moved to Jasło to live in the house of her brother-in-law, another of my Mother's brothers. Dyk had an unusual talent for inventing games. He used to disguise himself as an old man, an old lady, or an athlete, etc., and improvise monologues appropriate to these roles to the delight of my younger sisters and me. At one time wrestling was popular among the lads in our town to the point of mania. Its popularity flared up when a travelling circus came to town and their star wrestler Zbyszko Cyganiewicz,[39] said to be at one time a student of our Gymnasium, gave demonstrations of Greco-Roman wrestling. After each show he was followed home by a gaggle of us students, all dreaming of becoming famous athletes.

With Dyk I had no end of fun. In the morning, once the carpet had been rolled up and shoved under the table, we rode our bikes around the dining room table. We would use a *camera obscura* to form moving images in color of scenes in the market place, such as that of a cart pulled by a bay horse. We connected a thin piece of wire to the opposite poles of a small dynamo and heated it up to melting point. The most interesting games were always those invented by Dyk. For instance, his game called "The State" involved the use of lead soldiers supplemented by chess pieces. Each king had a retinue of ministers, an army, and a treasury consisting of colored weights representing priceless sapphires and rubies, as well as a palace built of Richter blocks,[40] of various colors—brick, cream, and steel-blue. They came in the form of prisms, cubes, arches, cupolas, and pyramids. Out of these blocks we built houses and towers, royal palaces and fortifications following the instruction manual, which provided not only full views but also precisely drawn cross-sections of structures one might wish to build. When all was ready the actual game began, often with Father participating enthusiastically. First the kings sent envoys and then paid state visits. These visits would give rise to conflicts either because a king might be offended by the modest gifts offered by his guest, or because the latter might be arrested on suspicion of intriguing against his host. Then a war would break out. Cannon fired lead shot and shattered the fortifications. Peace negotiations would begin prior to a ceasefire, and the defeated side had to disarm and pay tribute. Dyk was a master of diplomacy and I practically never won a war since our game had no well-defined rules and he would be constantly inventing new strategic

---

[39] Stanisław Jan Cyganiewicz (also called Zbyszko I) (1880–1967), strongman and professional wrestler in the US in the 1920s. In Poland he had been a noted intellectual, having studied music, philosophy, and law while growing up in Vienna. His brother Władek (also called Zbyszko II) (1891–1968) was also a noted strongman and wrestler.

[40] Toy building blocks of stone, produced from around 1880 by the businessman Friedrich A. Richter in Rudolstadt, Germany.

2 The Gymnasium 33

and diplomatic ruses. He was completely unscrupulous, both Machiavellian and Napoleonic.

Chess was highly esteemed in our circles. I began to play it very early with the seven-year-old Dziunia, daughter of my Father's close friend Mr. Mravinesies. (A footpath in the park in Krynica[41] was named after him in gratitude for what he did for that town as State Commissioner.) My sister Irena played chess from the age of five. We spent a great deal of time playing chess but played it badly. It was only when I became a university student that I stopped playing since I then came into contact with real chess players and realized how much time and effort would be needed for me to become even just an average player. But that was later; in the Gymnasium I had several chess partners. The most outstanding of these was Stanisław Adamski, junior to me but universally gifted. On becoming interested in chess he acquired a copy of Bardeleben's[42] guide. While he was too lazy to study all of it, his memory and overall grasp were so good that he played better than all of us. This applied also to his violin playing, duelling, stamp-collecting, artistic photography, and his approach to the game of taroc.[43] Except for stamp-collecting, all of these were fleeting passions with him. When Polish literature caught his passing fancy, he mastered Chmielowski's[44] textbook. For a very short period he was a mountain climber. He would drop a pursuit as soon as it turned to drudgery. Later he showed himself to be gifted in business, and, having a good musical ear, sang in a choir, danced the mazurka exquisitely, and became renowned for his dancing, having also the requisite good looks and correct posture for it.

A big town may have more things to entertain a child than a small one, but then the fewer diversions there are, the more thorough the child will be in extracting pleasure from them. Take, for instance, a magic lantern with a few pictures to scrutinize. The most exciting movie would not afford the inhabitant of a big city as much pleasure as our magic lantern gave us. Or take the page-flip bioscope, the prototype of a film projector. As one of us flipped the pages, we saw four men calmly begin playing cards, then start in to punching each other, then turn the table over in their anger....

* * *

Around the time I graduated from the junior to the senior Gymnasium, several significant events occurred. First, Principal Sienkiewicz died and his place was taken by Józef Słotwiński, a philologist, idealist, and patriot, and grandson of a director of the Ossolineum.[45] (I should mention that the aforementioned Stanisław

---

[41] Krynica-Zdrój, a spa town and winter sports center in the Beskid Mountains of south-eastern Poland, in what is now the Małopolska Voivodeship.

[42] Count Kurt von Bardeleben (1861–1924), German chess master. Best known for a game he lost to Wilhelm Steinitz in 1895.

[43] An old Italian card game played with the 22 tarot cards in addition to the standard 52.

[44] Piotr Chmielowski (1848–1904), a Polish philosopher, literary historian, and critic.

[45] An institute for the promotion of Polish science, combining a library, publishing house, and museum. Founded in Lwów in 1817 by Józef Maksymilian Ossoliński.

Adamski ended up marrying one of our new principal's daughters.) Second, my eldest sister[46] got married; the ceremony and reception took place in our house and lasted till morning, champagne flowing and music reverberating. The guests included all members of our extended family, from near and far.

Such events partly obscured for us the increased political tension in the world, although the ideological currents flowing ever more turbulently at that time certainly affected us. Some of my fellow students were reading *Promień* ("Ray of Light"), a socialist paper, while for others *Teka* ("Briefcase") was more to their taste. The latter were called "endeks".[47] Patriotic celebrations had both official and unofficial components, the latter involving, for instance, a crowd of students going to a gunpowder magazine, where a seventh-grader would recite Mickiewicz's plea "From Russian, Prussian, and Austrian slavery deliver us, O Lord," before torching the magazine. The explosion lit up the scene—including the gendarme standing by pretending not to notice. This behavior baffled me since I knew very well that the older people—the students' parents—were loyal to Austria, yet seemed nonetheless to approve of the rebelliousness of their sons. My friend Władek Korczak, son of the Commissioner of the *starostwo*,[48] once quoted approvingly words of his father to me: "Poland would give me nothing. I live on the pension Austria pays me!" Count Władysław Michałowski, the *starosta* at the time, once said in my presence: "Poland! God forbid! One big, vacuous pomposity!" Soon Korczak senior had his son enrolled in the applied military school in St. Pölten,[49] from which he graduated with the rank of Cadet and remained in active service in the Austrian army. In 1918[50] he joined the Polish army.

* * *

The time came to think about what work one would be doing. Father wanted me to be an engineer. Most of my friends were aiming at law school—a path leading to the courts, the Bar, local administration, the post of public prosecutor, the treasury, and other government ministries. One advantage of such a course of study was that one did not have to live in a university town while pursuing it: one could live in the country somewhere and be paid 50 crowns a month for acting as scribe for some lawyer or notary public, and pay a beadle in Lwów or Kraków just a crown to arrange for the relevant law professor to sign attendance. Of course, once a year such a student had to travel to the university in question to sit for examinations. In most cases these were passed on the second or third attempt, and at the end of the whole process the young man received a graduation

---

[46] Felicja Steinhaus.

[47] Acronym formed from the initials of *Narodowa Demokracja* (ND), the National Democratic Party, the main nationalistic, anti-Semitic, Polish political party.

[48] Regional administration.

[49] In Austria.

[50] When the Austrian, Prussian, and Russian empires had disintegrated, and Poland had once more become independent.

certificate called an *absolutorium*. Such a fledgling lawyer was quite the personage: considered an excellent match and a substantial member of society, with access to all the world had to offer. The cleverer among such young men exploited these advantages to the full. After completing court practice and moving to the practice of law, they were called "concipients". The passing of the law examinations entitled them to affix a tablet inscribed with their initials followed by the word "lawyer" on the door of their apartment, which opened to them the possibility of earning three times the salary of a commissioner in the *starostwo* or a teacher in a Gymnasium. Having read through several thousand pages on Roman, canonical, and Austrian civil and penal law, these young men had lost all desire ever to read anything else. This kind of education shaped strange individuals, some immortalized by the writer Adolf Neuwert-Nowaczyński in his novel *Wesele panny Ciompalanki* ("The Marriage of Miss Ciompalanka"). In his satire *Małpie zwierciadło* ("Monkey's Mirror"), a distorted mirror of relentless mockery, he skewered everyone from the ultraconservative Count Stanisław Tarnowski[51] down to advisers to the tsar, officials of the lower court, commissioners, clerks, and beadles. There was an unbridgeable gulf separating the clerks and beadles from our young lawyer types, since the latter had not only learnt in the offices of the *starostwo* and court to fold sheets of paper in two and write on them *p.u., exh., videat exp., reproducatur*, etc., but also knew secret truths that their future authority depended on, such as: *nemo iudex in causa sua; neminem laedo, dum ius meum quaero; lex posterior derogat priori; pater est, quem nuptiae demonstrant.*[52] These and a hundred other such nuggets of wisdom were trotted out at every opportunity and above all during court proceedings, and peasant and Jewish clients were greatly entertained by the eloquence of the young men who, in arguing the rights to a ridge between two holdings used words unintelligible to both disputants—who nonetheless paid gladly and handsomely for the uplifting performance. Successful provincial lawyers might have incomes ten times that of a court judge. In Jasło in 1905 there were four top lawyers, including my Father's brother, Uncle Ignacy.

Uncle Ignacy came almost every evening "to the market place", that is, to our house, for a chat. It was fascinating to observe these two so very different brothers. Father laughed readily, loved life, loved his home and family, good food and drink, good company, his fields and gardens, his brickworks, and the horses in his stable. He believed that greed and ambition, excessive zeal and social climbing are just as much obstacles to happiness as idleness, prodigality, incompetence, and readiness to associate with just anybody. He enjoyed good weather, a well-constructed building, or a beautiful apple tree in the garden more than a big bank account. Uncle was a very different kind of person. His ambition caused him to set store by the political

---

[51]Count Stanisław Tarnowski (1837–1917), Polish nobleman, historian, literary critic, and prominent public figure. Imprisoned by the Austrian authorities 1863–1865 for participation in Polish agitation for independence. Professor at the Jagiellonian University in Kraków 1871–1909.

[52]"Nobody should be judge in his own case;" "I hurt nobody when I pursue my rights;" "A later law supersedes earlier ones;" "The father is the one to whom the marriage record points."

goings-on in big cities, where war and peace are decided, where well-planned transactions may yield fortunes, and where one can hobnob with influential and titled people. Father was fond of his brother, but dismissed his raptures over the fashionable world with a smile. Uncle was eloquent, his arguments logical, and rhetoric persuasive. The world had a different aspect for him than for Father. For him the word "house" connoted "mortgage", or income from rental, or a contract of sale or purchase, while for Father the word suggested in the first place foundations and walls, the truss bearing the roof, and the people occupying its apartments.

Father evaluated people and objects the way he evaluated wine. I remember an occasion when Uncle tried to talk Father into buying a certain farm. Father went to look at the farm but turned back without seeing it because the condition of the access road was so poor: "If there is no decent approach then the farm is not worth a damn!" On the other hand, for Uncle a property stood for a bank account and stocks, and whatever property he owned he was always considering selling. Naturally, this difference in temperaments became an asset in any joint venture—and when it came to major transactions they did indeed combine their talents. If, say, it was a question of the purchase of some estate, Father would be the one to inspect and evaluate it, while Uncle examined the mortgage documents and drew up the contract of purchase and sale. They founded together a Credit Society, of which Father was president and Uncle chairman of the board. This company functioned for many decades, and was liquidated only upon the death of the brothers. As a result of his constant tripping back and forth to Kraków, Vienna, and Lwów, Uncle strained his nervous system and ruined his health. He would say to Father using words similar to those of the usurer in Horace's ode[53]: "O how happy you are living in a house and not a hotel, sleeping in your own bed, and not in a berth, and eating food prepared in your own kitchen and not in a restaurant!" But then after a week in Jasło he would become restless at the thought that big things were going on in Lwów or Vienna without his participation, and he would once again set off, ruin his digestion, and return.... He once tried to persuade Father to move to Lwów, where he had been offered the directorship of a major bank, but Father merely told his brother that he would rather be number one in Jasło than number ten in Lwów.

Years passed, and my view of objects and incidents observed as a child changed when I became sixteen or seventeen. Thus my enthusiasm for the Austrian army diminished greatly—largely on account of my grandfather on my mother's side, who was an ardent pacifist. I began to take an interest in what was written in the newspapers. The Credit Society subscribed to the Kraków paper *Nowa Reforma* ("New Reform") and the Viennese *Neue Freie Presse*, and Father would bring both papers home from the office. At that time whole columns of the Vienna newspaper were taken up with the Dreyfus affair.[54] This was a liberal paper then run by

---

[53]The poem in question, Horace's second epode, begins *Beatus ille qui procul negotiis*... ("Happy is he who far from the world of commerce...").

[54]In 1894 Captain Alfred Dreyfus, a young French officer of Alsatian Jewish descent, was convicted of spying for Germany and sent to the penal colony at Devil's Island. When it emerged

2 The Gymnasium

M. Benedikt,[55] a Jew, and at that time it had three aims. The first was to do a favor to Germany, which supported the paper, by envenoming the dissension in French public opinion between Dreyfusards and anti-Dreyfusards. (The extent to which this was indeed a boon to Germany became especially clear as late as the Second World War, when a France still divided politically and militarily offered little resistance to the invading German army in 1940, and the puppet Vichy government, deriving largely from nationalist and anti-Semitic circles still smarting over *l'affaire Dreyfus*, faithfully served their German overlords from 1940 to 1944.) Benedikt's second aim was to increase the readership of his newspaper by exploiting the interest of German-speaking Jews everywhere in reading about an innocent Jewish-French captain—that is, Alfred Dreyfus—who was banished to Devil's Island as a German spy as a result of the intrigues of Count Esterhazy, a colleague and a member of the general staff of the French army. Every Jew who knew German wanted to read the open letter *J'accuse!* written by Émile Zola in defense of an innocent Jew, published in translation in the Vienna newspapers, and also the testimony of Picquart[56] and Clam,[57] as well as other officers, indicating that the spy in question was in fact Esterhazy. Benedikt's third aim was to flatter the decaying Austrian imperial regime by comparing its army and judicial courts favorably with the corrupt French ones. This was of great importance since the imperial and royal army was at that time beset with scandals arising from the unmasking of the spy Major Redl, the discovery that a member of the general staff called Hofrichter had been trying to remove colleagues blocking his path to promotion by poisoning them, and the exposure of civilian agitators like Friedjung who had fabricated a Croatian plot. Needless to say, none of these revelations increased the prestige of the two-headed eagle. Although my own sceptical sense was not yet very developed, my grandfather's was even less so: he believed everything he read in the *Neue Freie Presse*. He was especially enthusiastic about Baroness Suttner's[58] pacifist book *Lay Down Your Arms!*. I did not share his enthusiasm.

Grandfather lived at 6 Wałowa Street in Tarnów. He had sideburns like those of the emperor Franz Joseph I, liked to wear a top hat, carried a walking stick with a white knob, and used to take me to the Strzelecki Gardens. This was a very beautiful park with a complicated sundial in the middle of a flowerbed which particularly appealed to me. On Grandfather's desk lay wooden printing blocks for reproducing

---

2 years later that a French major named Ferdinand Esterhazy was the culprit, and not Dreyfus, the evidence was suppressed, and false documents were drawn up by high-ranking military officials who wanted to frame Dreyfus. When the cover-up leaked out, the intense political and judicial scandal that ensued split French society. Dreyfus was eventually exonerated and reinstated as major in the French army.

[55] Moritz Benedikt (1849–1920), an Austrian newspaperman.

[56] Major Georges Picquart (1854–1914) brought evidence of fabrications to his superiors, and, when ordered to keep silent, leaked the information to the Dreyfusard press.

[57] Lieutenant-Colonel Armand du Paty de Clam (1853–1916), main accuser of Dreyfus.

[58] Bertha von Suttner, an Austrian peace activist, wrote *Die Waffen nieder!* in 1889.

drawings of cross-sections of a ring-shaped brick oven of his own design, whose novelty lay in having the exhaust gases from the kiln piped to a second chamber to dry the bricks prior to baking, before finally being allowed to escape through the chimney. Next to these lay models of roofing tiles, also of his design, with special grooves. Grandfather had lost a lot of money as a result of a suit he brought against Prince Sanguszko, for whom he had built a brick kiln. He claimed that the prince had not paid him enough, but lost the case. His only income came from his work as a representative of the Portland Cement Co. in Opole. His family came from Oświęcim,[59] before that from Piotrków, and before that Gdańsk; thus his ancestors had lived in Poland prior to the eighteenth century partitions.[60] (My Father's family, on the other hand, moved from Hungary to Galicia only at the beginning of the nineteenth century.)

My Tarnów Grandfather tried to make me loathe the army and militarism generally because he did not want me to become an officer. He told me that as a youth he had wanted to become a scholar, and that to this end he had studied Hebrew and Latin, but been forced to abandon his studies in order to earn his daily bread. He was tenderhearted, without any envy or hatred; he was just by nature and incapable of being argumentative. There were no spots or scratches or other blemishes on his body and he lived to be 83. I remember his advising me to rise before dawn and, alone in the quietness, witness the waking of the birds, who at first test their voices hesitantly, and then, as the sun rises, burst into a splendid joyful symphony.

Grandfather's fears were baseless since when still in the Gymnasium I decided on a future for myself that had nothing to do with things militaristic. Although Father wanted me to be an engineer, he did not pressure me—he knew how rapt I became when reading Jules Verne's novels and how interested I was in scientific discoveries. The library in our house was very modest, and I read more or less at random from it. It did contain the first volume, dealing with antiquity, of Szujski's[61] excellent *Literatura powszechna* ("General Literature"), which I read ecstatically many times over. Among my Father's acquaintances was a young engineer named Ludwik Silberman, who, in addition to qualifying in engineering at the Polytechnic, had also completed the mathematics program of the philosophical faculty of the university. When, at the first opportunity, I asked him where one might learn more mathematics, he told me that the university was the place for it. It was then that I made my decision.

The possibility of learning mathematics on one's own was somewhat remote. I once got hold of Dziwiński's[62] *Lectures on Mathematics* written for students at

---

[59] Auschwitz, in German.

[60] Partitions of the then Polish-Lithuanian Commonwealth by Russia, Prussia, and Austria in 1772, 1793, and 1795. With the third partition, Poland ceased to exist as a self-governing nation till 1918.

[61] Józef Szujski (1835–1883), Polish politician, historian, publicist, poet, and professor at the Jagiellonian University.

[62] Placyd Zasław Dziwiński (1851–1936), mathematics professor at Lwów Polytechnic. The *Lectures on Mathematics* (*Wykłady z matematyki*) is the work for which he is known.

2 The Gymnasium

the Polytechnic. While this fat volume impressed me tremendously, it contained so much material that I could not see where to begin; only much later did I find out that the chapters were ordered according to the letters in the author's name. Then Dr. T. Warchałowski, who worked for my Uncle as administrator, loaned me Bendt's textbook[63] on differential and integral calculus. I tried to read it and actually began to understand it. Then I came across *Analysis* by Burg.[64] In this book I encountered Taylor series, which I found fascinating. Thus I read without direction and without having more appropriate textbooks indicated to me. However, this may have been a good thing because if I had devoted myself too early to the study of mathematics, I would have neglected other things of importance.

The matriculation examination was now imminent—at that time still an elaborate ritual. It consisted of both written and oral examinations in mathematics, physics, Polish, German, Latin, Greek, and history. If one had excellent grades in any of these subjects from the preceding term, then the corresponding oral exam was waived, and in this way I avoided the orals in mathematics, physics, and history. In my last year—the eighth—I had been very busy with various things, and thus did not expect to have any oral exams waived—in fact I had made contingency plans in case the examination results were negative. I had been reading Schwegler's *History of Philosophy*[65] and learning English. My English teacher was an elderly Englishwoman with the looks of a fairy-tale witch. After living in our house for half a year, she decided to take a trip to Hungary. Since such a trip required changing trains at Nowe Zagórze, she thought of taking a preliminary excursion there to examine the timetable as it pertained to the rest of the trip. However, on arriving in Nowe Zagórze she was surrounded by a crowd of people who seemed to take her for a real witch—moreover an English-speaking one. She sought shelter in a store and, helped by the police, she managed to escape through the back door to the train station. My cousin Dyk had given me some dozen or so lessons using the Berlitz method. Her method was different: I would read an English text, with her correcting my pronunciation as I went along and explaining in English the meanings of the words I did not know. This method requires a great deal of effort on the part of the student but yields rapid results. The Berlitz method, which is based on the idea that children learn their native tongue by having objects pointed out to them and named, is not a very effective one. In the first place, it requires having a variety of objects or drawings of objects ready to hand, and, secondly, the procedure for assimilating abstract words is too long and complicated. Simple-minded people tend to exaggerate the difficulties arising from ignorance of the language when in a foreign country; the knowledge of the language sufficient for avoiding thirst and hunger is minimal, and can be acquired in less than a day. It may even happen that it

---

[63] F. Bendt, *Katechismus der Differential- und Integralrechnung*, Leipzig, Weber, 1896.

[64] Adam von Burg (1797–1882), professor of higher mathematics and mechanics at the University of Vienna.

[65] Albert Schwegler (1819–1857), German philosopher and theologian. Published *Geschichte der Philosophie im Umriss* in 1848.

is better to be without such knowledge—especially if it is not quite correct. This was brought home to me much later, on a hot day in Florence. In the square outside the Palazzo della Signoria, where stands a replica of Michelangelo's statue of David, we ran into an old woman with a younger one—probably her daughter—carrying a bag of fresh figs. I offered to buy the whole bag, using for this purpose the Polish word "figi", whereupon they both burst out laughing, gave us the fruit, and ran off without asking for payment. I later found out I had said something very improper in Italian.

In my last year at the Gymnasium, 1904, the Russo-Japanese war broke out over Korea and Manchuria. This was for us a most sensational development. I remember when we heard the news of the sinking of the Russian battleships *Tsarevich* and *Retvizan* in the harbor at Port Arthur. We—my immediate family—all sided with the Japanese. However, my cousin Ryszard, who at that time happened to be visiting us with his father, my uncle Marceli Frydman, had decided to enlist in the Russian army. He expounded to us the "realistic" view held in Viennese military circles, according to which the Russian colossus would quickly squash the Japanese squib and set up a Russian governor-general in Tokyo in place of the Mikado. But in Jasło, and in fact throughout Galicia, things Japanese became fashionable: the newest cufflinks were said to be in the Japanese style; a brand of collar was named "Banzai"; there arose a demand for lacquered products; and Japan was looked at through the eyes of Lafcadio Hearn[66] and Wacław Sieroszewski[67] as a land of brave soldiers, samurai warriors who put honor before life, geishas like Madama Butterfly, playing shamisens in little bamboo teahouses set in groves of cherry trees blossoming all year long. What neither my cousin Ryszard nor his father knew was that a new Prussia was already coming into being in the Far East, that German officer-advisors had introduced drills devised in Potsdam into the Japanese military barracks in Kobe and Nagasaki, and that the erstwhile samurai warriors and geisha girls preferred the smell of factory smoke to the scent of cherry blossoms. Little wonder that they were unaware of all this: on Ryszard's father's desk there lay not statistical reports, but the odes of Horace in the original Latin.

For the Polish cause a negative outcome for Russia in the war with Japan was desirable. The newspapers published ever more frequent reports—communicated by telegraph—of riots in Congress Poland and in Russia itself. We had no direct contact with the regions where outbreaks of violence occurred. However, I did come across a man who had avoided being impressed into the Russian army by fleeing from Congress Poland to Galicia, ending up in Jasło. It was springtime and the Wisłoka was high. We decided to help the carpenter Juryś go rafting. The crew was to include Stanisław Adamski, Ludwik Oberländer, Jan Węgrzyński, Juryś, the deserter, and

---

[66] Journalist and writer. Moved from the United States to Japan in 1900 at the age of 40, where he wrote books about Japan and published collections of Japanese legends and stories.

[67] Polish writer, active in the Polish Socialist Party, engaged in activities subversive to the Russian empire. Lived from 1858 to 1945.

me. When I asked the deserter why he had run away, he answered that the Russians had gone crazy and were sending their soldiers to a certain death.

The matriculation examinations were held on May 5, 1905, and were followed by five months' vacation. As there was nothing in particular to occupy me for such a long period, I was very grateful to Dr. Józef Kadyj when he offered to take me with him on an expedition to the Tatra Mountains[68] organized by his brother Henryk, a professor of anatomy. This was to be the first time I saw these mountains, which were several days' journey away. We crossed Goryczkowa Pass, skirted Krywań[69] through Cicha Glen to Podbańska, where we spent the night. If you hike through subalpine forests, then across mountain pastures, and through stands of dwarf pine, till at last you feel an unfamiliar wind blowing up the pass, see the lush sunwarmed grass of the southern slopes, and the Hungarian plain shimmering in the distance, radiantly blue, a plain guarded by Krywań, crouched like a dark lion, and hear below the sighing of a million spruce in the silence—if you do all this, then you will become the sworn enemy of the cable railway. Our group consisted of the professor and his wife, the doctor and his three children, a guide, and myself. (I note by the way that the professor's eldest daughter was one of the first female students to be admitted to the University of Lwów.) From Podbańska we descended to Lake Szczyrbskie, where, in an elegant restaurant with a Gypsy orchestra, the waiters looked askance at our hobnailed boots and heavy backpacks. The Hungarian summer resorts were western European in style, and Polish tourists were viewed by the Hungarians as savages. There was in this attitude a hidden component of dislike of their Slav neighbors, who were able to communicate so much more easily with Slovak highlanders than Hungarian gentlemen could. There were also many Budapest Jews looking more Magyar than the Magyars. More than 35 years later history would show that the Magyars did not reciprocate this fellow-feeling towards Jews, nor the Slovaks that of the Poles, and neither did Hungarian-Polish friendship prove to be more than pure fiction. But at that time the world wars were yet to come.... The only international quarrel then going on was the contest over Morskie Oko.[70] This started as a private dispute about the property line between lands belonging to Prince Hohenlohe and Count Zamoyski, but became an issue of jurisdictional, and even international, legality, because the Polish side cited grants by Polish kings and the Hungarian side those of their former kings. It came down to the question of where the Polish-Hungarian border had been before the first partition,[71] when Maria Theresa took control of Galicia and included it in Cislitawia. It followed that the dispute concerned the border between the Austrian and Hungarian halves of the Dual Monarchy, so that the opinions of neither Austrian nor Hungarian tribunals were decisive. Thus did the quarrel between Hohenlohe's German foresters and Polish highlanders become an international issue. The Austrian government

---

[68] The highest range in the Carpathians, forming a natural boundary between Poland and Slovakia.

[69] A mountain in the southwestern part of the Tatra Mountains with a curved tip, whence its name.

[70] "Marine Eye", the largest lake in the Tatra Mountains.

[71] In 1772.

appointed Oswald Balzer from the office of Galician Land Division in Lwów as its representative, and the Hungarian faction also chose a spokesman. These delegates in turn chose as chief adjudicator a Swiss historian by the name of Johannes Winkler to chair the International Court of Conciliation in Graz. Balzer convinced the tribunal that the Hungarians had interpreted the Latin text of a key document incorrectly, and won the case convincingly. Thus it was that the whole of Morskie Oko fell to Poland, of which it remains a part to this day.

Chapter 3
In the Capital Lwów

The vacation was long, and I occupied it mainly by preparing myself, a Jasło novice, for initiation into the ranks of what were then called in Galicia "academics"— a gross exaggeration, naturally. I spent a few weeks alone in Rabka[1] acquiring appropriate clothes and learning the elements of descriptive geometry, which I wanted to master at least to the level taught in the technical colleges, where it was a required subject. My cousin Dyk, my senior by a year, was already studying law in Lwów[2]. His mother had moved there to be with him, even though he still spent the vacations in Jasło. The transition from Gymnasium to university necessitated replacing the school uniform by a suit. Dyk directed me to the firm of Heilman and Cohn, where I bought a dark-blue outfit off the rack. Thus was I fitted out to appear in the Galician capital. Lodging had been organized beforehand: I was to be a subtenant "with board and service" in the flat rented by Dyk's mother at 10 Gołąb Street in Łyczaków. This had been easy to arrange since my parents and Dyk's widowed mother got on very well.

Dyk tried to impress me as a conqueror of women. In his home in Jasło there had been a nice-looking serving-girl who some years later married a craftsman. My cousin assured me that he had had an adventure with her like that in Żuławski's story *Impregnable Virginity*. I was sceptical about his account until I realised that if the parallel with Żuławski's story was accurate then it was he who was the conquered one.... It was easier to believe the story of a classmate, son of a village schoolteacher, who told me that *their* serving-girl had resisted him heroically till he offered her five dozen pears, the designated price of her virginity. Many of the older students frequented *The Red Inn* in Ułaszowice, just north of Jasło, whose atmosphere fit the description of a bordello in Zegadłowicz's novel *Nightmares*.

---

[1] A town between Kraków and Zakopane on the northern slopes of the Gorce Mountains.

[2] The capital of the Kingdom of Galicia and Lodomeria, the Austrian part of Poland, from 1772 to 1918 (the period of the partitions of Poland). Lwów, or L′viv in Ukrainian, is now in western Ukraine.

Fig. 3.1 View of Fredro Square, Lwów, as seen from St. Nicholas Street. The Scottish Café (see Chapter 7) was on the ground floor of the building on the right (Courtesy of the National Library of Poland, public domain.)

In Lwów I enrolled in the philosophical faculty, choosing to study mathematics and "exact" philosophy, a term invented by the Lwów University bureaucrats whose meaning was unclear. Naturally, like almost all first-year students I registered in Twardowski's[3] courses. Kazimierz Twardowski, from Skrzypna, descended from magicians, must also have been one—this is the only explanation of the hypnotic influence of his personality. He never sought popularity—rather the opposite, since he scheduled his lectures at eight in the morning, and in summer at seven. He scolded latecomers in public, insisted on accuracy and punctuality at his seminars, and ran his workshops with almost military discipline. Thus his behavior was such that, while in any other professor it would merely have generated antipathy, in his case it attracted an ever-increasing crowd of students of philosophy, law, mathematics, and the humanities. He was a graduate of the *Theresianum*,[4] a Gymnasium attended by the children of Austrian landed gentry and high government officials, where, after having first qualified as an actuary, he had settled on the study

---

[3] Kazimierz Jerzy Skrzypna-Twardowski (1866–1938), Polish philosopher. Founder of the Lwów-Warsaw school of philosophy.

[4] A private semimilitary high school for Vienna's bourgeois *élite*, founded by Maria Theresa in 1746.

Fig. 3.2 This 1898 photograph shows the Diet of the Province of Galicia and Lodomeria, then part of the Austro-Hungarian Empire. It became the main building of the Jan Kazimierz University of Lwów after World War I (Photo by Edward Trzemeski, Wikimedia Commons, public domain. Creative Commons Attribution/ShareAlike License.)

of philosophy under Brentano.[5] He was gifted musically, and actually composed music; I heard the music he had composed for Nietzsche's poem *O Mensch! Gib acht!*[6] He knew both ancient and modern languages. He prepared his lectures with meticulous care, and particularly stressed the importance of the history of philosophy. One learnt a great deal from him. For example, it was from him that I first heard of Lange's[7] *History of Materialism*,[8] which had been translated into Polish, and it was his mentioning Schopenhauer in his lecture on "scientific ethics" that led me to read that philosopher's works. Another philosopher, Professor Mścisław Wartenberg, gave lectures during my first academic year 1905/1906 on metaphysics and Kant's antinomy. I remember him asking in a loud voice, brooking no objections, the rhetorical question: "What could the letter t mean in Newton's equations of dynamics if there were no such thing as absolute time?!"

---

[5] Franz Brentano (1838–1917), influential German philosopher and psychologist.

[6] *O Man! Take heed!* Set to music by many others, including Gustav Mahler.

[7] Friedrich Albert Lange (1828–1875), German philosopher, pedagogue, political activist, and journalist.

[8] A standard introduction to materialism and the history of philosophy still in use well into the twentieth century.

I also heard about Ernst Mach,[9] and read his *Analyse der Empfindungen*[10] and *Populärwissenschaftliche Vorlesungen*.[11] Mach is an example of a popularizer who is not a vulgarizer. The leading representative of that illustrious set is Henri Poincaré.[12] In Poland there was Julian Ochorowicz,[13] but now there are virtually no such popularizing scholars here. We were lectured on mathematics by Józef Puzyna,[14] who had been a student of Fuchs[15] in Berlin—too bad he hadn't studied there earlier when Weierstrass was in Berlin! Many mathematicians of our epoch came to a sad end: the lives of Trzaska Kretkowski,[16] Rajewski,[17] and Kępiński[18] were ill-starred. On the other hand, Lwów could take pride in having among its faculty the first-rank physicist Marian Smoluchowski,[19] recruited from Vienna by his Theresianum class-mate Twardowski. The social sciences were in fashion then, so I attended the seminar of Stanisław Grabski[20] on the methodology of the social sciences. I must confess that I understood practically nothing, and was lost in admiration for my fellow student Ostap Ortwin for the way he got involved in learned discussions with the professor. Deciding that I should learn sociology, I borrowed Émile Durkheim's *De la division du travail social* from the library—to no avail. I could not understand why Durkheim considered the marriage customs of

---

[9]Austrian physicist and philosopher. In particular, he explained the puzzle of the behavior of rotating objects in terms of the background of stars in the galaxy. Exerted a major influence on the logical positivists. Lived from 1838 to 1916.

[10]"Analysis of Sensations".

[11]"Popular-scientific Lectures"

[12]Jules Henri Poincaré (1854–1912), preeminent French mathematician and theoretical physicist. Often called "the last universalist" in mathematics.

[13]Julian Ochorowicz (1850–1917), Polish inventor, philosopher, and theorist of positivism.

[14]Prince Józef Puzyna of Kozielsk (1856–1919), Polish mathematician. After completing his doctorate in Lwów, he went to Germany and attended lectures of Kronecker and Fuchs. Appointed professor in the University of Lwów in 1892.

[15]Lazarus Immanuel Fuchs (1833–1902), German mathematician. Fuchsian groups are named after him.

[16]Władysław Trzaska Kretkowski (1840–1910), Polish mathematician. Studied mathematics in Paris and Kraków. Habilitated in Lwów.

[17]Jan Rajewski (1857–1906), Polish mathematician and physicist. Professor at Lwów University from 1900.

[18]Stanisław Kępiński (1867–1908), Polish mathematician. From 1898 professor at Lwów University and Polytechnic.

[19]Marian Smoluchowski (1872–1917), Polish physicist. Pioneer of statistical mechanics, in a similar vein to Ludwig Boltzmann. Awarded a prize in experimental physics by the Vienna Academy of Sciences in 1908. Noted mountaineer.

[20]Stanisław Grabski (1871–1949), Polish economist and National Democratic politician. Ardent Polish nationalist. Retired from political life in 1926. Professor of Economics at Lwów University 1910–1939.

the Polynesians sociologically significant, but not their methods of waging warfare. In a nutshell, my sociological abilities were null.

* * *

In my first academic year I crammed without a break, there being nothing better to do that I could see. Although I had had a spell of taroc mania in sixth grade at the Gymnasium, I no longer felt like playing cards. I did not like imbibing alcohol, and as for dancing, I adhered to the maxim: *nemo saltat sobrius nisi forte insanit.*[21] My cousin Dyk was something else: he was sociable and an excellent dancer. Periodic dance parties had been held in our house in Jasło, and the dancing school *Karolina Witkay & Son* used to visit us once a year. But both there and in Lwów I was most of the time a mere passive onlooker at parties and fancy-dress balls. Occasionally, after dancing the night away, Dyk and the incomparable mazurka dancer Stanisław Adamski would wake me at seven in the morning to take me for a swim in the Wisłoka.

In Lwów I got to know who in the student body was politically active. There were two basic political factions, and everybody was caught up in the conflict between them. My cousin belonged to the organizing council of the progressive party, which called itself simply *Life*, and took his position very seriously, but I could never work up enough interest to attend even one of its evening meetings. Andrzej Niemojewski,[22] author of the cycle *Legendy* ("Legends"), used to give lectures at these meetings. However, *Legendy* was banned by the public prosecutor.[23] In particular, he advocated detailed critical analysis of Holy Writ, which made him *persona non grata* with the clergy, but *persona grata* with the progressives of *Life*. (As is well known, after World War I this same Niemojewski became a leading anti-Semite.)

The leaders of the *Life* movement were the Downarowicz brothers, and of the opposing camp Dmowski and Balicki, who were idolized by their followers. These two luminaries often attended meetings of *Life* to debate principles with their opposite numbers. The outcome of these debates, which went on all night, were heatedly discussed next day by the mass of common "academics". Among the latter there was a great variety of student types: anarchists, lovers of Sanskrit, perpetual students, those who lived off their winnings at a game of chance called "ferbel" played in the Central Café, impoverished young men who had mastered the fine art of surviving on 30 crowns a month, and law students eking out a living by struggling to spoon-feed less capable but richer young men with the contents of several thousand pages of Roman, canonical, and Austrian law. There were

---

[21]"Nobody dances sober unless he is quite mad".

[22]Polish writer and poet of the period of *Young Poland*.

[23]The ban was demanded by the church authorities. However, when certain members of parliament objected, Niemojewski's *Legendy* were allowed to appear in 1902 as *Tytuł skonfiskowany* ("A banned title").

students who wore bracelets, who pronounced *bardzo* as *bahdzo*,[24] students who were sycophants—doubtless in the hope of obtaining positive recommendations for high-level positions in the emperor/king's bureaucracy, soapbox orators, who on May Day exhorted crowds from rostrums covered with red cloth, and decadents who hung out in the Crystal Café in the Mikolasch Passage from early evening till dawn. Yes, Lwów was an interesting place! It also boasted many beautiful women, including some of my female fellow students.

My cousin once brought me hectographed[25] copies of a satirical student periodical published by the Jewish *élite*, golden lads whose pamphlets and parodies were aimed at proving before all how great their intelligence and sense of humour was. Twenty years later it was evident that these "café flowers" yielded no fruit, since none of them rose to be anything more than an average newspaper reporter—with one exception, namely Roman Tenner, a lawyer. He conceived the excellent idea of making so-called "war games" fairer by introducing the casting of lots weighted in proportion to the experience of those participating. He did not publish his idea and may not know to this very day the importance that accrued to it when scientific game theory[26] was invented at a later time and in another place.

At that time Lwów went through a period of feverish construction. Entire new streets of buildings came into being all at once, it seemed, and various Viennese and Galician banks financed the construction of multistorey structures on which laborers worked all day and night, so that the city was flooded with light from arc lamps placed high above the scaffolding. On Carl Ludwig Street[27] new stores appeared, with their show windows brilliantly lit till late at night. The sidewalks were thronged with people, at least half of whom were prostitutes, as one could tell from their inelegant but provocative style of dress. Of course, there were many strata in the profession, the loftiest being occupied by ladies who were driven by coach to the theater where they had their own boxes. Most of the clients of the denizens of that *demimonde* were army officers, who were in plentiful supply in Lwów. On Sundays crowds of them would gather at the corner of Academic Street to watch the ladies coming out of the cathedral after the 12 o'clock mass.

---

[24] *bardzo* means "very".

[25] Hectography is a printing process involving transfer of an original, prepared with special inks, to a gelatin pad pulled tight on a metal frame. The finished product is called a hectograph, or sometimes a "jellygraph".

[26] Game theory analyses conflictual scenarios where each party may choose any strategy from a prescribed set. In 1925 Steinhaus wrote a note setting out the fundamental notions of this theory. The note, entitled "Definicje potrzebne do teorji gry i pościgu", was published in the obscure review journal *Myśl Akademicka* in Lwów. An English translation entitled "Definitions for a theory of games and pursuit" appeared in the journal *Naval Research Logistics Quarterly*, 7 (1960), pp. 105–108. See also H. Steinhaus, *Selected Papers*, PWN, Warsaw 1985, pp. 332–336. (The French mathematician Émile Borel also published papers on game theory in 1921. However, the credit for developing and popularizing the theory is now usually given to John von Neumann, whose first paper on that topic appeared in 1928.)

[27] The main street of Lwów, renamed many times since.

The most popular theatrical attraction of the time was the operetta *The Merry Widow*, and its lead, Helena Miłowska, was the brightest star in the Lwów heaven. Lvovians were great fans of opera. My cousin Dyk, who was gifted musically, liked to play piano adaptations of the scores of the classical operas. The Italian tenor Dianni, who felt at home in Lwów, was *persona gratissima* to Dyk's family. The dramatic works of Ibsen and Maeterlinck were playing in the theaters then. The most popular play by a Polish playwright was Żuławski's *Eros i Psyche*, which the Lwów *élite* considered the last word in modernized and philosophically subtle Romanticism. It was now fashionable to view Mickiewicz—turned to stone and stood on a pedestal outside the George Hotel—as a writer for country squires, fit only for recitation by school children and at Gymnasium *soirées*. One swore allegiance instead to Słowacki,[28] for whom one of the theaters was named. The then youthful Juliusz Kleiner[29] had crowds—consisting mostly of females—attending his lectures on Słowacki, which were excellent even if his delivery was somewhat bombastic.

Literature was very much in vogue. It was in fact a highly creative era for European culture. One could see this in literature and painting, in music, although to a lesser extent, and—less easily since they are of a more technical nature—in the sciences. But in architecture a dreadful Secession prevailed, and this resulted in a blunting of the visual artistic sensitivity of the educated public. No one seemed any longer to appreciate the splendid Roman cathedral, the beautiful baroque *façades* of the old apartment houses fronting onto the Market Place, the lovely design of the Bernardine cloister and church, the group of structures making up the Church of St. Nicholas with its belfry and wall, the renaissance registry building at the rear of the Dominican Church, nor the glorious St. Jura Cathedral, with its curious Armenian alcove separated off with an inclined balustrade. No, Lwów was too progressive for that! It preferred the new multistorey Viennese buildings, stuccoed *façades*, the new train station—the largest in the whole of Austria-Hungary!—the electric street car, the arc lamps, and the boulevards. The byword was "comfort", and in time this word took on a special technical meaning, so that in Lwów one would routinely hear the phrase "an apartment with semi-comfort".

All in all, Lwów gave the impression of being unambiguously a capital city. A great many of its inhabitants were rich, having made money by dealing in Podolian wheat, Carpathian wood, or oil from Drohobycz.[30] This money was used to build or buy houses, or lost at the horse races, or frittered away at the tables of the *Casino*

---

[28] Juliusz Słowacki (1809–1849), Polish Romantic poet. One of the "Three Bards" of Polish literature. A major figure of the Polish Romantic period, his work often contains elements of Slavic pagan traditions, mysticism, and orientalism. Regarded as the father of modern Polish drama.

[29] Polish historian and literary theorist. Lived from 1886 to 1957.

[30] Podole (Podolia) is a historic region of Eastern Europe which belonged to Poland between the fifteenth and eighteenth centuries and again during the twentieth century interwar period. It is now in western Ukraine. Drohobycz was the name of a city near Lwów, now called Drogobych and belonging to western Ukraine.

de Paris while drinking champagne, or spent on carriages or rented boxes at the theaters. I was a mere onlooker at all of these goings-on, both passive and distant: I was not present at any of the receptions where strawberries imported from Italy were served in winter, or quail from God knows where, nor did I watch girls dancing the cancan, the crowning attraction and finale at the parties of the "golden lads".

* * *

My university studies were interrupted not only by the vacations but also by political events such as those involving the Ruthenians.[31] From time to time students of this nationality organized demonstrations demanding more Ruthenian university departments, a Ruthenian university of their own, the use of their own language in official dealings, and so on. (They also wanted even then to be called "Ukrainian", a right to which they themselves were of course fully entitled, but one—and this they failed to understand—not enforceable on other peoples, since these have their own customary name for them.... To take an analogy: it would not be reasonable for the Germans to insist that people of other nations call them "Deutsche".) Some Polish students organized counter-demonstrations, and during one of the resulting brawls the Ruthenians occupied part of a building on St. Nicholas Street and barricaded themselves in with benches. The Poles promptly took over the rest of the building, aiming to expel the invader, and a battle began in earnest. The conflict lasted three days. As weapons the combatants used sticks and burning logs which they hurled like grenades at the enemy. Occasionally an armistice would be declared for an hour or longer to allow for the removal of the wounded; during these breaks in the fighting, opposing combatants offered each other cigarettes, and chatted in a friendly manner. Neither Polish nor Ruthenian members of the Socialist Youth movement took part in the battle. A platoon of policemen was stationed on the ramp opposite the Church of St. Nicholas, but did not intervene. This is explained by the fact that the Rector of the university had jurisdiction over university territory, so that the police had to remain off campus unless he specifically asked for their help—which in this case he refrained from doing. Although "in its soul" the Austrian government sympathized with the Ruthenian cause and tried to meet some of their demands, it was unable to pursue such a policy of appeasement too radically because the prime minister Baron Gautsch did not want to incur the displeasure of a certain parliamentary faction crucial to the government's majority.[32] In any case, there were in fact several departments of Lwów University devoted specifically to things Ruthenian: the departments of the Ruthenian language, of Ruthenian history, of Ruthenian literature, and a few of the law departments.

---

[31] The Ruthenians were originally Ukrainian-speaking people dwelling in the deep, narrow valleys of the Carpathian Mountains. In the eleventh century, Ruthenia (or "Subcarpathian Ruthenia") was taken over by Hungary. A vigorous movement in the second half of the nineteenth century to save themselves from Hungarianization ultimately failed.

[32] At that time there were several conflicts going on between the very diverse nationalities that made up the Austro-Hungarian empire.

Chapter 4
Göttingen

In the spring of 1906 Stanisław Jolles, professor of descriptive geometry at the Charlottenburg Polytechnic in Berlin, came to Lwów with his wife on family business having to do with certain factories near Lwów owned by Mrs. Jolles. This lady happened to know my cousin's mother, so she and her husband paid us a visit at 10 Gołąb Street. The professor was soon informed that the subtenant—that is, I—wanted to be a mathematician, whereupon he immediately exclaimed: "Junger Mann, packen Sie Ihren Koffer und fahren Sie nach Göttingen!"[1] My Father's reaction to Jolles' "categorical imperative", when I came back to Jasło at the end of the 1905–1906 academic year, was lukewarm, but Mother favored the idea keenly—all the more so because she knew how eager I was. Thus it was that I transferred to Göttingen in the autumn of 1906. I travelled via Wrocław,[2] Berlin, and Halle, and, finally, Eichenberg. The Eichenberg–Göttingen train was a local one and almost empty; its only passengers were a few locals and some students wearing bright little caps. From the Bovenden station onwards I had a good view of the landscape of the place where I was to spend the next several years: low hills covered with deciduous forest, and in the foreground the steeple of the Johannis Church and some reddish roofs. The only colors were rust, brown, tan, and yellow: such were the land, the autumn leaves, and the roof tiles.

I took a room in a tiny hotel in Judenstrasse, and looked through the window onto the quiet street. I saw excited children amusing themselves with conical tops, which they kept spinning by means of little whips. The scene made me feel suddenly far from home: our children didn't play with this toy. A baker's wagon passed, its driver wearing a white cap. The harnesses were different from those in use back home, and the horses more of the cold-blooded breeds.[3] I recalled seeing such harnesses at

---

[1] "Young man, pack your trunk and go to Göttingen!"
[2] Breslau in German.
[3] The terms "cold-blooded" and "hot-blooded" as applied to horses, relate to temperament. "Cold bloods", such as draft horses and some ponies, are more suitable for slow, heavy work, while "hot

home in the show windows of stores selling children's toys, and realized then that these must have been made in Germany. For some reason this discovery displeased me.

Next day I went to the Rectorate, where the head beadle gave me a list of boarding houses. I ended up taking a room with board in a house owned by a Miss Thomas, at 82 Weender Strasse, next to the university auditorium, for 80 marks a month. Fräulein Thomas' table was frugal but sufficient. The term had not yet begun, so I acquainted myself with the professors beforehand by perusing the lecture timetable. It seemed to me then that people with such strange names as Hilbert,[4] Carathéodory,[5] Minkowski,[6] and Herglotz[7] simply had to be great scholars—and this turned out to be true even though my initial estimate had no rational foundations.

I had to learn how to speak German properly—for example, that it was impolite to greet someone without mentioning his or her name. Thus every morning I had to say "Guten Morgen, Fräulein Thomas!", and at noon, in the street: "Mahlzeit, Herr Lehmann!". That one shouldn't say simply "Guten Tag", was strange but true; however I didn't always observe such niceties.

My aunt had done some investigating, and discovered that Antoni Łomnicki,[8] who had a doctorate from the University of Lwów, was pursuing his academic career in Göttingen. It turned out that he was living nearby, opposite the auditorium, with his wife and one-year-old daughter Irena, and I paid them a visit. There were, in fact, many Poles living in Göttingen at that time: among the students, there were the brothers Władysław[9] and Wacław Dziewulski,[10] of whom the elder, Władysław, was an assistant in the department of astronomy,

---

bloods" have been bred for speed and endurance. Of course, all horses are warm-blooded in the biological sense.

[4]David Hilbert (1862–1943), German mathematician. One of the most influential and universal mathematicians of the nineteenth and early twentieth centuries. In 1900 he proposed a list of 23 problems which strongly influenced the course of much of the mathematical research of the twentieth century. Arrived in Göttingen in 1895 to lead the mathematics department.

[5]Constantin Carathéodory (1873–1950), Greek mathematician. First studied engineering in Belgium. Completed his graduate studies in Göttingen over the period 1902–1904 under the supervision of Minkowski.

[6]Hermann Minkowski (1864–1909), German mathematician of Lithuanian-Jewish descent. Invented the number-theoretic technique known as the "geometry of numbers", and formulated the special theory of relativity in geometrical terms, among other things. Joined Hilbert in Göttingen in 1902.

[7]Gustav Herglotz (1881–1953), German mathematician. Studied in Vienna and Munich. Completed his graduate studies in Göttingen under Felix Klein.

[8]Antoni Łomnicki (1881–1941), Polish mathematician, educated at Lwów University and the University of Göttingen. Appointed professor at the Lwów Polytechnic in 1920, where Banach worked as his *Assistent*. Murdered by the Gestapo in Lwów in 1941.

[9]Władysław Dziewulski (1878–1962), Polish astronomer and mathematician. Professor at the Stefan Batory University in Wilno, and then the Nicolaus Copernicus University in Toruń.

[10]Wacław Dziewulski (1882–1938), Polish physicist, editor, and sometime professor at the Stefan Batory University in Wilno.

and the younger, Wacław, was studying physics; Felicjan Kępiński from Piotrków, studying astronomy; and Leon Chwistek,[11] who arrived with his sister from Kraków somewhat later. Among the others were the astronomer Tadeusz Banachiewicz,[12] Włodzimierz Stożek,[13] Antoni Przeborski,[14] Kazimierz Jantzen, Moroński, K. Horowicz,[15] Janaszek, Sokołowski, Szadurski, the actuary Doerman, the medical doctor Ryder, the philosopher Rozenblum, the lawyer Marchlewski, the physicist Jan Króo,[16] the mathematician Żyliński,[17] the anthropologist Edward Loth, and many others. But they were far from being the only foreigners in Göttingen: there were students from America, England, Hungary, Japan, Russia, and even India. The reason there were so many Poles in Göttingen at that time had to do with the continuing strikes[18] at the schools and universities in the Kingdom of Poland.[19] It was noticeable that in Göttingen the young people—including the Poles—were more affluent than the students at the universities in Kraków and Lwów. This can be explained by the fact that in Germany, among the youth of the working or peasant class, only the exceptionally gifted completed high school. There were very few bursaries, so all but a few students paid the substantial tuition fees: university study was a privilege of the moneyed classes. The fees were used to pay a portion of

---

[11] Leon Chwistek (1884–1944), Polish avant-garde painter, theoretician of modern art, literary critic, logician, philosopher, and mathematician. Professor of Logic at the University of Lwów from 1929 to 1940, when, being sympathetic to Marxism-Leninism, he took political refuge in the Soviet Union, where he continued his scientific research and political activism. Believed that reality could be described only from multiple points of view: the physical, the popular, the phenomenal, and the visionary/intuitive. He became Steinhaus' brother-in-law. Born in Kraków, he died in Barvikha, near Moscow, in 1944.

[12] Tadeusz Banachiewicz (1882–1954), Polish astronomer, mathematician, and geodesist. Moved to Göttingen in 1905 following the closure of the Polish Warsaw University by the Russian government. At the Jagiellonian University of Kraków between the wars, he became a member of the Polish Academy of Sciences in 1922. A lunar crater and a planetoid are named after him.

[13] Włodzimierz Stożek (1883–1941), Polish mathematician of the Lwów school. Murdered by Germans in Lwów in July 1941 during the "Massacre of Lwów Professors."

[14] Antoni Przeborski (1871–1941), Polish mathematician. After beginning his academic career in Kharkov, Ukraine, he and his family managed, during the turmoil of the Russian civil war, to escape to Warsaw, where he held the position of Professor of Theoretical Mechanics till 1939.

[15] Kazimierz Horowicz (1884–1920), Polish statistician. After studying at the Russian University in Warsaw, obtained a doctorate in Göttingen. Took up actuarial mathematics on returning to Warsaw.

[16] Jan Króo (1885–?), Polish theoretical physicist and mathematician. Defended his doctoral thesis in Göttingen.

[17] Eustachy Żyliński (1889–1954), Polish mathematical logician. Member of the interwar Lwów school of mathematics. Joint supervisor with Steinhaus of Orlicz's Ph.D. thesis. During World War II worked in the underground Lwów University (1941–1944). Appointed Consul General to Ukraine in 1946. Head of the mathematics department at the Silesian Institute of Technology 1946–1951.

[18] The period 1905–1907 was one of upheaval in Congress Poland, stimulated in part by the 1905 revolution in Russia itself.

[19] That part of Poland ruled by Russia since 1815, also called Congress Poland.

the lecturers' and professors' salaries, and the system was such that those who taught required subjects were paid proportionately more than those teaching more recondite subjects. Actually, the system was much the same as in Austria, and hence also in Kraków and Lwów.

The German students were divided up into student "corporations",[20] some of which were distinguished by colors worn on ceremonial occasions and some not—so their members were called respectively "coloreds" and "blacks". The corporations consisting of "coloreds" formed the *All-German Association* (*Senioren-Convents-Verband*) and those made up of "blacks" the *Young Fellows*. The former were conservative, led by the scions of landed gentry. (Some conservative corporations—such as *Saxony* in Göttingen and *Borussia* in Bonn—restricted membership to sons of the aristocracy.) The Young Fellows' tradition, on the other hand, derived from the revolution of 1848, whose aims were universal suffrage and equal rights—especially for the middle class to which they belonged. This tradition they systematically celebrated—or drowned—in beer. The tradition of the "coloreds" was expressed through certain rituals: the downing of mugfuls of beer according to a precise schedule, the wearing of little colored caps and sashes, and *mensur*, a special style of duelling where opponents fenced with razor-sharp blades without wearing protective masks. Among the German students there were only a dozen or so "coloreds" all told, and their total number, including foreign students, was probably less than ten percent of the student body, but on the streets and in the beer halls they were highly visible, while in the lecture halls and seminars they were even less visible than this statistic might indicate. At academic ceremonies, inaugurations, and banquets, the "coloreds" always played a key role. In their shiny knee-boots, wielding unsheathed rapiers, and with faces beautified by *mensur* scars, they were idolized by the teenagers. They all knew by heart which color was that of Hanover, which of Brunswick, and which of Saxony, and some of them would even sign their names using a special corporate hieroglyph. In Göttingen every corporation had its own "house", and many members made use of the apartments in these houses, some of which, thanks to the contributions from *philisters*,[21] boasted pianos and libraries. Student life in a house with a beautiful garden, in the company of friends, and all paid for by rich parents, was far too pleasant for rapid completion of a program of study to be feasible. Membership in a corporation was expensive, costing about as much as was earned by a senior official with higher education—since in addition to the rent and everyday living expenses, there was the cost of the drinking-bouts and other kinds of carousing.

---

[20]*Studentenverbindungen*, similar to fraternities, but of older provenance, and including alumni (*Altherren* and *Hohe Damen*) once active in the corporation, who partly finance the corporation and help the active students in other ways. Some corporations' members wear distinguishing colored caps and ribbons, mainly on ceremonial occasions. Most student corporations date back to the mid-nineteenth century or earlier.

[21]That is, *Altherren*—members who had finished their studies.

Their playground of choice was Maria-Spring Wood, one stop by rail from Göttingen. It was customary for the students to rendezvous with girls there on Wednesday afternoons, when there were no lectures given. There was a pub there, and next to it an outdoor wooden dance floor on a stone foundation, with tables and benches arranged around it to form a sort of amphitheater. An orchestra played dance tunes by request at ten pfennig a time. By mid-afternoon the tables of German students were literally covered with empty beer bottles, testament to their drinking prowess. Some of the girls taking part in the revels were daughters of craftsmen and railroad workers, but most were *filiae hospitates*, that is, the daughters of landladies of the student boarding houses. There were few, if any, female students present at these dance evenings, since for the most part these were teachers completing their degrees by taking certain prescribed courses, and therefore older than the average male student.

In Galicia people were largely ignorant of German mores. Although Austrians were German-speaking, in Germany—and especially in Hanover—they were considered foreigners since their way of life was completely different. Later, when I had moved to a different boarding house, I discovered that the *filia hospitalis*[22] considered it a duty to help a student boarder on with his coat if she and he should by chance find themselves together in the anteroom after dinner—and in the absence of a *filia hospitalis* a female boarder was expected to take on this function. When one such young lady heard that I had bought a new pair of boots, she immediately came and took them from me in order to rub them with linseed oil to protect the leather from premature wear. At that time, German women—and for that matter also French, Swiss, and Belgian women—observed a tradition of prudence and thrift, and very often showed dismay at male untidiness and prodigality. They were especially outraged when foreigners displayed such manners. Constant swabbing and sweeping was the lot of both wives and daughters. The daughters of Göttingen professors scrubbed the steps of their parents' houses, and when they married they continued to do so in their own houses. On the other hand, servants had far more rights than in Poland; for instance, it was illegal for a servant to wax floors, and if a policeman happened to spy through an open window a servant using a floor polisher, he would note the name of the homeowner in order to impose a fine. Supper was usually eaten cold since the servants had to be free from seven in the evening. Walking along Planck Street around seven, one could see a light shining from the window of every garret, indicating that the housemaid was changing before going out for the evening. Such being the situation with hired help, only the better off, such as landlords of boarding houses, had servants. Most townsfolk—for instance, lower-rank officials—managed without, or at most employed charladies to clean for a few hours in the morning.

The town was divided into two parts. In the older, lower, part, there were several beautiful old cathedrals, streets of stores, and compact buildings of more venerable

---

[22] "Innkeeper's daughter", or, in a specialized sense, daughter of the house where a student had board and lodgings.

style, all surrounded by what had formerly been a fortified wall. The newer part of the town, outside the wall, was comprised for the most part of more modern houses with gardens. Many of the buildings in the old town dated from the sixteenth or seventeenth centuries. Some of these were of an exquisite wooden construction, whereby the exterior beams were elaborately sculpted and painted, and each storey protruded out further than the one below it—presumably to provide more space, this being circumscribed in olden days by the wall. In later, less dangerous, times the wall had been transformed into a sort of causeway or embankment, forming a beautiful raised boulevard surrounding the middle of the town, linked by green alleyways to the university garden, and providing a view from above of the many little private gardens. From the wall one could view the town park, adjacent to a small deciduous wood, and further out a vast forest through which one might wander for hours. By fiat of the university authorities, who wished to preserve peace and quiet, there were neither buses nor streetcars in Göttingen. They were in fact not needed; one was lucky to be able to roam the streets for hours talking till midnight amidst fragrant gardens—all the more so since there were no wardens in Göttingen, and each of us had a key to the house where he had his lodgings.

The group to which I found myself belonging developed a certain lifestyle. In the evenings we would gather at the residence of the Dziewulski brothers, a beautiful house near the Friedländerweg. This was not a boarding house, but one that they themselves owned and ran. More often than not they dined at home, and whenever one visited them one was sure of getting a glass of tea and often something to go with it. They also served as the group's bankers, lending money to the needy amongst us. Their house was the meeting place of Poles from the Kingdom of Poland, Galicia, Lithuania, and Ukraine, so we learned about the states of affairs in different parts of Poland or former Poland, and also smoothed out the provincialisms in our speech. However, the group had very few Poles from the region of Poznań[23] because these were being watched by the Prussian police and had to be careful. There was a rather exceptional young man from Poznań called Marchlewski, a lawyer—exceptional for his intelligence and broadmindedness, and for being different from most of the inhabitants of Greater Poland,[24] who spoke Polish poorly, had adopted German manners, but still retained a patriotic attitude. Then there was the likeable medical student Janaszek, of peasant background, and Jacobsohn, a Poznań Jew who considered himself a Pole—a very rare thing in Greater Poland. His great great grandfather's father had been doctor to Zygmunt August,[25] and had been ennobled

---

[23]Historic city in west-central Poland. From 1815 capital of the semi-autonomous Grand Duchy of Posen under Prussian hegemony, but, following the uprisings of 1848, reduced in status to the capital city of the Prussian province of Posen.

[24]The region with Poznań as its chief city, the core of the early Polish state.

[25]Sigismund II Augustus (1520–1572), Grand Duke of Lithuania from 1529, King of Poland from 1530, and King of the Polish-Lithuanian Commonwealth when Lithuania and Poland were united into one state in 1569.

by him. Jacobsohn later published his recollections of the German campaign against France in World War I.

At least once a week we would go to the café *National*. Kazimierz Horowicz, a student of statistics and actuarial science, was an *habitué* of the café, and knew by heart the prices of the several hundred items on the National's menu. He mingled freely with the English and American customers, and knew the actresses from the Göttingen theater; for us he represented a link to another, exotic world....

During their time studying in Göttingen, most of the Poles did not learn German well enough to converse freely in it. They all ate their dinner (*Privatmittagstisch*), prepared by the landlady, in their private boarding houses, and went to the café as a group, so made few German friends. Our Polish group even organized a secret Polish library. At first, we had wanted to establish a legal Polish association on the model of the Anglo-American Association, but the rector refused us permission, so we compromised with our secret library, changing its location from time to time. At one time the library was located in my apartment, and about two weeks later a man came to the boarding house and informed Miss Thomas that there was a secret Polish library in her house. So we changed its location yet again—but there were no consequences, and as far as we could discern the police did not investigate the matter further.

We went twice a week to the municipal swimming pool. It was very well run: the water was changed twice a week, and one had to take a hot shower before entering the pool. Wacław Dziewulski was the one who made sure we all made good use of the pool—in addition to our other pursuits such as playing tennis and going for walks.

The University of Göttingen was also called informally *Georgia Augusta* after the Hanoverian King[26] who founded it. It was divided into four faculties, one of which, the faculty of philosophy, had a subfaculty of agronomy. It was renowned for its mathematics center, and its faculty of medicine was also highly reputed. Among the university's most famous non-mathematical figures were the great physiologist Max Verworrn, the well-known expert on international law W. Baer, and the celebrated interpreter and biblical scholar Wellhausen; these and many others contributed to the university's fame, and attracted the best students not only from Germany but also from all over Europe, from America, and even from China and Japan. At the time of my sojourn in Göttingen, the students numbered somewhat over two thousand, some four hundred of whom were foreigners. Some of the houses on Prinzenstrasse had tablets affixed to them with inscriptions to the effect that for so-and-so many years princes of the Hanoverian dynasty then ruling England had lived there. The English influence on German habits and manners was very much in evidence, despite the absorption of Hanover into the Kingdom of Prussia after the battle of Langensalza in 1866. During my time in Göttingen, elections of representatives to the *Reichstag*

---

[26]The University of Göttingen was founded in 1734 by the Elector of Hanover, who was at the same time King George II of Great Britain.

were held. The *Welf*[27] party, which advocated non-recognition of Prussian rule, chose as its candidate a Welf by the name of Götz von Ohlenhusen.

Yes, the tablets fastened to many of the houses attested to the glorious past of the city and its university. They informed passers-by as to where and when the great had been present in the town: Bismarck, Heinrich Heine, Brentano, the composers Brahms, Spohr,[28] and Joachim,[29] the chemist Wöhler (the first to synthesize urea from inorganic substances), the philosophers Lotze and Hartmann, the physicist Everett, the statesman von Stein, the poet Tieck, the anatomist Teichmann (of Polish extraction), among many other famous names. In one of the town parks there was a monument to Gauss[30] and Weber[31] commemorating the building of the world's first telegraph line, linking the university to the observatory. The grave of Bürger,[32] creator of *Lenore*, is in the cemetery just outside the city gate leading to the village of Weende; there lies also Lejeune Dirichlet.[33] The Göttingen graveyards did not partake of any macabre or gloomy atmosphere: they were pleasant gardens used as shortcuts both day and night. Göttingen was also something of a summer resort town, whither people repaired from the cities for a stay in the country; the British Minister of War Lord Haldane, an alumnus of Georgia Augusta, spent his summers in Göttingen every year.

For me, as for a large number of my fellow students, the center of university life was the mathematics library. The mathematical tradition at Göttingen began, of course, with Gauss, "prince of mathematicians". Gauss made major, in some cases radically transformative, contributions to many fields: algebra, number theory, function theory, differential geometry, statistics, the theory of errors, geodesy, physics—in particular electrostatics—and astronomy, several of which had been in a sorry state before he resuscitated them. His notebooks are preserved to this day in the Göttingen astronomical observatory. At the time I was there, the Göttingen Academy of Sciences, in collaboration with other academies, was working on the publication of Gauss' complete works.[34] On Gauss' death, Dirichlet, the great number theorist (married to Mendelssohn's daughter, by the way[35]) became head

---

[27]The House of Hanover was a younger branch of the older House of Welf.

[28]Louis Spohr (1784–1859), German composer, violinist, and conductor.

[29]Joseph Joachim (1831–1907), Hungarian violinist, conductor, and composer.

[30]Carl Friedrich Gauss (1777–1855), considered one of the three or four greatest mathematicians of all time.

[31]Wilhelm Eduard Weber (1804–1859), famous German physicist.

[32]Gottfried August Bürger (1747–1794), German poet. Studied law at the University of Göttingen from 1768. His poem *Lenore*, by virtue of its dramatic force and vivid realization of the supernatural, made his name a household word in Göttingen.

[33]Johann Peter Gustav Lejeune Dirichlet (1805–1859), a preeminent German mathematician.

[34]C. F. Gauss, *Werke*, Volumes 1–6, Dieterich, Göttingen 1863–1874; Volumes 7–12, Teubner, Leipzig 1900–1917, Springer, Berlin 1922–1933.

[35]He was actually married to Felix Mendelssohn's youngest sister.

of mathematics in Göttingen, and upon his death, in turn, in 1859, Riemann[36] was appointed: Riemann, who, in his profound investigations of complex algebraic functions introduced the highly influential notion of multisheeted "Riemann surfaces", as they are now known, whose theory of non-Euclidean generalized surfaces or "manifolds" was later to provide the framework for Einstein's general theory of relativity, and who contributed substantially to the study of Fourier[37] series, a concept arising from the analysis of sound waves. This sequence of genii came to an abrupt end in 1866 with the death of Riemann, to be succeeded by a period of complete stagnation till the advent of Klein.[38] This great geometer was a creator, together with Sophus Lie[39] and Steiner,[40] of the German school of geometry.

By the time I arrived in Göttingen, Felix Klein had already realized his dreams: he was a *Geheimrat*, that is, privy councillor, a member of the Prussian House of Lords, and—most important of all—had the ear of Althoff,[41] which enabled him to make Göttingen into the world's leading mathematical center. But government subsidies were not enough for Klein: he founded a society for the support of pure and applied mathematics, and persuaded several rich industrialists from Hanover and further afield to join it, and to make large donations in support of the mathematics institutes at the University of Göttingen. His success was partly due to his emphasizing to them the applications of mathematics that would be forthcoming. However, there already existed an institute for applied mathematics in Göttingen, headed by Carl Runge,[42] who lectured on descriptive geometry and graphical statistics but was primarily interested in spectral analysis and atmospheric electricity. Another name among those listed as members of the applied mathematics institute was that of Listing,[43] author of the world's first article on topology, a subject very far from applications. There was also an institute of applied physics headed by

---

[36]Georg Friedrich Bernhard Riemann (1826–1866), German mathematician of genius. Made fundamental contributions to several areas of mathematics and physics.

[37]Jean-Baptiste Joseph de Fourier (1768–1830), outstanding French mathematician. Highly placed administrator under Napoleon. Proposed a mathematical theory of heat conduction, in connection with which he introduced the trigonometric series bearing his name.

[38]Felix Klein (1849–1925), German mathematician. Worked on function theory and the associated discrete groups, non-Euclidean geometry, and the connections between geometries and their groups of transformations.

[39]Marius Sophus Lie (1842–1899), Norwegian mathematician. Inventor of the theory of "continuous symmetry", that is, of "Lie groups", as they are now called, together with their "Lie algebras".

[40]Jakob Steiner (1796–1863), Swiss mathematician. Professor of geometry in Berlin from 1834.

[41]Friedrich Althoff (1839–1908), then departmental director in the Royal Prussian Ministry for Ecclesiastical, Educational, and Medical Affairs.

[42]Carl David Tolmé Runge (1856–1927), German mathematician, physicist, and spectroscopist. Co-developer of the "Runge–Kutta method" of modern numerical analysis. Took up the chair in applied mathematics in Göttingen in 1904.

[43]Johann Benedict Listing (1808–1882), German mathematician. First to introduce the term "topology" in an article of 1847. (Deceased long before the author's arrival in Göttingen!)

the aerodynamicist Prandtl,[44] and including the expert on electricity Simon[45] as a member. There was also a separate geophysical institute run by the celebrated Emil Wiechert.[46] The founding of these technical institutes resulted in a conflict with the Polytechnic in Aachen, where this was considered as exceeding the competence of Georgia Augusta and encroaching on the prerogatives of the polytechnics. However, the conflict was resolved somehow, and Klein—himself a great admirer of mechanics and coauthor with the eminent physicist Sommerfeld[47] of a book on the gyroscope—could continue his work in peace. Among the *Privatdozenten*[48] were Carathéodory, a qualified engineer and expert on mechanics and optics, and Herglotz, a Viennese mathematician who had had a seemingly universal education. Klein's favorite was Koebe,[49] who was investigating multivalued functions by geometric means, so that hardly any formulae or equations were to be found in his papers. The head of theoretical physics was then Woldemar Voigt,[50] and Max Abraham[51] and Paul Hertz[52] were *Dozenten* under him. The astrophysicist Schwarzschild[53] was at that time also a professor in Göttingen, with Brendel,[54] an expert on asteroids, working in the observatory. My friend Władysław Dziewulski was also working as an *assistent*—which shows that qualified foreigners had no difficulty obtaining appropriate university positions in Germany at that time.

---

[44]Ludwig Prandtl (1875–1953), German scientist. First to describe the boundary layer and its importance for drag and streamlining.

[45]Hermann Theodor Simon (1870–1918), who founded a division of applied electricity at Göttingen in 1907.

[46]Emil Johann Wiechert (1861–1928), German geophysicist. Appointed director of the Geophysical Laboratory at Göttingen in 1898.

[47]Arnold Johannes Wilhelm Sommerfeld (1868–1951), outstanding German theoretical physicist. Pioneered developments in atomic and quantum physics.

[48]Persons who have achieved the *Habilitation*, a degree somewhat higher than the Ph.D. qualifying them to give lectures, and opening the way to a regular university position.

[49]Paul Koebe (1882–1945), German mathematician. Completed his graduate studies at the University of Berlin in 1905. First to prove, with Poincaré, the "Uniformization Theorem" of complex analysis.

[50]A German physicist. Lived from 1850 to 1919. The term "tensor" with its modern meaning was introduced by him in 1899.

[51]German physicist. Studied under Planck, and worked as his assistant. Lived from 1875 till 1922.

[52]Obtained his Ph.D. in Göttingen in 1904 under the supervision of Max Abraham. Gradually shifted the focus of his interests from theoretical physics to the philosophy of science. Lived from 1881 till 1940.

[53]Karl Schwarzschild (1873–1916), German physicist. Found in 1915 the exact solution—the "Schwarzschild solution"—of the Einstein equations of general relativity in the case of an isolated non-rotating spherical body, latterly become important in connection with the physics of black holes. He died the following year from a painful autoimmune disease contracted while at the Russian front in World War I.

[54]Martin Brendel (1862–1939), German astronomer.

As a second-year student (counting my year in Lwów) it was difficult to appreciate the richness of the possibilities offered by Georgia Augusta. In addition to the above-mentioned mathematicians and physicists there was a group with David Hilbert as its spiritual leader. Klein had arranged for him to come to Göttingen from Königsberg, and in this he had made a brilliant choice. Hilbert was indeed a formidable scholar. His *Foundations of Geometry*,[55] which was awarded a prize by the Mathematical Society of Kazan,[56] served to remove all doubts concerning the validity of non-Euclidean geometries, and even, one might say, paved the way for an understanding of the theory of relativity.... Alas, the student finding himself in the midst of this saga was unable to appreciate the importance of the world-changing scientific breakthroughs being made all around him. But it was obvious to everyone that Hilbert was the greatest mathematical attraction in Göttingen. His first name, David, showed him to be derived from East-Prussian Quakers.[57] He was not only a great scholar but also a great man, as for instance his complete lack of interest in titles or medals demonstrates. Around him there were gathered such luminaries as Minkowski, an eminent expert in number theory and a laureate of the Paris Academy, Ernst Zermelo,[58] representing set theory and logic, and whose "axiom of choice" split mathematicians into two opposing camps, and Felix Bernstein, a student of the great Cantor,[59] who had anticipated the biologists in correctly formulating the laws of inheritance of blood groups—but who was probably the least popular of the professors.

Yes, getting a clear idea of the rich intellectual feast going on around me was difficult. In the course list, I crossed out a course on the history of philosophy, dictated syllable by syllable by an old man called Baumann, and then gave up philosophy of the old sort altogether. I decided to switch to applied mathematics as a minor subject. This included such topics as mechanics, analytic geometry, graphical statistics, numerical analysis, and geodesy. Thus I diligently attended classes at the Institute of Applied Mathematics on Prinzenstrasse, made drawings, and, under Wiechert's guidance, took measurements of polygons on the city streets with a theodolite. When on one occasion a double image of the sun appeared, he snatched the theodolite from us and measured the angle between the two images to

---

[55] This work gives a complete list of axioms for Euclidean geometry.

[56] Kazan' was the city in Russia where Nikolaĭ Lobachevskiĭ (1792–1856), one of the inventors of non-Euclidean geometry, had been a professor.

[57] Hilbert's family was actually Reformed Protestant, a common religious denomination in Prussia at the time.

[58] Ernst Friedrich Ferdinand Zermelo (1871–1953), German mathematician. Worked on the foundations of mathematics. In particular, proved that the Axiom of Choice implies that every set can be well-ordered. The Zermelo–Fränkel axioms of set theory have become standard. (Steinhaus later formulated a weaker version of the Axiom of Choice, his so-called "Determination Axiom".)

[59] Georg Ferdinand Ludwig Philipp Cantor (1845–1918), the inventor of general set theory, since become, in the form of the Zermelo–Fränkel axioms in conjunction with predicate logic, the standard foundation of mathematics. Defined both the transfinite cardinal and ordinal numbers, together with their arithmetics.

demonstrate to us that it was the same as the angle of diffraction of light rays in ice—or something of the sort. I also had astronomy as an elective course. Felicjan Kępiński and I attended lessons on the use of the instruments in the observatory, and under the direction of Professor Ambronn, we carried out computations using observations that had been made in Africa in connection with establishing the exact boundary between those parts of the Congo belonging to Belgium and Germany.

But for me the most important occupation was reading in the mathematics library. *Das mathematische Lesezimmer*,[60] a genuine source of pride for Klein, was located on the second floor of the main mathematics building on Weender Strasse, and so chosen that neither voices from the lecture rooms nor noise from the corridors reached it. For a payment of three marks,[61] a student obtained a key and an entrance pass valid for six months. The books, arranged on the shelves in alphabetical order, were removed and replaced by the students themselves. In a separate room Klein had placed ten thousand or so relatively recent dissertations—out of the many produced over the years. There was also a "B" section where the copies were the property of the university. The card catalogue was kept fully up to date. New issues of journals were ranged on separate shelves, and up in the higher reaches of the bookcases one could see complete sets of *Mathematische Annalen*, *Journal für Mathematik*, *Journal des Mathématiques*, *Acta Mathematica*, and the proceedings of various academies: the Paris Academy, the Berlin Academy, the new Munich Academy, the Viennese Academy—among many other Italian, French, English, German, and American periodicals. There were also sets of the complete works of Gauss,[62] Fermat,[63] Cauchy,[64] Hermite,[65] and others. Each seat had a lamp installed next to it, so that one could continue reading after dark. The lamp shades were made of asbestos, and one lit the lamps with a taper tipped with a copper-impregnated gas-absorbing sponge, which worked unfailingly. I spent many hours in this library; nowhere before or since have I seen a library organized so perfectly. It must have been used by—and helped to inspire—thousands of students from all over the world.

I recall an event that was interesting from a psychological point of view. After Minkowski's premature death, his post was taken over by Edmund Landau[66] from

---

[60]The mathematical reading room.

[61]The mark was worth $0.25 in 1905 US currency, so three marks would be worth about $18 today.

[62]By 1906, seven volumes of Gauss' collected works had been published; the final volume appeared only in 1933.

[63]Pierre de Fermat (1601–1665), French lawyer and amateur mathematician of genius. Credited with founding modern number theory. Made notable contributions to analytic geometry and the differential calculus. Best known for "Fermat's Last Theorem".

[64]Augustin-Louis Cauchy (1789–1857), outstanding French mathematician and physicist. Worked in many areas of mathematics. Laid the foundations of modern analysis in terms of limits and continuity.

[65]Charles Hermite (1822–1901), influential French analyst, algebraist, and number theorist.

[66]Edmund Georg Hermann Landau (1877–1938), German-Jewish number theorist and complex analyst.

Berlin. Landau assigned to me as an exercise the task of finding a putative error in Kőnig's[67] exposition of Gauss' law of quadratic reciprocity. The paper in question was so short that it was easy to locate the page that must contain the presumed mistake, and it seemed to me that indeed one of the formulae on that page did not follow from the preceding ones. However, there was no mistake at this place in the paper, and my inexperience, or possibly tiredness, had led me astray. I returned home and went to bed, and then, half asleep, suddenly saw in my mind's eye the page of Kőnig's article and the correct deduction of the formula I had doubted from the earlier ones. Early next morning I rushed to the library, opened the journal at the relevant page, but could no longer for the life of me see the connection between the critical formula and the preceding ones. I closed my eyes and tried to return to a half-conscious state, and after a while I again saw the required connection. I quickly opened my eyes to trap the thought like a butterfly and pin it to the page, and this time I succeeded! (I relate this for the psychologically inclined reader interested in the process of mental reasoning.)

As mentioned earlier, my apartment in Fräulein Thomas' house was close to one of the city gates, or, more accurately, to the spot where there had once been a gate in the city wall, in the direction of the village Weende. Long before my arrival in Göttingen, the gate had given way to a small building functioning as the city excise office. The excise officer, known to all as "Father Lotze", was responsible for collecting payments of excise duty on the produce the peasants brought to the city, sounding a trumpet in case of fire, selling postage stamps, sweeping the sidewalk in front of the office, and keeping the public toilet clean. For me it redounded to the merit of the city mayor—of whom I knew almost nothing—that he had been able to assemble such a set of responsibilities to keep Father Lotze busy. The art of administration depends not so much on new and all-encompassing ideas as on an ability to take care of day-by-day matters in an effective and frugal manner.

At this time, the works of Georg Cantor, the brilliant creator of set theory, were still relatively little known. Thus when the young Warsaw mathematician Wacław Sierpiński[68] discovered that the points of the plane can be coordinatized by single numbers,[69] he was surprised almost beyond words and wrote to his friend Tadeusz Banachiewicz in Göttingen, who went to the library, found the relevant paper by Cantor, and sent a telegram to Sierpiński with the words: "G. Cantor, Journal für

---

[67] Julius (Gyula) Kőnig (1849–1913), Jewish-Hungarian mathematician. After completing his mathematical studies in Berlin, he was appointed docent at Budapest University in 1871, and, in 1873, professor at the Technical University of Budapest, where he remained for the rest of his life. His paper on quadratic reciprocity appeared in *Acta Mathematica* in 1898. His son Dénes also became a distinguished mathematician.

[68] Wacław Franciszek Sierpiński (1882–1969), Polish mathematician. Made outstanding contributions to set theory, number theory, and topology. Three well-known fractals are named after him: the Sierpiński triangle, carpet, and curve.

[69] That is, as Cantor had discovered earlier, there is a bijection between the real line and the plane. The space-filling curves of Peano, Hilbert, and Sierpiński determine analogous *continuous* surjections, which cannot, however, also be injective.

Math., [year] ... [page number] ...." Our interest was piqued, and I, together with Antoni Łomnicki, began to study closely Borel's[70] *Théorie des fonctions*. At around this time Banachiewicz talked me into writing a popular article on squaring the circle for the Polish journal *Wszechświat* ("The Universe"). This was an exercise of more value to me than the readers of this useful journal, since it forced me to learn some history of mathematics.

We were so busy with our studies that we had little time left for a social life, but everything we observed about us, in particular during our strolls through the streets, brought home to us how different this world was from our Polish one. We were impressed, for instance, by the fantastic efficiency of the German postal service. If, during one of our walks lasting all night long, one of us set Banachiewicz a problem, and a few hours later, after the person who had asked the question had gone to bed, he found the solution, he would go with those of us still awake to the main post office and write the solution out on an express card. Ten minutes after he had dropped the card in a special box for such items, we would see a man on a bicycle taking the message to our friend's lodgings, where he was by now sound asleep. Another episode occasioned even greater admiration for the postal system of the Reich. Mrs. Antecka, the wife of Henryk Kołodziejski, had sent a telegram to Warsaw signed for with her maiden name, and by mistake the postal clerk had overcharged her. When the supervisor noticed the error, it fell to the post office to find the sender and reimburse her the amount by which she had been overcharged. However, the name "Antecka" did not appear in the Göttingen directory of residents, so postal clerks were set to checking letters from Warsaw to Göttingen in the hope that a letter addressed to someone by the name of Antecka would turn up—and indeed a few weeks later a letter addressed to Mrs. Kołodziejska-Antecka did arrive, and the postman triumphantly returned a sum of money in the amount of the overpayment to an utterly surprised lady.

I spent the long vacation at the end of my first academic year in Göttingen not back home in Jasło, but in the Belgian town of Blankenberghe, where my Mother and sisters were at the time. I travelled by train with a night's stopover in Antwerp. Blankenberghe was very different from Germany. The cries of newsboys: "Le Journal, Le Matin! Lisez L'Étoile Belge! L'Intransigeant!!..." woke me early in the morning. From an early hour the streets were full of gay and brightly dressed people. There was a porter at the hotel we were staying at whom I found interesting. He was a tremendous art enthusiast, and while the paintings of his own that he showed me were rather childish, I was moved by his expressions of admiration of the works of Rembrandt, Memling, Hals, and other masters, whose pictures, he said, could be found in galleries in Antwerp, Brussels, and decorating the walls of churches in Bruges. As a result of talking to him it was brought home to me that my artistic education was nil, and I decided that I had better acquaint myself with these formidable works of art. I anticipated that this would mean paying visits to

---

[70]Félix Édouard Justin Émile Borel (1871–1956), French mathematician. One of the pioneers of measure theory and its application to probability theory.

museums, palaces, and cathedrals without any reward of satisfaction of some kind, since my experience told me even then that these dark, strange paintings would make no impression on me, that I would not be able to feel their beauty. Thus I found myself in a difficult situation as far as art was concerned, since I knew that most educated people admired these paintings, while the rest at least pretended to do so, and that only cynics with the gift of exceptional moral courage would dare to put into words what I actually thought of classical art. So I tried to learn what one should look for in a painting, to develop a level of expertise sufficient to set against that of the established experts. However, such an educational program tends to prejudge the outcome in favor of those very experts, since the paintings thought to be worth looking at and the art books considered worth reading—like the musical works considered canonical—are all determined by the authority of those very experts, many of whom are simply snobs. So one is left with little choice.[71] It is strange that I never had any such doubts when it came to architecture. In Brussels I saw the Cathedral of St. Michael and St. Gudula, built in the Gothic style; the austere beauty of this cathedral, which impresses by its very shape and not by any ornamentation, needed no expert commentary. It is clear that my understanding of architecture was not innate since just a year earlier, in Lwów, I had been completely unconscious of the stone beauty all around me. I needed no one to instruct me in the beauty or grandeur of nature and architecture, yet painting was for me a closed book—in fact sealed with seven seals—and I cannot say that looking at hundreds of pictures officially recognized as great made much of a difference to me. The one positive effect it did have was that later, whenever I saw dreadful daubs in the shop windows in Jasło, I realized their ugliness.

Blankenberghe is one of the many seaside towns near where *La Manche* merges with the North Sea. Going southwards along the shore, you come to Wenduine, Mariakerke, and so on, until finally you reach Ostend, and going northwards, you come to Bruges-sur-Mer, and beyond it to Dutch seaside villages. From Blankenberghe the beacon of the Vlissingen lighthouse was clearly visible. We used to walk along the beach, especially in the direction of Bruges, where a harbor was being built. Bruges itself, made famous by Rodenbach's[72] calling it the "Venice of the North", and celebrated for its Gothic architecture, the brick spire of the Church of our Lady, the thirteenth century belfry housing a carillon that rang every hour, and its Béguinage,[73] impressed me less than other historic towns I was to see later. In Blankenberghe we met a Russian general, whose daughter, a young beauty, had been invited to balls given by the tsar himself. She declared to me that if the Poles

---

[71]Over the last decades of the twentieth century an alternative movement developed, called "visual culture", whose proponents attempt to discard preconceived ideas of what is or is not "art", and examine every visual object by its own lights, or how it impinges on the viewer directly, unfiltered by mere stuffy tradition—as opposed to a healthy critical tradition.

[72]Georges Raymond Constantine Rodenbach (1855–1898), Belgian symbolist poet and novelist.

[73]Building used by *Béguines*, members of lay sisterhoods of the Roman Catholic Church, consisting of religious women who sought to serve God without retiring from the world.

continued to agitate against Russian rule in the Kingdom of Poland, Russia would cede the territory to the Prussian emperor Wilhelm II,[74] who would know how to introduce order.

The manners of the Belgians were stiff and provincial, and they were not as tidy as the Dutch. There was a large business class, which dined well on lobster, shrimp, and tiny sea snails called *moules*, among more mundane foods. That they were politically clearheaded was shown by the fact that despite possessing profitable colonies in Africa, they were not interested in acquiring a fleet of warships—realizing, no doubt, that a fleet capable of defending those colonies would cost more to build and maintain than the revenue from said colonies. The Belgians thought of themselves as representing a northern branch of Latin culture,[75] and in all things looked to Paris as their model. There was an express train running from Brussels to Paris that got one there in under three hours. It was Belgian engineers who built the Paris *Métro*, and likewise the Warsaw streetcar system, whence the Polish term *belgijska buda* ("Belgian caboose") for streetcar. Not to be outdone by Paris's *Bois de Boulogne*, Brussels has its *La Cambre*, a beautiful suburban wood. On Sundays its lawns were literally covered with people picnicking, and the restaurants nearby were also packed. The public toilets were inadequate, however, as was clear from the immense line-ups of the distressed of both sexes. Overcrowding was in evidence everywhere in Brussels, an abomination to someone from a country with a much less dense population. During the long summer evenings the suburban streets were full of *promeneurs*, and before almost every gate there were seated lace-makers, creating from white thread their famous *point-de-rose* lace, then in demand by women the world over—and by smugglers, being worth its weight in gold. At that time, Flemish[76] in Brussels played the same role as Ruthenian in Galicia: foreigners were unaware of its existence. During our stay we never heard Flemish spoken; even the fishermen spoke French, addressing everyone with the familiar "tu". Their women walked barefoot on the beaches, and, although their feet were shapely, they themselves were by and large not attractive. Many of the old Belgian women had such luxuriant hair on their faces that they looked as if they had moustaches and beards; anyone from our part of the world would have been simply appalled at the sight.

For the next few years my studies alternated with summer and winter vacations—of almost equal length—spent at home in Jasło. My cousin Dyk, who was studying law in Vienna and Graz, and my friend Ludwik Oberländer, one year junior to him in the law program in Vienna, also came home for the vacations. From Ludwik

---

[74]Frederick William Victor Albert of Prussia (1859–1941), the last German Emperor (Kaiser) and King of Prussia, ruling from 1888 to 1918. Dismissed Bismarck in 1890 and launched Germany on a bellicose new course in foreign affairs culminating in his support for Austria-Hungary in the crisis of July 1914 that led to World War I.

[75]This doubtless applied only to the Walloons, the inhabitants of the southern French-speaking region of Belgium, and not to the Dutch-speaking Flemish.

[76]A language very close to Dutch.

I learned of many Danubian marvels, in particular, of a relatively new Viennese periodical called *Die Fackel* ("The Torch"). My clever friend brought me a few issues of this strange magazine, which was not financed by advertisements and did not guarantee regular publication to its readers. The Viennese intelligentsia initially thought that *Die Fackel* must be the product of a gifted decadent, a leftist journalist who, late at night, in a café, dashes off satirical pamphlets, directing his sarcasm at the bourgeoisie, while spending their subscription monies on alcohol. Of course, we speak here of none other than Karl Kraus. This brilliant cynic made it clear to the victims of his barbs that they should be proud to be subscribers to *Die Fackel*. He was considered a literary figure, one similar to Peter Altenberg[77]—acknowledged as a born poet—with an admixture of something of Harden,[78] mocked by Kraus for the baroque style of his periodical *Die Zukunft* ("The Future"). This was a preliminary estimate, based on the early issues of *Die Fackel* from the close of the nineteenth century. Initially Kraus accepted articles by other writers, but from 1911 till 1936, the year of his death, he was the sole contributor. (Not long after his death Vienna was occupied by the most imbecilic of armies commanded by the most obscene of evildoers.[79]) Prior to 1905, *Die Fackel* had concentrated primarily on local Viennese matters, which were of little interest to me. Luckily I did not begin reading the magazine till 1906, otherwise I might have been put off and forgone a 50-year-long boon of richly rewarding reading matter. Anyone who has read the New Testament closely will have been struck by the way in which minor, seemingly unimportant, facts are enumerated concerning Jesus, which, taken together, seem to authenticate his identity. The proper way to assess a text is as follows: before extending moral credit to the author, assuming beforehand that he knows how to write and what to write about, before assuming that a passage defying comprehension represents failure on the part of the reader, that a logical error is not a slip of the author's but an integral part of the text, that a stylistic infelicity demonstrates lack of education in the reader and not the writer—to repeat: before we extend this tremendous moral credit to the author, we must determine whether his work is distinguished by the occurrence of any marvels emerging from his prose incidentally, much as Jesus' identity emerges from an accumulation of minor facts.

Kraus' writings contained many such marvels—so many, in fact, that even a 19-year-old student unable to read German fluently, a provincial young man without aesthetic criteria but with a predilection for mathematics, could spot them in the slim issue of *Die Fackel* with its red cover, given to him by his friend with the admonishment: "Read this; this man should be a member of our Club of Crazies!" The latter term was our favorite name for our group, since, under the influence of the book

---

[77] Viennese writer and poet, a key figure in the rise of early modernism in Vienna.

[78] Maximilian Harden (born Felix Ernst Witkowski) (1861–1927), influential German journalist and editor.

[79] Hitler's Austrian *Anschluss* in 1938.

Physiologie des Schwachsinns ("The Physiology of Feeblemindedness"),[80] we—half jokingly—tried to see in ourselves and others the weakening of will and thought considered to be the mark of the decadent—a decline in will power whereby one felt compelled to enter a café and stay there till closing time, to squander the money sent from home instead of using it to pay the tuition fees, and never to read a book to the end—in short, a state of mind inimical to the way of life judged reasonable by the unimaginative older generation. The inclusion of Kraus in this club—without his knowledge, of course—was intended as a mark of recognition. However, in fact, although Kraus had very little in common with the class of merchants, lawyers, and hack journalists, he had even less in common with the young decadents. Both groups appear in his satirical operetta *Literatur oder Man wird doch da sehn* ("Literature or We Shall See About That"),[81] where each of the two groups converses in the language peculiar to it, and although they seem to be in disagreement they actually agree. Kraus' epigram "When young he had intended going into business, but ended up going into literature" pillaries both groups at one stroke.

The "red magazine" was one of a kind in many ways, but first and foremost for its lack of advertisements. In my earlier days, my Tarnów Grandfather had told me that the text of a newspaper is merely an addendum to the advertisements, which form the real basis for the newspaper's existence, and that the editorials represent decoys for attracting subscribers and readers. Advertisements are the bread and butter of the publisher, since their sponsors pay handsomely for them, whereas the news articles and columns have to be paid for. The cover of each issue of *Die Fackel* carried the warning: "We do not accept advertisements, reviews, or manuscripts. Postage enclosed with these will be used for philanthropic purposes."

And then there was the language of *Die Fackel*: every article, every gloss, every response seemed sculpted in stone.[82] There was none of the language of journalists or merchants, no officers' solecisms, none of the Yiddish loan words used by the *Neue Freie Presse*, no Berlin jargon. German speech, freed from nineteenth century affectations, professorial pedantry and neo-syntax, as well as Nietzsche's demagoguery and the telegraphic literary style of proponents of the twentieth century, became lean and light, muscular and strong. Syntax, punctuation, rhythm, and even the layout of the text on the page were flawless.

Finally, there was Kraus' polemical style. This was something entirely new, relying on careful quoting of whole pages from other sources to demonstrate a subtle point. Occasionally, *Die Fackel* would print photographic reproductions of cuttings from newspapers the magazine wished to satirize. The paper *Neue Freie Presse* was often the butt of Kraus' scorn, but other papers also provided

---

[80]This may be *Über den physiologischen Schwachsinn des Weibes* ("On the Physiological Feeblemindedness of Women"), published in 1900 by the German neurologist P. J. Möbius (1853–1907), who believed that excessive thinking makes women ill.

[81]Performed and published in 1921.

[82]One of the main points of Kraus' writings was to show the great evils inherent in seemingly small errors, including linguistic errors or carelessness.

fodder for the magazine. One such was the *Österreichische Handelskorrespondenz* ("Austrian Business Correspondence"), which printed an advertisement stating that: "'Golgotha,[83]' attractive for shop windows, is available," and quoting the cost of installation and a business address. Kraus reproduced the advertisement in *Die Fackel*, and gave the name of the newspaper in which he had found it. The editor then wrote him a letter demanding that "On the basis of Article 19 of the Press Law, *Die Fackel* should publish a correction stating that 'It is not true that such an advertisement appeared in the *Österreichische Handelskorrespondenz*, but that, on the contrary, no such advertisement appeared in the newspaper...'." Kraus did more, publishing the entire text of the letter from the editor of *Österreichische Handelskorrespondenz* from vignette to signature, and adding: "I have thus fulfilled the request." But then he continued with a rather overheated piece of sophistry: "The advertisement in question could not have appeared, since if there is anything that cannot be used to attract buyers, it's 'Golgotha'! I am therefore forced to believe the editors of the *Österreichische Handelskorrespondenz* when they say that such an advertisement should not, could not, and did not appear. However, this is not the first time I have reprinted impossible, and hence non-existent, things. After I've reprinted them, I don't believe what I see. The present case is precisely of this sort. This must all be just a ghastly dream. The only thing for me to do is to pick up a copy of the issue of *Österreichische Handelskorrespondenz* dated ... of the year 19..., to see if indeed on page..., column..., line..., there appears an advertisement stating that 'Golgotha', attractive for shop windows, can be purchased for ... crowns, from the firm ... in ...."

This was beyond arguing since every reader of *Die Fackel* could, in almost every café, ask for a copy of the cited issue of the *Österreichische Handelskorrespondenz* and verify that the advertisement was indeed there. Thinking about Kraus' unanswerable apostrophizing, I understood that of course the editors of the *Österreichische Handelskorrespondenz* had not intended to commit a blasphemy, and that after reading the reprinted advertisement in *Die Fackel*, they were unable to bring themselves to believe that their paper had advertised the attractiveness of Golgotha. This episode taught me that a literal quote, or, better, a photograph of the original text, could entail a revelation. I understood that words contain within themselves unfathomable philosophical mysteries. In the case in question, Kraus' polemic had achieved the ideal objective: not *écraser l'infâme*, but to correct a blasphemy.[84]

Such were my reasons for admiring Karl Kraus. The reader curious as to more official estimates of him may consult *The Encyclopaedia Britannica*.

*　*　*

The second time I travelled from Jasło to Göttingen, I went through Dresden rather than Berlin. This afforded me the opportunity of seeing such historic

---

[83] The traditional western Christian name for the site of Jesus' crucifixion. Also called Calvary.

[84] A hundred years later, it all seems more like a storm in a teacup.

landmarks as Brühl's Terrace, the Zwinger Palace, the Dresden Manor and the *Frauenkirche*, and all of the architecture referred to by Baedeker as "late Italian renaissance accommodated to the climate of Saxony." In Germany, Saxons have something of a reputation for modesty. And indeed I noticed that the Saxon railroad conductors did not affect the barking of an N.C.O., and that the people seemed cheerful and readily engaged in conversation—and that the girls were pretty. However, the so-called "flower coffee"[85] was simply bad; the thimbleful of milk they added was enough to deprive it of all color and taste, which it didn't have much of to begin with. The Saxon towns I saw were surrounded by allotments with summer cottages on them, where industrial workers rested on their days off by toiling in their tiny gardens. There were a fair number of Poles living in both Leipzig and Dresden, and a large contingent of Englishmen, attracted by the low rents, prices in Germany being generally lower than in England. In those days, when travel abroad was not impeded by passport or foreign exchange regulations, Germany was an attractive place for a foreigner to stay since public services were run extremely efficiently—it was almost as if they were operated by unfailing automata. The postal and rail systems functioned flawlessly, the streets were sewered and kept well lit and swept, access to libraries and reading rooms, museums and art galleries was open to all, there were excellent and frequent concerts given, an abundance of bookstores and clean restaurants—though the food served in them was often second-rate. But then the bakeries sold top-quality baked goods, the butcher shops sold sausages the worst of which were at least edible, and the groceries fine cheeses; there were cheap smoked salt-water fish, delicious chocolate, and a variety of fruits to be had. A foreigner could rent comfortable quarters in Germany and live peacefully and eat better than at home. I found only one item of food reminding me of home: *Krakauer Wurst* ("Sausage à la Kraków"), which in Poland we called tenderloin sausage. It was not imported from Poland, however, since the importation of ground or minced meat was forbidden in Germany. I remember that on one occasion my parents sent me some *pâté*, and the customs office in Göttingen called me to pick up the parcel, which I had to open in the presence of customs officials. When they saw the *pâté*, which was of a chocolaty color, they asked me what it was. When I told them, they informed me that such food items could not be imported. I then told them that they could keep the *pâté* as a present, but they explained that they were not allowed to accept things the recipients had to pay customs duty on, and that they therefore had to burn the package. Finally, one of them cut a tiny piece of *pâté* off with the tip of a knife, tasted it, and asserted that the stuff was not *pâté* but chocolate. Not being very quick on the uptake, I insisted it was *pâté*, but the officials outvoted me, and sent me home without my having to pay duty.... Only when I reached home did it dawn on me what a thickhead I'd been.

Around that time in Germany, single-family dwellings and larger brick villas were springing up. More and more, such villas were being built in the style promoted

---

[85] In German *Blümchenkaffee*, referring to brewed coffee so thin one could see through it the flower placed at the bottom of the cup.

by Muthesius,[86] their roofs steeply mansarded with a low overhang, and tiled with corrugated red tiles. But there were some that were built of artificial grey stone blocks, and tiled with sheets of slate; their dun color called for climbing roses or grape vines with clusters of grapes, and these were used unsparingly. The interiors of German houses were also rather dun, lacking both tastefully colored carpets and attractive furniture, which was compensated for to some extent by luxury glass and porcelain sculptures of *Gallé* glass,[87] Belgian glass, Copenhagen porcelain, and, inevitably, Dresden porcelain. Such items were all available for purchase in Göttingen.

At some point everyone was seized by bibliomania. My mentor Horowicz introduced me to a Mr. Kugelman from Hamburg, who invited us to his lodgings. He was a student, and, like most students in those days, changed universities several times over the course of his studies. Mr. Kugelman had a large personal library which travelled with him. It took up every wall of a huge room, and even then the books were two deep on some shelves. He had first editions of Verlaine's poetry, books bound in vellum, volumes made of parchment, books with covers dating from the Renaissance, facsimiles of famous old books, amateur editions autographed by their authors, editions printed on hand-made "Japanese-Emperor" paper, prints made with a hand-press, and so on. Mr. Kugelman was clearly not very interested in the contents of his books. Sometimes he read a few pages, but it was clear that he viewed the contents of his books as subsidiary to the cover and the material and form of their pages. This biblio-aestheticism had spread like wild-fire in Germany, with booksellers and publishers feeding the frenzy as hard as they could. Hence for a while a book in Germany became something different from a book in France, say. Thus the grotesque situation came about that while in Göttingen's bookstores there was displayed a profusion of splendid volumes bound in parchment white as snow, you could not for love or money get hold of an issue of *Die Fackel*—which had just 600 or so subscribers in the whole Reich, despite the fact that it was very cheap yet contained priceless intellectual material assembled with scrupulous editorial care.

At first sight, German students seemed different from foreign ones. Their trousers were without cuffs, and in equable weather they walked about coatless, while on frosty days they wore thin winter coats. Their hair was invariably cut, carefully brushed, and stuck flat to their heads. They seemed to eat very little, but drank a lot, and eating anything while drinking beer was disapproved of. They seldom took lodgings in boarding houses accommodating foreigners. After a year in Miss Thomas' boarding house I moved to the house of a Mrs. Gericke at 1 Planckstrasse, where there was always congenial company—the absence of which had made me feel somewhat lonely on Weender Strasse. Among my fellow boarders there

---

[86]Hermann Muthesius (1861–1927), German architect, publicist, and diplomat. Promoted ideas of the English arts and crafts movement in Germany, and influenced the later movement of German modernism in architecture.

[87]Émile Gallé (1846–1904), French artist in glasswork, revolutionized the art of glass sculpture by combining his own ideas with such ancient techniques as enamelling, cameo, and inlaying.

was a certain Mr. Lange from the city of Plauen in Saxony. He was a handsome and elegant young man, presumed to be studying political economics. He let on that since his Munich days he had been a member of one of the most exclusive student corporations. His room was full of beautifully bound books; he pretended to be reading Baudelaire, but spoke French badly. He smoked opium-impregnated tobacco *à l'américaine*, wore heliotrope ties, and liked to repeat that "one has to have an occupation, but it must not be allowed to degenerate into work." The female students living in the boarding house, who were a few years older than us, used to visit Mr. Lange's room gladly in my company; he would recite Nietzsche to us, and we would admire the beautifully framed reproductions from *Jugend*[88] on the walls. However, the idyll came to an abrupt end when Mr. Lange suddenly upped and left Göttingen without paying his last month's rent. It gradually emerged that he had devised an economic system for himself enabling him to live reasonably comfortably on relatively little. The stores in Göttingen extended credit to students quite readily, especially to members of student corporations, who usually paid their bills on a monthly basis. Mr. Lange had capitalized on his highly presentable appearance to make purchases on credit of a variety of articles from various stores, which he then sold to fellow students and others for ready cash. I later found out that he had used his system before in Plauen on a large scale, and had had to flee to Switzerland to evade prosecution.

In my new boarding house there also lived at one time a chemist from Frankfurt called Rosenbusch, a Greek called Chardalias, an Italian lecturer by the name of Albano, and an American called Durand-Edwards, among others—but the lodgers were always changing. I became friends with Mr. Louis Durand-Edwards, who hailed from Cleveland, and who was preparing himself for a consular career. He was a happy young man through whom I came to know American culture from the best side. He decided to take charge of my physical education, and train me in long-distance running. I remember our first run well: it was the beginning of winter, there was a little snow visible in the fields, and the puddles were half-frozen. One evening he commanded me to don tennis shoes and a sweater—but no waistcoat—and to keep up with him. After running along the street uphill for a few minutes—nothing to write home about so far!—he suddenly turned into a fallow field close to a wood, and I had to follow him. I felt ice-cold water seeping into my shoes each time my foot sank into the half-frozen soil. I didn't at all feel the cold, but I did soon begin to feel a stitch in my side. Edwards yelled at me sharply to keep running and that if I didn't I'd catch pneumonia. I felt sure that my time had come to die, as it comes to everyone, and that this would therefore be the last as well as the first run of my life. After a little the stitch ebbed away: that was the first miracle. We ran all the way home: that was the second miracle. Edwards undressed me in my room, rubbed me down with a towel, massaged my legs skilfully, and let me eat some supper and go

---

[88] An art magazine of the time that helped popularize in Germany the new "art nouveau" style of art, architecture and applied art that peaked in popularity around the turn of the twentieth century. *Die Jugend* means "The youth".

to bed. When I got up next morning, I felt like a newborn babe, and from then on we ran together at least once a week. Our runs were over 5, 6, and sometimes even 12 km distances, and one December we ran 26 km through snow—so that our speed was somewhat reduced. Neither Edwards nor I was interested in professional sports or following the times of world-class runners.

Edwards was what might be called very American. There were a few other Americans in Göttingen, one of whom was a student of art history who—so Edwards informed me—worshipped European art and science, and believed that the beauty of European cathedrals and the millennial European artistic tradition were incomparably greater than anything on offer in American culture—whose barbarism became apparent as soon as the comparison with the European variety was made. I saw that young man once, and, although I did not have a chance to talk to him, I must say that his facial expression—if this can be ascribed a general type—looked "European". However, Edwards regarded him as having betrayed his country. It was not that what he was claiming was wrong, but that he should be affirming things American—like a true American pragmatist such as Edwards—rather than denigrating them. Edwards told me a great deal about the United States, and convinced me—not dialectically but in actuality—of the value of the type of person considered ideal there. He claimed that an American upbringing inculcates straightforwardness, loyalty, a sense of responsibility for oneself, physical fitness, combines respect for the elderly with freedom, makes one strive to be independent of assistance, and teaches one to apply objective criteria in all matters, just as is normally done in sport. Durand-Edwards was a graduate of a good American college, and knew, for instance, who Edgar Allan Poe was. He explained to me why it was that the U.S. was having a great deal of trouble with its envoys and ambassadors, who were often chosen from among bank directors and factory owners, people with no notion of European sensitivities and rituals. A man accredited as ambassador to the Kingdom of Serbia must be sufficiently in control of himself to avoid clapping the king on the shoulder and saying: "I like you, Sir, very much, because, Sir, you're a darned good old king, but I could buy your palace with its inventory and grounds without noticeably affecting my balance sheet for the year!" He told me that Americans don't hold much with ceremonial orders, and related the story of the American envoy, who, observing at palace receptions the envoys of other countries resplendent in their sashes, stars, and crosses, went to a jeweller's and bought himself some diamond pins, brooches and medallions, which he wore pinned to the breast of his tailcoat at the next palace ball. A European might better understand the psychology of the American envoy by imagining a European invited as guest of a minor black African king wishing to show respect for his host's customs. Edwards said he valued Poles as good workers, but the issue of Polish independence was for him one of the great mysteries of the disordered Balkans—by which he meant eastern Europe, perhaps.

On entering my room one day, I found a young man in a kimono and moccasins stretched out on my couch reading a newspaper and eating my apples. The guest greeted me as if he had known me from childhood, lay back down and informed me that Mrs. Gericke had told him that he might see if I minded his staying with me till such time as a room should be freed up. This was Rex Benson, a nephew of the Governor-General of Canada, a 20-year-old lieutenant-lancer in the British army. I had a sudden insight that the manner in which my room had been summarily annexed by the army of his Britannic Majesty may indicate how—though of course on a much larger scale—the British empire augmented its dominions—but I made no protest. Although his behavior was unheard of as far as I was concerned, it came so completely naturally to him that I found it impossible to be offended.

Rex Benson, called "Baby" by Edwards, was handsome and well-built, with a small head and beautiful features. He and Edwards had plenty to talk about together concerning sport in general and gymnastics in particular. Both would jump from the first floor into the garden and return to the room through the window by clambering up the molding. They told me that in England and America such skills were considered part of a general education. Benson had been sent to Germany by some politician or other to assess the mood of the German people—thus he was actually conducting military reconnaisance, although legally so.

Another fellow lodger was Mr. Seifert, a teacher in a local Gymnasium. He was also a reserve officer in a guards regiment, and was as proud of this as only a German teacher can be of having an officer's rank. Naturally, he owned copies of all infantry regulations, and Baby borrowed them to study, Seifert being only too happy to oblige. Benson was a happy person, and could often be heard singing French and English cabaret songs to his own piano accompaniment. He showed me a diary he had compiled while on a trip in Canada, consisting of short descriptions in pencil, sketches of people he had met, tickets and invitations glued to the pages, and photographs and autographs. This amusing way of recording one's personal history was then unknown on the continent. Baby had served in a cavalry regiment since he turned 17, and was not interested in women, but—and this is even more interesting—despite his cutting such a dashing, elegant, splendidly male figure, they reciprocated this indifference. In general, young Englishmen seemed to avoid women as if they considered them dangerous. I was never able to discover how they had acquired this deep knowledge of the gentler sex so early on in their lives.

The Germans resembled the English in many ways. There was a certain group of students from the larger German cities who were unable to accustom themselves to the serious atmosphere prevailing in Göttingen, and for relief they would go from time to time to Kassel,[89] where there was a popular theater, cabarets, and girls. All this was but a two-hour express train-ride from Göttingen. There was also a beautiful art gallery boasting more than a dozen Rembrandts—said to represent the largest such collection outside the Hermitage—as well as many other Flemish and Italian paintings. A short distance from the city was the former palace of the prince-elector, surrounded by a spacious park, beautifully landscaped, with waterfalls—but, taken altogether, somehow lacking charm. On the other hand, the woods and rivers seen

---

[89]The capital of the electorate of Hesse-Cassel from 1567 until it was annexed by Prussia in 1866, when it was merged into a new Prussian province called Hesse-Nassau. In the early nineteenth century, the brothers Grimm lived in Kassel and collected and wrote most of their fairy tales there.

from the train between Göttingen and Kassel were very beautiful. The small town of Münden, near where the rivers Werra and Fulda flow together, was exquisite, especially in springtime, when it was blanketed with cherry blossoms, and we went there several times. In winter we organized hikes to the villages Schierke and Wernigerode in the Harz Mountains, taking sleds with us to ride down the kilometers-long trails through the forest. In Schierke we watched ski and bobsled competitions held on specially constructed ramps and tracks. Late one evening, we lost our way trying to get to the Brocken,[90] which plays a role analogous to that played in Poland by the Łysa Góra.[91] Our leader, Wacław Dziewulski, took care of the train timetables and our supplies of sausage. The Harz was very beautiful—except for the smell of bad coffee that was wafted even to the mountain tops, and the crowds of people, apparently all cognizant of the maxim "Frühes Kommen sichert guten Platz",[92] which I had seen on a sign in a wood near Dresden. But it was not difficult to avoid the masses—since they came only in the holiday periods—and moreover the inns were cheap, clean, and comfortable. We also visited the old towns of Goslar and Hildesheim, where the streetcars were not manned by conductors, so that as one entered one had to drop two ten pfennig pieces in a glass container near the front door. Every now and then the driver would lean back and turn the glass container upside down so that the coins fell into a box below. I thought to myself that if an honor system of this sort were introduced into Poland, the management of the streetcars would soon have enough buttons to open a button store. Once we returned to Göttingen in a fourth-class compartment with benches ringing a small iron stove. The other passengers were Harz woodcutters, who, after listening to our Polish conversation for a considerable time, finally asked politely what German dialect we were speaking. While written German was the same as the German taught in school, most uneducated people spoke a regional dialect, so that a Frisian fisherman would not be able to communicate with a Tyrolean shepherd. Despite the fact that Hanover is considered a region where standard German is spoken, the speech of the peasants in the neighborhood of Göttingen sometimes sounds like English. In Gymnasiums in Hanover, students were taught to pronounce the letter "s" in such words as *Spitz, Stein,* and *Sprache* as in the English words "sleep" and "stand", while Bavarians, Austrians, and, generally speaking, people living to the south of the Danube all pronounced the "s" of *Spitz, Stein,* and *Sprache* the "shushing" way, like "sh" in the English "ship". The social class divisions are also, of course, associated with differences in speech—and not just between rail passengers. When I was a student, in both Poland and Germany you could sense the antagonism of ordinary working people towards those from the propertied classes. A young man in flannels on his way to play tennis, racquet in hand, had to skirt the workers' district if he did not want to hear remarks about idle spongers, and in large towns such as Düsseldorf and

---

[90] The highest mountain in the Harz range.

[91] The Bald Mountain, second highest peak of the Świętokrzyskie (Holy Cross) Mountains in south-central Poland.

[92] "The early comer gets a good spot."

Cologne, a lady in a fur coat would risk being spat at or having mud thrown at her if she ventured on foot into the factory district.

Hamburg was also not so far away, so Sierpiński,[93] Banachiewicz, and I once went there on a three-day trip. I felt that Hamburg was perhaps the most beautiful city I had seen in Germany. Near the center of the city was an artificial lake formed by the damming of the Alster, a river flowing into the Elbe at Hamburg, and around the lake one saw the great buildings of this Hanseatic[94] port. This lake was connected to another further from the city center, of such a size that one could not see the end of it from the bridge separating it from the smaller one, and ringed by the villas and gardens of rich Hamburgers.[95] Yachts and smaller pleasure boats on the lakes lent a special charm to the city, as did the northern Gothic cathedrals. Lining the canals were granaries of old darkened brick, and the harbor itself was like another city, thronged with great ships, and busy with little steamers carrying tourists, merchants, sailors, and wharf laborers from one wharf to another. Here one could hear all the world's languages—those of blacks and Chinese in particular— but it was all Dutch to me. I took a ride in a ferry, gawping at the huge cranes and the gigantic transatlantic liners. One evening I went to a restaurant near the harbor, where the guests all seemed to be merchants. They were all red-faced, spoke loudly, and guffawed, while wolfing steaks washed down with red wine. Then they smoked Havana cigars, and pinched the bottom of a buxom waitress who took it all in her stride. Here was none of the frugality and modesty of your German provincial! The food tasted good, and there was plenty of it. This was not Prussia, but the *Free and Hanseatic City of Hamburg*, as, by historical tradition, it was officially called.

Initially, I had some difficulty understanding the class structure of German society, since the Polish class system did not match the German one. In Poland, city dwellers who were members of the intelligentsia—in various professions—or clerks of various ranks who travelled first-class by train, made up our "Class I". In Germany this class was numerically much smaller, and would be designated there as forming "Class II", especially since they actually travelled second-class on the trains. So who travelled in the carriages bearing the Roman numeral "I"? Well, I only saw this numeral on long-distance trains, and passengers on such trains were seldom German. And since in Germany the mass of city folk were of the type who travelled third-class, they would naturally form "Class III" in Germany. But I had intimate knowledge of the third-class passengers. These were craftsmen, mechanics, drivers, teachers, railroad workers, storekeepers, students, business agents, shop assistants, salesmen, technicians, dentists, and so on. They all resembled one another in many ways: each had an instruction manual or textbook of some sort to bone up on, each had a small briefcase or bag that could easily be carried in one hand,

---

[93] Sierpiński came to Göttingen in 1907 for a few months' study.

[94] The Hanseatic League was an economic alliance of trading cities and their guilds maintaining a trade monopoly along the coast of northern Europe, lasting from the thirteenth to seventeenth centuries.

[95] Residents of Hamburg.

none of them hired a porter or ate in a station restaurant, many carried with them bread and liverwurst or frankfurter sausage, and at the larger stations many drank beer at the snack bar, and sometimes ate sausages with bread and mustard on paper plates. Quite a few read newspapers, and were prepared to talk about politics, and then they expressed political views that mixed democracy and nationalism. When I asked my first landlady Miss Thomas, a member of this class, what income tax is for, she replied: "For the army." And what good is the army? "To fight the French." And why fight the French? But this she couldn't answer.... Their understanding of democracy was as a complex of tenets: everyone must serve in the army; everyone must pay taxes; one must not—God forbid!—order coffee in a restaurant, because it costs twice as much as a cup of tea; one should travel third-class because it's just as comfortable as second class; the minister for railroads should respond to a complaint sent by postcard by resolving the issue and sending the answer promptly; trading for profit, that is, business, is not to be considered in any way inferior as an occupation to the professions acquired in higher institutions of learning; and every salesman should strive to attain the ideal of becoming a *Grosskaufman* ("Wholesaler").

In second-class carriages the travellers were army officers, bureaucrats with official titles, such as *Assessor, Landrat, Studienrat*,[96] and so on, as well as members of student corporations, *Junkers*,[97] and factory owners. These people loved to use the word "noble" to such an extent that it appeared even on their furniture, inkpots, and ashtrays—which demonstrates the linguistic barbarism characteristic of Germany after 1871.[98]

Ladies travelling second-class typically dressed in the English style: austere navy blue outfits and American shoes. Trinkets were eschewed except possibly for a brooch of amber or old gold. This fashion suited blonds especially. While travelling in their second-class carriages, members of this class read the illustrated magazine *Jugend*, which, while posing as liberal, was in its sly way anti-Catholic, anti-individualistic, and anti-Polish. In his history of three French generations, Bainville[99] correctly stated that in the imperial united Germany that came into being in 1871, a Germany where politics and education had become standardized nationwide, basic tenets of liberalism were understood in a very idiosyncratic way. Thus when the *Reichstag* readily voted credits to build a fleet, it was the businessmen and exporters rather than *Junkers* and peasants who reaped the benefits, when they voted to spend more on general education in Germany, it was because more officers were needed for the armed forces, and when they agreed to invest more in the German colonies in Africa—the need for which Bismarck said he couldn't understand since they were merely a financial burden—it was for the sake of the

---

[96]Respectively assistant judge, district administrator, principal of a Gymnasium.

[97]A *Junker* was a member of the landed nobility of Prussia.

[98]The year Germany was unified into a politically and administratively integrated nation state, with Wilhelm of Prussia as Emperor Wilhelm I of the German Empire.

[99]Jacques Bainville (1879–1936), French historian and journalist. Well known as a Germanophobe. A staunch monarchist, active against Dreyfus.

colonial class. The Hohenzollerns,[100] with their version of the dream of a German empire, imposed without apology the tradition of a Prussian Sparta on the whole of Germany, turning it into a land of tillers of the soil and army officers, administered by a frugal bureaucracy, under the spiritual guidance of a puritanical protestantism. After the Prussians took over, so to speak, their serene highnesses the prince of the Principality of Reuss Younger line and the duke of Saxe-Coburg-Gotha,[101] for example, became the butt of primitive jokes—like those about mother-in-laws, dachshunds, or absent-minded professors that used to appear in the pages of the popular magazine *Fliegende Blätter*.[102] These were shoddy attacks on those who had had only the appearance of power: in united Germany they no longer had any say concerning the army, the railroad, the post office, taxes, or anything else. However, these princelings, still living quietly in their palaces, represented the Germany that had been, the Germany idealized by Mme de Staël, the Germany of *The Grand Duchess of Gerolstein*,[103] who when asked by a poor coal merchant if she needed any charcoal, turned to the footman following behind with the order: "Tell this man that the answer is 'no!'" This is a different style of arrogance altogether from the famous *Immerfestedruff*[104] telegraphed by the Crown Prince of Prussia in late 1913 to the commanding German general faced with an uprising in Zabern, Alsace[105]: "Go ahead! Stick it to them!"

There was a satirical German weekly called *Simplicissimus* which first appeared in Munich in 1896, but was too strong for the above second-class citizens. Its contributing editor was Ludwig Thoma and its leading caricaturist-cartoonist the Czech Řezníček. In 1898, Kaiser Wilhelm's objections to being ridiculed on the cover led to the magazine being suppressed: one offending drawing represented the Kaiser by a big red bulldog that had slipped its chain and was sitting on railroad tracks, causing a locomotive to rear up on its hind wheels in fright. Another cartoon, caricaturing Wilhelm II's colonial policy, depicted an old Prussian fort being besieged by black African Herero,[106] with anxious white faces visible behind the palisade, waiting for the Herero to breach the gates. Text: "Why do we have

---

[100]The House of Hohenzollern is that of the dynasty of electors, kings, and emperors of Prussia, Germany, and Romania, originating in Swabia in the eleventh century.

[101]Former small states in Germany.

[102]"Flying Leaves", a humorous, illustrated German weekly, published in Munich between 1845 and 1944.

[103]A French *opéra-bouffe* by Jacques Offenbach, ridiculing court favoritism and military paraphernalia.

[104]*Immer feste druff* means "Always strongly at them", or, essentially, "Always give it to them!". Also the title of a satirical poem by Karl Kraus.

[105]Alsace-Lorraine had been annexed by Germany at the time of the Franco-Prussian war in 1871, and was a source of continuing resentment for the French.

[106]In January, 1904 the Herero people of German South-West Africa—present-day Namibia—rebelled against German colonial rule. The rebellion was put down, and the Herero were driven into the desert where many died of thirst.

to put down the uprising of the Herero in Africa?" "Because otherwise the blacks may reach Potsdam and put an end to slavery there." Much later, in 1923, a German lady who had returned with her photographer husband from South-West Africa to her native Leipzig, explained to me that after the defeat of the Germans in 1918, the British had handed the colony over to South Africa,[107] but that the Boers had made life unendurable for the German colonists. When I asked her about the Herero people, she answered that the blacks can't live without the whites because they have nothing with which to get a living.

– And how did they live before?—I asked.
– They had cows.
– And what happened to the cows?
– We took their cows away from them a long time ago to punish them for rebelling.

I then recalled the cartoon in *Simplicissimus*, and a brief history of that poor country flashed before my eyes: first the cows were enslaved by the Herero, then the Herero by the Germans, then the Germans by the Boers....

In his book *L'Allemagne*, the French journalist Jules Huret composed a lively and witty portrait of Germany in the first ten years of the twentieth century. The book caused a fuss in Göttingen over Huret's charge that the Göttingen "mandarins" gave no lectures on the latest advances in, for example, mathematics, but merely presented the standard material required of most students for the appropriate degree. However, Huret was under a misapprehension stemming from his ignorance of the German system of *Privatdozenten*, which did not exist in France. It was these "lecturers" who usually lectured on advanced topics beyond what a future Gymnasium teacher, for instance, needed to know in order to pass the requisite examinations. Being more experienced pedagogically, it fell to the professors to instill the elements of analysis, algebra, and geometry in those students whose aim was to pass the state examinations required of mathematics teachers in Gymnasia— and that is why the professors usually lectured to a full auditorium. It was suspected that Huret's source was the astronomer Brendel, whose wife was French, and as a result of the to-do he had to leave Göttingen. Of course, Göttingen mathematics professors greatly respected their French *confrères*, and recommended their students read the textbooks on analysis by Jordan, Picard, and Goursat, and Serret's[108] textbook on algebra, among others. Klein was an anglophile, and advised us to learn English in order to be able to read the mechanics texts by English authors such as

---

[107] To administer as a Mandated Territory on behalf of the League of Nations.

[108] Marie Ennemonde Camille Jordan (1838–1922), remembered for the Jordan curve theorem and the Jordan normal form of a group, among other things. Charles Émile Picard (1856–1941), known in particular for the "Picard group" of a linear differential equation. Édouard Jean-Baptiste Goursat (1858–1936), famous for his *Cours d'analyse mathématique*. Joseph Alfred Serret (1819–1885), remembered in particular for the Frenet–Serret formulae of the differential geometry of a space curve.

Routh and Lamb.[109] On the other hand, the physicists learned experimental physics largely from Russian works, and the geophysicist Wiechert recommended to us the German translation of the text *Geofizyka* by Rudzki,[110] a professor at Jagiellonian University in Kraków. Thus in order to get Göttingen University more in perspective, it is important to realize that the instruction there relied quite heavily on non-German textbooks.

By agreeing insanely to be guided by an insane Hitler and renounce the contributions of German Jews to German science, the Germans were condemning their science to a secondary ranking in Europe. Erasing from the book of *Gloria Germaniae* the names of such mathematicians as Minkowski, Landau, Schwarzschild, Toeplitz,[111] Born,[112] and Courant,[113] would reduce the scientific status of Göttingen to that of a little known provincial university.

Georgia Augusta benefitted greatly also from the presence of foreign scientists: the Greek Carathéodory, the Austrian Ehrenfest and his Russian wife,[114] the Dane Harald Bohr,[115] the Poles Banachiewicz and Sierpiński—and many others. Outstanding foreign scientists such as Henri Poincaré and Marian Smoluchowski also spent short periods in Göttingen. Albert Michelson,[116] a Nobel laureate in physics, also spent several months in Göttingen in 1911. Some of these visits

---

[109]Edward John Routh (1831–1907), English mathematician. Contributed to the systematization of the mathematical theory of mechanics. Horace Lamb (1849–1934), English mathematician. Headed the Mathematics Department of the University of Adelaide from 1875 to 1885, when he took up a chair in Victoria University (now the University of Manchester). Two of his influential textbooks are *Infinitesimal Calculus* and *Higher Mechanics*.

[110]Maurycy Pius Rudzki (1862–1916), Polish geophysicist. Established a chair of geophysics at Jagiellonian University in Kraków in 1895. Author of "Geophysics".

[111]Otto Toeplitz (1881–1940), German-Jewish mathematician working mainly in functional analysis. Appointed extraordinary professor at the University of Kiel in 1913, moving to Bonn in 1928. Emigrated to Palestine in 1939.

[112]Max Born (1882–1970), outstanding German physicist and mathematician. Nobel laureate for physics in 1954.

[113]Richard Courant (1888–1972), German-American mathematician. Studied at the universities of Breslau (Wrocław), Zürich, and Göttingen, where he eventually became Hilbert's assistant. Emigrated from Germany to the US via Cambridge in 1933. Obtained a professorship at New York University in 1936, where he founded a highly successful institute of applied mathematics (since 1964 named the Courant Institute of Mathematical Sciences). Coauthored with Hilbert *Methods of Mathematical Physics* and with Herbert Robbins the popularization *What is mathematics?*

[114]Paul Ehrenfest (1880–1933), Austrian-Dutch physicist and mathematician. Contributed to statistical mechanics and quantum mechanics. Studied under Ludwig Boltzmann. A severe depressive, he committed suicide in 1933. His wife Tatyana Alekseevna Afanaseva (1876–1964) was also a mathematician and collaborated with her husband in his work.

[115]Danish mathematician and star soccer player. Invented the field of almost periodic functions. Brother of Niels Bohr. Lived from 1887 to 1951.

[116]Albert Abraham Michelson (1852–1931), American physicist known for his work on the speed of light, and especially for the Michelson–Morley experiment. First American to receive the Nobel Prize in any science (in 1907).

were financed from a fund set up by the industrialist Wolfskehl.[117] In particular, he bequeathed 100,000 marks to the first person to solve Fermat's Last Theorem, but after World War I this sum was reduced by hyperinflation to practically nothing.

The publication of Einstein's special theory of relativity in 1905 generated much turbulence in the air over the river Leine, the stream that flows through Göttingen. It was presented to the Mathematical Society by Hermann Minkowski, who had been able to recast it in a completely geometrical form. He prefaced his talk by saying how great Albert Einstein's scientific achievement was, but adding that the mathematical education of the young physicist was rather modest, justifying this with the words "I can say this in complete confidence since he learned his mathematics from me in Zürich." Not long after this lecture Minkowski died from a burst appendix, not operated on in time. Just before his death, he expressed his disappointment in not being present to read the solution of Waring's problem[118] recently obtained by Hilbert—thus solving a problem proposed more than a century earlier. Not even imminent death can dampen the scientific enthusiasm of such people! This is not the only example of its kind: Hilbert said that if he could rise from the dead in 200 years, his first thought would not be to ask what social or technological progress there had been, but what had been discovered about the zeros of the zeta function[119]—"because that is not only the most interesting unanswered mathematical question, but the most interesting of all questions...."

At that time, in view of Minkowski's geometric reformulation of the special theory of relativity Göttingen considered that theory a triumph flowing ultimately out of Hilbert's *Foundations of Geometry*. Einstein's theory encountered objections in Göttingen: Wiechert, and especially Runge, refused to acknowledge it. However, their reasons were much more sophisticated than the nonsensical ones one could read about ten years later in various little pamphlets. One has to admit that attempts to popularize difficult scientific theories among the masses, generally speaking, end up spreading a great deal of confusion rather than genuine knowledge. Einstein never asked Moszkowski[120] to become his "court journalist", representing him to the masses, and neither Moszkowski nor certain contributors of popular articles

---

[117] Paul Friedrich Wolfskehl (1856–1906), German industrialist with an interest in mathematics.

[118] In 1770, Edward Waring asked if for each positive integer k there exists an associated positive integer s such that every sufficiently large natural number is the sum of at most s kth powers of natural numbers.

[119] This is the function defined by

$$\zeta(s) = \frac{1}{1^s} + \frac{1}{2^s} + \frac{1}{3^s} + \cdots$$

for real $s > 1$, continued analytically to the complex plane. First considered by Euler in 1740, later by Chebyshev for all real $s > 1$, and, finally, by Riemann, who showed that the above function $\zeta(s)$ is defined for all complex s with real part greater than 1, and can be analytically continued to a function defined for all $s \neq 1$. In 1859 Riemann published a paper establishing a close relationship between the zeros of the "Riemann zeta function" and the distribution of primes.

[120] Alexander Moszkowski (1851–1934), Berlin humorist and journalist.

on science to the *Berliner Tageblatt*[121] were able to facilitate understanding of his theory—for the very good reason that it is fundamentally inaccessible without knowledge of some nontrivial mathematics. What their efforts in that direction *did* accomplish was the fostering of a desire to calumniate Jews as spreaders of pseudoscientific theories.... Realizing that this was a popular trend, the newspapers soon began publishing articles linking a negative view of relativity with a negative view of Jews.

Minkowski's successor Edmund Landau was never happy when physicists came to the meetings of the mathematical society because he was not fond of physics and did not like talking on a topic unless he had a perfect understanding of it. He had an epigrammatic talent. When set theory's star began to ascend, a popular book entitled *The Essence of Infinity*, by Kurt Geissler, appeared. When I asked Landau if he had read Geissler's book, he replied: "Kurt Geissler was at one time the resident tutor of Count Arco in Berlin, and one fine morning he absconded to Italy with the count's wife and money from his safe. Thus since his behavior in the realm of the finite was so odious, I don't expect him to behave any better in the realm of the infinite...."

Landau never seemed to tire. He was of a robust build, always in a good mood, and always ready to discuss mathematics, be it in a seminar, in the street, at home, in a restaurant, in the morning or in the evening. He was very rich. He was married to the daughter of the famous medical scientist Ehrlich[122] from Frankfurt, who, together with the Japanese Hata, discovered the drug Salvarsan 606. His private library was as large as any seminary's. He had been a student of Pringsheim[123] in Munich, and had imbibed precision of argument from him. Landau was inordinately concerned that a text be absolutely correct, and submitted his papers to multiple corrections, both by himself and others. When the Warsaw journal *Prace Matematyczno-Fizyczne* ("Mathematical and Physical Works") omitted the period after his name on the title page, he insisted that the sheet be reprinted. He talked disparagingly, even contemptuously, of mathematicians who published papers containing errors or incomplete proofs or hypotheses, considering them dishonest people. His skill at finding mistakes was so great that the Ministry of Railroads always sent him the new timetables, knowing he would detect errors where no one else could. He liked Sierpiński's papers very much because their precision was up to his discriminating standards. After all this, the reader will not be surprised to learn that he was also an excellent chess player.

All of these famous professors were very diligent in carrying out their duties. For instance, they never skipped lectures, were punctual, and did not remove books from the reading room. On the rare occasions when Minkowski did actually take a book from the reading room—of which, after all, he was co-director—then he, together

---

[121] German liberal newspaper published from 1872 to 1939.

[122] Paul Ehrlich (1854–1915), medical scientist. With Sahachiro Hata he developed Salvarsan, the first effective treatment for syphilis, in 1910. Nobel laureate for medicine in 1908.

[123] Alfred Pringsheim (1850–1941), German mathematician and patron of the arts. Worked mainly in complex analysis. Great admirer and patron of Wagner.

with others, would have to listen to Klein's public reprimand, clearly directed at him although he would not be named explicitly. People used to a different, more relaxed, social climate did not feel very comfortable in this very German one. This applied not only to foreigners but also to the Viennese—who were in fact officially regarded as foreigners by the Germans. German officialdom was consistent in identifying a person's nationality with his citizenship, so that Poles from the Poznań area[124] were for all official purposes treated as German. In Mrs. Gericke's boarding house there lived for some time a student by the name of von Strauch, a German from the Baltic region, who had graduated from a German Gymnasium in Dorpat, spoke only German, and considered himself German. Yet his fellow students all referred to him as *der Russe* ("the Russian"), since someone who was obliged neither to pay taxes nor serve in the army of the Reich could not possibly be German.

Another episode bringing home to us Poles the otherness of some German ways: Once Mrs. Gericke obtained some tickets to a ceremony of the association of *Burschenschaften*[125] in the great hall adjacent to the town park, so we went there and seated ourselves in one of the loges. In the center of the hall there were rows of small tables at which sat *Burschen*[126] wearing little colored caps and sashes. They were drinking beer and singing loudly in accordance with the rigorously observed beer-drinking ritual. A girl selling roses was walking about among the tables. When one of the students bought a rose, I thought he was going to present it to one of the teenage girls in the loges, all of whom were so obviously enraptured by the sight of the drinking brotherhood. But to my surprise he offered it instead, with a tremendous show of deference, to a comrade—an event that was to be repeated some dozen times in the course of the evening. I understood this to signify that the German valorous ideal was not that of a mediaeval knight, women being for Germans weak vessels worthy only of contempt. The true German crusader feels at home only within the walls of his castle, which at the same time serves as his guardroom, his drinking den, and his male cloister—so women are excluded on three counts.

The Göttingen mathematical faculty did not engage only in theoretical work. Every few weeks Klein and Hilbert organized flights by balloon with the aim of carrying out meteorological observations, especially those having to do with atmospheric electricity. Runge also took part in the expeditions when there was a lunar eclipse to observe.

* * *

In 1910, I transferred for the summer term to Munich. There I attended Pringsheim's lectures, but not many other mathematics lectures. I attended Seeliger's lectures on astronomy, partly as a means of getting to know other Poles, and through them the town and its environs. There were many Poles in Munich then, and they had formed an officially recognized Polish association, at whose headquarters I met

---

[124] At that time belonging to the part of Poland ruled by Prussia.

[125] Student corporations of a special type, inspired by both liberal and nationalistic ideas.

[126] Lads, members of the *Burschenschaften*.

the astronomy student Kazimierz Jantzen. I lived in a room on the fourth floor of the house at 66 Arcis-Strasse, and from my window I was often able to see Venus in her guise as the morning star. On the wall opposite the window there was a picture of a Tyrolean rifleman being given a cup of water by a girl. In the morning while still half-asleep, the picture would sometimes seem to be of something else—for example, the head of an old man—and this vision would persist for a dozen or so seconds after I had awoken. My *faux-tableau* was assembled out of parts of the actual picture, such as one sees in picture puzzles purposely drawn so as to allow different interpretations. What is even more interesting is that my phantom picture was different every morning, but was always utterly clear, and always made up of elements comprising the rifleman.

The Polish colony in Munich was susceptible to cases of wounded honor. I was once enlisted as counsel in such a case, and calculated that each of the people involved wasted scores of hours on that single case, one of several such occurring each semester. Finally, the problem was solved when the elders of the association introduced a system of substantial fines for all transgressions. After a few months of this regime, affronts to honor had magically disappeared.

In May I went with the Munich Polish colony on a hike in the vicinity of Königsee,[127] and from there to Garmisch-Partenkirchen,[128] and on to the Watzmann peak.[129] Our company consisted of the ladies Mesdames Dziewulska and Majewska, and the gentlemen Messieurs Horowicz, Hedinger, Jantzen, Cieszyński, L. Suchowiak, T. Rotwand, Jan Fryling, and others. We attempted walking barefoot in the snow in the Alps in May, and convinced ourselves that it was indeed possible to do so quite comfortably for hours, without really feeling the cold. Once we even organized a trip to view Halley's Comet. I recall P. Rozenblum from Wilno,[130] who stood out in our colony, and who took me to the suburban *Mittagstisch*[131] favored by Poles. There I met a student who was all agog over Freud's theory of sexual repression, which he had learned about either in Vienna or Switzerland somewhere. Rozenblum and his *fiancée* loosed the psychoanalyst on me as soon as an opportunity arose. The young man told me that sexual dreams and unconscious sexual drives form the true basis of all our conscious views and thoughts, and that the coming into being of the latter can be fully explained in this way. So I asked him if Freud's theory itself could be explained in that way. At first he did not understand what I was getting at, but finally he twigged and was duly crestfallen. Thus Rozenblum, who had hoped to bask in the glory of my being scooped on psychoanalysis, basked instead in the glory of my put-down of the poor young Freudian—although he surely should have foreseen that things would take this turn.

---

[127] A lake in Bavaria, near the border with Austria.

[128] A resort town in Bavaria.

[129] Third highest mountain in Germany. Not far from Berchtesgaden.

[130] Vilnius in Lithuanian.

[131] Restaurant serving a midday meal.

I did not know then Karl Kraus' aphorism: "Psychoanalyse ist die Krankheit, für deren Therapie sie sich hält."[132] This would have done equally well in vanquishing the young Freudian, and, indeed, reduces all discussions on that theme to this one brilliant epigram.

I felt free to pay a visit to Professor Pringsheim, since my Tarnów Grandfather had known his father very well at some time in the past in Silesia. Pringsheim *père* had founded a railroad where the wagons were drawn by horses, and had become very rich as a result. Pringsheim told me that his father had told him many stories about my Grandfather. When I told him that I was studying in Göttingen, he said that in Göttingen they knew very difficult things but didn't know the simplest things. He gave as an example of this the case of a Göttingen dissertation in which the Fourier double integral theorem was systematically attributed to Hilbert—which was, of course, not Hilbert's fault but that of the doctoral candidate.

I spent the 1910 summer vacation in Jasło as usual. One evening, on returning from a trip to Gorajowice, my sisters described strange violet lights they had seen in the sky to the north. Next day the newspapers reported disruptions in radio communications, so I felt sure they had seen the Northern Lights. I noted down the time and date of the phenomenon, and when I returned to Göttingen in autumn told Wiechert about it. He was very surprised that the Northern Lights should be visible from such a low latitude.

During my last year at Göttingen, 1910/1911, a great variety of interesting things occurred. One of these was my befriending of Albert A. Michelson, who had been invited to spend part of that academic year in Göttingen, and who had become famous for the experiment he carried out with Morley in 1895 showing that the Earth's motion around the Sun had no optical effect whatsoever, which seemed to prove that the mysterious medium called the "luminiferous aether", which was supposed to participate in the propagation of light, did not in fact exist. Lorentz[133] and Einstein then proposed different theories relating the motion of matter and light, in order to accommodate the seemingly paradoxical conclusions emerging from the Michelson–Morley experiment. Michelson was staying at Mrs. Gericke's house, so I got to know him rather well. He was born in Strzelno, in the region of Poznań. His father had been an innkeeper, and in the 1860s had emigrated to the United States with his son, then just a few years old. Eventually Michelson completed an engineering degree and joined the navy. He loved to draw, and he told me how this passion had once landed him in trouble. During a voyage whose purpose was the training of the navy cadets in navigation, they were allowed shore leave till sundown whenever the ship was in port. Michelson's fellow cadets went to town, got drunk,

---

[132]"Psychoanalysis is that mental illness for which it fancies itself the cure."

[133]Henrik Antoon Lorentz (1853–1928), Dutch physicist. Proposed that moving bodies contract in the direction of their motion, in order to explain the result of the Michelson–Morley experiment, whence "Lorentz transformations". The same explanation had been advanced earlier by the Irish physicist George Fitzgerald, so the supposed contraction is also called the "Fitzgerald–Lorentz contraction." Joint Nobel laureate for physics, with Peter Zeeman, in 1902.

and came back on board late, but young Michelson found a solitary spot and set to drawing the scene before him, becoming so engrossed that he forgot the time and, like the rest, returned late. The captain considered the cadets' excuse that they had been too drunk to keep track of the time as a justification he could understand, but Michelson's excuse that he had been so absorbed in sketching that he had forgotten to check his watch, he regarded as absurd. The result was that he was not allowed to leave the ship for two weeks.

A short time later, Albert bid the navy adieu and began to study physics on his own. He soon became famous as a manufacturer of diffraction gratings, which are metal mirrors engraved with parallel grooves so improbably precise and dense that each millimeter-wide strip might contain several hundred grooves. His apparatus for producing the diffraction gratings was an automaton operating a sort of minuscule plough with a tiny cut black diamond as blade, chosen by means of a magnifying glass from among others for its perfect natural crystalline point—the ideal engraving tool. The force pressing this crumb-sized diamond onto the metal plate was provided by the diamond's weight—but even this tiny force was sometimes too great and had to be reduced using a counterweight. The process whereby the little plough furrowed the metal lamellae was fully automated. The whole apparatus was set up on a floating base in Michelson's basement laboratory in order to eliminate as far as possible any movement, and a fixed temperature was maintained by means of a rheostat that turned the lights on or off in response to any change in temperature. Eventually, after introducing gradual refinements of his apparatus, he was able to make lamellae with grooves several thousand to a millimeter. He had a few of these with him, and demonstrated how the sunlight reflected from one of them produced a spectrum a few meters wide. Even without collimators or lenses, one could see dark red stripes indicating the presence of iron in the sun. Before working on his apparatus for making these marvellous mirrors, Michelson had designed and constructed his interferometer, a precision optical instrument that split a beam of light and directed the two beams along different paths before recombining them to produce an interference pattern. This instrument was accurate enough to detect differences in the lengths of the two paths as small as a millionth of a millimeter. He told me that if you attached one mirror of the interferometer to a large rock set in the sturdy concrete wall of his basement laboratory, and pressed the rock gently with your finger at a place a meter from the mirror, you would observe a shift in the interference fringes, showing that the mirror had moved back ever so little! On one occasion, the precision of the interferometer was the cause of an unexpected little contretemps. A female graduate student was trying to measure a certain magnetic effect in his laboratory, and Michelson had warned her beforehand to be very careful not to bring any metallic objects into the laboratory. The experiment wasn't going as expected, but she assured him that she had neither watch, rings, nor keys on her person. Finally Michelson guessed that her shoelaces were threaded with metal wires, and to her great embarrassment she had to admit that he was right....

Michelson had used his interferometer in the famous experiment in which he, together with Edward Morley, compared the speeds of light in the direction of the Earth's motion with its speed in the opposite direction. The two speeds, and in fact the speeds of light in various inertial frames, were all equal, a fact of profound significance for our understanding of physical space, and the starting point for the theory of relativity. It was for this experiment that he received the Nobel Prize. He told me that the manufacture of a certain crucial screw for the interferometer had required six months' meticulous work. When at last the screw had met the appropriate requirements of precision, a specialist fitter and turner in the laboratory, who happened to be Swedish, accidentally dropped it on the stone steps, and, even though no damage could be seen even under a powerful lens, Michelson considered it necessary to make another one.

At the time I met Michelson in Göttingen, he was professor of experimental physics at the University of Chicago, and, being a Nobel laureate, had a privileged position there. He said he gave two lectures a week "if it wasn't raining". His wife and three little daughters were with him in Göttingen. He was then 58 years old, but could still play an excellent game of tennis. He was very easy and natural in company. Haeckel's[134] book *Kunstformen der Natur* ("Art Forms of Nature") delighted him, and he told me how he had once taught a half-course where all he did was show drawings and photographs illustrating the amazing similarities of form between living and inanimate nature. He showed me a beautiful, and very simple, experiment: if you let fall a drop of ink into a glass of water, the drop first forms a vertical thread in the water, ending in a little ball, which then changes to a ring from which two more threads of ink descend, each of which in turn repeats the process, and so on. Within a few seconds a whole upside-down tree of threads of ink has formed in the glass. I never found out what physicists or chemists have to say on this beautiful phenomenon. I would say that Michelson himself was a naturalist rather than a physicist. It seemed that for him acquiring mathematical knowledge was as difficult as the construction of an optical instrument would be for me. Just after I obtained my doctoral degree, Michelson suggested I go with him to Chicago as his mathematical assistant. I turned him down, however, because I had already had enough of living in exile.

There was a large group of Japanese students in Göttingen, with whom we were on friendly terms. We once invited them to our quarters and presented them with a book of reproductions of Polish art. They were not very receptive to this art, saying that apparently only Wyspiański[135] knew how to draw. When they then sang their national anthem for us we had difficulty keeping straight faces, but when we reciprocated by singing the *Warszawianka*,[136] they laughed so hard they all but fell off their chairs.

In 1910 another American, a Professor Porter from Austin, Texas, lived for some time in Mrs. Gericke's boarding house. Thus I had ample opportunity for getting to know Americans. Their approach to life is more proactive than that of Europeans;

---

[134]Ernst Haeckel (1834–1919), eminent German biologist, naturalist, and philosopher.

[135]Stanisław Wyspiański (1869–1907), famous Polish playwright, painter, and poet.

[136]The Song of Warsaw, written in 1831 on the occasion of the November Uprising against Russian rule in 1830–1831.

in fact, I would say that this is the main difference between the peoples of the new and old worlds. The Americans will have no truck with pessimistic philosophizing. They prefer acting to talking. They have a capacity for enjoyment that is rather childlike—adult Americans even seem to enjoy pillow fights! The youthful among them are afraid of women, like to drink, and like to break things when drunk: during the celebration of their Independence Day they smashed all the signs on the edge of town advertising Rohn's restaurant. They like wearing new clothes and look good in them, and are less money-conscious than Europeans, with a devil-may-care attitude to their possessions. Their musical sense is superior to that of the English, as is shown by the fact that they are good dancers. When they reach the age of 15 they consider themselves grown up. The young are polite to their elders without deferring unduly to them. They are subtler than the Germans; for instance, they understand irony, which is inaccessible to Germans. I had observed that on the trains, German conductors always allowed that a passenger was right if he behaved like a noisy brute when airing grievances, while they assumed that a passenger who did so calmly must be afraid of them. The average German dislikes logical reasoning since it is at odds with his world-view. They have an exaggerated respect for Anglo-Saxons: the Berlin newspapers reporting on the behavior of the crew of the Titanic during the 1912 catastrophe concluded that "the English are better people than us."

Fig. 4.1 Part of the first and last pages of the certificate of studies awarded to Steinhaus in 1910 in Göttingen (Courtesy of the HSC Archive, Wrocław University of Technology.)

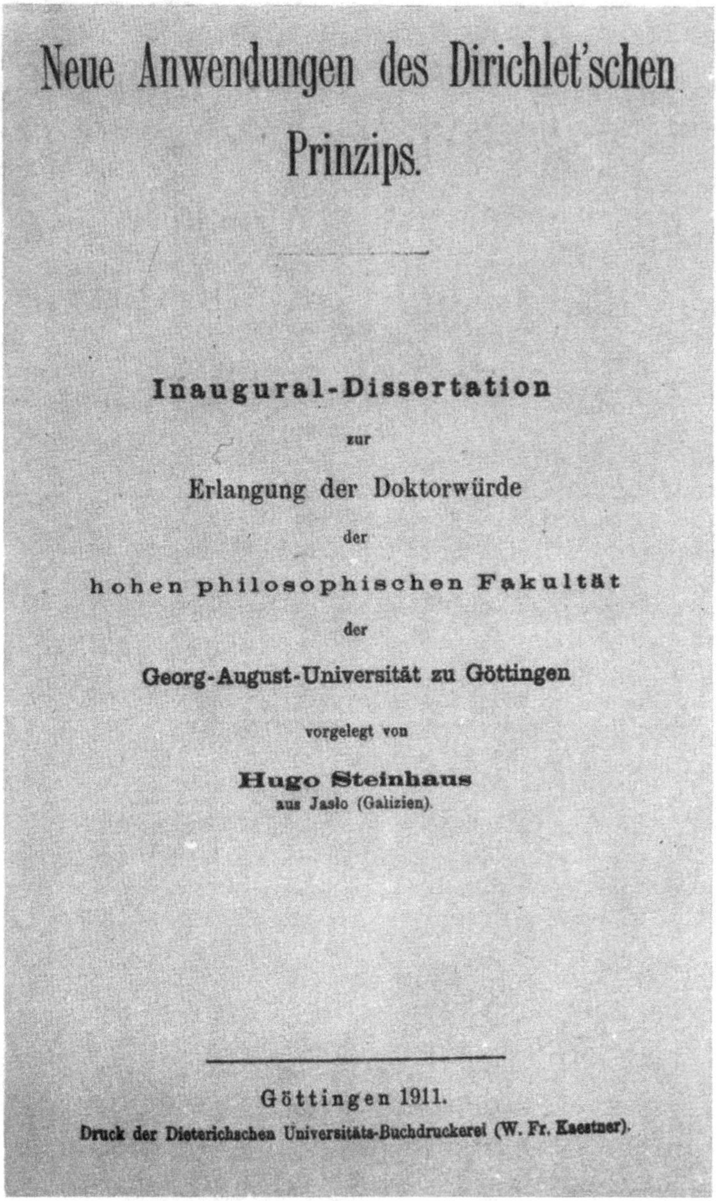

Neue Anwendungen des Dirichlet'schen Prinzips.

Inaugural-Dissertation

zur

Erlangung der Doktorwürde

der

hohen philosophischen Fakultät

der

Georg-August-Universität zu Göttingen

vorgelegt von

Hugo Steinhaus

aus Jasło (Galizien).

Göttingen 1911.

Druck der Dieterichschen Universitäts-Buchdruckerei (W. Fr. Kaestner).

Fig. 4.2 Title page of Steinhaus's Ph.D. dissertation (Date of the oral examination: 10 May 1911. Referee: Herr Geh. Reg.-Rat. Prof. Dr. Hilbert.) (Courtesy of the HSC Archive, Wrocław University of Technology.)

Chapter 5
The Return Home

I passed the final examination for my doctorate on May 10, 1911. The examiners were Hilbert, Runge, and Hartmann. Having thus completed my studies, I went to Lake Lugano in the Italian-speaking region of Switzerland, where I stayed for two weeks at the *Hôtel du Lac*. It was perfectly located, and very peaceful, the guests being for the most part elderly couples putting to good use the money they had so labored to accumulate over the years. Each couple sat at its own little table looking bored—presumably they were no longer capable of pleasure at the sight of the Alps and the lake. Most interesting in Lugano are the small villages along the lakeshore—Oria, for example. The houses are all of stone, and are accessed from stairs carved in stone, rising from the shore. From Lugano I went to Milan. The Gothic style of the Milan Cathedral is awe-inspiring, with its thousands of spires and thousands of steps, with its rooftop balustrades, where each support of the handrail is capped with a head in low relief as meticulously carved as if it were crucial to the whole edifice. This fine work misleads one into thinking the interior should be equally rich, but instead it's dark and austere. There it impresses by the massiveness of its columns and the *chiaroscuro* of the enormous enclosed spaces, relieved by the suppressed play of colored light from the stained glass windows. High up in the south wall one can spy an aperture, and down on the parquet floor there is a brass inlay on the same meridian of longitude, so placed that at true local noon time there appears a momentary flash of light at a spot on the north wall. I wanted to buy a *Borsalino*[1] hat, a popular Italian hat with a moderately wide brim, so I went to the biggest hat shop I could find. There they looked at me disdainfully as they informed me that they did not sell such hats.... To this day I remain baffled by this event....

From Milan I travelled to Venice, which I won't describe since, like the character Bukacki in the *The Połaniecki Family*,[2] I would have to say that "I spied under the

---

[1] A company that produced hats, known particularly for its fedoras. Established in 1857 by Giuseppe Borsalino.

[2] A novel by Henryk Sienkiewicz.

Ponte di Rialto an eggshell floating next to a piece of orange peel. Suddenly a wave carried the eggshell closer to the orange peel and they swam away together"—but then, despite Sienkiewicz's great authority, I fear no one would believe me. For me the most interesting thing about Venice was its architecture. Venetian palaces have a style of their own: the marble is black in some places from the action of water, and retains its whiteness in others, and this tends to emphasize the noble lines of these buildings, reflected in the black waters of the canals. St. Mark's Basilica must be one of the strangest churches in the world: only in this place, lying between the East and the West, could a church be built combining both Italian and Byzantine elements, the marble slabs of one of whose walls—the one facing the Doge's Palace—are of all the pastel colors of the shells on the Lido Beach. Before the basilica stand the famous Horses of St. Mark[3] taken away by Napoleon, and returned after his fall. The interior is decorated with mosaics on a gilt background suggestive not only of eastern hieratic primitivism but even the perversion of the East. Built over relics of St. Mark brought from Byzantium,[4] this church is the chief symbol of Venice, a city whose crest features a lion holding the Gospel of St. Mark.

On returning to Poland, the first thing I did was attend the First Meeting of Polish Physicians and Naturalists in Kraków, at which I gave a presentation in the mathematics section. Not being used to scholarly meetings, I listened attentively to many of the lectures. The best of the general lectures was Romer's[5] talk "On Landscape in Geography and Art." I had a long discussion with Stanisław Zaremba senior[6] about the axiom of completeness introduced by Hilbert as one of the axioms of Euclidean geometry. Zaremba wanted to reformulate this axiom in the following way: "We include in geometry all and only such objects that satisfy the axioms previously listed."[7] Eventually Zaremba saw that he could not convince me, so proposed an armistice in the form: "We declare peace since I cannot refute his arguments nor he mine." This discussion, which took place in the coffee house *Bisanza*, was provoked by my note on the concept of limit that had appeared in *Mathematische Annalen*.[8]

---

[3]Installed in about 1254. Believed by some to have once adorned Trajan's Arch. They were taken to Paris by Napoleon in 1797, and returned in 1815.

[4]They are usually said to have come from Egypt.

[5]Eugeniusz Romer (1871–1954), outstanding Polish geographer and cartographer. Founder of modern Polish cartography. Professor at the University of Lwów and the Jagiellonian University.

[6]Stanisław Zaremba (1863–1942), one of the internationally best-known Polish mathematicians of the time. Professor at the Jagiellonian University from 1900. Co-founded the Polish Mathematical Society in 1919.

[7]It is unclear whether he means that the axiom of completeness follows from the other axioms, or something else.

[8]Bibliographical data on all of Steinhaus's publications as well as reprintings of the more important ones can be found in: H. Steinhaus, *Selected Papers*, PWN, Warsaw 1985. See also the following selection of his papers in Polish: H. Steinhaus, *Między duchem a materią pośredniczy matematyka* [Between Spirit and Matter There Exists Mathematics], Wydawnictwo Naukowe PWN, Warszawa–Wrocław 2000.

5 The Return Home

Upon arriving in Jasło, I was greeted at the corner of Third of May Street and Kościuszko Street by the poet Ludwik Eminowicz,[9] who asked me what I had seen in Germany. Indicating the spot where we were standing, I replied: "The Germans will come here"—and my prediction was fulfilled three years later.... But in our family, a local matter such as, say, my uncle's candidacy for the position of parliamentary representative of the Bełz district caused more commotion than the threatening clouds of international discord. In Galicia at that time a discernible rift had appeared between the Kraków and the Podole factions of the conservative party. In order to prevent the Vienna parliament from breaking into a great number of nationalistic splinter groups all at each other's throats, and so rendering the normal working of the House of Representatives impossible, the government, at the Emperor's urging, introduced a law to the effect that the election of representatives to parliament should be general, equal, and by secret ballot. Thus in Galicia, in particular, in compliance with the new law, the system of curial voting[10] to the Lower Chamber was to be replaced by the system "one man, one vote." The governor of Galicia, Michał Bobrzyński, was in favor of the change. While he and Leopold Jaworski,[11] the best mind in the conservative party, were not unduly troubled by the idea of general enfranchisement, the Podolians, that is, the owners of traditional manorial properties east of the San River, had every reason to fear the loss of political power resulting from the admission of the Ruthenian masses to the ballot boxes. Uncle Ignacy had the support of the Kraków faction, and therefore of the government and its deputies, but the Podolians refused to agree to the Kraków faction's recommendations, and put forward the local landowner Count Starzyński, a law professor at the University of Lwów, as the candidate for the Bełz district. In many of the small towns in that region—such as Bełz itself, Rawa Ruska, Sokal, and so on—many of the potential voters were Jewish, and it was thought that their votes might very well turn out to be significant, given the split among the conservatives. In Bełz there lived a celebrated miracle-working rabbi, and my uncle had to pay him a courtesy visit. Knowing that my uncle was a non-observant Jew, the rabbi asked what he thought of the traditional Jewish customs. My uncle's answer: "I'm not familiar with them, so intend to leave such matters to people who know the Holy Scripture, that is, to people like you, whose opinion must be decisive in such matters." Although this would normally have been enough to reassure the pious man, there was another source of doubt: Stańczyk's[12] candidate Starzyński, *homo*

---

[9] Polish poet and translator. Lived from 1880 to 1946.

[10] That is, whereby the electorate is subdivided into groups each with one vote. From the Latin *curia*, in early Roman times signifying a section of the people—a tribe, more or less.

[11] Władysław Leopold Jaworski (1865–1930), Polish lawyer, conservative politician, and professor at the Jagiellonian University of Kraków.

[12] Stańczyk (1480–1560?), the most famous court jester in Polish history. Became a symbolic figure in Poland, with patriotic associations. A conservative political group formed in Galicia following the collapse of the 1863 January Uprising called its political platform *Teka Stańczyka* ("Stańczyk's Portfolio"), whence the name of the group: *stańczycy*.

regius,[13] was, as always, counting on the rabbi's support. Although, it is true, the *starosta*[14] had assured the rabbi that Dr. Steinhaus was the candidate favored by Governor Bobrzyński, this wasn't convincing enough. He suspected that the Polish gentlemen had quarrelled, and that neither in Bełz nor even Lwów would he be able to discover the truth as to who was the more worthy of his support. Such being the case, he sent his secretary to the emperor in Vienna, and, surprisingly, the secretary in his gabardine[15] cloak found his way to the chancellery and obtained the required information. The officials in the chancellery doubtless consulted with the governor in Lwów, who would, of course, have confirmed in confidence that the government favored the parliamentary candidate Dr. Steinhaus in the election.

Głos Jasielski ("The Jasło Voice") was a weekly that was put out in Jasło for a few years around that time by an engineer named Kostkiewicz. It was a publication of modest ambitions with national-democratic leanings, embellished, of course, with local gossip. "Steinhaus", the name by which I have the honor to be known, appeared in that rag several times: my Father and uncle were described there as tyrants who did whatever they pleased. The journalist who had written these words in the hope of perpetrating an insult was mistaken, since attacks of this kind provide free publicity that ends up merely increasing the influence of prominent local personages. After my uncle had been elected to the parliament in Vienna, a large crowd of young people gathered at the train station in Sokal to show their displeasure with him and the government he represented. However, the disturbance that ensued when he arrived there was confined merely to some shouting.... I have to admit that my uncle was a consummate diplomat. For example, if a certain *starosta* was inconvenient to him in some way, he would not resort to saying anything negative about him to his superiors in Lwów, but do the exact opposite, with the aim of having the *starosta* transferred elsewhere by way of promotion to a better posting. He would explain that the *starosta* represented "a force which is being wasted in the provinces."

In the Fall of 1911, I went into the army to do my compulsory service. The barracks of Fortress Artillery Regiment No. 5, named for Freiherr von Rouvroy,[16] were near the regimental depot in Montelupi Street in Kraków. The training school for us would-be soldiers doing our year-long national service was nearby. I did not find the gymnastics, knee-bends, runs, and drills too tiring, but the compulsory early rising, polishing of shoes and buttons, the dismantling and assembling of guns,

---

[13] "The king's man"

[14] Of Bełz, presumably.

[15] A tough, tightly woven fabric used to make overcoats, trousers, etc. May once have been the material of preference for garments worn by Jews; in *The Merchant of Venice*, Shakespeare has Shylock talk of his "Jewish gabardine."

[16] *Festungsartillerieregiment Freiherr von Rouvroy Nr. 5*, named for the Austrian general J. Th. von Rouvroy (1727–1789).

the making of beds and folding of coats, having to listen to a sergeant's idiotic instructions, and the chicanery going on all about me—all this made the military repugnant to me beyond all limits. The barracks were at one and the same time like a prison, a hospital, a boarding school for small children, a pub, and a training ground. The corporal whose job it was to teach us how to aim the heavy field guns told me that I would never understand how a Vernier scale works. The captain who taught the art of fortification was a pedagogical genius. He lectured using the Socratic method, by asking questions, and I admired him for this, since this approach to instruction requires a very thorough knowledge of the subject. But I had very soon had enough of military theory and practice, and wrote complaining letters home. It turned out that acquiring a heart condition costs much more than curing one, but that it can be acquired for a price—there are ready suppliers even of that commodity.

So I soon found myself in a hospital room with two beds. The other bed was occupied by a cadet from Łobzów,[17] who, after learning that I had a doctorate in mathematics, said scornfully: "You don't claim to know more mathematics than our teacher in the military school for cadets, do you?" In the hospital I also met an uhlan who had become deaf in one ear as the result of a slap administered by his sergeant-major: he had misheard the command for a half turn as that for a full turn. He hadn't submitted a complaint, he said, because apart from this incident, the sergeant-major was a good man. The uhlan knew that when an ordinary soldier comes to the divisional headquarters with a problem, then the officers will be interested in hearing about what's going on in his squadron—in particular, how the NCOs are behaving themselves. However, he had kept quiet about the sergeant-major's treatment of him.

I had something to read in the hospital—a French version of *Don Quixote* with woodcut prints. When the chief staff doctor came to examine me in the company of two assisting physicians, he looked with interest at *Don Quixote*, but the two assistants seemed indifferent. But when he noted a difference in my pulse rates when measured using different wrists, they readily confirmed this extraordinary phenomenon. I returned to the barracks with a certificate releasing me back to civvy street. I presented my beautiful full-dress parade shako[18] to Corporal Korbulak, an acquaintance from that episode.

The matter of my military service having been settled to everyone's satisfaction, I went to Lwów—partly to spend a few months near my sisters, who were studying at the university. They were living in a boarding house called *Litwinka*. I took a room in a boarding house called *Polonia*. One of my fellow boarders was Stefan

---

[17] A village close to Kraków (now become part of that city).

[18] A tall cylindrical military cap, with a metal badge embossed with the Austro-Hungarian imperial double-headed eagle in front, and a pompom on top.

Mazurkiewicz,[19] who was then working towards a doctorate under Puzyna. It was said of him that he would pace back and forth in his room for hours. He smoked constantly, and disliked opening windows. Another of my fellow boarders was Professor Weyberg,[20] who looked more like an actor than a scientist. One of the sons of my landlady was a government official in the department of industrial concessions.[21] He assured me that in Warsaw the water utility draws water from the canals and by means of excellent chemical filters is able to deliver water as pure as spring water. I expressed my scepticism to him, with the result that he declared to one and all that if I thought I would obtain my *Habilitation* in Lwów, then I was laboring under an illusion. At that time I met at Wacław Sierpiński's a young Stanisław Ruziewicz,[22] fresh from completing his doctorate, so at a stage in his career similar to mine. I myself was just beginning to do what one might call genuine mathematical work: I read Lebesgue's[23] *Leçons sur les séries trigonométriques*, and as a result managed to write a note which Sierpiński edited—so that it ended up being in his style—and had published in the *Sprawozdania Towarzystwa Naukowego Warszawskiego* ("Reports of the Warsaw Scientific Society"). All the same, I was just a private scholar without a definite occupation. I accompanied my sisters to balls, visited the members of the family of my cousin and uncle, and played tennis in Jasło in the summer. The next year I lived for a time in a boarding house in Kraków, or rather, to be more accurate, in the house of a Mr. Lepszy and his family on the border between Kraków and Łobzów. Mr. Lepszy was an iconographer, but to this day I don't know what the term means. My sisters lived a few houses away at Professor Ciechanowski's[24] place.

The year was 1913, but it did not seem that the end of the world was nigh. The First and Second Balkan Wars[25] had even come to an end. My uncle used to say at that time "We are choking", meaning that it was becoming more and

---

[19] Stefan Mazurkiewicz (1888–1945), influential Polish mathematician. Studied under Sierpiński and numbered Borsuk, Kuratowski, Knaster, Saks, and Zygmund among his students. Worked in mathematical analysis, topology, and probability. Spent most of his career at the University of Warsaw.

[20] Zygmunt Weyberg (1872–1945), Polish crystallographer and mineralogist.

[21] Concessions allowing exploitation of forest resources and the lease of state-owned industrial enterprises.

[22] Stanisław Ruziewicz (1889–1941), Polish mathematician. A student of Sierpiński. One of the founders of the Lwów school of mathematics. On July 12, 1941, he was arrested and murdered by the Gestapo, along with other Lwów professors.

[23] Henri-Léon Lebesgue (1875–1941), French mathematician. His generalization of the Riemann integral revolutionized integration theory.

[24] Stanisław Ciechanowski (1869–1945), originally Dean of the Medical Faculty of the Jagiellonian University of Kraków.

[25] In the First Balkan War, in 1912, Bulgaria, Greece, Montenegro, and Serbia attacked the Ottoman Empire, ending the latter's five-century rule over most of the Balkans. The Second Balkan War (1913) was essentially a squabble over the spoils between Bulgaria and the other victors of the First Balkan War, with Romania and the Ottomans intervening to make territorial gains.

5 The Return Home

more difficult to find good opportunities for investing capital. But on the other hand getting credit was not very difficult. When my Father and uncle, together with Loewenstein, Gorayski, Stojowski, Macudziński, and Wiktorowa, formed a consortium in order to purchase a parcel of farm land in the region of Polanka with a large steam-operated brick kiln on it, the banks loaned them a million crowns. They bought the property from Count Goetzendorf-Grabowski, who was half-German, and he lived extravagantly for a time on the proceeds, eventually ending up as a horse-breaker in the Renz circus.[26] There was also a manor house on the property, which the new owners rented to Stanisław Klobassa, who had also gone through a fortune,[27] for one crown a month. The brick-kiln was managed on behalf of the consortium by its youngest member, Kazimierz Macudziński, who was soon faced with an emergency: the mechanic got drunk and forgot to replenish the steam turbine's oil-pan, and friction from the motion of the large fly-wheel melted the bearing ring. As a result the wheel began to lean, threatening to knock down the adjacent wall. Mr. Macudziński managed to stop the engine in time, and when he had it dismantled and understood the cause of the near-disaster, he immediately fired the mechanic. Being aware of the vengeful nature of some people, he tried to ensure that the mechanic was kept away from the factory. He telephoned to Bern for a fitter, who duly refitted the engine, but when after starting it he observed the action of its two pistons, he ordered it stopped again. One had to be an experienced mechanic to note something amiss in the rhythm of the two pistons.... Another moment, and the cylinder that should have been at low pressure would have exploded since the valve admitting steam into the cylinder was closed when it should have been open, so that action of the piston in the other cylinder caused pressure to be constantly exerted on the trapped steam in the first. The Swiss machinist took apart the timing mechanism and showed Mr. Macudziński the small wheel regulating the opening and closing of the cylinders by means of grooves in its plane. It turned out that after he had taken the engine apart after the accident, the original mechanic had had enough time to smear graphite on one of the grooves and make another one with a few strokes of a file. This technical trick might have caused an accident with no one being the wiser.

In Kraków that year, I wrote four notes on Fourier series. Zaremba presented them to the Academy[28] and they were published in 1913. (For a long time I kept Zaremba's letter in which he expressed the wish that my French were better. There was a mistake in this letter, which was written in Polish: he misspelled the word *życzenie* as *rzyczenie*.[29]) I joined a rowing club and learned to row, and even raced a few times in a rowing eight. Among the members were Miss Bujwidówna and

---

[26] A circus founded by the Dutchman Arnold van der Vegt in 1911, at one time featuring Tom Mix as the star attraction. The name was changed to "The Herman Renz Circus" in 1923.

[27] See Chapter 1.

[28] The Polish Academy of Learning (before 1919 the Academy of Learning), founded in Kraków in 1872 and merged in 1951 with other institutions to form the Polish Academy of Sciences.

[29] The papers in question were written in French. The Polish word *życzenie* means "wish".

her sister, Zoll Jr.,[30] the Jentyses, and so on. The boats had to be carried from the boathouse, which was located on an elevated part of the river bank without suitable access to the water, to a river dock in town. Changing into and out of our rowing gear, and marching to town and back took more time than the actual activity of rowing on the Vistula. The whole outing took half the day, and most of one's energy for the whole day. I conclude that whoever wants to practise a sport seriously must give up his real work. Since the latter option didn't appeal to me—and in any case I would never have become more than a third-class oarsman—I limited myself to the tennis courts.

In 1913, my parents took my sisters and me on a trip to Italy. We first went to Venice, where we stayed in the Hotel Bauer. There they served Viennese food, but the smell of fish being fried in olive oil was wafted to us as through the dining room window as we ate. I would very much have preferred the fried stuff of the poor street vendor to the veal cutlets fried in grease by Mr. Bauer. My sisters and I made an excursion to the Lido in a small steamer, and on the return trip we noticed a gentleman in an elegant grey outfit with opera glasses on a strap around his neck, accompanied by a blonde as fashionable as a model from the latest *Vogue*. They were seated in the stern, and, when the ferry approached the *Riva degli Schiavoni*,[31] they rushed to the prow and began taking turns looking through the binoculars with exaggerated expressions of delight on their faces. When my sisters and I turned round to find out what they were looking at, we saw standing on the quay a wicked-looking character, and nearby a camera man recording the whole scene. After we had berthed, the whole troupe set off—with us close behind—for the *Piazza San Marco*[32] to be filmed feeding the pigeons against the background of St. Mark's Campanile, newly restored. The whole thing must have yielded a cheap little flick! This episode prompted me to reflect on the autonomy of the phenomena of real life. When the cinema was invented, it was at first thought that it might serve to record the unfolding of natural phenomena. However, the new medium has from the beginning devoted itself instead to the recording on film of motions that have been planned and learned beforehand, and movies of naturally evolving or moving organic or inorganic nature comprise a vanishingly small part of cinematography.[33] Cinematographers don't care to let us see what is happening out there in reality; they form a powerful clique forcing millions to view things invented by the hirelings of big business, who—and this is truly scandalous!—insist on being called *auteurs* and *artistes*. Radio is a similarly manipulative medium—and the press also, which has given up reporting the observed bare facts of the matter and arrogated power to itself by actively putting its own constructions on the facts.

---

[30]Fryderyk Zoll the younger (1865–1948), Polish jurist.

[31]Venice's main waterfront boulevard.

[32]St. Mark's Square.

[33]A hundred years later, perhaps the balance has been restored somewhat with the many excellent nature films now appearing on television.

5 The Return Home

Those were the pioneering days of aviation. I had first seen a plane take off when I was in Munich; the pilot was Lindpaintner,[34] and, on landing, his plane got smashed to pieces. For a long time I kept a wooden splinter from the wreckage as a memento. Nobody predicted that nations would soon be making their people sweat blood to produce flocks of high-flying metal monsters that would be used to reduce palaces, thousand-year-old cathedrals, and hundreds of towns to so much rubble, forcing people to hide in basements and cellars. But at the time of which I am writing, airplanes were the work of carpenters and locksmiths, and we thought they heralded an era of unrestricted international travel, that everyone would be as free as a bird, so to speak....

From Venice we travelled to Rome via Florence. In Florence we stayed at the *Hôtel des Américains*, which had been recommended to us by a young Hungarian lady, Cécile de Tournay, whom we met on the way. It seemed that she found us very congenial, so we chatted very amicably with her. She told us that she made a living from writing fiction. After a little while she confessed that when she saw our faces covered with red blotches she had thought to commiserate with us for some dreadful disease that had struck the whole family. She calmed down when we explained that these were just the traces of mosquito bites. Later, when we were sitting at our small table in the hotel dining room, we could not help overhearing a conversation right next to us between an elegantly dressed Italian lady and an Austrian aristocrat. When the Austrian announced pompously that a certain lady with a long noble pedigree would soon be arriving in Florence, the Italian lady said loudly: "Sie stinkt aber!.[35]" I gathered that the lady was descended from the Doge Mocenigo,[36] but this didn't stop her expressing her dislike of another woman in this blunt fashion.

In Rome, of all the works on display in the many churches and museums we visited, I was most impressed by the Medicean Venus[37] and The Wrestlers.[38] We also joined a crowd of a few hundred pilgrims who were being given an audience by the pope. When the pope appeared on the balcony accompanied by the priest Count Sampero, the Italian children present began to chant loudly "E'viva il Papa!" without kneeling, and the pope raised a finger in a gesture of admonishment. Before leaving, we bought some postcards in a little shop adjacent to the Vatican. There were some depicting that same Count Sampero, a dashing, black-eyed Italian, and when the saleslady saw us admiring his likeness, she said in heavily accented French: "C'est

---

[34] Otto Lindpaintner (1885–1976), German aviation pioneer.

[35] "But she stinks!"

[36] The Mocenigo family was prominent in Venice for more than four centuries. Several of its members served as doge.

[37] A statue formerly standing in the Villa Medici, carved by the Athenian artist Cleomenes in the second century BC.

[38] A famous Roman sculpture in marble after a lost Greek original of the third century BC, now in the Uffizi collection in Florence. (Several copies have been made of the sculpture, so perhaps the author saw one of these in Rome—or perhaps he actually saw it in Florence.)

joli, le Comte Sampero!" At this, a priest, probably one of the pilgrims, gave us a disapproving look.

From Rome we took the coastal train to Nice, where we stayed for some time. It was autumn, but the sea was still warm enough to bathe in. Once, even though quite close to shore, I almost drowned: the undertow caught me and despite my swimming as hard as I could towards the shore, I was swept out till I grabbed hold of an iron stake poking out of the water. An Englishman swimming close by advised me to dog-paddle instead of using my usual stroke since that is less tiring and one doesn't need to make dangerous pauses.... I imitated him, and after some time managed to reach the shore safely. My Mother had been watching me from the promenade the whole time but hadn't suspected I was in any difficulty.

Of course, we made a side trip to Monte Carlo, and also to Grasse, where they grow carnations to make perfumes. Everywhere we went we—my sisters Irena and Olga and I—did a lot of walking. Once, when walking through the suburbs of Nice, we came across a barred cell set in a wall, which contained a live bear, and in the yard behind we could see a fake stall with sausages and legs of ham made out of *papier-mâché* laid out on the counter. It soon dawned on us that this was a depot of the French film company Pathé—confirming for me once again that the cinema is devoted to combining objects in the most unlikely ways: the stall with its fake comestibles was doubtless intended to be turned upside down in some comedy or other.

From Nice we went to Paris, where we stayed in the *Hôtel Cambon* on Cambon Street, and did all the things tourists do in Paris. After a few days, my parents and I returned home, leaving my sisters in a boarding house. However, I returned the following spring, and took a room in a *pension* at 36 Rue Gay-Lussac, with a view to doing some mathematics. I met Henri Lebesgue, and attended lectures by Borel and Picard.[39] At that time, Henryk Lauer[40] and Aleksander Rajchman,[41] both from Warsaw, were visiting Paris. Once, while we were strolling in the Luxembourg Garden, I talked to Rajchman about trigonometric series, telling him, in particular, of Riemann's famous paper[42] on the subject, written almost 60 years earlier. Even today one comes across the ahistorical expression "Riemann's localization

---

[39] Charles Émile Picard (1856–1941), French mathematician. Known mainly for his work in complex analysis and differential equations.

[40] Henryk Gustaw Lauer (1890–1939), Polish mathematician, economist, and social activist. A leading figure in the Polish Communist Party 1920–1922. Victim of the Stalinist purges.

[41] Aleksander Rajchman (1890–1940), Polish mathematician. Worked at the University of Warsaw from 1919 to 1921. Obtained his Ph.D. under Steinhaus at the University of Lwów in 1921. Habilitated at the University of Warsaw in 1925, he was a *Privatdozent* there till 1939. In the late 1930s lectured for a time at the Collège de France in Hadamard's seminar. Arrested by the Gestapo in April 1940, he perished in the Sachenhausen concentration camp later that year.

[42] A paper on the properties of a function given in the form of a trigonometric series. It was originally part of Riemann's *Habilitation* dissertation.

5 The Return Home

principle",[43] used as if this was how Riemann himself had phrased it. But it's all my fault, because this is how I expressed it to Rajchman, and he was unaware that I had made the name up.

I especially enjoyed going to the teahouse in the Luxembourg Garden, but sometimes when at a loose end, I would go to a café—such as the one next to the Corn Exchange—and try eating their oysters, snails, and various other molluscs. One of the strangest of Parisian streets was the narrow *Rue Mouffetard*, where the store fronts were wide open onto the street, and everywhere—in the store windows, and outside in the street—lay baskets of all kinds of produce: fish, vegetables, slabs of meat, oranges and apples,.... The accumulated detritus, like the moraine deposited by a glacier, made no impression of foulness or litter.

The Parisians like to stroll in the middle of their streets, so that, having been tamped down for centuries, they form an excellent base for the laying of asphalt. Like the Italians, Parisians live an important part of their lives in their streets, yet at the same time they have a way of maintaining order unobtrusively which is hard for foreigners to grasp. The rotating brushes of mechanical street sweepers swept the dirt and rubbish off the curb into the gutter, whence it was washed into open sewage drains by streams of water from hydrants.

Paris is many towns combined. On the Left Bank the institutions of higher learning are clustered: the Sorbonne, the Faculties of Law and Medicine, and close at hand the *Hôtel de Cluny*, parts of whose domes and walls brought to mind Julian the Apostate.[44] In the Cluny Museum the crown of Pepin[45] was on view. There were Renaissance coffers of black oak, looking as if they were carved by Colas Breugnon,[46] and arrayed on top of these ivory caskets, enamelled reliquaries, and the wine goblets of mediaeval knights—including the Knights Templar[47]—whose vestments hung in showcases and whose armor, helms, and swords decorated the walls. The best way to visit this museum is to do so on impulse when you just happen to be passing by. It was never crowded, even on those—quite frequent—days when admission was free. Through the windows one glimpsed the garden and greensward, fenced off by a railing from the *Boulevard St. Michel*, with its smooth but sticky

---

[43]This states that if a periodic function with period 2π is integrable over the interval $[-\pi, \pi]$, then for each point x in that interval, the convergence of the function's Fourier series at x depends only on the behavior of the function in an arbitrarily small neighborhood of x.

[44]Flavius Claudius Iulianus (331–363), Roman emperor from 355 to 363. Took command of the western provinces in 355 as Emperor in the West. On the death of Constantius, Emperor in the East, in 360, he was proclaimed Emperor at the *Thermes de Cluny* in Paris, whose Gallo-Roman ruins form part of the present-day *Musée de Cluny*. A non-Christian, he attempted to revive traditional Roman religious practices.

[45]Pepin "the Short" (714–768), king of the Franks from 752 to 768. (Possibly the museum guide misled the author, since the three crowns in question are those of seventh century Gothic kings. It is known for certain that one of these was the Visigothic king Reccesvinthus.)

[46]The title character of a novel by Romain Rolland (1866–1944), a sculptor.

[47]A Western Christian military order endorsed by the Roman Catholic Church in 1129, and active in the crusades.

asphalt and busy traffic. You can cross the boulevard and sit in the Luxembourg Garden, warm yourself in the sunshine, and look at the charming children skipping rope so gracefully near the pond. Close by is the *Palais du Luxembourg*, seat of the French senate, where it is made very clear that the palace is not in the jurisdiction of the Prefect of the Seine,[48] but in that of the *commandant* of the senate—whose job doubtless represents the perfect sinecure. Here we were able to visit the Luxembourg Museum, a small Romanesque building full of the most beautiful marble sculptures, and paintings by Degas and other impressionists—reminding one that at that time these painters were still regarded as revolutionary, so not exhibited in the Louvre but on the Left Bank, where the young have been in charge for centuries.[49]

If you proceed upwards along the *Boulevard St. Michel*, you eventually come to the Paris Observatory, while if you walk the other way, and turn left when you reach the Seine, so that you're walking in the direction of flow of the river, then, after passing the headquarters of the publisher *Gauthier-Villars*,[50] you'll find yourself standing before the dark façade of the *Académie Française*. In front of the *Académie* stands a statue of Voltaire, in belated homage to the great rationalist. The Parisians aver that for some time after his death, he was honored by the academicians with a statue hidden behind the door—thus continuing the tradition of striving to ignore him that they had practised during his life.

On the bank of the Seine above its stone bulwark one finds, then as now, the stalls of second-hand booksellers, replete with old books, etchings, sketches, illustrations of English scenery, and so on. We once found a copy of Stendhal's diaries, but when I tried to buy them, the vendor asked me to come back in a few days since he was in the process of reading them himself. Also on the Left Bank, near *Saint-Sulpice*—made famous by Huysmans'[51] novel *Là-Bas*—was a street full of stores selling religious accessories: devotional articles, holy pictures, and theological and religious literature. Near the Panthéon there is the Ste Geneviève Library and the exquisite church Saint-Étienne-du-Mont with its famous rood-screen crossing the nave like a bridge, supported by spiral staircases on either side. The house where Calvin[52] lived is also to be found in this quarter, where most of the houses are very old four-storey stone structures. The multitudes of cats visible in the courtyards apparently enjoyed almost Egyptian veneration.

In Paris the mud dries quickly, doubtless because the sewerage system is so good. The feet of all those myriad human beings that walked in the past and continue to walk along *Rue Saint-Jacques*—a street following the ancient *Via Romana*—so compressed the ground that the pavement has a perfectly hard, unyielding base.

---

[48] Now called the *Préfet de Paris*.

[49] Much of this collection was moved to the *Musée d'Orsay* in 1986.

[50] No longer at this location.

[51] Charles-Marie-Georges Huysmans (1848–1907), French novelist, most famous for his novel *À rebours*. His work expresses a disgust with modern life and a deep pessimism.

[52] Jean Cauvin (or John Calvin) (1509–1564), highly influential French theologian and pastor during the Protestant Reformation. Founded the system of Christian theology later called Calvinism.

5 The Return Home

In spring the Parisian streets smell of sun, moisture, scents from the flower stalls overflowing with blooms—and of gasoline from the cars. The many young girls—students, models, and *midinettes*[53]—were dressed with becoming modesty. There were several Polish and Russian restaurants, in one of which someone said they heard the following order given to the waiter: "Koleżanko Szapiro, pozhaluĭsta minya a sztikele von diesen farszirte poisson!"[54]

It was easy to idle away the time in Paris. A short stroll down a gently sloping boulevard and you're at the Seine, where the streets in the neighborhood of the *École Polytechnique* are named after Monge, Ampère, and other scientists. Down on the *Quai de Suède*, not far from where the great mass of *Nôtre Dame* rears itself up, one might muse one's life away leaning on the hand rail. Just below are the barges that seemed to have always been moored there. The bargeman's family lives in isolation, doing their mundane chores, hanging out their washing, frying fish, indifferent to the hubbub around them called Paris.

I often crossed the bridges to *L'Île Saint-Louis*, to the *Quai d'Orléans* on the south bank of this island in the middle of the Seine. That is where the Polish library was, formerly the Parisian abode of the Czartoryski family,[55] and later Władysław Mickiewicz, son of Adam Mickiewicz. But I was more interested in the street market in little animals there: canaries, tiny green parrots in cages, tortoises, and other zoological marvels. Further on, now on the Right Bank, one came to *Les Halles*.[56] There, in vast sheds, were great mounds of meat, fish, vegetables, oysters, lobsters, artichokes, olives, apples, oranges, grapes, and piles of other foods, both familiar and less so: bananas, pineapples, fresh figs, coconuts—all that grew in France, Algeria, and perhaps even Indochina. The produce was transported from the environs to the marketplace on two-wheeled carts drawn by massive draught horses with little straw hats on their heads.

If one should wish to visit the outlying museums and theaters, embassies and cafés of Paris, then one must perforce avail oneself of the *Métro*, the world's eighth wonder. It consisted then of ten lines, around a hundred stations, and carried about half a million passengers a day. At noon, during the dinner break, the throng in the *Métro* was stupefying, since every Parisian who worked in the center wanted to eat his *dîner* at home and at the prescribed time, namely 12:30. The French don't seem to be aware of the freedom in the allocation of time enjoyed by the barbarians: they all eat dinner not later than 13:30, and a lighter evening meal around 19:30. Theatrical plays begin at 20:00, and end around 22:00, and from midnight on, only foreigners roam the city. Tall stories about Parisian night life arose from the fact

---

[53] Parisian seamstresses or salesgirls.

[54] "Colleague Shapiro, could I please have a piece of this stuffed fish!" The sentence mixes Polish, Russian, Yiddish, and French words.

[55] The leading noble family of eighteenth and nineteenth century Poland.

[56] Till 1971 the central wholesale marketplace of Paris.

that along *Les Grands Boulevards*[57] there are a dozen or so spots open all night and diligently frequented by the tourists. However, after midnight the *Métro* does not run, neither are there cabs or buses, the shop windows are shuttered, and the Parisians are all asleep—because they must be at work by eight or nine in the morning. Their eight-hour work day is divided up rigorously: if you should happen to enter a restaurant at three in the afternoon, say, then you may well be directed to the manager since they will assume you're a supplier wanting to discuss business.

Northern Frenchmen differ from the southern variety; in their restraint they resemble Germans. They are xenophobes. They like their privacy. The administrative and officer class of France is made up mainly of northerners. The southern Frenchmen, on the other hand, are brunettes, lively gesticulaters, lack punctiliousness, and probably don't make as good soldiers as the northerners. From these the cadre of French politicians and journalists is drawn.

---

[57]Especially on the *Boulevard de Clichy* in the district of Pigalle, where the *Moulin Rouge* was built in 1889. And then there is Montmartre.

Chapter 6
The Life of a Private Scholar

Somewhere around that time I got to know Professor Jan Sleszyński[1] in Kraków. He refused to write the *kreska* over a soft "s",[2] but this unorthodox orthography—which may have had something to do with "borderland phonetics"[3]—was not his only eccentricity. For Stanisław Zaremba he represented a cult figure from the time when he was his teacher, and that is why he brought this retired professor from Odessa to the Jagiellonian University. Sleszyński was a born logician. He was closer to Shatunovsky and Kagan,[4] and other such Russian investigators of the foundations of mathematics than to Western mathematical logicians. With Russian intransigence he asserted that a concept or statement that is not absolutely clear and comprehensible is without value for science. On another occasion he told me that the exclusive source of human misfortune and wrongdoing is the failure to tell the truth. In his lecture on irrational numbers that I attended, everything was stated with the utmost clarity. He regarded proofs by contradiction as manifestations of ignorance of the logical "principle of reversibility." When he attempted to paint the

---

[1] Ivan Vladislavovich Sleszyński (1854–1931). Studied under Weierstrass in Berlin, and held a professorship in Odessa from 1882 to 1909. His translation, with commentary, of Couturat's *L'algèbre de la logique* greatly influenced the development of mathematical logic in Russia. Thanks to the recommendation of Stanisław Zaremba, appointed ordinary professor at the Jagiellonian University of Kraków in 1919.

[2] In Polish, a stroke (*kreska*), somewhat like an acute accent, over a consonant indicates that it is "soft", that is, palatalized—pronounced towards the front of the mouth, in the vicinity of the alveolar ridge—in the case of "ś" like a softened English "sh".

[3] That is, of the region bordering on Poland and Ukraine.

[4] Samuil Osipovich Shatunovsky (1859–1929), Russian-Ukrainian mathematician. After enduring great difficulties due to poverty, finally obtained a position at Odessa University in 1905. Worked on the axiomatic foundations of mathematics, among other things. Beniamin Fedorovich Kagan (1869–1953), Russian-Ukrainian mathematician. Head of the Department of Differential Geometry at Moscow State University from 1922. Worked on the foundations of geometry, in particular. Edited the complete works of Lobachevsky.

Fig. 6.1 Hugo Steinhaus at around 30 (Courtesy of the HSC Archive, Wrocław University of Technology.)

Fig. 6.2 Hugo with his cousin Władysław (Władek) Steinhaus, in the uniforms of Polish legionnaires, standing in the Trojanówka market, September 24, 1915 (Władek died from shrapnel wounds on September 30, 1915. This photograph appeared in his posthumously published "Diary of a Legionnaire".) (Courtesy of Agnieszka Mancewicz.)

trees in the garden he could see from the window of his apartment, he expressed surprise at the fact that the seeming arbitrariness of the entanglement of the net of branches requires the artist to proceed according to intuitive rules which are difficult to formulate clearly, and if this vague recipe is not adhered to, then the resulting representation will not resemble the crown of a tree. My idea of a machine not guided by reason that could output *all* the digits in the decimal expansion of the square root of two appealed to him. His grey patriarchal beard, love of nature, Socratic scrupulousness in reasoning, and "borderland" phonetics—all these together with his radical simplifications of human affairs and uncompromising atheism, made him one of a kind. He would assert that probability was nonsense, and would use his unworldliness to support this view, among other similarly extreme ones. For example, when challenged to explain why the casinos in Monte Carlo cannot but prosper, he answered that they succeeded because they didn't know the rules of roulette.[5] Today I feel that Sleszyński was right. During a discussion of the *numerus clausus* rule[6] as it applied to the Department of Philosophy of the university, he asked: "If the quota of Jews had already been met, and then Jesus Christ applied, would he be admitted?" In the gallery of rationalists his portrait would surely be prominently displayed near that of Leo Tolstoy.

As I mentioned earlier, in 1913 a few of my notes were presented to the Kraków Academy by Zaremba, and published in the *Biuletyn Akademii Umiejętności* (*Bulletin international de l'Académie des sciences de Cracovie* or Bulletin of the Academy of Learning). Zaremba, a gifted mathematician who had done significant work—on potential theory, in particular—was much influenced by the French school, in fact by all things French. Thus he greatly admired the French way of life, the political forms of the *Troisième République*, French cuisine, and actually spoke and wrote French better than Polish. This being so, it is surprising that he did not remain in his adopted *Patrie*[7] but returned to the real one.

He was very demanding of the students: passing his examination for prospective mathematics teachers was considered a major achievement. He was extremely stubborn and uncompromising, and this sometimes occasioned fallings-out with colleagues: for example, he started a feud with Natanson[8] lasting a lifetime, and Banachiewicz also became averse to him. As I mentioned earlier, when he proposed that we agree to disagree on a topic prompted by my note on the notion of limit, I considered this outcome a resounding success.

My life as a private scholar, interspersing games of tennis and rowing sessions on the Vistula with scholarly pursuits, may have become tedious if it weren't

---

[5] Meaning, perhaps, not just the basic rules of the game but the theory of games.

[6] A rule for limiting the number of students admissible, sometimes used to impose religious or racial quotas—in the present case on the number of Jews.

[7] Zaremba graduated from the Sorbonne in 1889, and stayed in France until 1900, when he took up a position at the Jagiellonian University in Kraków.

[8] Władysław Natanson (1864–1937), Polish physicist and mathematician.

Fig. 6.3 Wedding photograph of Leon Chwistek (in the uniform of a Polish legionnaire) and Olga Steinhausówna, December 14, 1916 (Courtesy of Agnieszka Mancewicz.)

for the historical developments then impending. The first ever international tennis tournament held in Kraków took place in 1913, with me acting as chief umpire. However, the next tournament, in the early summer of 1914, was rudely interrupted when suddenly an order was promulgated forbidding the participation of officers of the imperial army: Franz Ferdinand,[9] Archduke of Austria-Este, had been murdered in Sarajevo. However, one of my cousins, a young man of 17, felt that history's machinations should not be allowed to disrupt such an important matter as a tennis tournament, and organized a local tournament in Jasło. My uncle, his father, claimed that war was inevitable because the Stock Exchange somehow sensed it in the

---

[9]Heir presumptive to the Austro-Hungarian throne. His assassination precipitated Austria-Hungary's declaration of war against Serbia.

atmosphere. Exuding great confidence, he also predicted an Austrian victory and the emergence of a paradise following the war, to which my Father answered: "Yes, indeed. It will be paradise because, having lost even the clothes from our backs, we'll all be walking around naked!"

And it came to pass! The Austro-Hungarian ultimatum to Serbia—and here the dualism meant more than just the two heads of the eagle symbolizing emperor and king, because the Hungarian businessmen had even earlier been pushing for war with Serbia in order to restore the import of Serbian pigs[10]—and its stridency in both form and content, the support of Berlin and the rejection of the British proposal of mediation, and then the declaration of war against Serbia—these events followed one another with lightning speed.... The emperor announced the war to the people of the Empire-Kingdom with the words: *Ich habe diesen Krieg nicht gewollt*,[11] and his sincerity was then amply illustrated by the mobilization posters which quickly appeared everywhere. Then on August 1, 1914, Germany declared war on Serbia's ally Russia, whereupon the latter's ally France rushed to support it in turn, followed by Britain, so that the British-German and Russian-Austrian conflicts erupted in the space of a single week—demonstrating the genuineness of the various accords and alliances! The effect of the world-wide conflagration to ensue was visible from our window onto the marketplace: instead of the vendors' sitting each at his stall, they now congregated in groups trying to understand events outside the narrow framework of their experience of life. The elders of our community could be seen sitting on the bench in front of the pharmacy *Pod Gwiazdą* (At the Star) earnestly discussing the baffling situation, in particular, the pressing matter of Poland. The *Naczelny Komitet Narodowy* (NKN) (People's National Committee) issued a proclamation containing the warning: "Whoever waits cautiously for the game to end, won't win," and urging people to join Piłsudski's[12] Legions. In this somewhat nebulous situation Polish patriotism did not expose politicians to the threat of a military trial and imprisonment since Piłsudski appeared to be siding with the emperor-king. This exalted monarch, who was secretly pleased at the death of Franz Ferdinand because he had not liked him and considered his wife of inferior rank, who did not allow the couple to be buried in the traditional burial place

---

[10] In 1906 Austria-Hungary had imposed a customs blockade on imports from Serbia as punishment for Serbia's attempting to evade economic and political control by the Habsburgs. The blockade was lifted in 1909. Prior to the dispute, most of Serbia's exports had gone to the empire, with pork being the major export. The pigs were transported to Hungary for fattening, and most of them returned to Serbia for slaughtering and processing.

[11] "I did not want this war."

[12] Józef Klemens Piłsudski (1867–1935), long-time proponent of the cause of Polish nationalism. When war broke out in 1914, he seized the opportunity of leading Polish troops against Russia in a loose alliance with the Austrians, although his real aim was Polish independence. From mid-World War I, Piłsudski was an important figure on the European political stage, being largely responsible for Poland regaining its independence in 1918. He was Head of State 1918–1922, First Marshal of Poland from 1920. Informal leader of the Second Polish Republic following the so-called May *Coup d'État* of 1926.

of the Habsburgs, and did not even attend the funeral, was to become the Polish king—such were the bold ambitions of the NKN for an Austrian-Hungarian-Polish monarchy.

We did not have long to wait for the first blows to fall: Lwów very soon found itself smack in the front lines. Piłsudski's so-called Eastern Legion was withdrawn from the front, and ended up in Jasło together with thousands of refugees, so that the streets resounded to the Lwów accent and humor. It was summer and the war was perceived as a mere escapade. The soldiers painted the words *Schnellzug nach Petersburg*[13] on the sides of the railroad wagons. Various local committees brought flasks of tea and coffee, and sandwiches and cookies to the station for the soldiers, and train after train passed through heading north and east, each stock wagon carrying around forty soldiers who had sworn fealty to a double-headed double-crowned eagle, emblematic of the monarch and the crown prince, and to "live and die as brave soldiers should" by reason of the emperor's noble blood. Count Skarbek, the political patron of the Eastern Legion, also arrived in Jasło; he belonged to the Podolian faction of the conservative party, which, together with the National Democrats, wanted to prevent Poland from fighting on the side of the Central Powers, whose defeat they both wished for and predicted. However, Poland's situation was such that the Law of the Excluded Middle did not apply to the statement "Either the Germans will win or the Russians will win." Even though they were political opponents, my uncle Ignacy had to allow Skarbek to put up in his house. Skarbek managed to so manipulate matters—with the approval of the NKN and almost all of the Polish representatives in the Vienna parliament—that the Eastern Legion dissolved itself, many of the troops joining Piłsudski's formation in Kraków. Piłsudski himself did not exactly have the confidence of the powers-that-be. Who was he, after all? He did not figure in any official plans for conducting the war, was not a reserve officer, seemed to have no definite profession, no one knew his family, no Galician count had ever heard of him, and there was nothing in the newspaper files to enlighten an enquirer.

The sensation of fear that fate was dealing one a poor hand was now exacerbated by the obvious fact that the Austrians were being defeated. Auffenberg near Lwów and Dankl[14] near Kraśnik had discovered that the Russian artillery used live shells, and that the dirty-brown uniforms of Russian foot soldiers were more practical on the battle-field than the red trousers of the Austrian uhlans, and also that the only thing that happened to cavalry charging trenches—which they tried out near Satanów—was that they mostly perished "as brave soldiers should" without even reaching the first line of barbed wire. By the time the harvest had been brought in, there were clear indications of defeat. Thus a German *Landwehr* soldier from

---

[13] Express train to St. Petersburg.

[14] Count Moritz Freiherr von Auffenberg (1852–1928), a general in the Austro-Hungarian army. Count Viktor Dankl von Kraśnik (1854–1941), career Austro-Hungarian officer.

Woyrsch's[15] corps, which had been pushed back to Rzeszów, loudly proclaimed to all and sundry in the marketplace that the game was up, and that he was going back home to Katowice. Then there was a lull. Then a column of cars carrying senior officers, escorted by the gendarmerie, passed through, heading west; this was the high command escaping to Nowy Sącz. A major and his adjutant were lodged in our house; he was in charge of an infantry unit armed with machine guns. We were advised to leave, and allocated horses to transport our belongings to the station. We took only a very few things, abandoning all else: furniture, most of our clothing, and household wares and implements—crockery, bedding, shotguns, harnesses, and so on. Father took the books of the Credit Society, a small hamper containing bottles of vodka and other spirits, eau-de-cologne, a chunk of salami and a round of Swiss cheese. We travelled in a third class rail-car through Sącz to Preszów[16] in Hungary, and after a short stay there, transferred to Budapest, where Father had relatives, and where my Tarnów Grandfather and his daughter had found refuge. While we were there my Grandfather caught a cold, and what with his age—eighty-one—and the turmoil of moving, this was enough to finish him off. My Father then decided we should move on to Vienna. There he rented a room at 12 Wipplingerstrasse from which he continued to manage his Credit Society, paying his clients reasonable interest rates on their deposits. For us, his family, he first rented an apartment in a villa in the suburb Währing. In the same house there lived a Dr. Fröschels, a disciple of Freud, who taught deaf-mutes to communicate with one another.

This was a sad autumn. I went with my sisters to the grounds of Schönbrunn Palace,[17] the refuge of the sublime melancholic, the mausoleum where a monarch as petrified as his empire dithered, close to death himself, but still having the power to send others to their deaths—me, in particular. I was sent back to Kraków, where I put up at the quarters of one Bolek R., a gendarme in the Legions. He advised me to join the gendarmerie, but that dignified way of avoiding the worst didn't quite appeal to me. I went to the recruiting office of the Legions, where they assigned me to the Military Department of the NKN. Soon Kraków found itself close the front lines—or rather the front lines found themselves close to Kraków—so the head of the department, Władysław Sikorski,[18] went to Vienna with me in tow. My family had by then moved to an apartment in a house at 33 Währingerstrasse, where it so happened that Władysław L. Jaworski, a professor of

---

[15]Remus von Woyrsch (1847–1920), Prussian lieutenant-general. Called out of retirement in 1914 to command the Silesian Landwehr Corps on the Eastern Front.

[16]Now Prešov, in Slovakia.

[17]Then the imperial summer residence, now a major Viennese tourist attraction.

[18]Władysław Eugeniusz Sikorski (1881–1943), Polish military and political leader. Agitated for Polish independence before World War I. Fought with distinction in the Polish Legions during World War I, and in the Polish-Soviet War of 1919–1921. Prime Minister of Poland from 1922 to 1923 and holding important posts subsequently, he fell out with the *Sanacja* government in 1926. Prime Minister of the Polish Government-in-Exile and Commander-in-Chief of the Polish Armed Forces during World War II until his death.

law at the Jagiellonian University of Kraków, was also staying. At that time, he was leader of the conservative wing of the Polish Circle,[19] and was generally regarded in Vienna as the wisest of the Polish politicians. I was assigned to the office of the presidium of the NKN as a translator of non-Polish documents, answering directly to its head Count Władysław Michałowski. Jaworski was a disciple of Machiavelli: he thought that people were either stupid or dishonest, and that real politics consists in taking advantage of human failings in order to promote one's own ends. Yet his own, personal, ends were not at all selfish: he was certainly a Polish patriot, and in him this inclination trumped his conservatism. He was also broad-minded, understanding very well that people like Piłsudski appealed to the minds and hearts of the masses more powerfully than the cleverest of the clever in the Polish Circle. Although he invoked supernatural forces in his writings, he was basically a rationalist—but intelligent enough to understand that in Poland rationalism was unpopular, and that a wise man can perhaps tell the whole truth to Michał Bobrzyński or Wojciech Dzieduszycki,[20] but not to Abrahamowicz or Hupka. It was like a chessboard on which he moved the pawns of his political world. He talked in a hoarse voice because he had a piece of tubing in his throat from an earlier operation. Every morning he was driven in a chauffeured car from 33 Währingerstrasse to 9 Neutorgasse, where the NKN had its offices. All sorts of people came there: politicians from the Polish Circle, muckrakers, journalists, legionnaires seeking financial aid, Galician busybodies, philanthropic ladies, half-Polish half-aristocrats, but above all those wanting to avoid a trip by boxcar to the front. I even saw Piłsudski himself there once, and Sławek[21] several times. But they did not regard the NKN as an institution to be taken seriously, knowing full well that political manoeuvring by itself would never detach Poland from Austria. On the other hand, it might be detached by more violent means, since Austria was under threat not only from her eastern foe, but also, more insidiously, from her northern ally.

The mood at 9 Neutorgasse was not a happy one. The whole of the population of the Kingdom[22] and at least half of that of Galicia were opposed to the Central Powers. The brutality exhibited by the Germans in Kalisz,[23] and the sight of the

---

[19] In 1860, the government of the Austrian Empire was reformed by establishing a dual monarchy together with a bicameral State Council, to the lower house of which representatives were sent from the autonomous regional parliaments of the empire. The Polish faction in the State Council achieved official recognition in 1867 as the "Polish Circle."

[20] Bobrzyński was then a minister for Galicia in the Austrian government, and Dzieduszycki a government appointee to the upper house. The point here may be that the representatives Dawid Abrahamowicz and Józef Hupka were members of non-Polish minorities, Jewish and German-Silesian respectively. However, according to other sources Dawid Abrahamowicz was a Polish conservative politician of Armenian origin.

[21] Walery Jan Sławek (1879–1939), Polish politician, army officer, and activist. A close aide to Piłsudski. Later, in the early 1930s, he would serve three times as Prime Minister of Poland.

[22] The part of Poland ruled by Russia.

[23] Near the beginning of the war, German officers had ordered the razing of the Polish town of Kalisz.

Austrian army retreating in Galicia, contributed to this alienation. The Viennese, on the other hand, saw in the crowds of refugees a threat to their coffee, cream, and buns, and the Austrian diplomats, who—as Karl Kraus put it—"lie to the journalists in the evening, and believe those same lies in the morning when they see them in print," regarded the Slavs as *the* great misfortune of Austria. It was they who started the sinister joke circulating that "the war between Austria and Russia is over Galicia, and the loser will have to take it." The military command in Vienna had observed the large number of young men working in the offices of the NKN, and when Daszyński[24] reproached Jaworski on this score, all of us shirkers promptly submitted requests to be sent to the front. Jaworski had needed these requests merely for show, and refused all of them. However, the surrounding atmosphere having becoming unpleasant, a few of us office workers did opt to transfer to the front line formations of the Legions.

The boarding house *Atlanta* run by Herr Brandt at 33 Währingerstrasse was not a lively place. Some of the older boarders—Jaworski, my uncle, Hupka—played bridge. There was also a Mrs. Askenazy who lived there with her daughters, the American consul—incidentally, the only person who seemed concerned at my imminent departure—, an Irish girl who was the *fiancée* of a young medical student called Gröer who used to pay her visits, a Mr. Maliniak from Warsaw, the only cheerful member of the older generation of boarders, and a philologist from Kraków called Mr. Cięglewicz, a translator of Aristophanes, who played chess with me and was unable to hide his irritation when he lost, even though that happened rarely. The physicist Hevesy,[25] who later discovered the chemical element hafnium, used to come for dinner, as did the secretary of the Albanian Postal Ministry, who had come to Vienna before the war to order postage stamps, and stayed there. He told me that in all of Tirana, the capital of Albania, there was at that time just one mailbox.... A Mr. Wołyński also used to drop in; he impressed me very much with his theoretical knowledge. Several of these people were citizens of enemy states, so had to report each week to the police, but apart from this they were allowed to live in peace.

My sisters Irena and Olga took me in a taxi to the railroad station, whence I travelled, in civvies, to Piotrków. I was supposed to go to Jeżów, but that town was under quarantine because of an outbreak of typhoid fever, so I took advantage of the ten days I was forced to stay in Piotrków to transform myself into an artillerist, to which end I girded my loins with a heavy, elaborately styled sword that looked as if it had been retrieved from a theatrical rubbish heap. In a café I met up with my cousin Władek, who had been assigned a special duty: a Russian captain of gendarmes had somehow been left behind when the Russians abandoned Piotrków—probably intentionally, in order to join the Legions with the aim of gathering intelligence. He had reported to the Austrian headquarters in Piotrków, and was now being kept under surveillance—by Władek, who had been ordered

---

[24]Ignacy Ewaryst Daszyński (1866–1936), Polish politician and journalist. He would become prime minister of Poland in 1918.

[25]George Charles de Hevesy (1885–1966), Hungarian radiochemist. Nobel laureate in 1943.

to accompany him everywhere. The captain was a burly officer, typical of the Russian gendarmerie; he wore knee-boots of high quality and riding breeches. He looked with pity on the soldiers of the Polish Legions in Piotrków: that haphazard collection of volunteers, a large proportion of which were teenagers, lacking NCOs and barracks training, must have struck the tsarist officer as something like a crowd scene from a Viennese operetta. The Piotrków notables, lawyers, and the majority of the townspeople had a similar view of the spectacle. But there were those inhabitants who willingly sent their sons to join this bedraggled army formation, with its one cannon serviced by a platoon of Piotrków artillerists.

The first artillery regiment of the Polish Legions was stationed in Jeżów, and when the quarantine was lifted, I went there to report to Major Jełowicki. It was rumored that this officer had organized an uprising in Albania on behalf of Prince Wilhelm von Wied, among other adventures. He wore a white uniform with silver buttons and was followed about by a retinue consisting of an aide-de-camp called Beck,[26] the standard bearer of the division, who was to become minister of foreign affairs in postwar Poland, the divisional doctor, a bugler, and a few mounted soldiers who were also leading horses. The commander of the battery that was being formed, and which I was to join, was Jan Maciej Bold, later promoted to inspector of artillery, and its officers were: the aide-de-camps Mazurkiewicz and Gumiński, later to become governor of the Kraków district, and the artist Sichulski, who wanted to buy my theatrical sword from me. I refused to part with it. This staff was billeted in the manor house of the Zakrzewskis in Rozprza, where the sixth battalion of the infantry regiment in which my cousin Władek was serving was also stationed. It could never be said that the officers in charge of our platoon knew anything about artillery, but we had a young NCO assigned to us by the name of Frank who did know something. He had managed an equestrian school in the Austrian army, so was an expert on horsemanship—but we didn't have horses to spare for him to demonstrate his skill. At last we were assigned an NCO called Bura to instruct us in the art of field artillery. He was a professional officer, a graduate of the cadet school, well brought up, and disdainful of us legionnaire amateurs. We were also assigned an Austrian NCO named Cejnar, a Czech by nationality, and several Hungarian horse leaders, experts at having their mount lead six other horses pulling a piece of field artillery.

Life with the platoon was rather monotonous. We were billeted in a large barn, and I slept next to Zygmunt Janiszewski,[27] who had obtained his doctorate at the Sorbonne. His dissertation, entitled *Sur les continus irréductibles entre deux points*, attracted the attention of Henri Poincaré, justifiably considered the greatest mathematician of the first decade of the twentieth century. Janiszewski

---

[26]Józef Beck (1894–1944), subsequently a Polish statesman, diplomat, and army officer. Close associate of Marshal Piłsudski. Polish Minister of Foreign Affairs 1932–1939.

[27]Polish mathematician. Of great personal integrity, he refused to swear an oath of allegiance to the Austrian government. He was the main force behind the creation, after World War I, of the world-renowned Polish school of mathematics. Lived from 1888 to 1920.

had dedicated the dissertation to Marc Sangnier, leader of the French Christian Socialists. Unfortunately, he soon left our battery, being reassigned to the command of a unit in Warsaw when the Germans occupied that city. A few years later he told me that he had been met at the Warsaw train station by the eminent geometer Max Dehn,[28] with a car to take him to his quarters. I can imagine their mutual joy at meeting a kindred spirit—so much so that it's a wonder the car actually reached its destination without accident.

The food in the battalion was excellent, and rice, Swiss cheese, chocolate, and wine were distributed to the soldiers in generous quantities. There were quite a few members of the intelligentsia in the battery: the physician Studencki, always cheerful, Słobódzki, later known (under the pseudonym Kozar-Słobódzki) as a composer of songs, the chemist Zwisłocki, who later married the daughter of President Mościcki,[29] Zięborak, whom I would meet a few years later as a chemistry *Assistent* in the University of Lwów, and a young man called Sarnecki who hailed from the vicinity of Piotrków. However, while the military ate well, the local people were going hungry. Girls from nearby villages would come to the wood adjacent to the battery ostensibly to gather mushrooms, but really counting on the soldiers to give them some of their bread, chocolate, or canned goods.

On the whole the local youth avoided us, but there were a few exceptions who did seek to join up with us. There was a young man, for example, a handsome blue-eyed blond, and a born bandit, who decided to join the battery, but since army discipline was not to his liking, he spent much of his time under guard in the guard house—actually a shed in the backyard of a peasant's hut that he would have had little trouble escaping from had it not been for the guard posted there around the clock. Once I was assigned the midday shift of guard duty, which involved taking the prisoner for a half-hour walk. I ordered him to walk ahead of me, and was grateful to him for not attempting to grab my gun or run away.... Clearly, he was obedient merely from a feeling of friendliness towards me.

After several months of waiting, the field guns arrived. There were four of them, bronze-plated according to an old recommendation by the Austrian general Uchatius.[30] Since the barrels were fixed to the gun carriages, they did not recoil independently, so that after each shot the whole cannon would jump back and had to be re-aimed. Our first firing exercise was preceded by a *promenade militaire* of some 20 km. In the artillery caisson for which I was responsible, there was a single compartment for sacks of gunpowder, shrapnel shells, and grenades, and another where I placed the metal box containing the detonators. At the end of the *promenade*

---

[28]German mathematician. Student of Hilbert. Famous for his work in geometry, topology, and geometric group theory. Solved the third of Hilbert's 23 problems. Fled Germany in 1939, eventually ending up in the US. Lived from 1878 to 1952.

[29]Ignacy Mościcki (1867–1946), Polish chemist and politician. President of Poland from 1926 to 1939.

[30]Franz von Uchatius (1811–1881), Austrian artillery general and inventor—in the field of cinematography, in particular.

I opened the caisson and saw that the detonators, thoroughly shaken by the long ride, had all gone off. If even one sack of gunpowder had been in the same compartment, the whole battery would have been sent to kingdom come. I could perhaps lay the blame on our superficial training. The firing exercise went well, however.

It was time to go to war. A whole train was placed at our disposal, with flatcars for the field guns, goods wagons for the horses, and passenger cars for us. Although the journey from Piotrków to Kowel[31] took five days, we had ample food, hot water from the locomotive for tea and shaving, and in September 1915 it was still not cold—so we were comfortable. In Kowel we were billeted in a house that had belonged to a Russian general. There we found large quantities of maps of Port Arthur and Manchuria, and evidence of looting and of books having been used as fuel in the stoves. But doubtless the house had been plundered even earlier. I met a Jewish cabinet maker who lived nearby, and went to his place to wash, shave, and make chocolate—which was easy, since all you had to do was melt a stick of *Suchard*[32] in hot milk.

The town itself was devastated. We set up our field guns on a nearby meadow. Mrs. Dąbrowska, a Polish woman, lived not far away, and she spent the whole day frying schnitzels, which we bought from her in great quantities: I sometimes ate four of them between breakfast and lunch. Pasted up on store fronts along the main street there were trilingual posters exhorting the locals to "get out of the way of officers and bow to them...."[33]

We soon set off again, with the horses pulling the guns and the caissons. When we arrived in Maniewicze, we saw a Russian reconnaissance plane fly overhead. Standing there in the wood at the edge of a sandy road, I watched as a German infantry regiment marched sluggishly by. When I was told that this was the eighty-second regiment, whose home base was Göttingen, I called out: *Weender Strasse ist schöner!*[34].... They all immediately turned their heads to the right as if I were a general reviewing his troops in a march-past—clearly they were very surprised by such an unlikely ejaculation. We continued on our way, and when night fell, found ourselves in the middle of a wood. It was a beautiful clear autumn night, the sky studded with stars. Somewhere ahead there was a suspicious glow visible against the night sky, and after a few more hours of marching it was realized that we were lost. The officers consulted me as to the appropriate direction to take because they saw that I was forever gazing at my compass. Since the glow remained puzzling, the NCO Fyda and one other were ordered to investigate, and they moved off cautiously through the woods, leaving us waiting in a clearing in the forest for their return. After an interval, two helmets glinting quite beautifully in the moonlight

---

[31] Kowel being situated then in the part of Poland ruled by Russia. Now Kovel', a town in the far northwest of Ukraine.

[32] Swiss chocolate, as produced by Philippe Suchard (1797–1884), Swiss chocolatier and industrialist.

[33] The text in the Polish original is a travesty of Polish.

[34] "Weender Strasse is more beautiful!"

emerged from the impenetrable black of the forest wall. These turned out to belong to two German soldiers, who, when they saw us, asked *Wo geht man hier nach Homburg vor der Höhe?*[35] They were on leave, and in a tearing hurry to get home to Homburg.... They explained that the glow was coming from a fire accidentally started by German fusiliers. In fact, during the war many villages were burned to the ground by accident, as it were, the soldiers not caring to take the trouble to extinguish the fires.

We passed the night in the forest. Even though I had lain on a blanket, when I awoke my leg was stiff from the cold ground. Sometime in the morning we at last entered the war zone: it was recognizable from the ceaseless booming and the desolation all about us. I no longer rode on the wagons, since I had been told that in the war zone it was forbidden to ride on horse-drawn wagons carrying field guns and caissons, although the others continued to do so. After some time, when I was beginning to be left behind, but had a good view of the road stretching ahead, the battery suddenly put on speed, so I began running after it as hard as I could in my heavy boots. Then quite close to me on the sand a black bush of smoke suddenly sprouted up out of nowhere and I heard a whistle and a thump. I tried to run faster but the distance between me and the battery kept on increasing. Several more shells landed in the sand and then the firing ceased; presumably the battery had passed out of the line of sight of the Russian observers. When I finally caught up with the battery, they had found a suitable emplacement near a village called *Bolszoje Miedwieże*,[36] and it was there that for the first time we fired our big guns in earnest. Nearby stood Major Rylski of the sixth infantry regiment, and with him my cousin Władek, who had the job of dispatch rider, and other officers—all of whom seemed to believe that the elevation of our field guns was high enough. It was already getting dark, and when I went over to a caisson to fetch a shell, I omitted to take a flashlight, and picked out a shrapnel shell[37] by mistake, set to explode after travelling a relatively short distance. When the clang of the shell bursting in the air was heard, followed by the whistle of the shrapnel fragments, everyone scampered for cover. It was all my fault: I had picked out a shell marked with a "V" for *vortempiert*,[38] in Austrian artillery dialect.... So I caused a stampede, with Ensign Mazurkiewicz shouting "They've found us!" But it was soon realized that it was our own shrapnel. I dutifully reported my mistake to my NCO, who promptly jumped down my throat: "Shut up!" I understood: he didn't want our battery to be compromised in front of the officers and foot soldiers or even our own comrades in

---

[35] "How do you get from here to Homburg-by-the-heights?" *Bad Homburg vor der Höhe* is a town in Hesse, Germany, on the southern slope of the Taunus mountain range.

[36] Meaning "Big Bears". However, this name, which sounds Russian more-or-less, and the alternative *Niedźwiedzie Większe* given by the author, are likely inexact. (The Polish word for "bear" is *niedźwiedź*.)

[37] An anti-personnel shell filled with small shot or pieces of metal designed to explode in flight and scatter its contents.

[38] Meaning that the detonator is timed for short ranges.

the battery. The deployment of the guns was declared dangerous, and was changed. There were other mistakes made, none of which was reported in the official history of our battery. For example: some way off there was a railroad flag stop occupied by Russians and we aimed our guns at the spot where, according to our map, it should have been located. But the map was wrong, and we hit a shed behind which a battalion of Austrian dragoons had dug in. Spurred on by our shelling, they moved forward willy-nilly and captured the railroad stop.

Some days later we were redeployed to a new position on a prominence just outside Podczerewicze. There we fired on the bell tower of the church from which a Russian scout was directing onto us the fire of a platoon of light artillery; its commandant seemed to have it in for us. Right next to our position there stood a detachment of Legion infantry firing their rifles in unison to the order of their commanding officer as if it were a drill. They sneered at us for ducking down behind our gun shields when shrapnel shells burst. During this action, Major Rylski, his aide-de-camp—a female disguised as a male—and a Legion gendarme by the name of Zielony[39]—whose niece I married two years later—observed the proceedings from not too far away. The horses and artillery caissons stood safely more than a hundred meters further back. After a volley or two, the Cossack field guns found our range, and shells began exploding no more than a few dozen meters from our position, so it was decided to abandon it. The horses were brought up, and while I was trying to hook the gun carriage onto the spike of the caisson to be towed away, a shell exploded right next to the horses. Luckily, it had struck loose sand, so that the force of the blast tended upwards out of the pit made by the impact, rather than outwards. By the time I had collected my wits, the horses and caisson were gone and my field gun left in the emplacement! I looked around. We, that is, my little artillery unit and our gun, were alone. The senior staff, including the female aide-de-camp, who had been seized by a fit of sobbing, had gone with the others, so we, the artillery unit of cannon No. 4, of which I was in charge, also abandoned said cannon and rejoined the battery. But then we—five young men and I, of which Lieutenant Maciejowski led the horse pulling our caisson—had to go back up to retrieve it. The Russkies had their sights trained on the field gun, just waiting for us to return.... However, this time we were able to hook the gun carriage onto the spike at first try, and by the time the first shell hit we had moved some dozens of meters lower. We had to return again for the boxes of shells and the gun-aiming instrumentation that had been left behind. When this had been done, Brigadier Grzesicki arrived to verify that all was in order, and I allowed myself a perverse pleasure: contrary to battlefield etiquette, I saluted formally and reported in an official tone that all the equipment had been successfully removed. In fact, the only thing left at the emplacement was a book my Mother had sent me—Romain Rolland's *Jean Christophe*, which I have

---

[39] "Zielony" was the Legion pseudonym of Leon Grünwald, an uncle of Stefania Szmoszówna, who became the author's wife in August 1917. Zielony was a man widely known in society for a variety of talents: he was an excellent bridge player and a specialist in the settling of issues relating to offended honor, that is, fighting duels.

never read to the end even to this day. By this time it was ten in the morning, and, exhausted, we all fell asleep on the grassy floor of the valley where we had arrived. I slept till sunset.

On another occasion our position was too narrow for four field guns to be deployed, so mine remained idle at some distance in the woods behind. I went on foot up to the emplacement, only to be sent back with some order or other on horseback. As I was riding through a meadow without a thought on my mind, suddenly something caught at my throat, and when by reflex action I moved my head backwards, my cap was torn from my head. I then understood that the culprit was the wire of a field telephone. As I approached the wood where my cannon stood, I encountered a Legion officer apparently resting his horse, which was standing nearby. He engaged me in conversation, in German, and asked how things stood with us. My suspicions being aroused, I later reported him as a Russian spy, but have never discovered who he really was. There was a conjecture that it was Lord Sutherland, an Englishman who had volunteered to fight in the Legions.

Then we undertook a retreat lasting a whole day and night during which we fired very effectively on the weir near Wołczeck. Foot soldiers who had retreated earlier had left behind hundreds of blankets, knapsacks, canteens, and greatcoats, so we could gather up such items to our heart's desire. During the nighttime portion of the retreat, my unit got mixed up with an Austrian column whose general loudly cursed us. We were later chided by Lieutenant Bold for not understanding that we might have given him as good as we got since he wouldn't have been able to do anything about it in the dark.

On yet another occasion, when the battery was resting in a warm and quiet glade in the woods, the commandant assigned me to nighttime guard duty, somehow knowing that I would not fall asleep. Nearby there was a barn where an Austrian patrol was spending the night—a few dragoons and their horses. Was I responsible for or answerable to them? In fact, I was answerable only to the fragrant summer night during which the nightingales seemed to be trying to outdo one another in the beauty of their song. So I was satisfied with my night of pleasure, and the commandant was also satisfied. But the Austrians were less satisfied. Although they had slept well, three of their horses had been stolen during the night: it was not nightingales but Cossacks who had been whistling so sublimely to one another.

Finally, we arrived back in Kowel where we had started, and were allowed some weeks rest. A few days later, the commandant informed me that my cousin Władek was in hospital, mortally wounded. He had been riding near Kukły to deliver a dispatch and had been hit by two pieces of shrapnel, one piercing a lung, and the other damaging his spine. He was in a state of delirium, saying such things as: "This is not the place for an officer. My place is at the front!" Uncle Ignacy arrived in time to find his son still alive, and bid him adieu.[40] My uncle was able to arrange a leave of absence for me, and we returned to Kraków together with the coffin of Władek. There Mr. Srokowski, editor of the democratic paper *Nowa Reforma* (New

---

[40] Władysław Steinhaus died on September 30, 1915.

Reform) delivered a funeral oration which our family found offensive. Władek had left a written record of his experiences entitled *Pamiętniki legionisty* (Diary of a Legionnaire). From the comments of the publisher of this diary I learned that he was to be promoted posthumously to the rank of ensign. The Germans also honored him by awarding him the Iron Cross, and after the war a street in Lwów was named after him.

I did not return to the military. Even with the support of the Jasło *starosta*, representations to the authorities to obtain exemption would normally have been fruitless, but my Mother, without anyone's knowledge or support, went alone to an official in the Ministry of National Defense in Vienna, and was herself surprised when, after listening to her request, the official decided the matter with the single word *enthoben*.[41]

In Jasło my Father was living like a bachelor, having returned from Vienna in early 1916 after the breakthrough in the "Gorlice line"[42] by the Austrians with the help of the Germans under the command of Mackensen.[43] Mother and my two sisters had remained in Vienna. I found two of my old friends still in Jasło: Janek Adamski and Ludwik Oberländer. The Russian army had left some preserves behind, and one could also buy bread—even decadent white bread—and other edibles in the tavern. Such were the times that, walking though the market place on a summer evening, one could hear the detonations of heavy field guns coming from the direction of Podhajce: this was Japanese artillery on the Russian side firing on Ottoman Turkish infantry fighting on the Austrian side!

My farewell to arms could not be played out indefinitely in Jasło. My uncle once again exerted his influence, this time to obtain a position for me in the head office of the Agency for National Reconstruction, set up by the government with the mandate of making the most necessary repairs and restorations of farms, handicraft workshops, and factories. Thus from July 1916, I lived in Kraków in a boarding house at 9 Karmelicka Street. I was assigned to the technical section of the Agency for National Reconstruction, which was headed by Counsellor Ingarden. In addition to the full-time, permanent civil servants, I met many of the contract workers, mostly engineers. In my new position I had the opportunity of learning technical terms such as "exhibit', "official reminder" ("O.R."), "make a fair copy", "approve", *videat*, *reproducatur*, and many others, together constituting the essence of bureaucratic thought. I had no sense, however, that I was to be among the last few people to be exposed to this arcane terminology, and that within the space of a little over two years this bureaucratic argot was to vanish from the earth together with the imperial-royal monarchy of Austria-Hungary. There were enthusiastic connoisseurs of the art,

---

[41]"exempt"

[42]The Gorlice-Tarnów offensive of 1915 started as a minor offensive to relieve Russian pressure on the Austro-Hungarians, but resulted in the total collapse of the Russian lines, and their withdrawal into Russia.

[43]Anton Ludwig August von Mackensen (1849–1945), German field marshal. One of Germany's most successful military leaders in World War I.

such as, for instance, the *starosta* Gaspary, the aesthete of the little green table—his desk—where a collection of stamp pads and red, black, and blue pencils sharpened to an ideal point was carefully arranged. When Mr. Gaspary prepared a document, the artistic conventions were followed to the letter, and the result was so gorgeous that a petitioner receiving one, even if in the form of an official refusal, must surely have felt honored by the amount of stylistic, graphical, and coloring work that had been invested in his case: "...in accordance with the letter of the i. and r. governor circulated 3 III 1916 I. COK1947/16, insofar as said circular letter anticipates the furnishing of aid only to such of those injured parties whose workshops are located in that part of the country occupied by the i. and r. military forces, and *a contrario* prohibits the allocation of aid to citizens of the i. and r. monarchy who, while admittedly at present relocated in the liberated region of Galicia, have their permanent domicile in the region occupied by the Russian forces...." Although a petitioner from Buczacz[44] might discern in this a refusal of his case, he could not but be honored at the tremendous effort that had evidently gone into the framing of said refusal.

Although Kraków was still formally organized somewhat like a military fortress, one could still go for walks in the parkland on the outskirts. During one such walk, I overheard the words "...Lebesgue measure..." spoken in a conversation between two men seated on a park bench, so I approached and introduced myself to the two young mathematical neophytes sitting there. They introduced themselves as Stefan Banach[45] and Otto Nikodým,[46] and told me that Witold Wilkosz,[47] on whom they heaped lavish praise, was also of their company. From then on we met regularly, and since at that time Władysław Ślebodziński,[48] Leon Chwistek, Jan Króo, and Włodzimierz Stożek were also living in Kraków, we soon decided to organize a mathematical society. As one of the initiators of the project, I made my room available for the meetings of our society, tacking a square of oilcloth on a wall as blackboard. When the Frenchwoman who managed the boarding house saw this, she was terrified: what would the proprietor say? I calmed her by telling her that the owner was a son-in-law of my uncle, so would make allowances, but in

---

[44] A town in Podole, now part of western Ukraine, 135 km southeast of L'viv.

[45] Stefan Banach (1892–1945), Polish mathematician. Self-taught prodigy, and founder of modern functional analysis. Later to become a founder of the Lwów School of Mathematics and one of the most influential of twentieth century mathematicians. Steinhaus claimed that Banach was his "greatest scientific discovery".

[46] Otto Marcin Nikodým (1887–1974), Polish mathematician. Connected with Warsaw University in the interwar period. Emigrated in 1946, arriving in the US in 1948.

[47] Witold Wilkosz (1891–1941), Polish mathematician. After completing his doctorate in 1918, taught in private secondary schools. In 1920 defended his *Habilitation* at the Jagiellonian University in Kraków, after which he joined the faculty there.

[48] Władysław Ślebodziński (1884–1972) studied physics and mathematics in Kraków. Pioneered differential geometry in Poland. From 1942 to 1945 was in various concentration camps, including Auschwitz. Professor at Wrocław University and Polytechnic from 1945.

this I was completely wrong. Mr. L., the owner, adopted the stance of a hard-nosed landlord unmoved by the lofty goal the "blackboard" was supposed to serve. But we persevered, and the society expanded—the first ray of light of this kind in Poland. Some years later, Mr. L. was awarded some sort of medal for providing a locale for the infant society, but I'm sure he never understood that his greatest merit in that respect was not so much his contribution of the walls of his house, as the holes the tacks holding our "blackboard" up had made in one of them, contrary to his will....

At about the same time, I received a visit from Puzyna,[49] who was then a mathematics professor at the University of Lwów. The upshot of this meeting was that in March 1917 I went to Lwów to deliver my lecture for the *Habilitation*,[50] that is, my "inaugural" lecture. I then applied—successfully—for a transfer from the head office of the Agency for National Reconstruction in Kraków to the branch office in Lwów, anticipating that I would be granted my *Habilitation*.

On August 9, 1917, my wedding took place in Kraków, and my wife and I spent a few weeks' honeymoon in Zakopane.[51] We hiked in the mountains, sleeping in abandoned huts, and bathing in mountain pools; there were very few people about on account of the war. On one of our walks we bumped into two professors from Lwów who informed me that I had indeed been granted *venia legendi*[52] at the University of Lwów.

We were lucky with our accommodation in Lwów since my cousin Dyk's mother was spending the wartime in Jasło, so we could live in her furnished apartment at 10 Gołąb Street in Łyczaków for some time at least. My life now became split between boring, tiresome, and pointless sitting around in the office of the Agency for National Reconstruction, and my duties as docent[53] and *Assistent*. I did manage to attract a few students to my lectures on Lebesgue's theory of integration, but of course most young men were serving in the army. The following spring, in March 1918, there was a mass demonstration against the Treaty of Brest-Litovsk,[54] and for the termination of the war. Tens of thousands of people marched through the town, and despite the fact that many streets were left empty, there was no looting.... We

---

[49] In 1908 Puzyna proposed Sierpiński for a professorship, with the result that in that year one of the world's first courses in set theory was given in Lwów. He also arranged for Steinhaus to take his *Habilitation* in Lwów, thereby establishing him in that city.

[50] Bestowing the right to give lectures at the University of Lwów.

[51] Town lying at the foot of the Tatra Mountains of southern Poland, a range in the Carpathians. A center of "highlander" culture. Called "the winter capital of Poland".

[52] Permission to lecture.

[53] Temporary lecturer aspiring to a professorship.

[54] A peace treaty between the new Russian Soviet Federated Socialist Republic and the Central Powers, marking Russia's exit from World War I. Since there was no mention of Poland in the treaty, its signing caused riots and protests in Poland, and the final withdrawal of all support for the Central Powers.

took part in the demonstration, and there I happened to meet Wacław Sierpiński,[55] who had managed to return to Poland via Siberia and Sweden.

The Agency for National Reconstruction was not happy with me. Counsellor Rozwadowski expressed his dissatisfaction with me in a report containing the words: "He does nothing, offering as excuse his university responsibilities." This was true, but at least I did not take my duties at the university lightly—or rather was simply unable to do so since Professor Puzyna had fallen ill and I was the sole representative of mathematics in the whole university. Representations were made to the military concerning this extraordinary situation, and soon an official letter arrived from an office of the Ministry of National Defense—which, I should mention, was not the same as the Ministry of War, believe it or not. The upshot was that I was permitted to relinquish my position in the Agency, without regret on either side.

Events now came thick and fast. In April 1918, my wife gave birth to a daughter, and as soon as feasible I sent her with the child to Jasło, to her parents, remaining myself in Lwów. I put most of our things in storage, shut up the house, and moved into the boarding house *Anuta*, on Copernicus Street. It turned out that Mr. Pieniążek, the landlord, was the paternal uncle of one of my fellow soldiers in the battery. And that's where I found myself at the end of World War I—and almost immediately right in the middle of a new Polish-Ukrainian war[56] that broke out in November. There was one day when the Ukrainians had unquestionably won possession of the city, but then Polish forces began attacking Ukrainian patrols in the vicinity of the railroad station—the legionnaire-mathematician Felsztyn having the honor of taking the first Ukrainian prisoner. Soon a line had been established dividing Lwów into Polish and Ukrainian zones; it ran along Ossoliński and Słowacki Streets and cut right through the Jesuit Garden. My boarding house happened to be in the Ukrainian zone. The unceasing staccato of machine guns accompanied the booming of a Ukrainian battery perched up on "High Castle".[57] Normal activities ceased, the stores were all closed with their blinds rolled down, the electricity supply was shut off, and the Ukrainian commandant of the city ordered the city's gates closed. Bullets fell on the iron roofs from morning till night with a frightful clatter, a continual reminder that a war was going on—a war following immediately on another war!

---

[55] Sierpiński spent the years 1914–1918 in Russia, having found himself and his family caught up there in 1914 when war broke out. He spent the war years mainly in Moscow working with Nikolaï Luzin. He returned to Lwów in 1918, but was shortly thereafter offered a position at the University of Warsaw. (The author's claim that he returned via Siberia is dubious.)

[56] An armed conflict lasting from November 1, 1918 to May 22, 1919, between the forces of the Second Polish Republic and the West Ukrainian People's Republic over control of Eastern Galicia following the dissolution of Austria-Hungary. The "Battle of Lwów" (or "Defense of Lwów"), in which the local Polish civilian populace participated, took place as one of the several fronts of the war.

[57] One of the hills of Lwów on top of which there once stood an old castle, much repaired and rebuilt, now in ruins.

Among the few boarders still living in *Anuta* was a Captain J., a Russian, who had participated in the Russo-Japanese war in 1905. He had many interesting stories to tell about Koreans and other peoples of the Far East. Every day he risked his life crossing over to the Polish zone—not for military reasons, but because he was carrying on a liaison with a lady living in that zone. There was also a Count Komorowski, an oil man, who passed the time playing cards with Mrs. Linhardt, sister to Tadeusz Rittner,[58] and our landlord. Finally, there was Mr. Decykiewicz, who had formerly been a high-placed Austrian bureaucrat. He was a Rusyn,[59] an advocate of compromise, liked reading Anatole France in the original French, and was very objective in his views. Every day he went to the Ukrainian command post of the town, where he had obtained a high position as a civilian administrator. He complained that various "Cossack hetmans", as he called them, from both zones, were completely impervious to the voice of reason, and that this absurd war would totally destroy a country that had just been half destroyed in the previous war. He said that if it weren't for the Jews remaining, the whole of Galicia would soon not be worth a brass farthing....

The situation in Lwów augured nothing good, and after tolerating it for three weeks, I decided to flee to Jasło. Mr. Decykiewicz wangled me a pass stamped with the image of a lion, and I was put in touch with the well known actor Dante Baranowski,[60] who was in Lwów with his wife and child, and also wanted to go west. Around November 21, we took advantage of a one-day truce between the warring parties in Lwów, and travelled by means of a peasant wagon—actually the driver was Jewish—to Żółkiew. At the railroad station there we encountered a rabble of soldiers and a few officers all waiting for the train. The officers made an attempt to introduce order among the soldiers, but these merely scowled, otherwise ignoring them. Of course, at this time there was a political vacuum in that part of Poland: there was no Polish, Austrian, or Ukrainian governance of any kind in the region, although a Ukrainian system of rule was beginning to take shape. Our driver suddenly declared that he would go no further, although we had paid him ahead of time. My fellow passengers called over a Ukrainian militiaman, who obligingly threatened our wagon driver with his bayonet. The driver took hold of the bayonet gingerly, and after arguing with the Ukrainian for some time, finally agreed to drive on. At some point we blundered into a Ukrainian military outpost, and all around us Ukrainian soldiers suddenly rose up out of the dark flourishing their bayonets *à la* Rinaldo Rinaldini.[61] They confiscated our thermos flask as an item of military significance, but the image of a lion stamped in our passes was sufficient for them

---

[58] Polish dramatist, writer, and literary critic. Lived from 1873 to 1921.

[59] The Rusyns form an Eastern Slavic ethnic group, also called Ruthenians. A mountain-dwelling people of the eastern Carpathians.

[60] Venice-born actor and director in the theater, for instance in Zakopane and Tarnów. Lived from 1882 to 1925.

[61] A fictional, noble-minded Corsican bandit, the hero of *Rinaldo Rinaldini, der Räuberhauptmann*, written by Christian Vulpius around 1800.

to let us proceed. We passed through the deserted streets of Rawa Ruska with their boarded-up stores, and continued to Jaryczów, where we spent the night. There only Jews remained, and by five in the morning they had all gathered in the synagogue; it was hard for me to imagine how they passed the days otherwise in that desolation. At last, not far from Jarosław, at the further end of a bridge over the river San, we came to a Polish military outpost, and were able to continue on by train. While the train was stopped in Rzeszów, we heard gunfire, but it turned out to be just young railroad workers, soldiers, and other youths who had looted a munitions store and were shooting off the rifles in the air for fun. When finally I arrived in Jasło, dirty, hungry, and with a four-days' growth of beard, my wife and her parents had difficulty recognizing me.

I remained in Jasło from then till 1920. I was not called up, because I was just beyond the recruiting age at that time. Lecturing in Lwów being out of the question, I managed to get a job with the gas pipeline firm of Gartenbarg, Waterkeyn, and Karpaty, which had constructed a pipeline from Męcinka, near Krosno, through Jasło, to the refinery at Glinik Mariampolski, near Gorlice, and supplied methane from the mineshafts in Męcinka to the refineries and also to the small towns strung along the railroad that ran parallel to the pipeline. They produced large amounts of cheap odorless gas of high caloric value, and in that region one could hardly find a dwelling heated by coal, wood, or coke. I worked in an office in a refinery in Niegłowice, near Jasło. The man in charge was an engineer by the name of Aleksander Dietzius, the best engineer I have ever known. My role was that of mathematical expert, and we got on swimmingly for as long as MacGarvey, the head of the firm Karpaty, did not interfere. His aim was to have his gas company declared politically neutral.... I thus had a twofold pretext for resigning: in protest against the intentions of a millionaire from Vienna, and my indispensability at the University of Lwów, in view of the fact that Sierpiński and Janiszewski were now teaching at Warsaw University, and Puzyna had died. The docent recently habilitated in Lwów was therefore shirking his duty by not returning there to hold the mathematical fort!

And there's a sense of equity involved when a poorly paid lecturer in mathematics can thwart the aims of a millionaire, albeit on a rather modest scale.

Part II

Chapter 7
In the University Town Lwów

When I arrived in Lwów, I found that the things I had left in storage —under the less than watchful eye of a Mr. Zawadzki—had been stolen, and that there was no room to be had in my own house, so I had to rent a basement flat. I eventually lost the lawsuit over our possessions, and it also turned out that the contract of purchase and sale of the house was not legally valid because the seller, who died a few months following the transaction, was not the legal owner of the whole property. I had now to validate the contract, in particular by obtaining the signatures of the owner's heirs. All I could do was persevere in the face of these difficulties.

Changes had taken place at the university in my absence. My colleague Zygmunt Janiszewski had died in 1920 of the terrible Spanish flu that had swept the world in the aftermath of World War I, and, as I mentioned earlier, Józef Puzyna had died the year before. In his place Żyliński, from Kiev, whom I had met in Göttingen, had been appointed professor. Although he and I did not always see eye to eye, on my recommendation Stefan Banach was made *Assistent* to Antoni Łomnicki at the Polytechnic. It was Janiszewski who had founded a mathematical society in Lwów, later to become a branch of the Polish Mathematical Society.[1] Its earliest members were Janiszewski, Krygowski,[2] Żyliński, Łomnicki, Banach, Ruziewicz, and I, with Dziwiński[3] attending quite rarely. Our little mathematical society flourished. It did not take long for Banach and Ruziewicz to achieve their *Habilitation*. Although Lwów University was growing scientifically, nonetheless the dominant mood in the town was one of postwar social chaos and bedlam: frenzied dancing and playing

---

[1] *Polskie Towarzystwo Matematyczne* (PTM). This was to grow out of a mathematical society formed in Kraków in 1919. (See Chapter 6 for more on the initial stages of the Mathematical Society in Kraków.)

[2] Zdzisław Krygowski (1872–1955) was then a professor at the Lwów Polytechnic, where he had been rector in 1917–1918. He moved to Poznań University in 1920.

[3] Placyd Dziwiński (1851–1936), Polish mathematician. Worked in geometry and the history of mathematics.

Fig. 7.1 Interwar Poland (Map courtesy of Carolyn King, Department of Geography, York University, Toronto.)

games, tripping hither and thither, libertinism with the accompanying temporary liaisons and betrayals—in a word, the release of the steam of freedom that had been bottled up for years, the reward for the wartime years of oppression and greyness. All this was characteristic of the years 1920–1923.

In 1920, following on the death of Janiszewski, the first issue of *Fundamenta Mathematicae* appeared in Warsaw; this had been his idea: an international mathematics journal published in Poland, and dedicated to one area of mathematics, namely set theory and its applications. Janiszewski had reasoned that set theory had so many adepts in Poland that such a journal would have at least as many Polish contributors as foreign ones. The project succeeded brilliantly, but, alas, its author did not live to see even the first issue. Today the first few volumes of the journal are in such demand that it paid to have them reprinted several times.

The mathematics lectures were all delivered in a lecture room in the physics building on Długosz Street, and among the students who attended them were J. Schenck, Władysław Orlicz,[4] Herman Auerbach,[5] a Mr. Homme, and two ladies,

---

[4] Władysław Roman Orlicz (1903–1990), Polish mathematician. Worked mainly in topology and functional analysis. In 1937 he was appointed professor at Poznań University, where after World War II he created a highly reputed center for functional analysis.

[5] Herman Auerbach (1901–1942), Polish mathematician. Professor at Lwów University. Remained in Lwów all his life, until sent to the Bełżec concentration camp in 1942, where he perished.

Fig. 7.2 Group portrait taken on the occasion of a visit by Ernst Zermelo to Lwów in 1929. Seated (*left* to *right*): Hugo Steinhaus, Ernst Zermelo, Stefan Mazurkiewicz; standing: Kazimierz Kuratowski, Bronisław Knaster, Stefan Banach, Włodzimierz Stożek, Eustachy Żyliński, and Stanisław Ruziewicz (Courtesy of the HSC Archive, Wrocław University of Technology.)

Fig. 7.3 Inside the Scottish Café in its heyday (Public domain.)

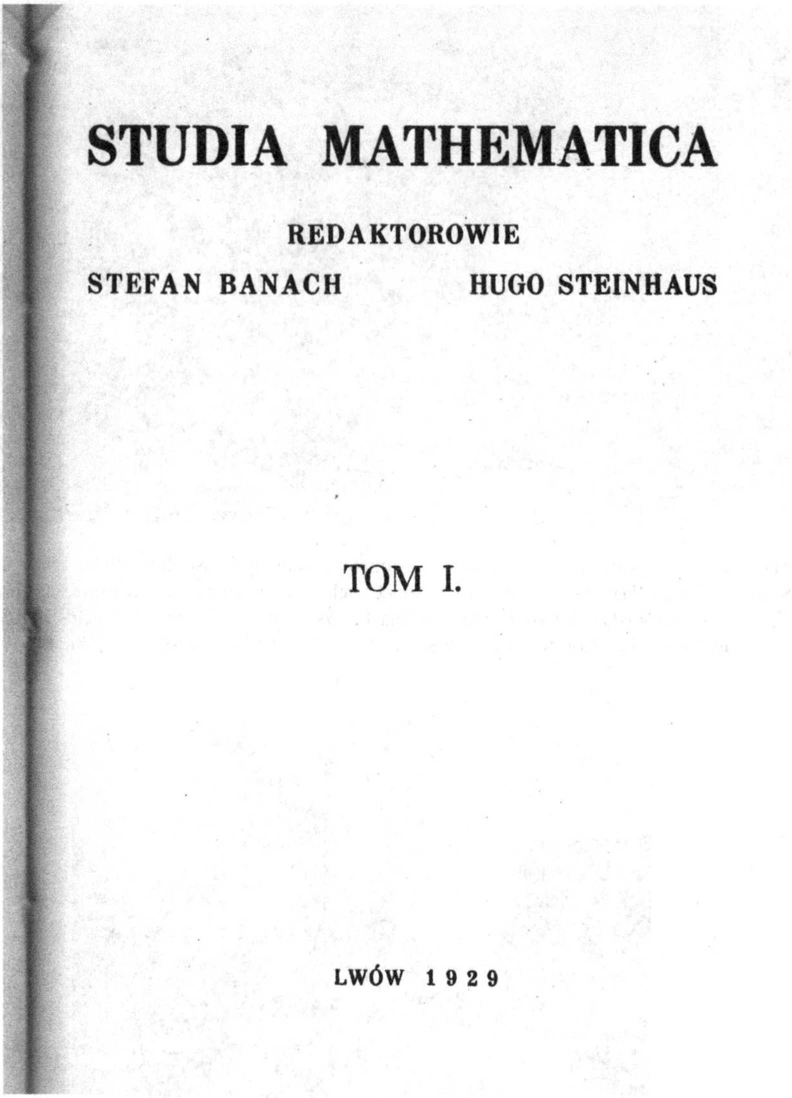

STUDIA MATHEMATICA

REDAKTOROWIE

STEFAN BANACH HUGO STEINHAUS

TOM I.

LWÓW 1929

Fig. 7.4 The title page of the first volume of *Studia Mathematica* (Courtesy of the HSC Archive, Wrocław University of Technology.)

> Le journal „Studia Mathematica" publie des Mémoires et Notes du domaine de Mathématiques pures et appliquées. La Rédaction poursuit le but de grouper autour de ce journal les recherches concernant l'analyse fonctionnelle et tout ce qui s'y rattache.
> Les manuscrits doivent être adressés à l'un des Rédacteurs:
> Stefan B a n a c h, Lwów, ul. św. Mikołaja, 4.
> Hugo Steinhaus, Lwów, ul. Kadecka, 14.
> Après avoir été examiné par les Rédacteurs ils seront pourvus d'une date d'entrée, qui sera imprimée à la fin du Mémoire en question.
> Les travaux doivent être dactylographiés en français, allemand, anglais ou italien. Tout caractère différent du type courant doît être marquê au crayon en couleur.
> Les épreuves corrigées avec tout le soin possible doivent être renvoyées sans aucun délai à M. H. Steinhaus, Lwów (Pologne), ul. Kadecka, 14, (tel. 6490). Les changements du texte causant des dépenses additionnelles seront executés aux frais de l'auteur. On fournit gratuitement 50 tirages à part; pour ce qui dépasse ce chiffre MM. les auteurs seront chargés des frais d'impression. Le prix du volume est $1^1/_2$ $ à l'étranger, 12 zł. en Pologne. Un volume contiendra au moins 200 pages.
> Les versements (par mandat international ou par chèque [P. K. O. Nr. 154 127, Studia Mathematica]), les tirages à part et les périodiques destinés pour les „Studia" doivent être adressés à l'„Administration de Studia Mathematica, L w ó w, ul. św. Mikołaja 4, Uniwersytet".

Fig. 7.5 The inside back cover of the first volume of *Studia Mathematica* (Courtesy of the HSC Archive, Wrocław University of Technology.)

Miss Krzysikówna and Miss Schoenówna. They were all very keen to learn as much mathematics as they could, so I gave them the keys to the seminar library, which they made good use of, causing virtually no damage to or loss of the library's inventory.

At this time, I was partially dependent on my wife's parents and my own to eke out support for my family of three, since the pay of an *Assistent* was relatively meager. For this reason, my wife and daughter were then living with her parents in Jasło, but I was able to visit them often during the frequent breaks in bouts of lecturing. Hypnosis was then a fad, and my sister-in-law Hela discovered that she had a hypnotic gift, especially when her friend Ewa served as medium. On one occasion, following my request, they arranged a *séance* in our room. I had thought of a few experiments ahead of time to rule out any deception: I asked Hela to suggest to Ewa during the hypnotic trance that she forget her name on awakening, and also that she take Volume II of a book of reproductions of paintings and sketches by Leonardo da Vinci when she left. (Both volumes were lying on the table.) When Ewa was woken from the trance, I—having in fact just been introduced to her—asked her for her name, whereupon she turned red in the face and seemed unable to reply. Then she went to her handbag, took out her visiting card, and read off her

Fig. 7.6 Hugo Steinhaus with his niece Alina Chwistkówna in Jasło in 1936 (Courtesy of Agnieszka Mancewicz.)

name. A little later, she said her goodbyes and prepared to leave, but, although we did not stop her, she lingered—even though she had said her mother was waiting for her. She turned to Volume II of the Leonardo book, and looked through a few pages, and her face brightened as she asked me if she might borrow the book. I said dryly that I could lend her Volume I, but not Volume II, which I was in the middle of reading. She then began to insist, because—as she put it—only Volume II interested her. Since we didn't know each other at all well, this insistence was actually quite improper; here one could see clearly the effect of the hypnotic suggestion. Finally, Hela told Ewa of the suggestion she had made during the trance, whereupon Ewa immediately ceased her demands and seemed to entirely lose interest in the book, even though I now offered to lend it to her for an extended period of time.

I relate these experiments partly because I see in them proof that a sense of awareness that one is acting freely is no guarantee of actual freedom of will: after all, Ewa's behavior showed clearly that she had this awareness in choosing the book

7 In the University Town Lwów

Fig. 7.7 Professor Leon Chwistek with wife Olga and daughter Alina in Lwów, 1934 (Courtesy of Agnieszka Mancewicz.)

in question—she felt she had made this choice independently of our wishes—and yet we, that is, Hela and I, knew beforehand which book she would demand.

* * *

Yes, things were livening up in Lwów, scientifically speaking. Banach defended his doctorate and not very much later his *Habilitation*. I attended the seminar run by Twardowski devoted to reading Russell.[6] Among the other attendees were Ajdukiewicz,[7] Zawirski,[8] the Gromskis, and Czeżowski.[9] At that time, Lwów was agog over the "new" poetry, and of the new poets I met Józef Wittlin,[10] who was then translating *The Odyssey* into Polish.

---

[6]Bertrand Arthur William Russell, 3rd Earl Russell (1872–1970), prominent English philosopher, mathematical logician, historian, and social critic. The work in question may have been *Principia Mathematica* by him and A. N. Whitehead, in which they attempted to derive mathematics from basic postulates of logic.

[7]Kazimierz Ajdukiewicz (1890–1963), Polish philosopher and logician of the Lwów-Warsaw school.

[8]Zygmunt Zawirski (1882–1948), Polish philosopher and logician. Later held professorships in Poznań and Kraków.

[9]Tadeusz Czeżowski (1889–1981), Polish philosopher and logician of the Lwów-Warsaw school of logic.

[10]Józef Wittlin (1896–1976), Polish expressionist poet, essayist, novelist, and translator. His most highly regarded work is the 1936 novel *Salt of the Earth*. Lived in the US from 1941.

The question of decent quarters was causing me difficulties. At last an apartment had become vacant in the house I owned, but I had nothing to furnish it with and it didn't make sense to buy furniture, since, not being a professor, my future in Lwów was uncertain. So I let the apartment temporarily to a judge by the name of Antoniewicz. Then when I was appointed "extraordinary professor" in 1920, and tried to get the judge to vacate the apartment, he invoked a law protecting tenants, and I had to wait another year for a judgement to come down, and then to be executed.

So I took a leave of absence for a term and went with my wife to Göttingen, leaving our daughter behind with my wife's folk. At that time Polish money was more stable than German money, and the exchange rate correspondingly favorable to us, so my salary sufficed for a tour there. In Berlin I phoned Leon Lichtenstein,[11] a professor there, and he came to see us at our hotel, where we soon warmed to one another. He told me I should meet Professor I. Schur,[12] so I went to the university and met several people, including Schur and Bieberbach.[13] Schur, and especially his wife, a Russian Jewess, hated Poles and made me feel their enmity. However, my friendship with Lichtenstein lasted for the rest of his life, that is, for the next 10 years.

Immune from the galloping inflation, we travelled on to Göttingen. I went straight to the house of my former landlady Mrs. Gericke, but as she no longer took in boarders, we put up instead at a Mrs. Lotheisen's. In addition to a few German students, she provided room and board to two Chinese, a few Hungarians, and a star of the local German operetta, a Miss Chandler from New York. When once she asked the Chinese students if they liked the operetta, they answered politely that, of course, they liked it a great deal, except for the frequent fits of shouting by the singers on the stage. It turned out that such was the effect on them of choral pieces. They had strips of white paper hanging in their room with Chinese characters drawn in special red ink on them. They explained to us that the characters so executed combined calligraphic beauty with the beauty of the thought they expressed. Naturally, such a combination is somewhat beyond the ken of a European. The meals at Mrs.

---

[11] Leon Lichtenstein (1878–1933), Polish-German mathematician. Worked on the calculus of variations, partial differential equations, and potential theory. A founder, and the first editor, of *Mathematische Zeitschrift*. Abandoned his position at the University of Leipzig on the accession of Hitler to power in 1933, and returned to Poland, where he died in Zakopane.

[12] Issai Schur (1875–1941), Jewish mathematician. Became professor at the University of Berlin in 1919, remaining there till dismissed by the Nazis in 1935. Among his students were Richard Brauer, Bernhard H. Neumann, and Richard Rado. Worked mainly in the theory of group representations. Among other things, introduced the "Schur multiplier" of a group, shown some 40 years later to be the second cohomology group of the group over the complex numbers.

[13] Ludwig Georg Elias Moses Bieberbach (1886–1982), German mathematician. Known in particular for his work in complex analysis. In 1916 formulated a famous conjecture on the condition on a holomorphic function's Taylor series for it to map the unit disc injectively into the complex plane, settled by Louis de Branges in 1984. A Nazi sympathizer, was involved in the repression of Jewish colleagues in the 1930s—including Edmund Landau and Issai Schur—and co-founded a "German mathematics" movement and journal.

Lotheisen's boarding house were wretched. Occasionally we were served just a thick soup and a thin rhubarb compote; one had to admire the students for being able to work on such a diet.

At that time, physics was flourishing in Göttingen, in particular, Franck[14] was a rising star of the field. However, I found the sort of physics they were preoccupied with[15] unattractive. I was then working on probability and its applications to statistical mechanics, and at that time the only person interested in this topic was Artin.[16] I spent our time in Göttingen leafing through the mathematics journals, and showing my wife the sights in Cassel and Hann. Münden.[17]

We moved on to Hamburg, where a new university had just been opened. We went together to Hamburg's ethnographic[18] museum, well worth seeing. Then, while my wife explored the town and the port, I went and talked to Ostrowski, Blaschke,[19] and especially Rademacher,[20] who lived in Blankenese on the bank of the Elbe, some way downstream from Hamburg—a beautiful place brimming with roses. We sat in a small café from which we could see the Elbe, seemingly as wide as a sea.

At the port we saw a quite unusual sight. A ship had just been built that rode upright in the water only when laden, so when first launched its watercocks would need to be opened to provide enough ballast to keep her upright. However, the engineer in charge of the launching had forgotten to do this, with the result that the ship had capsized, drowning forty seamen. What we saw was this newly built ship being cut into pieces by means of blowtorches to salvage the iron that had gone into her hull. These blowtorches were run on hydrogen, and I also saw in Hamburg how they used them to help in loading barges with scrap iron: a crane would grab

---

[14] James Franck (1882–1964), German physicist. Appointed Professor of Experimental Physics in Göttingen in 1920, and, together with Max Born turned Göttingen into an important center for quantum mechanics. (In particular, Robert Oppenheimer came to further his studies with this group.) Nobel laureate for physics in 1925. Moved with his family to the US in the 1930s to escape Nazism.

[15] Quantum mechanics.

[16] Emil Artin (1898–1962), Austrian-American mathematician. Came to Göttingen in 1921 to pursue post-doctoral studies in mathematics and mathematical physics with Courant and Hilbert. During his year in Göttingen worked closely with Emmy Noether and Helmut Hasse.

[17] Cassel (Kassel from 1926) is a town in northern Hesse. Hannoversch Münden is a town in Lower Saxony at the confluence of the Fulda and Werra rivers, notable for its old houses, some dating from more than 600 years ago.

[18] Ethnography is the branch of anthropology where peoples are studied via their cultural practices and artefacts.

[19] Alexander Markowich Ostrowski (1893–1986), mathematician. Born in Kiev, he studied under Hensel in Marburg, and after World War I wrote his doctoral dissertation in Göttingen. Worked as Hecke's assistant in Hamburg from 1920, where he completed his *Habilitation* in 1922. Wilhelm Blaschke (1885–1962), Austro-Hungarian differential and integral geometer.

[20] Hans Adolph Rademacher (1892–1969), German mathematician. Worked in mathematical analysis and number theory. Dismissed from his position at the University of Breslau in 1933 for publicly supporting the Weimar Republic, he moved to the US, obtaining a position at the University of Pennsylvania.

a huge tangle of scrap iron which a man standing on the barge cut into manageable portions with a blowtorch before they were deposited in the barge.

From Hamburg we travelled to Kiel, in Schleswig-Holstein, to visit Otto Toeplitz, who had been one of my teachers in Göttingen and was now a professor at the University of Kiel. Although Kiel was a port, it was a relatively quiet, provincial sort of place, and one could stroll about the town and indulge in peaceful reflection. Toeplitz received us very cordially, and since he had just then received back from his students exercises he had set them at his seminar, he asked me if I would like to correct them—occasioning some surprise amongst the students when they perceived that the corrections had been made in the hand of someone other than their teacher. Toeplitz invited us to dinner, and I suggested to my wife that she take along some sliced ham and smoked herring that we had in our hotel room. This turned out to be the right thing to do since at that time conditions in Germany were such that even a professor could not afford to provide a dinner for three guests—the third being Mrs. Radbruch, wife of the Reich Minister of Justice, a post that had not existed before the war. We asked Toeplitz and her if they knew of a good seaside resort where we might spend a couple of weeks, a holiday we had been very much looking forward to, and Madame Radbruch suggested Laboe, where Kiel Bay opens to the sea. So we took a ferry to Laboe, traversing Kiel Bay from shore to shore, but although there was a small beach area there, it was not sufficiently open to the sea for us, and we continued along the shore looking for a spot where the sea horizon spanned at least 150°. And eventually we came on just such a place. It was called Stein, and consisted of an inn, large and very clean, and a few fishermen's huts, and there was a footpath leading down to the seaside. We decided to stay at the inn, and to rush back to the hotel in Kiel, check out, and transport our things to Stein. As we were leaving, we were lucky enough to meet someone with a car who, we hoped, might be heading for Kiel. We explained our intentions to him with the aid of a considerable amount of gesturing, and a little later, after he had completed some errand or other, he caught up with us and drove us to the hotel. We quickly paid the bill and went with our bags to the port, where, to the hooting of the siren, we just managed to throw them and ourselves on board the last ferry for the day. In Laboe, we were met by a horse-drawn wagon sent by the innkeeper, as an exceptional favor, to convey our luggage, as the only alternative was to go on foot. All this rushing about in order to save a few złotys!

Stein belonged to the so-called *Schöne Pfarre*[21] region, raised high above the sea, with rich, fertile farmland, meadows surrounded by rose hedges, with succulent moist grasses on which the cattle grow splendidly fat. The milk and butter were as good as the Danish varieties. Each morning, ravenous after the skimpy meals served in Göttingen and Kiel, we each consumed several buttered rolls and drank four cups of coffee made with milk. There were very few guests staying at the inn at any one time. There were some businessmen from Hamburg, and an "admiral", which was the nickname we gave to a man from Frankfurt who always sported the cap of

---

[21] Meaning "beautiful parish".

some yacht club, and whose family each day filled the inn with shouted orders for preparing and lowering a small boat—which he called a "canoe"—into the water. We got to know a former sergeant who told us about secret organizations plotting the overthrow of the Republic.[22] He said they were masterminded by former aristocrats in cahoots with high-level former imperial military personnel, who foregathered in the castles of the old nobility to receive oaths of allegiance, washed down with champagne, from various supporters of plans for a vengeful coup—but that no one took these people seriously. Professor Toeplitz visited us once, and I discussed my work on additive operators[23] with him. Relaxing after these demanding rigors, he told me of a serious error made during the war involving the great Tirpitz,[24] that was well known in Kiel. It seems that when planning submarine warfare, Tirpitz had presented to the Kaiser estimates of the productivity of British and American shipyards which omitted the tonnage produced by neutral states on order from Britain. When a certain naval officer noticed the error and asked that his report be presented directly to Kaiser Wilhelm, Tirpitz vetoed it, thus persisting in furnishing his own side with disinformation—invaluable to the entente powers. Incidentally, one could see along the shore of Kiel Bay the ruins of forts that had had to be destroyed in accordance with the Treaty of Versailles.

The first leg of our journey home was by electric train from Kiel to Leipzig, which ran on rails resting on drained peat bogs, the peat evidently not being commercially profitable. Travelling east, we were surprised by the number of towns we passed through with Slavic names—whose meaning and origins were a matter of indifference to most Germans, of course.

In Leipzig we were met at the station by Neder, Lichtenstein's assistant,[25] who had found an inexpensive furnished room for us. The son of our landlady was a "bible student", and put into practice his belief that one had a moral obligation to show Christian love for one's neighbor in the details of daily life. Thus he would never refuse to do a favor. He told us that in the factory where he had once worked his fellow workers had started by laughing at him, but ended by forming a sect of disciples. If, for instance, he happened to see a sealed and addressed envelope lying on a desk, he would go to the trouble of finding the correct stamp for it. When his wife complained that her new shoes were too tight, he put them on himself and walked in them all night in his study to widen them for her.

---

[22] The Weimar Republic, a parliamentary republic established in Germany to replace the Imperium. It was beset with many difficulties, some resulting from the harsh penalties imposed by the Treaty of Versailles, and including a period of hyperinflation. It ended, in effect, in 1933, with Hitler's accession to power.

[23] That is, assumed to be just additive, rather than fully linear?

[24] Alfred von Tirpitz (1849–1930), German admiral. Considered the founder of the German Imperial Navy.

[25] In 1922 Lichtenstein left his position in Münster to take up a professorship at the University of Leipzig. He had been professor at the Polytechnic in Berlin-Charlottenburg 1919–1920, transferring in 1920 to Münster.

We returned home through Silesia[26] which we found in a state of ferment. The borders having not yet been stabilized, the factories were paralyzed, the trains ran at random, and crowds of people seemed to be roaming aimlessly through the streets of the towns. We were close to running out of money, and in Katowice, after a day without food, we ate a fried herring washed down with beer, paid for with our last few coins. It was one of the finest dinners I have ever eaten!

The year was 1923, and the fourth volume of *Fundamenta Mathematicae* had just appeared, containing my paper on the probability of convergence of an infinite series whose terms are weighted, that is, are probabilistic. This was a problem not considered by anyone earlier, and which I had thought of one day while walking along Kościuszko Street in Jasło. In the second section of this paper I had asked a question eventually answered by Khinchin,[27] who received the prestigious Stalin Prize for his solution in 1941. Some years later, Willy Feller,[28] who was raised in Zagreb, studied in Göttingen, and taught in Stockholm, expounded the general treatment of probability as a subdiscipline of measure theory, and my work was subsumed under that approach. The same volume of *Fundamenta Mathematicae* contained a paper by Antoni Łomnicki with the same basic idea as Feller was to have, but with the problem of "multiplicativity" stated in an unclear way; a precise version of the concept is included in that of "stochastic independence",[29] formulated much later. It was in this year also that I was appointed ordinary professor at the Jan Kazimierz University of Lwów.

In Lwów at this time I met a Mr. Blumenfeld, who worked as manager of a chemical plant but had imbibed a fair amount of mathematics and philosophy as a student at the University of Vienna. He introduced me to Jakub Parnas,[30] an eminent chemist, and professor of medical chemistry at the university.

It was around this time that I wrote a little book of popular mathematics called *Czem jest, a czem nie jest matematyka* (What mathematics is, and what it is not),[31]

---

[26] A region of present-day southwest Poland, rich in mineral and natural resources. Ruled by Prussia and then Germany from the late eighteenth century to the end of World War I. The eastern part was transferred to Poland after that war (with an Austrian remnant going to Czechoslovakia) and, after World War II, most of the rest also became part of Poland.

[27] Aleksandr Yakovlevich Khinchin (1894–1959), celebrated Soviet probabilist.

[28] William Feller (1906–1970), Croatian-American probabilist. Born in Zagreb to a Polish-Jewish father and a Croatian-Austrian Catholic mother. After evading the Nazis for some years from 1933, he arrived in the US in 1939, ending up at Princeton University.

[29] Based on the notion of independent events in a probability space.

[30] Jakub Karol Parnas (1884–1949), prominent Polish-Jewish-Soviet biochemist. Head of the Department of Medical Chemistry in the University of Lwów 1920–1941. He collaborated with the Soviet regime from the time of its invasion of Poland in 1939. Moved to the Soviet Union in 1941, where he was made a member of the Academy of Sciences and served as Director of the Institute of Biological Sciences in Moscow 1943–1947. In January 1949, in the course of a Stalinist purge, he was arrested by the NKVD on the false charge of spying for the West and imprisoned in the Lubyanka, where he died.

[31] A popular book in Poland up to World War II.

which was eventually published by the firm owned by Altenberg. Mr. Altenberg was very friendly towards his authors: when I once asked him for an advance, justifying my request by mentioning the costs associated with Sylvester night,[32] he paid me at once. That was later, when we were going through a dancing phase, and danced a great deal. The main promoter of new dances in Lwów was then Mr. Sklepiński, a pharmacist, who, despite his heavy build, was an extremely good dancer.

The early 1920s was a time of extreme fluidity. New joint-stock companies were continually being formed and exploited the instability in currencies to show fantastic profits. And then to service the new state of Poland a multitude of new civil service positions needed to be set up and filled. Social traditions, fashions, dances—everything was in flux. This fluidity and lack of tradition sometimes had strange consequences. For example, the ministry responsible for universities, whose employees had been drawn from who knows where, demonstrated a complete failure to understand basic principles of law, let alone how universities function. On one occasion, a student who had been refused admission to our university by the relevant dean, appealed the decision to a committee of the faculty, which confirmed the dean's decision, whereupon the student appealed to the ministry. It would seem that in this case the ministry should have adopted the stance of a supreme authority of last resort and either overruled the faculty committee's decision and admitted the applicant, or explained to him that that committee represented the highest authority in such matters. Instead, however, the ministry informed us that we should hold another ballot to vote down our previous decision!—which we refused to do, naturally. A similarly curious thing happened in the Polytechnic. There was a new regulation relating to tertiary institutions according to which the docents in each department had to be represented by two delegates to the department's council. However, since the regulation did not specify how the delegates were to be chosen, one department decided that its own council would nominate the two delegates. Yet surely they should have been chosen independently of the departmental council so as to truly represent the docents' interests! The department in question might be likened to a parliament that pretends to be representative yet chooses its own members.

However, such anomalies and setbacks notwithstanding, overall things were developing, even flourishing, as they should. Around us we could see that some sort of evolutionary progress had been made amid the confusion—for instance, the confusion created by the student strikes, which even then we had to endure, although they were still largely kept under control. The very word "strike" would not seem to have any meaningful application in the context of an academic institution, and from its misuse alone one could infer that those involved in the disruption were in some kind of trance leaving them unable to reason. The suspension of lectures at the demand of certain students is not analogous to a strike as normally understood, but rather to a lockout in response to workers' demands. The parallel between universities and their students on the one hand, and factories and their workers on the

---

[32]New Year's Eve, named after Pope Sylvester I.

other is false. A professor is not an employer for whom the students work for pay—on the contrary, it's the students who pay for the services of the professors. Thus if the word "strike" is to be applied correctly in the university context, it is really only the professors who are in a position to strike for improved conditions or higher pay—from the state or at the cost of increasing student fees. Thus the Göttingen Seven[33] protest was closer to being a true strike than so-called student strikes, where what is actually taking place is the exploitation of the extraterritoriality of universities for purposes having nothing to do with their academic activities. The strikes I witnessed at that time were motivated purely by student political organizations' wanting to bring their aims to the attention of a larger forum. The strikers never made demands for the improvement of teaching or other conditions related to their studies. In any case, refusal to attend lectures was not likely to result in their demands being met since ultimately this merely increased pressure on the financial resources of the parents, rather than those of the professors or the state. A student "strike" becomes dangerous only when their idiocy begins to infect the professors, members of parliament, and government ministers—which is, in fact, what the strikers were counting on, and here they did not miscalculate. At that time most student strikes were going on outside Poland; in our country they would ultimately have led to the liquidation of the universities.

In the department of mathematics students of exceptional ability soon turned up—such as Schauder, Orlicz, Halaunbrenner, and Mazur.[34] Mazur was somewhat unsystematic in the way he did mathematics. Once when I heard that he had solved a difficult problem concerning the summation of infinite series, I asked him if he would show me his solution written up. He kept putting me off until finally it became clear that he had found an error in his proof—but he didn't give up, and in the end did produce a correct proof, fulfilling his own prediction, so to speak.

Collaboration was the order of the day in our group: Orlicz and Mazur, Sierpiński and Ruziewicz, and Banach and I—these were the pairs who most frequently wrote papers together. Żyliński and Ruziewicz wrote a textbook together, and later Stożek, Banach, and Sierpiński would collaborate in writing textbooks for the Gymnasiums.

---

[33] A group of seven professors from Göttingen University who, in 1837, protested against the annulment of the constitution of the Kingdom of Hanover and refused to swear an oath of allegiance to the new king Ernest Augustus.

[34] Juliusz Paweł Schauder (1899–1943), Polish-Jewish mathematician, working mainly in functional analysis, partial differential equations, and mathematical physics. Collaborated with Jean Leray in Paris from about 1932 to 1935. Appointed senior assistant at the University of Lwów in 1935. Known, in particular, for the concept of a Schauder basis for a Banach space and the Leray–Schauder principle providing a means of establishing solutions of partial differential equations based on a priori estimates. Murdered by the Gestapo in 1943. Jacob Halaunbrenner (1902–1943). Murdered by the Gestapo in France in 1943. Stanisław Mazur (1905–1981) was supervised by Banach for his doctorate. Became one of the leading experts on functional analysis at the University of Lwów.

Thanks to the influence of Professor Bartel[35] at the Lwów Polytechnic, a department had been created there dedicated to general theoretical training of students, to which Kazimierz Kuratowski,[36] from Warsaw, was appointed professor, with the result that the Polytechnic became a highly reputed center for mathematics. He pushed for new assistantships to which first Kaczmarz and Nikliborc,[37] who came from Kraków, were appointed, and then Mazur and Orlicz, among others. There were only two students who obtained their doctorate in mathematics there: Jan Antoni Blaton and Stanisław Marcin Ulam.[38]

In 1925 I was given leave for another term and once again used it to travel with my wife—this time to Paris. We stayed in a cheap and well run *pension* owned by a Monsieur Ruben, situated in the Gobelins District, far from the city center. At the time, Stefan Banach, Professor Lehr-Spławiński, and Professor Kuryłowicz[39] and his wife were also staying there. It happened that Professor Parnas and our friends from Lwów Emil Bader and his wife were also living in Paris just then.

I remember an outing with the latter three to Fontainebleau to visit the *château* where Napoleon had spent the waning days of his reign. However, instead of taking a tour of the *château*, we spent a great deal of time in an excellent restaurant that Parnas had found, and then went at his urging on a several-hours-long walk through a deciduous wood nearby, where, just as painted by the French landscape artists, the leaves really were of a familiar delicate green, and the air also was suffused with those certain soft shades of color.

---

[35] Kazimierz Władysław Bartel (1882–1941), Polish mathematician and politician. Member of the Polish parliament from 1922. Prime Minister of Poland five times between 1926 and 1930. In the 1930s wrote a monograph on perspective in European art. Murdered by the Nazis in 1941 after refusing to form a Polish puppet government.

[36] Kazimierz Kuratowski (1896–1980), prominent Polish mathematician specializing in logic and set theory. Gave an alternative axiomatization of topology to the one now standard. A leading representative of the Warsaw school of mathematics. Appointed to the Lwów Polytechnic in 1927.

[37] Stefan Kaczmarz (1895–1939) obtained his doctorate and *Habilitation* at Lwów University. Vanished during World War II. (There are claims that he was murdered by the NKVD at Katyń in 1940.) Władysław Nikliborc (1889–1948) also obtained his *Habilitation* at Lwów University.

[38] Jan Blaton (1907–1948), Polish physicist. Stanisław Marcin Ulam (1909–1984), Polish-Jewish-American mathematician. Went to the US in 1938 as a Harvard Junior Fellow. Returned to Poland on the eve of World War II, but eventually managed to regain the US. Participated in the Manhattan Project. Well known for his work in various fields: number theory, set theory, ergodic theory, and algebraic topology.

[39] Tadeusz Lehr-Spławiński (1891–1965), Polish linguist specializing in Slavic languages. Jerzy Kuryłowicz (1895–1978), Polish historical linguist. One of the greatest of twentieth century scholars of Indo-European languages. Professor at the University of Lwów from 1928. After World War II held professorships at the universities of Wrocław and Kraków.

One day Montel, Lebesgue, and Hadamard[40] took Banach and me to dinner at a restaurant on the *Quai d'Orsay* where we ate a delicious meal. Unfortunately, Hadamard could stay only briefly. I remember Montel and Lebesgue asking about the anti-Semitic movement in Polish universities.

I took the opportunity of learning a little about the Paris mathematicians whose work was of greatest interest to me. One of these was Szolem Mandelbrojt,[41] originally from Warsaw. He had married a Frenchwoman, taken out French citizenship, and had served in the French army. He was a gifted mathematician, and was soon appointed *maître de conférences* in Clermont-Ferrand, and on the retirement of Hadamard was made his successor at the *Collège de France*. He invented what was then a new type of seminar, namely one devoted to the discussion of recently published works.

Mandelbrojt liked to frequent, in the company of the rest of us Poles, the *salon* run by Madame Kasterska,[42] a novelist, and the wife of the Romanian mathematician Sergescu,[43] who later made a name for himself with his research into the history of French mathematics. In particular, he discerned in the writings of scholars at the Sorbonne in the Middle Ages theses reminiscent of concepts of nineteenth century set theory—a discovery which to some extent rehabilitated scholastic mathematics. Among the other *habitués* of Madame Kasterska's *soirées* there were a Mr. N. Ottman, who had lost a leg in the defense of Lwów against the Ukrainians, and was married to a beautiful raven-haired lady, Mrs. B., an artist, and Mr. J. Mirski, also from Lwów. Madame Kasterska was active politically, professing herself a Polish royalist. She carried on an intense correspondence on political matters with people in Lwów, and even asked us if we would take some of her letters with us when we returned there, on the grounds that the post was not reliable. In this and in other ways it seemed that she felt that now, in 1925, things were much the same as in 1825. Generally speaking, the whole company of Poles in Paris affected a kind of nostalgia for Poland, as if they had left the country for good to escape the invader, when they were for the most part essentially just office workers on holiday spending their salaries abroad.

I must say a few words about M. Ruben, our landlord. He was a Frenchman from the south of France, and in a state of constant awareness of his provenance. He would explain to anyone willing to listen that the conflicts between the Northerners

---

[40]Paul Antoine Aristide Montel (1876–1975), French mathematician. Worked in complex analysis. Jacques Salomon Hadamard (1865–1963), eminent French mathematician. Made major contributions to number theory, complex analysis, differential geometry, and partial differential equations. The first, with Charles-Jean de la Vallée-Poussin, to prove the "Prime Number Theorem".

[41]Szolem Mandelbrojt (1899–1983), Polish-Jewish mathematician, working mainly in classical analysis. He studied under Hadamard, and was an early member of the Bourbaki group. In 1938 appointed Hadamard's successor as professor at the *Collège de France*. During World War II he was at the Rice Institute in Houston.

[42]Maria Kasterska (1894–1969), Polish writer. Earned a doctorate in French literature at the Sorbonne.

[43]Petre Sergescu (1893–1954), Romanian mathematician and historian of science.

and Southerners in France had lasted for centuries and cost millions of deaths. In his apartment he had a history of these troubles published around 1870, whose author, a Southerner, referred to Northerners throughout as "they", reserving the first person plural "we" for those from Provence. All the same, he belonged to the Society of Friends of the Louvre, made up of local patriots whose subscriptions were used, year in and year out, to buy new paintings or sculptures for the gallery. He was nothing like your rapacious hotelier and seemed satisfied with the rather basic prices he asked.

Back in Lwów life took on some regularity. We usually spent the summers in Jasło, where we enjoyed a rather active social life, on top of the tennis tournaments that were organized there every July and August. There was someone there who felt it one of his responsibilities to keep the town from falling asleep: Antoni Zoll, the *starosta*, who would keep everyone awake day and night by playing opera arias *fortissimo* on his gramophone or singing them himself at an open window. This type of eccentric *starosta* had been thought to have vanished with Count Michałowski. Zoll's rooms were crammed with phonograph records, copying machines, jars of pickles, and a multitude of other objects. He had his own special way of carrying out his duties. Once the town police arrested an elegantly dressed gentleman arrived from nowhere who couldn't speak Polish and whom they suspected of being a Russian spy. The *starosta* took an interest in the matter, and, after visiting the stranger in his cell, declared that the town jail was not an appropriate place to incarcerate such an upstanding gentleman. So he had him transferred to a room adjoining his own bedroom, but took care to lock the suspect in whenever he left his apartment. He had the distinguished gentleman looked after in his apartment, and dined with him each evening. One day, while he was walking with Janek Adamski through the market place, he suddenly rapped his head with his knuckles and said:

— I must go back home.
— Why?
— I forgot to serve my guest his usual cup of black coffee!

This was all done with the utmost sincerity and seriousness. The probability of anyone under arrest meeting with such treatment must be almost zero—except possibly in England.

Zoll was a close friend of Didur[44] and was often visited by Kiepura.[45] Members of the local aristocracy were also wont to call on him. He was game for anything, happy to oblige—in signing the papers authorizing the award of prizes to the winners of the tennis tournaments, or in any other matter making for a better community. It would be wrong to think that he was somehow shirking his duty in being so accessible to all. His attitude to his responsibilities was rather that of a

---

[44] Adamo (Adam) Didur (1873–1946), talented Polish bass opera star. Performed in Europe and at New York's Metropolitan Opera.

[45] Jan Wiktor Kiepura (1902–1966), Polish tenor and actor. Emigrated to the US in the late 1930s to evade Nazism. His brother Władysław (1904–1988) also sang in the opera.

noble distributing largesse to his subjects than that of a petty official. He impressed people more by his other-worldliness, his disdain for formality, and by his many talents, than by his former official Austrian title of "King's Councillor." On retiring, he took lodgings in the house of the engineer Dietzius near Niegłowice. He kept himself occupied by picking up nails from the road so that they wouldn't puncture the tires of passing cars, collecting flat stones to play "ducks and drakes[46]" with, and catching moths by night. His collection of moths was indeed very beautiful.

Sometime in the mid-1920s we started spending the summer vacation differently. It began when our friend Ignacy Blumenfeld talked us into going with him and his wife to Czarnohora.[47] For ourselves, we chose the resort town of Worochta there as our base, and, guided by the Blumenfelds and Stanisław Vincenz,[48] we learned to explore a landscape that was completely new to us. The most beautiful place within striking distance of Worochta is the valley of the river Prut where it flows through the village of Foreszczenka. We started from the Worochta sawmill from which there issued a narrow-gauge railway used for transporting wood. We seated ourselves in the small open wagons and off we went, struck dumb by the sun's dazzle, the scents, and the twittering and rustling from the forest hemming us in. We were heading for Howerla, where Vincenz had a cottage. After we left the little train to continue on foot, it soon became so fiercely hot that I immersed my daughter Lidka in the Prut— as Thetis did to Achilles[49]—to prevent her from overheating. We entered a shady, almost tropical thicket, then walked along a contour of a steep slope dense with raspberries, and lastly through an old, virgin forest. We met nobody on the way, and at certain junctures used a compass to get our bearings—and did manage to end up as planned at Vincenz's place when it was already quite dark. His cottage, a little wooden house whose only singular feature was a small room lined with Swiss pine—presumably because it was resistant to a certain fungus—was on the bank of the river Bystrec. We met his neighbors the Gołyczs, and Ołeksa Maksymiuk, a true Hutsul,[50] with a traditional knowledge of the timbers of the region and how to work them, and an excellent hunter of mountain deer—and it seems he wrote poetry into the bargain. Vincenz told us that when they had heard someone address him as "Doctor Vincenz", Ołeksa had taken him aside and told him that "people should always tell the truth when talking with a priest, lawyer, or doctor," and that he knew

---

[46]The pastime of skipping flat stones over the surface of a stretch of water so that they bounce as many times as possible.

[47]"Black Mountain". A mountain range in the Eastern Carpathians. Part of Ukraine since the end of World War II.

[48]Stanisław Vincenz (1888–1971), Polish novelist and essayist, lover and connoisseur of the Hutsul people of the Eastern Carpathians, and the thought and art of Ancient Greece.

[49]According to the Greek myth Achilles' mother Thetis, a nymph, holding her son Achilles by a heel, dipped him in the river Styx in order that the sacred waters render him invulnerable. However the heel that remained dry was left vulnerable.

[50]The Hutsuls (or Huculs) are an ethnic group of highlanders of mixed Rusyn and Wallachian origin, living in the Eastern Carpathians.

for a fact that there were doctors in the city who could prescribe drugs to restore vigor in a man even if, like him, they are over sixty. Vincenz had then started to explain that he was not a medical doctor, and that in any case "the state of natural celibacy that frees a man from a thousand troubles with women is surely a most desirable one." "Yes," Maksymiuk had answered, "You are right, *Pan* Doctor, but if a young woman were to make a polite request, then it would be impolite to refuse her!"

Vincenz later wrote a book about Czarnohora titled *Na wysokiej połoninie* (*On the High Uplands*),[51] in which he describes, in particular, how Gołycz's wife decided to divorce her husband and return to her home town Żabie on the gaily flowing Czeremosz.[52] It seems they had owned their house on the Bystrec jointly, so she hired woodcutters to saw it into exactly two halves, and had her half moved to Żabie, leaving the other half in place, gaping wide on one side. During our stay in Czarnohora I came to admire the way the little Hutsul ponies[53] moved so agilely. Once when out walking, I was watching a Hutsul pony cross a bridge over the Bystrec, when suddenly a plank gave way. Instead of striking out with its hooves in fright as an ordinary horse might, the Hutsul stopped dead and delicately extricated its leg from the gap. One can load a Hutsul with a quintal[54] or more and lead him over terrain covered with logs and stumps: he will climb up on a log and then balance there for a few instants with all four hooves occupying a space the size of a chessboard, before nimbly stepping down—a different species from other horses altogether, one might think. On a hike we took to Pop Iwan[55] I met an Austrian officer who owned property near Vienna, but who was so fond of the little Hutsul ponies that he came every few years to Czarnohora just to renew acquaintance with them—which was no longer so easy since now he would need a passport and a visa to enter Poland.

There were many bulls roaming through the Czarnohora, sometimes representing a real danger to tourists. Stefan Kaczmarz told me that he and two others were hiking in the Gorgany,[56] when they were forced to take cover behind some trees and wait for several hours before a bull would allow them to proceed.

Once, for a change, we went instead to Zakopane,[57] staying in a guest house called *Albion*. Together with my colleague Tarski[58] and a few other people we went

---

[51] English version published in London in 1955.

[52] A tributary of the Prut.

[53] A small breed of pony from the Carpathians, noted for their great endurance.

[54] A hundred kilograms.

[55] "Ivan the Priest", the third highest peak of the Czarnohora range.

[56] A mountain range in the outer Eastern Carpathians adjacent to the Czarnohora range.

[57] A town lying at the foot of the Tatra Mountains, a range in the Carpathians.

[58] Alfred Tarski (1901–1983), Polish logician and mathematician. A leading light of the Lwów-Warsaw school of logic. Held a professorship at the University of California at Berkeley from 1942. One of the most outstanding mathematical logicians of the twentieth century. Famous in particular for the "Banach–Tarski paradox".

on a most delightful hike. We first drove to the Biała Woda,[59] and from there walked across the Koperszady,[60] which separates the Bielskie Tatras from the High Tatras, where we encountered three bison escaped from the Hohenlohe Zoo.[61] They stood motionless a hundred steps from the path, their bodies perpendicular to us, and their big shaggy heads turned to gaze steadily at us like guards on alert. Thousands of cattle feed in the Koperszady, and the hiking paths are sometimes obstructed by herds of them.

The best part of the hike was the return trip through Rohatka.[62] The Staroleśna Valley is surrounded by black, granite peaks covered with lichen, and in the sunshine they glowed with a mixture of colors: emerald-green, violet, and lemon, under a sky of autumn-blue—summer was by then coming to an end. By this time we were in Czechoslovakia, and I took it on myself to boldly lead our group back to Poland without going through the formalities at the customs outpost. We pulled it off: neither gendarme nor customs officer was anywhere in evidence along the route we took. One of our party, a Mr. Wilder, later put this experience to use: having run up a great many debts in Kraków, he escaped to Czechoslovakia by this very route.

In the summer of 1927, when we were once again in Worochta, we awoke one morning to see that the Prut had become a lake. There had been a tremendous cloudburst in the mountains and a vast amount of water had poured down the mountain sides. In the past the dwarf mountain pines[63] would have prevented the worst, but some firm, doubtless employing bribery, had got around the prohibition against harvesting them, and set up a primitive distillery for processing the sap from the dwarf pines on the sides of Mariszewska.[64] As a result, a flood of water rushed unimpeded down the mountainside, turning it into a cascading mass of rocks, mud and water, sweeping everything before it; the distillery shed, installations, and boilers vanished without trace, ground up by the plunging torrent. Four workers were killed, but a woman with a baby who happened to have been some distance higher and away from the main force of the avalanche, survived. There were no trains to evacuate people since the flood had washed away the bridges and undermined the tracks. It was vital for me to get to Lwów since the first Polish Mathematical Congress was shortly to take place.[65] The road out of Worochta had been half washed away by the flood, so, a few days later, when the river had receded somewhat, we had to go on foot for a certain distance to the nearest functioning station. As we walked along we saw some curious things, such as a stretch of

---

[59]"White Water", a certain mountain stream.

[60]A branch of the Jaworowa Valley.

[61]A zoo in the then private park of Prince Hohenlohe in the Tatra Mountains.

[62]A pass in the Tatra Mountains. The eastern slopes of the pass descend to the Rohatka Valley, an upper branch of the Staroleśna Valley.

[63]Low, shrubby pine trees. A sort of sap from the buds and cones can be boiled down to a concentrate and combined with sugar to make "pine syrup."

[64]Or Maryszewska, a nearby mountain.

[65]It began on September 7, 1927.

essentially intact railroad embankment denuded of rails and ties; even the imprints left by the ties were still visible. The embankment was well above the level of the Prut. What had happened was that the river Żeniec, a nearby tributary of the Prut, had become blocked with logs and boulders to such an extent that its waters had overflowed onto the railroad track and, lifting off the rails and ties at one point, had deposited the end of a length of track in the Prut. And then the Prut, having thus got hold of a bit of track—the end of a ribbon, so to speak—proceeded to tug more of it off, until it had torn off a few hundred meters of the ladder-like strip, which it then wrapped around the pillars supporting the bridge a few kilometers lower down. Further along we saw that only half remained of the road cut in the rocky cliff above the Prut; after a few days of the raging Prut's smashing boulders into the cliff face, half of the road had caved in. The train we caught in Podleśniów had to cross the bridge over the Dniestr very slowly because the water was almost up to the bedding. Over the vast lake there circled various kinds of gulls, storks, common terns, and other waterfowl I didn't recognize—all attracted by the flood and the easy pickings it brought with it.

The congress in Lwów was a great success. After it was over it occurred to me to found a new international mathematics journal[66] based in Lwów, so I broached the idea to Banach, and we took the first steps towards its realization. There was an immediate objection from the Polish Mathematical Society (PTM)[67] in Warsaw, even a threat to take us to court. Motivating this opposition was the fear that the proposed journal would be reliant on financial support from the relevant ministry and thus cause the grant-in-aid to *Fundamenta* to be reduced. This might have been a valid objection if *Fundamenta* had accommodated the contemporary Polish production in all branches of mathematics. However, this was certainly not the case, since from the very beginning that journal had been devoted to set theory and neighboring fields, so I didn't take the objection very seriously. Eventually, we did indeed obtain financial support from the government, but it took some time for the first issue to come out.

In those years many foreign mathematicians visited Poland; mathematicians came to visit us in Lwów from France, Germany, Britain, America, and elsewhere, among them John von Neumann, Zermelo, Ayres, Whyburn, Ward,[68] and Lebesgue. Thanks to the influence wielded by Bartel, who, in particular, obtained for the PTM free transferable first-class train tickets, visits to Kraków and Warsaw by foreign mathematicians were much facilitated. He was also able to obtain grants towards

---

[66] The first issue of *Studia Mathematica*, with Banach and Steinhaus as editors, appeared in 1929.

[67] Founded in 1919. Stefan Banach, Franciszek Leja, Otto Nikodým, Stanisław Zaremba, and Kazimierz Żórawski were among its founding members.

[68] John von Neumann (1903–1957), Hungarian-American mathematician. Made major contributions to many fields. Regarded as one of the greatest of twentieth century mathematicians. Frank J. Ayres, Jr. (1901–1994), American mathematician. Gordon Thomas Whyburn (1904–1969), American mathematician. Morgan Ward (1901–1963), American mathematician.

the cost of a dozen or so Polish mathematicians' attendance at the International Mathematical Congress in Bologna in 1928.

In 1928 I went to France again, primarily to take the cure at Vichy. On the way I stopped over in Leipzig to see Lichtenstein, who was ever more preoccupied by questions of cosmology. He had laid out for himself a substantial program of research, and had made considerable progress with it. He was critical of what he called the onesidedness of Polish mathematics, yet it was just at this time—as was pointed out later in conversations with Zeltner[69]—that Polish mathematics was undergoing a wider, many-faceted development. He strongly recommended that if I went to Paris I get in touch with Meyerson,[70] whom he admired as a truly great philosopher, in contradistinction to Ernst Mach, whom he denigrated as a "trivialist." Meyerson had begun serious scientific work only when he was fifty, taking particular interest in questions of scientific methodology. Among his works were an interesting history of chemistry, the two-volume *De l'explication dans les sciences*, and *Identité et réalité*.[71] When I did finally visit him he was busy writing another brilliant opus, but was capable of working only two hours each day because—as he himself said—he had so many ailments the doctors called him a "perambulatory clinic". While I was in Paris, there was a celebration honoring the 50 years of scientific activity of Émile Picard, which I attended as representative of the Jan Kazimierz University of Lwów. I had been commissioned to find a theoretical physicist to fill a vacant position in Lwów, there being no one suitable in Poland, so I visited Paul Langevin[72] to get his advice. He suggested I approach Maurice de Broglie,[73] but in the event I didn't, and to this day do not know if de Broglie ever heard of the suggestion.

In Vichy I stayed at the *Hôtel de Genève*, by no means luxurious but very convenient and comfortable; a four-week stay, including spa treatment and the services of a doctor, cost me the equivalent of just 100 dollars—an important consideration although some might accuse me of stinting myself, perhaps. The lady who owned the hotel told me that the manager of the chain of large "Palace-Hotels", owned by a partnership called *Vichy-État*, was a certain Aletti, an Italian from Trieste. He had once received a telegram from an American in Paris, enquiring about the cost per day in Vichy of the best available suite, including a room for his

---

[69] This may be Hermann Zeltner (1903–1975), German philosopher.

[70] Émile Meyerson (1859–1933), Polish-born French chemist, epistemologist, and philosopher of science.

[71] *Explanation in the Sciences* and *Identity and Reality*.

[72] Paul Langevin (1872–1946), French physicist. One of the founders of the *Comité de vigilance des intellectuels antifascistes*, an antifascist organization created in the wake of the riots by supporters of the French far right in 1934.

[73] Louis-César-Victor-Maurice, 6ième duc de Broglie (1875–1960), French physicist. Not to be confused with his younger brother Louis-Victor-Pierre-Raymond (1892–1987), who introduced the idea of electron waves.

chauffeur, a garage, room service, medical expertise, hydropathic[74] procedures, tips, and so on. Aletti had an appropriate apartment for 200 francs a day, but decided to take advantage of the situation by adding to the cost substantial extras for the room, for the chauffeur, for serving breakfast in the suite, for the services of a doctor and a masseuse, for the use of a golf course, and so on, managing to raise the price in this way to 500 francs a day. However, since the dollar was then worth 25 francs, this brought the cost to a mere 20 dollars per day, and when this price was quoted to the American, he decided not to take up the offer on the grounds that it seemed too low.

Vichy was then a health resort frequented by the rich of America, France, and Britain, by Indian maharajas, and French missionaries returned from China and Indochina—all, it seemed, victims of overindulgence, with diseased livers, spleens, or gall bladders. In the center of the town, near the main pavilions where the cure was administered, there was a casino where the favorite game was *La Boule*,[75] in which the odds of losing are five to nine, worse for the gambler than roulette. Till midnight there was a line of limousines parked in front of the casino to take the players back to hotels not much further than 150 paces away. One of the missionaries commented that it is not so much what the Bolsheviks were doing in Moscow, but what the owners of these limousines were doing right here that provided the communists with their most effective propaganda.

There was a dam on the Allier River that was used to raise the height of a certain stretch of it to be used for regattas, beyond the river there was a splendid golf course, and there was also a race course operating in season, all for the benefit of the rich Americans and Britishers—not to mention the few thousand *cocottes* who swelled the population of Vichy in season.

Of course, the poor sufferers from malfunctioning liver or spleen were unable to partake of the excellent French fare on offer, sharing instead the fate of rich people the world over who pay not so much to eat as to fast. In my hotel, everyone had a card from a doctor with a description of his or her diet, or *régime*. Meals were served at precise times, and there was a single waiter to serve the 30 or 40 guests. Just prior to a meal, food items not needing to be served hot were arranged on side tables, as were, of course, jugs of water, carafes of wine, glasses, flasks of vinegar and oil, and cutlery. Every few minutes the waiter would go up to the little opening in the wall of the dining room and call out *"Trois entremets*[76] A!"*, meaning three portions of diet A, and the required dishes would quickly appear, brought up to the dining room by means of a lift. After dinner I would usually go and sit on a bench in front of the hotel to watch the passers-by. Once the hotel owner pointed out to me a hefty workingman pushing a two-wheeled handcart laden with vegetables, saying that he was a former Russian major who had escaped from the Bolsheviks.

---

[74]Hydrotherapy, formerly called "hydropathy" or "water cure", involves the use of water for pain relief and treatment of illness.

[75]A game similar to roulette, but simpler and faster.

[76]An *entremets* is literally a dish between main courses.

The only things worth looking at in the center of the city were the displays in the many store windows, especially those of the jewellers, in one of which I saw huge chunks of rock from which there seemed to have grown veritable colonies of crystals of amethyst or pink quartz. These were the so-called "stones of Auvergne". It was unpleasant walking in the center of the city since the automobile traffic was dense, the roads narrow, and the sidewalks often non-existent, so that often one had to step into the gutter to avoid being hit by a fender. However, the roads, mostly cobbled, were well maintained.

Of course, Vichy was famous for its few dozen hot springs of various temperatures and mineral content, some of which spouted geysers now and then. The waters from some of these springs were channelled into baths furnished with hydropathic[77] installations, including a machine for performing underwater massages, which I tried out a few times. It is curious how much of medicine seems to depend on fashion. Dr. Raymond, the doctor assigned to me at the spa, put questions to me and prescribed a diet altogether different from those of our doctors at home. I doubt that our doctors had even heard of the laxative—magnesium sulphate—that Dr. Raymond prescribed. But the cure was excellent, in addition to which, being without much company there, I was able to write up the work that was later to appear in Volume I of *Studia Mathematica*.

On the return trip I travelled from Paris to Trier[78] in a passenger train carrying French soldiers returning to the Rhineland at the end of their leaves. Their duty there was to enforce the occupation.[79] Since the main purpose of this particular train was to transport these soldiers back to the current political hotspot, it ran express to Trier, so there were not many civilians aboard. Furthermore, since my date of departure had not been noted and there was no passport control at the border, I was able to spend two weeks in Germany instead of the permitted three days. Naturally enough, I did not register my presence anywhere in Germany, and, after examining the *Porta Nigra*,[80] I travelled on to Bonn, where I visited Toeplitz and Hausdorff.[81] Then I went to Göttingen, where I looked up Courant, and, at a reception at his house, met a number of members of a new generation of European mathematicians: Hans Lewy, Otto Neugebauer, Emmy Noether, Willy Feller, the brilliant Hungarian mathematician Gábor Szegő,[82] and others. When Karl Kraus's name cropped up at

---

[77] That is, hydrotherapeutic.

[78] German city on the Moselle.

[79] That is, the occupation of Alsace-Lorraine by France in accordance with the Treaty of Versailles.

[80] "Black Gate", a large Roman city gate in Trier, so named in the Middle Ages because of its darkened sandstone.

[81] Felix Hausdorff (1868–1942), German-Jewish mathematician. Considered one of the founders of modern topology. Committed suicide with his wife to avoid being sent to a concentration camp.

[82] Hans Lewy (1904–1988), German-American mathematician. Known chiefly for his work in partial differential equations. Otto Eduard Neugebauer (1899–1990), Austrian-American mathematician and historian of science. By interpreting clay tablets of the ancient Babylonians, revealed that they knew much more mathematics and astronomy that had been previously thought. Amalie

this gathering, of those present only Willy Feller endorsed my enthusiasm for him—a situation that was to repeat itself, in somewhat altered form, some nine years later. On the topic of the imminent International Congress of Mathematicians in Bologna, Courant expressed the opinion that Germans should attend it, an opinion shared by Hilbert and Landau.[83]

* * *

Slowly but surely Lwów was changing. The grassy meadow on the east side of Kadecka Street, opposite our house, where one used to be able to lie down and doze and where the occasional hare could be seen, was no more, since new houses were being built there. Burglaries were becoming more and more common. Insurance for household items was as high as in Mexico. The garden behind the house was level with our second-floor apartment, and burglars and vandals used it as a thoroughfare. One morning my wife was dismayed to see that all the roses—of which there had been dozens the evening before—had been plucked from the rosebushes. This stimulated me to invent an electric alarm system to forestall future invasions of our property, which I had made for me by the caretaker's son, an electrician. It consisted of four wires strung one above the other on insulators and running horizontally around the garden at levels of 30, 60, 90, and 120 centimeters above the ground. At intervals short wires ending in hooks were attached to the top three wires, cut to lengths a little less than 30 centimeters in order that they not quite reach the wire below that from which they hung. The idea was that if in the dark an intruder were to brush against the wires or push them down, then one or more of the hooks would catch onto the wire below, causing an alarm to sound and an electric bulb to go on and light up the garden—and what's more giving the interloper a decent shock, reduced to 60 volts since the light drew off some of the power. Sometimes I even had a miniature cannon rigged up to fire a warning shot. These acoustical, optical, and tactile surprises sufficed, as later experience demonstrated, to repel a great number of trespassers—all the more so since often when the alarm sounded I would go out onto the porch and fire a revolver at a mound of earth.

In the autumn of 1928 we went to the ICM in Bologna. Others from Lwów attending were Antoni and Zbigniew Łomnicki, Stożek, Ruziewicz, Żyliński, Banach, Chwistek, Kaczmarz, Nikliborc, and Schauder.[84] There were also present Hilbert, Weyl, Kellogg, Landau, Hadamard, Lichtenstein, de la Vallée-Poussin, Serge Bernstein, Khinchin, Mirushov, Haar, and Dénes Kőnig,[85] among many other

---

Emmy Noether (1882–1935), German-American mathematician. Made ground-breaking contributions to abstract algebra and theoretical physics. Considered the greatest female mathematician of all time. Gábor Szegő (1895–1985), Hungarian-American mathematician. Moved to the US in the mid-1930s.

[83]The reluctance of some German mathematicians to attend the ICM in Bologna in 1928 had to do with a quarrel going back to World War I, for more details on which, see below.

[84]For bios of these delegates, see earlier footnotes.

[85]Hermann Klaus Hugo Weyl (1885–1955), all-round German mathematician and theoretical physicist. Early member of the Institute for Advanced Study in Princeton. Oliver Dimon Kellogg

well known mathematical personages. Notably absent was Vito Volterra,[86] who had moved to Paris following the fascist coup in Italy. Some German mathematicians were boycotting the congress in order to register their protest at the ten-year interval following the Treaty of Versailles during which the Germans had essentially been snubbed by the international mathematical community. They had evidently forgotten about the "Manifesto of the Ninety-Three", in which 93 prominent German scientists, scholars, and artists declared their unequivocal support of German military actions during the early part of the war. Later Planck[87] defended himself by saying that he had signed the Manifesto without knowing its contents.[88]

Bologna is a town in a class all by itself. Many of the older buildings are of brick, but the seams between the bricks are worn so smooth that it is as if they were formed from a single block of material. Many of the streets are lined with porticos or arcades, some of great length: the San Luca arcade is some 4 km long. In the middle of the town stands a very beautiful mediaeval Town Hall, with a "victory bulletin" carved in its facade giving fantastic numbers of Austrian prisoners of war and field artillery supposedly captured in 1918 at the Battle of Vittorio Veneto.[89] What had happened in fact was that Count Károlyi[90] had called his troops home, the Austrian army had followed suit, and, abandoning their weapons, they had all set off over the Alps. Not satisfied with the exodus, however, small Italian units in army vehicles pushed their way through the disorganized throng of departing soldiers—most of whom were now in any case without arms—and, setting up roadblocks at certain points, took as many of them prisoner as they could—quite unnecessarily, since the Italian general staff was already in Vienna.

At the foot of Bologna's spectacular double Asinelli Tower was a smallish open area where we spent a great deal of time standing around little stone tables. The towers both lean, and, according to a fifteenth century etching of Bologna, it used to boast dozens of leaning towers. It is interesting that no one today knows—we asked

---

(1878–1932), American mathematician. Obtained his doctorate in Göttingen under Hilbert. Charles-Jean Étienne Gustave Nicolas Levieux, Baron de la Vallée-Poussin (1866–1962), Belgian mathematician. Best known for proving, at the same time as Hadamard, the Prime Number Theorem. Alfréd Haar (1885–1933), Hungarian mathematician. His doctorate was supervised by Hilbert in Göttingen. Haar measure, Haar wavelets, and the Haar transform are all due to him. Dénes Kőnig (1884–1944), Jewish-Hungarian mathematician. Wrote the first textbook on graph theory.

[86] Italian mathematician and physicist. Well known for his approach to solving integral equations. Lived from 1860 to 1940.

[87] Max Planck (1858–1947), German physicist. Founder of quantum theory. Awarded the Nobel Prize in 1918.

[88] In 1915, Planck publicly voiced his disagreement with parts of the Manifesto, and in 1916 signed a declaration against German annexationism.

[89] A battle marking the end of the war on the Italian front, and the dissolution of the Austro-Hungarian Empire.

[90] Mihály, Count Károlyi (1875–1955), Hungarian statesman. Appointed Hungarian prime minister on October 31, 1918, by Emperor Charles I of Austria, in recognition of Hungary as a separate state.

7   In the University Town Lwów   155

the Bolognese professors about this—whether they were purposely built that way, as in Pisa,[91] or if they leaned subsequently to being built.

Galvani and Volta[92] were sons of Bologna, but there was nonetheless a philistine air about the place—perhaps connected with its celebrated baloney. Northern Italian cuisine differs from the southern variety; in Bologna the main meal seemed to be broiled chicken served with a salad of tomatoes and peppers.

Paintings from the Bolognese school were on exhibition in the art gallery, but I saw no masterpieces there. The complex of religious buildings around the *Piazza Santo Stefano* was the most interesting part of Bologna architecturally speaking; it consists of seven churches and sanctuaries arranged attractively around the large open space of the square. According to tradition it was built in the fifth century on top of a temple to the goddess Isis, now below street level.

Bologna is the capital of the region of Italy called Emilia-Romagna. On the whole it is beautiful—as indeed are almost all Italian cities. There was a splendid new swimming pool surrounded by a stone amphitheater, built by Mussolini's fascists.[93] We had run into Blackshirts earlier when we went through passport control on entering Italy. They were armed with revolvers, and their attitude to us foreigners had been a sort of primitive suspiciousness, despite our having been invited by the Italian government to a congress held under the distinguished patronage of King Victor Emmanuel III and His Excellency Benito Mussolini. They were gruff and impolite—like the gendarmerie everywhere, only more so. In the town they hung about the market place making sure that the trolleys departed on time. It was explained to us that before the fascists came to power, the trolley drivers often used to smoke and chat at the terminus for 20 minutes or more beyond their departure time, and these delays had been impossible to eliminate. It was only the fascists who were able to "introduce order" into public transport in Italy—by means of clubs.

We went on an excursion through Emilia-Romagna with the Knasters,[94] and I formed a clear impression of the countryside, which consisted largely of quadrangular fields separated by ditches lined with trees festooned with grapevine, and extending in all directions. The fields themselves were like luxuriant grassy market gardens of pumpkins, melons, tomatoes, and all sorts of other vegetables, catering to one of the most densely populated regions of Europe.

At the congress itself there were a few issues needing to be settled. For example, a certain Mr. Grużewski had proposed to lecture on a result of his to the effect that there are no two-sided surfaces having the trefoil knot as boundary, and a group

---

[91] In fact, the Tower of Pisa was intended to stand vertically, but began to lean in the early stages of its construction in 1173 because of a poorly laid foundation and loose substrate.

[92] Luigi Galvani (1737–1798), Italian physician and physicist. Pioneer in bioelectricity. Count Alessandro Giuseppe Antonio Anastasio Volta (1745–1827), Italian physicist. Developed the first electric cell in 1800.

[93] Benito Mussolini had become Prime Minister of Italy in 1922 by means of a *coup d'état* by his National Fascist Party.

[94] Bronisław Knaster (1893–1980), Polish mathematician. Worked in point-set topology.

of us descended on him at his hotel demanding the details of his proof. In fact, a year later it was shown by Frankl and Pontrjagin that there is such a surface.[95] We also managed to settle positively issues pertaining to our projected journal *Studia Mathematica*, and we entered into discussions concerning the publication of a series of monographs under the general heading *Mathematical Monographs*.[96]

As part of the lighter side of the activities, the attendees and their families were taken by bus on an excursion to Ravenna, where we saw the tomb of Theodoric[97] and also of Galla Placidia,[98] which was even more impressive: there the light penetrated underground through a thin covering of pink marble, and the Byzantine mosaic immediately transported one to the strange world of the early Middle Ages. The San Vitale Basilica and the other churches of Ravenna were for me like memorials to the struggle of various trends and cultures for the soul of Europe—a game whose outcome was determined only a thousand years later. Ravenna had originally been built on piles on a series of small islands in a marshy lagoon with direct access to the Adriatic, but today it is some miles from the sea.

For lunch we were then bused to a *pineta*, a pine grove, where the trees were sparse enough to be counted. In Italy every such grove has a special name and is considered a place of pilgrimage for tourists—probably because the Romans denuded northern Italy of trees for their buildings and galleys. Among the umbrella pines tables had been set up for the few hundred of us guests. I happened to sit next to a Mr. C. Karpinski,[99] an American geometer, and I asked him if he knew anything about the author of *Laura and Philo*.[100] He said he was related—a fact of interest, perhaps, to literary historians which would seem not to be well known. It is another little known literary fact that the character Selim Mirza of the eponymous

---

[95] See: F. Frankl und L. Pontrjagin, "Ein Knotensatz mit Anwendung auf die Dimensionstheorie" (A Theorem on Knots with an Application to Dimension Theory), *Mathematische Annalen* 102 (1929), pp. 785–789. Here it is shown that in fact every simple closed polygon in three-dimensional space is the boundary of some orientable surface. Such surfaces were later called "spanning surfaces" for the knot represented by the polygon.

[96] In Polish: *Monografie Matematyczne*. This was intended to be a mathematical series of high-quality research monographs in areas of mathematics where Polish mathematicians were prominent. Prior to 1939 ten volumes in the series appeared, authored by such notable mathematicians as Banach, Saks, Sierpiński, Kaczmarz, Steinhaus, etc.

[97] Flavius Theodoricus or Theodoric the Great, king of the Ostrogoths (471–526). Ruler of Italy from 493 to 526.

[98] Aelia Galla Placidia (392–450), daughter of the Roman Emperor Theodosius I, and consort of Ataulf, king of the Goths, and then of Constantius III, Western Roman Emperor.

[99] This may be Louis Charles Karpinsky (1878–1956), American historian of mathematics and cartography.

[100] Franciszek Karpiński (1741–1825), the leading sentimental Polish poet of the Enlightenment. *Laura and Philo* was one of his most popular love poems.

novel by Sienkiewicz[101] was modelled on Bruno Abakanowicz,[102] whose integraph, manufactured by the firm *Coradi* in Zürich, is known only to mathematicians interested in the more practical side of their subject. Perhaps I might also mention the following somewhat obscure fact: the Lobachevsky family, to which the Russian geometric genius belonged, has a Polish branch, some of whose members could be found at the time of writing in Małopolska. And the contemporary poet Antoni Słonimski[103] has two mathematical ancestors whose names appear in histories of computing mechanisms. One of these was Abraham Stern,[104] who was elected to the Warsaw Society of the Friends of Science, and the other is the namesake of our contemporary man of letters.

But to return to the ICM: its last gasp, so to speak, took place in Florence, whither we were transported via the new electric railroad cutting across the Apennines. We stayed on the *Via dei Calzaiuoli*[105] and had plenty of time to explore Florence and visit some of its galleries since there remained only the final plenary session of the congress. The most anticipated talk of the session was that of Birkhoff,[106] the great American topologist, who completed the proof of Poincaré's last theorem. His lecture was on the mathematics of aesthetics: using pictures of Greek and Etruscan vases, he attempted to derive mathematical criteria for the beauty of a curve. After 15 minutes of listening to his arguments, a Frenchman sitting next to me muttered: "*C'est d'une puérilité déconcertante.*[107]"

But Americans are not afraid of running the risk of attracting ridicule. They organize beauty competitions and reward the young woman whose bodily proportions come closest to those of the *Venus de Milo*, they are prepared to weigh a dying man to calculate the weight of his departing soul, and they run for shelter when the radio reports that the Martians are invading Earth.[108]

---

[101] Sienkiewicz's novel *Selim Mirza* appeared in 1876.

[102] Bruno Abdank-Abakanowicz (1852–1900), a Lithuanian-born mathematician, inventor, and electrical engineer. Among his many inventions was the "integraph", which, given the graph of a function, draws the graph of its integral.

[103] Antoni Słonimski (1895–1976), Polish poet, translator, and journalist. Known for his devotion to pacifism and social justice. Converted from Judaism to Roman Catholicism. One of his great grandfathers was Abraham Stern (Sztern).

[104] Polish-Jewish scholar and inventor. One of his inventions was a computing machine, perfected in 1817. Lived from 1769 to 1842.

[105] The most centrally located and one of the most elegant streets of Florence.

[106] George David Birkhoff (1884–1944). Best known for his Ergodic Theorem. Made significant contributions also to number theory, graph theory, and theoretical physics. The title of his talk at the ICM in 1928 was "Quelques éléments mathématiques de l'art". His paper completing the proof left unfinished by Poincaré was entitled "Proof of Poincaré's geometric problem" (published in *Trans. Amer. Math. Soc.*).

[107] "It's a piece of disconcerting childishness."

[108] On October 30, 1938, Orson Welles broadcast an adaptation of H. G. Wells' *The War of the Worlds* in such a way that it sounded like a news bulletin about an invasion in progress. The hoax caused general panic in the US.

Florence is celebrated for its geographical location—but not just because it is situated on the Arno, with its beautiful stone *Ponte Vecchio*,[109] on which clustered little stalls stuck to the pillars like swallows' nests, where jewellers sold rings fashioned in the manner of Cellini,[110] complete with a secret compartment under the emerald for poison. A stanza from *La Divina Commedia* carved in the stone of the central column reminded us that Dante was as moved as we were at how the stone arch of the bridge soars over the eternally flowing stream—and young Englishmen, apparently oblivious of the softly glowing Tuscan evening, strained at the oars of a rowing eight as if the river were the Thames and not the Arno.

No, that is not the whole explanation for Florence's reputation, even if one includes the *Palazzo Vecchio*,[111] the vast *Palazzo Pitti*, the splendid specimens of Greek sculpture from the time of Praxiteles—such as the Uffizi Wrestlers, more beautiful even that the Medici Venus—and the gardens south of the Arno. No, it is above all what one sees when one gazes from Fiesole[112] towards the Florentine hills, and what Goethe found so moving: the mildness and sweetness of a landscape smoothed down by 3000 years of culture. There is not a stone there that has not been worn smooth by being continually touched by human hands.

Italian women are not especially beautiful, generally speaking, but opposite our hotel there was a tobacco store in which sat a young woman with an ideally oval face, with a complexion of matte gold, and with black almond-shaped eyes. She gazed steadily—it seemed for hours—at the storekeeper, her *fiancé*, who appeared to take this homage for granted.

Some of the younger Polish participants decided to make a side trip to Venice on the way home from Florence. They later told us the following story: they had met a young Venetian girl in the street who offered to act as guide, which greatly pleased the young Poles. When it came time for them to depart, they asked her how they might recompense her. She said she wanted no money, but that a pair of stockings would be appropriate. As time was running out, they suggested they give her the money for her to buy the stockings herself, and, since she didn't know the exact price, they gave her 20 *lire*. Two weeks later, in Lwów, Zbigniew Łomnicki received a letter from the girl with money enclosed, the excess of 20 *lire* over the cost of the stockings.

We also went to Venice, spending a whole week at the Lido on the *Via Quattro Fontane*. Since our hotel was closer to the lagoon than to the sea, the price for a large room was reasonable, and furthermore the more comfortable in that in Venice proper one can't sleep for the mosquitoes. We rode up the elevator in the *Campanile di San Marco* to look at the view over the lagoon at sunset, one of rare beauty.

---

[109] Old Bridge.

[110] Benvenuto Cellini (1500–1571), Italian goldsmith, sculptor, painter, musician, and soldier.

[111] The Romanesque, crenellated fortress palace dating from the early fourth century and now functioning as Florence's Town Hall.

[112] A small town in the mountains just to the northeast of Florence.

However, our real aim was Verona. On the train taking us there we met a garrulous Italian businessman, who, when he learned that we had been at a mathematics conference, asserted baldly that he regarded such conferences as humbug: mathematics being a finished subject, the sole purpose of such meetings could only be to travel, eat, drink, and be merry at government expense. Although in our compartment there was also a chemistry student with a more informed opinion, I could not help feeling how one simply cannot rely on the friendly attitude of the masses for the promotion of learning—in particular, of science. Simple-minded people generally believe learning necessary because illiteracy, and perhaps lack of knowledge of foreign languages and of the law, can cause difficulties for them. The slightly more educated regard the study of Latin as unnecessary because it's a "dead language", and that all of mathematics has been done, so there's an end to it. More examples: A local councillor and director of a private Gymnasium in Lwów by the name of Petelenz once declared to me that it was a waste of effort teaching Polish children songs, poetry, and literature, and that they should instead be taught a useful trade and have imparted to them the basic information needed in daily life. When I once took the liberty of telling an engineer called Lauterbach, a man a little older than I, that the journal *Krytyka*[113] to which our club subscribed contained an interesting article, he all but jumped down my throat, declaring that "if more people in Poland did a proper job of work, things would be better." He evidently believed that in other countries—say Czechoslovakia, Denmark, Britain, and Holland—everyone "works", rather than engaging in such "non-work" as publishing articles in literary journals.

Mr. Henry Ford[114] should have been crowned king of all these utilitarians, since for him not only Latin, but also history, mathematics, and philosophy were dispensable.

However, if I'd thought to ask him, the Italian businessman would probably have admitted astronomy as a useful pursuit, since he would doubtless have considered weather forecasts as useful, and these depend largely on astronomical techniques and observations—or he may simply have felt something of the awe inspired by the tremendous distances between the stars as revealed by astronomy. But perhaps not: I have known engineers who surely knew much more about astronomy than this businessman, yet still regarded it as a luxury science. All this muddle in the minds of the half-educated, half-intelligent stratum of society derives from the simple fact that *learning—serious learning—is not for everyone*. For the great majority of people the scientific attitude, the scientific method, and the accompanying wonderment, remain inaccessible. The same goes for poetry: I knew people who regarded poetry

---

[113]"Criticism"

[114]American automobile industrialist. Quoted in the *Chicago Tribune* of May 25, 1916 as saying: "History is more or less bunk. It's tradition. We don't want tradition. We want to live in the present and the only history worth a tinker's damn is the history we made today." Also said reading muddled his brain. Between 1919 and 1927 published *The Dearborn Independent*, a weekly newspaper promoting his anti-Semitic views.

as a school exercise for teaching young people good style—whereas I would have thought the aim was a completely different, much more immediate one.

Frigyes Riesz[115] had recommended a hotel in Verona to us, but it proved beyond our resources, so we had to shorten our stay. It was old-fashioned, with antiques by way of furnishings, and the guests looked like members of the old nobility come to dwell in their traditional cavernous, poorly lit rooms, as in an ancient castle.

Verona has two squares of glorious perfection: the *Piazza delle Erbe* and the *Piazza della Signoria*. Around the church of Santa Maria Antica stand the tombs of the Scaligeri,[116] Gothic shrines enclosing sarcophagi in the form of knights sculpted out of white marble, smiling with the smile of the northern conqueror. Two of them bore the names Canis I and Canis II (*canis* meaning "dog"). These tombs stood under marble canopies and seemed to be almost perfectly preserved. The fortified castle, the *Castelvecchio*, was still surrounded by a moat, and nearby there was a recently rebuilt Roman amphitheater from an earlier era. There was also a beautiful old church, named for St. Zeno, a black bishop who, according to tradition, introduced Christianity to Verona in the fourth century. Some of Verona's streets reminded us of Kraków, and the spacious squares and barracks-like buildings were reminiscent of Vienna.

From Verona we travelled on to Desenzano on Lake Garda, where we took a ferry to Gardone Riviera,[117] for a stay of two days. We found Lake Garda a rather melancholy place, and in fact one of its islands was said to be the model for Böcklin's[118] *Isle of the Dead*. Along the shores of the lake there grew palms and agaves. We continued on by means of a narrow-gauge railroad along the northern shore of the lake, passing through Riva del Garda and Arco to Merano; the vineyards were so thick with vine and so close to the track that the grapes almost burst against the windows of our carriage. We saw, cut out of the rock, a steep road from Riva del Garda up along the cliff to the extraordinary *Tagliata del Ponale*, an elaborate Austro-Hungarian fort high up above the lake—the scene of much carnage during World War I, now in ruins—and continuing beyond to the Val di Ledro. We were heading for the Brenner Pass,[119] but on the way spent a night at the inn *Zur Post* in the (formerly Austrian) South Tyrol, which surely hadn't changed much since the time Goethe spent a night there. In the lobby Italian and Austrian customs officers played cards, and the inn was hemmed in all around by a towering spruce forest.

---

[115]Hungarian mathematician. Made fundamental contributions to functional analysis. Lived from 1880 to 1956.

[116]The noble family ruling Verona during the thirteenth and fourteenth centuries.

[117]A lakeside resort in Lombardy, with hotels, a promenade, impressive Botanical Gardens, and *Vittoriale degli Italiani*, a former residence of the poet Gabriele d'Annunzio.

[118]Arnold Böcklin (1827–1901), Swiss symbolist painter. Best known for his five versions of *Isle of the Dead*.

[119]A pass through the Alps between Italy and Austria, one of only a few, and for this reason much fought over.

In Innsbruck we again interrupted our journey homewards. Seen from the railroad high above it, the Inn Valley is like a uniform green carpet, with little villages clustering here and there, and the gleaming river cutting through it all, giving the impression of a land of idyllic happiness, paradise on Earth. In the town, when we went to buy an aluminum lunch-box, there in the store stood the storekeeper in leather shorts revealing his bare knees, and with a little Tyrolean hat perched on his head with a huge feather stuck in the hat-band. We had to pay quickly and run outside to avoid offending him by bursting out laughing.

Since our train to Salzburg was leaving late in the evening, we decided to join an excursion by cable railway to view the Alps in comfort—proving, however, merely a waste of money. Although, it is true, ascending through a thousand meters in a quarter of an hour affords a certain amusement, whoever thinks a ride along a cable railway will enable him to commune with the mountains is mistaken. At the stations along the cable railway there were no "mountains" as one would like to have them before one: there we saw mainly empty sardine cans, discarded newspapers, and advertising billboards, and smelled bad coffee—and in saying this, I am not agitating for the preservation of unblemished "virgin" nature, as the romantic phraseology has it. No, to experience the pleasure of being in the mountains one needs to have endured an arduous climb along narrow paths, and the slow erasing from memory of the down-to-earth atmosphere of life in the valleys, with the stuffiness of the towns, the noise of the crowds, and the densely concentrated masses of people obtruding on one's well-being—all of which the cable railway seemed to drag along with it from beginning to end.

In Salzburg, a beautiful town in the style of Austrian baroque, we visited a museum dedicated to "the fatherland and the military," where the cast-off trousers of Austrian generals were solicitously preserved, and on large wall maps the sites where Austrian arms had won glory were indicated. It is somewhat surprising that, despite all these glorious victories, the borders of Austria should finally have shrunk right down to those of the little "Republic of Austria" as defined in Saint-Germain-en-Laye.[120]

In Vienna we were met by my wife's sister and her husband, who took us to Bratislava,[121] where he was director of the oil refinery *Apollo*. Bratislava retained the atmosphere of a pre-war Austrian city—albeit with a hint of provincialism—with elegant cafés, its old castle, and period customs that had ended everywhere else in 1918, but here were cultivated assiduously by a few counts from the Ballhausplatz,[122] as it were, by a handful of ladies formerly belonging to the court,

---

[120] The Treaty of Saint-Germain-en-Laye was signed in 1919 between the Allies of World War I and the new Republic of Austria. The treaty defined the terms of the dissolution of the Austrian Empire—in particular the considerably constricted new Austrian borders—and specified war reparations to be paid to the Allies.

[121] Now capital of Slovakia. Situated very close to Vienna.

[122] A square in central Vienna on which the official residence of the former Austrian State Chancellor, today the Austrian prime minister, is situated.

still stubbornly travelling about in carriages reminiscent of the pre-war elegance of the traffic in the Prater,[123] by rich Hungarians who preferred Austrian company to the Czech sort—and, last but not least, by the Polish colony. The members of the latter group, encouraged by the Polish consul, saw fit to belittle the Czechs when in the company of Slovaks. The political goal behind this behavior was the fracturing of Czechoslovakia,[124] to be attained by flattering the Slovaks at the expense of the Czechs, thus undermining the cohesion of that newly formed country. Slovakia was considered by the Polish powers-that-be as a would-be nation deserving Poland's friendship; actually they had their collective eye on a disputed region of northern Slovakia. Just how misguidedly self-serving this policy was, became clear in March 1939 when the Slovaks did indeed form an independent state,[125] serving as a base from which to launch an attack on Poland, the land of "the wise in hindsight".

What was most beautiful in Bratislava was the Danube, which is very wide there. Since 1918 the Danube had been under international control,[126] so the Czechs and Slovaks had only very partial jurisdiction over it. A certain Frenchman attempted to exploit this somewhat anomalous situation by setting up a café on a lighter moored to the bank, reasoning that he would not be liable for taxes since the river was not within the fiscal purview of Czechoslovakia. However, his project came to grief when the municipality refused to supply his establishment with electricity.

In Bratislava, signs were all in triplicate: first Slovak, then German, and last Hungarian—but not Czech, even though Czechoslovakia was officially bilingual. A professor in Prague might deliver his lectures in Slovak without needing to obtain permission to do so. In those days, probably the only notable Slovak living in Bratislava known outside the country was Stodola.[127]

On my return to Lwów I was swamped with work. In particular, our journal *Studia Mathematica* required a great deal of preparatory editorial work. For one thing, there were then no typesetters in Lwów capable of composing a mathematical text up to internationally accepted standards. Such established conventions as having algebraic symbols italicized, and standardized abbreviations—such as log, sin, lim—in Roman font would have to be drummed into our typesetter. And then I had to decide on the style of type, of the headings of each section or chapter, the form of running heads, the format of page references, and of references to cited works. I brought much pedantry to these tasks: I wanted authors' names to be in

---

[123] A large public park in Vienna's Leopoldstadt.

[124] A new state formed in 1918 from regions of the defunct Austrian Empire.

[125] A few days later Hitler declared the Czech part (Bohemia and Moravia) of Czechoslovakia a German protectorate, and Slovakia became a puppet state of Germany, taking part in the Polish campaign of September 1939 as allies of Germany.

[126] In 1918 the International Danube Commission had been set up, replacing the earlier European Commission. The parties to the commission were the riparian states, dominated by Germany.

[127] Aurel Boleslav Stodola (1859–1942), ethnic Slovak and pioneer in the area of thermodynamical technology and its applications. His book *Steam and Gas Turbines*, first published in English in 1927, became a basic reference in the field.

standard form, that is, I did not want some to appear with initials and others with first names in full, and I wanted a single form for place names—for example, not "Warszawa" in one place, and "Varsovie" in another, French, context.

Articles were acceptable for publication in *Studia* in any of four languages: French, German, English, and Italian. In the initial stages at least, most papers were in French, somewhat fewer in German, a few in English, and only very sporadically did we get a paper in Italian. As might be expected, there were recurring difficulties with authors over corrections because some were too tardy in making them while others were too superficial. There was also a lot of work involved in organizing and carrying out the scheme whereby *Studia* was exchanged for other Polish and foreign periodicals. We ended up exchanging *Studia* with over a hundred and thirty other journals—which naturally then had to be sorted and catalogued. It was considered inappropriate to have assistants[128] do such work. However, Ludwik Sternbach[129] willingly took it on, and did it conscientiously and intelligently, and, like everyone on the editorial staff, without pay. He also kept the books, that is, the record of expenditures, and each year submitted to the ministry a statement of account attached to a request for further funding. Initially we had been funded by the National Culture Fund and then by the Ministry of Science and Higher Education. When the great depression began to affect Poland, in 1930, Professor Bartel managed to wangle a subsidy from the oil company *Małopolska*. There was a certain amount of revenue from sales abroad: we received checks in dollars, Swiss francs, German marks, and what have you. Out of an edition of 500 copies, about 200 were used to fulfill our exchange agreements, 200 sold abroad, and 100 at home, mainly to bookstores. The booksellers often made things difficult for us since many of them seemed to be of the general opinion that a publisher should be happy to supply copies of a book and leave the matter of when or even whether payment would be made entirely to the bookseller. Some booksellers, having exhausted the patience of a publishing firm, and faced with a customer wanting a book put out by that firm, would turn to another source to obtain the book. Thus we were often asked if we could supply issues of *Fundamenta Mathematicae*, but Bronisław Knaster was on to their game and advised us to demand payment in advance.

Some of the Varsovian editors of *Fundamenta Mathematicae* demanded a commitment from us not to publish papers on set theory,[130] but this was largely unnecessary since in Lwów the main mathematical interests were linear operators, probability, and other not especially set-theoretical areas. We were thus very surprised when one of those making the request then sent us a paper on set theory—which we eventually had to reject.

Mathematics was developing fast in Lwów. The department already had four professors and three docents, and at the Polytechnic Professor Bartel was ardently

---

[128]That is, lecturers.

[129]Ludwik Sternbach (1905–1942), Polish mathematician. Studied at the Jan Kazimierz University of Lwów. Worked as Banach's *Assistent* and as a Gymnasium teacher. Killed by the Nazis in 1942.

[130]Since *Fundamenta* was dedicated to publishing papers in that area.

pursuing his ambition of having the theoretical sciences taught there. As already mentioned, by using his influence with the government he had been able to found a General Studies Faculty at the Polytechnic, where mathematics was taught by such stars as Kuratowski, from Warsaw. Antoni Łomnicki and Włodzimierz Stożek—whose mathematics chairs were located in the departments of engineering and mechanics—also taught in the new faculty. Bartel himself taught descriptive geometry, and Rubinowicz[131] theoretical physics. In addition to all that, Banach and Ruziewicz sometimes gave guest lectures for the students there.

It was not clear what graduates of the General Studies Faculty would pursue by way of a career. There was a proposal that they teach mathematics in technical high schools, but these were still at the planning stage. However, a good education is always a good thing, independently of its advantages for getting a livelihood. And some graduates of that faculty went on to distinguish themselves: for example, Jan Blaton, who became director of the State Department of Meteorology, and Stanisław Ulam, who eventually taught at Harvard and participated in the Manhattan Project at Los Alamos. Soon, however, in the aftermath of the collapse of the stock market in America, retrenchments had to be made, curtailing the program somewhat; in particular, Kuratowski returned to Warsaw and Rubinowicz transferred to the university.

Our Mathematical Society developed and flourished, sometimes in rather unexpected ways. We met once a week, and sometimes as many as three papers were read at a meeting. We were an active group, and the number of new results of some significance obtained in a year by its members amounted to several dozen. We also regularly got together informally in a café called *The Scottish Café*, which thus became the scene of passionate mathematical discussions. These meetings had been going on for some time when Banach's wife hit on a clever and original idea: she bought a thick, sturdily bound exercise book, to be kept in the café, in which anyone so wishing might record a question, problem, solution, or what not, appending their name, signature, and the date of the entry. This book was called *Książka Szkocka* (*The Scottish Book*),[132] and was eventually to have an impact on mathematics far beyond the borders of Poland. It soon contained a few hundred problems, next to many of which a prize for the solution was indicated, to be awarded by the poser, ranging from a glass of beer or black coffee to a whole meal or a live goose. Some offered such extravagant rewards as *fondue à la crème* to be served in Geneva. News of the book soon spread beyond Lwów—especially since many of the problems had been posed by Banach, Mazur, and Ulam.

---

[131]Wojciech Sylwester Piotr Rubinowicz (1889–1974), Polish theoretical physicist.

[132]Following World War II, *The Scottish Book* was brought to Wrocław by Banach's widow. A copy now resides in the library of the Institute of Mathematics of the Polish Academy of Sciences in Wrocław. A typed English version can be viewed at the website: http://banach.univ.gda.pl/pdf/ks-szkocka3ang.pdf. A published English version is also available as: R. Daniel Mauldin (ed.), *The Scottish Book: Mathematics from the Scottish Café*, Birkhäuser Boston, Boston, MA, 1981. The tradition started by Lwów mathematicians was continued for many years by Wrocław mathematicians in the form of *The New Scottish Book*.

7 In the University Town Lwów

One could sense how rapidly mathematics was changing. One noticed, in particular, how specialization was becoming more and more narrow; for example, Mazur was at that time working almost exclusively on summability theory, Auerbach on differential geometry, Schauder on partial differential equations, Kaczmarz and Orlicz on orthogonal series, and Nikliborc on celestial mechanics and stationary solutions of the many-body problem.

Collaboration seemed to be on the rise. I wrote a paper on the uniform boundedness principle[133] with Banach, Banach and Tarski collaborated on their famous result that a ball in Euclidean space can be subdivided into a finite number of pairwise disjoint pieces that can be rigidly reassembled into two copies of the original ball,[134] Mazur and Orlicz also collaborated, as did Łomnicki Jr. and Ulam, Ruziewicz and Sierpiński, and so on. If in those days one were to join the names of every two collaborating mathematicians by a line segment, one would have obtained a graph in which Polish mathematicians were connected to mathematicians of every mathematically significant country in the world.

One could begin to distinguish mathematical "schools". There was the Banach school, working on linear operator theory, Chwistek's school of mathematical logic, and Bartel's group at the Polytechnic, concentrating on the relation between descriptive geometry and artistic representations of space—in this connection studying Italian etchings from the Renaissance, and publishing a beautiful book on them. For my part, for some time I consulted my friends about launching a seminar in mathematical physics. Finally, yielding to Dr. Blumenfeld's advice, I decided to go ahead with it. Together with Banach and with input from Dr. Blumenfeld, we ran the seminar for a year. Although it was not a resounding success, it did have the effect of making our students better educated.

Then there was the administrative side of our work. We were constantly being bombarded from above by memoranda concerning changes and reforms, and by ballots to do with polls of one kind or another. For example, as a result of the introduction of the trimester system and a master's degree program, we found ourselves at one point having to assess large numbers of students all at the same time in a single large room with just a few blackboards. This being exhausting for both us and the students, we soon went over to a different system whereby each of us examined small groups of students in his office. The protocol was such that each student had to submit to two such examinations, by different faculty members, each examination consisting of one written and two oral components. The results were then compared to determine the acceptability or not of the student. This system had the advantage that it incidentally tested the ability of each faculty member to make a proper assessment of a student. After some teething problems, we soon reached a situation

---

[133]This, also known as the Banach–Steinhaus theorem, states that a pointwise bounded family of continuous linear operators from a Banach space to a normed space is uniformly bounded.

[134]This is the celebrated "Banach–Tarski paradox". The "pieces" are exceedingly complicated subsets of points scattered throughout the ball. The proof uses the Axiom of Choice in an essential way.

where less than ten percent of cases required any discussion, and those where there was drastic disagreement—an estimate of very good by one faculty member and unsatisfactory by the other—made up only about one percent of all cases.

We also had to administer entrance exams. At some point our department had decided it could admit only a limited number of students, admission being determined by an entrance exam. It happened once that the mother of a female applicant whom I had failed in the entrance exam came to me to express her puzzlement over the fact that her daughter had been a distinguished mathematics student in the Gymnasium, often helping those of her classmates who had actually passed the exam. I asked her to send her daughter to me so that I could explain to her how little a student loses as a result of delaying one's studies by a year provided he or she works hard. For, in any case only a tiny percentage of students finished their studies in the prescribed standard number of years, most not managing to prepare themselves in that time for their final-year examinations. The lady let herself be persuaded and sent her daughter to me. The young lady admitted that her real aim was to study architecture at the Lwów Polytechnic, and when she had been rejected there, she had applied to the Mathematics and Natural Sciences Division of the university. To see how developed her taste in architecture was, I asked her which building in Lwów she liked most. It turned out that the building housing the Ritz Café was for her the most appealing, a big shed in the style of the Secession building[135] in Vienna, weighed down with heavy decoration in dark concrete. When I asked her what she thought of the Latin Cathedral,[136] she had to confess that she didn't even know where it was. This was hard to comprehend, since she had spent all of her Gymnasium years in Lwów, and the fact that she was Jewish and therefore paid less attention to Christian churches hardly justified her ignorance in my eyes. In my role as custodian of the Association of Jewish Students, I had had a lot to do with such young people. They were certainly no worse than average as people—certainly far better than the thuggish types who, armed with knives and knuckle-dusters, periodically attacked them. However, by and large their upbringing had been narrow, their culture suffered from lack of fertilization from outside, and they had no political nous whatever. They represented a third of the student body, but since they seldom had access to bursaries or other financial assistance, their contribution in fees represented half the total—of which most, about ninety percent, went to the construction and maintenance of academic buildings and dormitories to which they were denied access. On graduation, positions in the civil service or other administrative bodies would largely be closed to them, as would most teaching positions, and, furthermore, since the Faculty of Medicine admitted only about ten percent of all Jewish applicants, very few could expect to qualify as doctor. Their situation being similar at the Polytechnic, it is hard to comprehend why, instead of inferring that they should give up their studies, or at

---

[135]Built in 1897 for exhibitions of the Vienna Secession, a group of artists including Gustav Klimt, Koloman Moser, Max Kurzweil, and others.

[136]The Archcathedral Basilica of the Assumption of the Blessed Virgin Mary.

least present their case clearly *urbi et orbi*,[137] all they did was have their complaints printed in the Zionist press, with specific cases of discrimination itemized, and including appeals to humanitarian and democratic principles—failing to grasp that in actuality such principles played a very minor role in the society in which they were embedded. They didn't know any history, or they would have taken a lesson from the student revolutionaries in the Kingdom of Poland[138] in 1905, and organized private instruction for themselves, either openly or clandestinely, as the Polish students had done back then.

But a few of the shrewder among the Jewish students did in fact organize coaching sessions—not for themselves, but for other students needing to prepare for the finals, so that for a few hundred złotys a student could get help in preparing for any examination, or, for that matter, buy himself a set essay towards the undergraduate degree, or, if cost was no object, even a doctoral dissertation. When, inevitably, a fake piece of work of this sort was detected at the Polytechnic, the resulting professorial hue and cry quickly became muted when a discreetly phrased note from the "Rector" of the underground "Academy of Sciences" made it clear that the sons of several of the professors at the Polytechnic had been awarded engineering degrees largely on the basis of work purchased at that "Academy". In this way the Jewish contingent had established a convenient *modus vivendi* with precisely those gentile students who regularly ejected them from the lecture halls. Thus a small coterie of the practical-minded among the Jewish students secretly sold knowledge in written form to those of the Polish youth distinguished by their ready cash rather than a desire for learning.

* * *

In connection with the foregoing, it should be mentioned that the university's Philosophy Faculty was then split into two: a humanistic half and a mathematical and scientific half, with a corresponding difference in professors' titles. It may also be appropriate to indicate how the professoriat was behaving itself in Lwów in those days. To some extent this depended on the law regulating tertiary educational institutions, an initial version of which had been passed by parliament in July 1920. This had two particular merits: it had been written by a professor, so took into account the realities of academic institutions, and it was short, so we practically knew it by heart. However, then the government drew up a new statute on universities, parts of which were in conflict with the law as it still stood. Following the initiative of Professor Negrusz,[139] a dozen or so of us examined the new statute in detail, and found many places where it came into conflict with the unabrogated old

---

[137] "not just to the city (of Rome) but to the whole world"

[138] Or Congress Poland, the northeastern third of Poland ruled by Russia from 1815 to 1915, where from 1905 to 1907 there occurred many armed insurrections of Polish workers, peasants, and students against the Russian controlled government—the wider ramifications of the Russian Revolution of 1905, subsequent to Russia's defeat by Japan.

[139] Roman Negrusz (1874–1926), Polish experimental physicist.

law. We brought an action against the statute, and the Administrative Tribunal came down in our favor, much to the government's chagrin. Then in September 1930, the Brześć issue[140] came to the fore: the imprisonment of certain parliamentarians in the Brześć citadel on the River Bug, where they were beaten and starved. When a large number of professors, myself included, signed a petition protesting this episode, the government reacted by passing a new set of laws relating to tertiary institutions, allowing them to abolish or reorganize departments as they chose.

But worse than this loss of control was the general undermining of professorial authority in society at large that had been taking place for some time. In addition to their salaries, professors were being paid substantial supplementary sums for administering exams and giving specially commissioned lectures—and their prosperity was by no means always justified by scientific work of any significance. Many of them were heavily involved in political agitation, either by means of articles published in the dailies, or organization of political mafias of one kind or other. Still others devoted themselves to the acquisition of administrative power, running divisions or departments as their own private fiefdoms. Sad to say, there were scandalous things going on behind the walls of academia: professors appointing their mistresses to assistantships—or, conversely, turning their female assistants into their mistresses—, assigning commissioned lectures to themselves when these were often the only means of subsistence for docents, and skipping many of their regular lectures simply from being unable to cope with a multitude of self-imposed duties. They also discreetly fanned the flames of the students' anarchistic impulses, especially of those of the radical-nationalistic sort, and used their chairs as soap-boxes from which, without fear of punishment, they could attack the hated "Sanation[141]" government. In their weeping and wailing over the threatened loss of "academic freedom" they were being extremely hypocritical, since—as many of the more disinterested faculty members clearly saw—in practice this academic autonomy came down to a licence for right-wing radical youth to beat Jews with clubs, interrupt lectures, and organize student "strikes" every few days without intervention by the police—which to my mind represented a danger to society emerging from below. The rage of the politically active youth was especially directed at Jędrzejewicz,[142] the Sanation minister of education, who, together with his colleagues in the government, was afraid to lay any blame on them, suggesting instead that student unrest was being fomented by the professors—in response

---

[140]In 1930, an aging Marshal Piłsudski was called upon to end the chaotic political situation in Poland. Among his first actions on assuming power was to arrest several former members of the opposition and imprison them in the Brest Fortress (formerly known as the Brest-Litovsk Fortress; the Polish name of the city is Brześć Litewski), built by Russia in the early nineteenth century.

[141]Following on Piłsudski's *coup d'état* of May 1926, a Polish political movement calling itself "Sanation" (Piłsudski's watchword, from the Latin word *sanatio* for "healing"), and led by close associates of Piłsudski, came to power on a platform of the primacy of national over democratic interests.

[142]Janusz Jędrzejewicz (1885–1951), Polish politician and educator, one of the leaders of the Sanation movement, and a former aide to Piłsudski. Prime minister of Poland from 1933 to 1934.

7 In the University Town Lwów   169

to which the sanctimonious professoriat invoked the holy catch-all of academic autonomy, deploring the threat to Polish learning represented by the imprisonment of a thuggish faculty member who had thrown a female colleague down the stairs. Thus the university as a genuine institution of higher learning was under attack both from above and below, as a result of which it was becoming impossible for it to exploit in a calm and systematic fashion the potential afforded by the excellent work of a dozen or so capable and devoted people. The division of the Philosophy Faculty into two halves led to my appointment as Dean of the Mathematics and Natural Sciences Division. I remember that when soon afterwards I went to see vice-minister Pieracki in Warsaw on university business, he refused to see me because I had signed the Brześć petition.

That was also the time when I received a visit from Professor Gröer,[143] a Lwów pediatrician, whom I had met in 1915 in Vienna when he was an assistant to Pirquet,[144] and used to come to the boarding house *Atlanta* at 33 Währingerstrasse to visit his *fiancée*. Pirquet had been using inoculations of Koch's[145] tuberculin to diagnose tuberculosis in children: inoculation of a tubercular child resulted in a rather more substantial red blister in the area where the tuberculin had been pricked through the skin, than in the case of a healthy child. Gröer had begun measuring the sizes of the blisters and had discovered a systematic correlation between the diameter d of the blister and the concentration c of the solution of tuberculin, indicating a concentration of 10^{-c} milliliters per milliliter of water. For each patient Gröer wanted to plot points (d, c) in the coordinatized plane and join them with a curve showing the relation between d and c characteristic of that patient—which required giving the poor patient a dozen or so subcutaneous inoculations of serums of differing concentrations. His idea was that then one might draw diagnostic conclusions from the shape of the patient's (d, c)-curve. I advised him to use logarithmic paper for plotting the graphs, then obtainable from the firm of *Schleicher and Schüll*, where both axes are calibrated logarithmically, such paper being, of course, more useful when the relation between the logarithms of two quantities is simpler than that between the quantities themselves. It turned out that the relation between the logarithms of d and c was linear, that is, for each patient the plot of the points (d, c) on the logarithmic paper was a straight line. It followed that the "allergy curve" of a patient, that is, the (d, c)-curve, could be determined from just two inoculations.

---

[143]Franciszek Gröer (1887–1965), Polish pediatrician and professor at the University of Lwów. Manager of the Lwów Opera 1931–1933, and well known for his photography. Director of the Institute of Mother and Child in Warsaw 1951–1961.

[144]Clemens Peter Freiherr von Pirquet (1874–1929), Austrian scientist and pediatrician, known for contributions to immunology and bacteriology. Devised the test for tuberculosis involving the inoculation of a suspected tuberculosis case with tuberculin, an extract taken from the tuberculosis bacterium, to see if there is an allergic reaction caused by the prior presence of that bacterium.

[145]Heinrich Hermann Robert Koch (1843–1910), German physician. Isolated the tuberculosis bacillus in 1882. Nobel laureate in physiology or medicine for 1905.

Gröer considered that this so-called "allergy law" must be important in connection with diagnosis and monitoring of tuberculosis in children, but it was still not clear how. He saw as basic that he first needed to assign each allergy curve two parameters determining it, and had even decided on names for these parameters beforehand: "reactivity" and "sensitivity". I suggested that he take as parameters simply the axis intercepts of the straight-line plot on the logarithmic paper. Thus the present state of a tubercular patient was now characterized by two parameters, namely R (reactivity) and S (sensitivity), which could be determined easily by means of two inoculations. The values of R and S on any particular day yielded a point (R, S) representing the patient's state on that day. By plotting the points (R, S) each day—this time on ordinary squared paper—over the course of a month, one obtained a curve indicating the development of the disease in the patient over that month. After making these measurements in several patients, and plotting their (R, S)-curves over a month, it became clear that the curves of patients who were getting better differed in a striking and uniform way from those of patients where the disease was worsening. In this way we gave added depth to the old Pirquet test, at the expense of making two inoculations each day rather than just one.

One problem with Gröer's test was that the diameter of a blister was frequently difficult to measure accurately, since it was often not exactly circular, so we decided to use, as a more precise indicator, its area. Following my advice, Gröer measured the area of a blister by placing a small square of glass on top of the blister, etched with a square grid or lattice of points with adjacent points a distance 3.16 mm apart. Then by counting the number of cells of the grid over the blister and multiplying by 10, one obtained an estimate of its area in square millimeters—since $\sqrt{10} \approx 3.16$.

After these results of Gröer and his collaborators (Kochanowski, Chwalibogowski, and others) were published, a rejoinder by the Warsaw pediatrician Mieczysław Michałowski[146] appeared in *Pediatria Polska*. The author, although generally speaking a decent fellow, on this occasion perpetrated a piece of nonsense: instead of pointing out the weaknesses in Gröer's test—the difficulty in measuring the size of the blister accurately, the fact that so complex a phenomenon as a blister surely depends on many factors having nothing to do with the presence of tuberculosis bacteria, the complete inapplicability of the test in adult cases of the disease, and the fundamental issue of the absence of any alternative diagnostic procedure for providing a comparative validation of the test—instead of such serious objections, he made such ridiculous statements as: "Ordinary mathematics cannot be applied to the biological phenomena encountered in medicine." Well, that is at least a coherent point of view, so I might have forgiven him if he had not also written such

---

[146]In fact the author of the rejoinder was Mieczysław Michałowicz (1876–1965), Polish physician, social activist, and politician. Taught pediatrics at Warsaw University from 1920 to 1931. (The clash of opinions was recorded in *Pediatria Polska* 1935, Vol. 15, Fasc. 6, containing the proceedings of the Fifth Congress of Polish Pediatricians in Łódź, November 1–3, 1935, at which Franciszek Gröer gave a lecture entitled "Zasady patergometrii i allergometrii" (The principles of patergometry and allergometry) (pp. 407–419), and Mieczysław Michałowicz responded in the discussion following the lecture (pp. 421–423).)

things as: "The four-dimensional mathematics of Łukasiewicz and Leśniewski[147] is more suited [to medicine]." I knew these logicians well, but to this day I still have no idea what Dr. Michałowski could possible have had in mind in referring to them in this way. There were two other articles on this theme in the issue of *Pediatria Polska* in question, one by a Mr. Jan Grączewski, which was a mine of ignorance—the author did not seem to grasp that $a = b$ implies $\log a = \log b$—and the other, by another wise man whose name I have forgotten, which was a congeries of nonsense sprinkled with big names such as Einstein, Planck, and Eddington, and terms such as "quanta", "electrons", and "stars".[148] It resembled the bag of Ali the Persian as described by the Kurd who stole it in the Arabian tale *Ali the Persian and the brazen Kurd*.[149] If I remember correctly, Stanisław Mazur wrote rebuttals to these articles in *Mathesis Polska*,[150] which was treating them far more seriously than they merited. I recall a favorite proverb of one of our servants in Jasło: "There are a tremendous number of fools in the world, but one doesn't notice since they don't tend to go around together."

I might insert here a memory of an episode involving my daughter, who was then ten years old. She complained that her teacher had given her the homework assignment of measuring the length of the Vistula by means of a length of thread placed on a map of Poland, but that she found this method of estimating the river's length impractical. It was for this purpose that I invented my longimeter,[151] a version of which the publishing house *Książnica Atlas* used at one time to include with the maps they produced.

To return to my decanal ordeals: I was kept very busy, for instance by having to attend meetings of the Division, which began at five in the evening and often went till ten, after which we would adjourn to the George for a late supper.

---

[147] Jan Łukasiewicz (1878–1956), Polish logician and philosopher. Pioneer investigator of multi-valued logics. Born in Lwów. Stanisław Leśniewski (1886–1939), Polish mathematician, logician, and philosopher.

[148] These were published as commentaries arising from the post-lecture discussion: Jan Grączewski, "Uwagi matematyka do odczytu o patergometrii i o allergometrii" (Remarks of a mathematician concerning the lecture on patergometry and allergometry) (pp. 433–438), and Józef Marzecki, "Matematyka i medycyna" (Mathematics and medicine) (pp. 426–432).

[149] A tale told by Ali the Persian to the sleepless Caliph Harun al-Rashid, about how a Kurd stole his bag, and when asked to describe its contents before a judge, enumerated at great length an extremely improbable set of items, including carpets, tables, a pregnant cat, two beds, a female bear, a lioness and two lions,....

[150] *Polish Mathesis*. A monthly devoted to the science of numbers and methodology, first published in 1926.

[151] The "Steinhaus longimeter" consists of a transparent sheet on which three congruent square grids of side 3.82 mm are marked, two of which are rotated with respect to the third through 30° and 60° respectively. Placed on a sheet of paper on which a curve is drawn, the number of intersections of the curve with the lines of the longimeter gives the approximate length of the curve in millimeters.

Krzemieniewski, Rogala, Kulczyński, Ruziewicz, and Tokarski[152] came regularly, others more sporadically. Following each of these divisional meetings I had to see that the decisions taken were put into effect by dispatching appropriate memoranda. There were also various committees whose meetings had to be presided over by the Dean, and twice a week there was the prescribed ritual of receiving student petitions. By way of example of the sort of problem that arose, I might mention the search for a new head of the Department of Geography following on the resignation of Eugeniusz Romer. Romer had recommended Mieczysław Limanowski from Wilno,[153] but he refused. As our second choice we invited Stanisław Pawłowski from Poznań to come and discuss the possibility of taking up the position. Having done so, Pawłowski returned to Poznań and fell silent, neither turning down nor accepting our offer. (I later found out that he had been using the offer as a bargaining tool for securing the post of Rector at the University of Poznań.) Eventually Julian Czyżewski[154] was appointed.

The meetings that I found most unpleasant were those of the university senate, attended by the Rector and Prorector and the Deans and other representatives of the university's five divisions. Apart from the fact that I hardly knew these people, what made the meetings difficult for me was the attitude of the members, so different from that of my close colleagues. One of the special areas of competence of the senate was that of student issues, a recurring source of trouble. Even if there were no current cases formally before the senate, inevitably a submission would be received at the very last minute from some such student movement as that calling itself "Regeneration", made up of Christian students of a puritanical bent, requesting the senate to see to it that all advertising billboards and movie posters depicting scantily clad or completely naked women should be taken down, since such scenes were disturbing the young male students, and preventing them from paying due attention to their studies. I remember taking part in the resulting discussion, proposing that the submission be relayed to the provincial administration as the competent authority—but no one supported me. Is there anything more futile than the idea that a young man who, having seen a girl's naked *derrière* on a movie poster downtown somewhere, and being unable to concentrate on studying as a result, should have someone submit to the senate on his behalf a request that he be shielded from that paper Aphrodite?

However, the Rector at that time,[155] Professor Krzemieniewski, was in fact worthy of that high post. Unlike others, he did not bow to pressure from the

---

[152]Seweryn Krzemieniewski and Stanisław Kulczyński were from the Department of Botany (specializing mainly in floristics), Wojciech Rogala was in the Department of Geology, and Julian Tokarski was head of the Department of Mineralogy and Petrography.

[153]Now Vilnius, capital of Lithuania.

[154]Julian Czyżewski (1890–1968), Polish geographer. Studied and worked at the University of Lwów between the wars.

[155]The academic year 1931–1932.

paedocracy.[156] It was during his tenure that demonstrations took place protesting the death of a student by the name of Wacławski in Wilno,[157] and the murder of another called Grodkowski in Lwów. In dealing with student unrest of this sort he favored being completely open about the situation. I remember the occasion when the "little" senate—consisting of him, the Prorector and the Deans—came to a protest rally by students and certain faculty members at the main university entrance. The organizers of the rally, student representatives and certain faculty members with many years of political demagoguery behind them, were seated at a table ready, as it were, to engage in debate with us, the members of the little senate. However, there was no debate, in the sense of reasoned argument or two-way discussion: they launched a tirade of criticism of the ministry, claiming that learning was under threat from the government since it intended—so they maintained—to by-pass the divisions by arrogating to itself the function of making academic appointments to the university. But behind their avowed solicitousness for learning, there lay the ulterior motive of getting anti-Semitic resolutions passed concerning such appointments. Then from the mass of students assembled just outside came the cry "Death to the Jews!" and the official monitor of the rally, Professor Longchamps,[158] was forced to intervene to restore decorum. Among the student representatives at the table was a girl by the name of Piepes-Poratyńska, who, having long been exposed to the racist lucubrations of her fellow students, seemed surprised that I did not show some sign of obeisance before her.

The continual pressure from the anti-Semitic element in the university eventually led to an arrangement whereby every few weeks the student body would designate "days without Jews," that is, days when no Jewish students were to be allowed on university grounds. What then inevitably happened was that some professors cancelled their lectures in protest on those days, following which the Rector was essentially forced to close the university for a certain time—so that what had begun as a "strike" of sorts became a "lockout". During one such lockout, I happened to be leaving a university building on Kościuszko Street, where caretakers were guarding the gate on the Rector's orders, when a group of students stormed the gate—thirsty for knowledge, perhaps. One of them grabbed a caretaker around his middle, spun him round as if dancing the mazurka, in this way managing to get past. I went up to him and asked for his name. He answered: "I won't give you my name. Let the police arrest me!" When I told him that the police had no jurisdiction in the university, but that I, as Dean of the Mathematics and Natural Sciences Division, certainly did, he bowed and gave me his student card, indicating that he was in fact a first-year student at the Polytechnic. I immediately took the card to the Rectorate. Next day

---

[156]That is, rule by a child or children.

[157]Wacławski was killed during an anti-Semitic uprising by Polish students in Wilno.

[158]Roman Longchamps de Bérier (1883–1941), Polish lawyer and professor. Last Rector at the Jan Kazimierz University of Lwów. Murdered by the Nazis in the Massacre of the Lwów Professors.

the Prorector, Professor Edmund Bulanda,[159] asked me if I agreed that the card be returned to the student, his mother, in a pitiful state of despair, having implored him to overlook the incident. I replied that I had no personal stake in the matter, and that it was for the Rector to decide, not me. I later found out that Bulanda returned the card without bringing the matter before the Rector.

It was getting to the point where the so-called "youth movement" made it impossible for the university to remain open for more than a month at a stretch. The professors blamed the continual unrest on the government's policies, and the government in turn explained to the *Sejm*[160] that it was the professors who were instigating the disturbances as a way of carrying on a covert operation against the ruling Piłsudski camp. On the other hand, the students themselves justified their actions before public opinion in terms of the holy watchwords of patriotism, maintaining that it was necessary to eject Jews from the university since the Jewish element threatened Polishness, religion, and virtue, and was a hotbed of communism, bolshevism, cosmopolitanism, and pacifism. The police, happy to avoid responsibility, adopted an attitude of excessive correctness towards the territorial autonomy of the university, apparently feeling no call of duty to intervene during excesses that threatened life and limb and the destruction of public property. Society as a whole, not liking to ponder complex issues, seemed unable to reach consensus as to what should be done, but I must say that one section of it, the lower classes—laborers, servants, country women, and so on—regarded the students' pranks as unpardonable licence. Their frequently heard comments might be summed up as follows: "Instead of being grateful that they don't have to do hard physical work and can better themselves by studying at their fathers' expense, they do mischief."

As Dean I also had to deal with the Ruthenian[161] problem. On one occasion a student from this group asked me if the chemistry curriculum would suffice for making explosives, and on another, a police detective demanded, through my secretary, that I let him copy the names of all "Ukrainian" students from the records of enrolment. I instructed her to tell him that the university communicates with outside authorities only through the Rectorate, and that, just as we, when requesting information from the district offices, do not send a clerk to the secretary of the district council, so too, if the town *starosta* wants information from us, he should either call the Rector, or write to him directly. Then if the Rector so informs me, I will allow the police agent to copy the students' names. It cast an interesting light on police methods and procedures that the agent never returned. Naturally, I told the Rector of the incident, and he approved my position.

---

[159]Edmund Jan Bulanda (1882–1951), Polish archaeologist. Appointed Rector of Lwów University in 1938. When the university was closed by the Nazis in 1941, he continued as Rector of the "underground University of Lwów" for the duration of the war.

[160]The traditional lower house of the Polish parliament.

[161]Although this term has a variety of meanings, here it refers to people of what was south-eastern Galicia and the eastern Carpathians, who spoke Ukrainian and called themselves Rusyns. The issue in question had to do with their desire to be given separate national status.

7 In the University Town Lwów

I seldom had to deal with police, but the few times I was involved with them were all very different from one another. Our young racing car enthusiasts conceived the brilliant idea of organizing a race through Lwów streets along a route modelled on the Monte Carlo circuit. A great deal of effort went into choosing Pełczyńska, Stryjska, and Kadecka Streets to form the circuit, the erecting of judges' boxes along the route, and waking everyone up at the crack of dawn with the roar of muffler-less racing cars practising on the three-kilometer circuit. On the day of the race I invited a few guests to watch from our balcony. Some time before the start I was visited by a detective who had come to ask about some or our tenants. The conversation went as follows:

– Do you know the Potylickis?
– Yes, they have been living here for about a dozen years.
– I can't tell you the reason for this inquiry as it's a secret.
– Good.
– But I will tell you something.
– Yes?
– It's about the car race.
– Yes?
– It seems the Rusyns[162] intend to throw a bomb into the street during the race.

To which I replied:

– Yes, but that would not be to their advantage since this is an international race with Italians and Germans participating, so throwing a bomb would cause an international incident that would reflect badly on them.
– Indeed, you are right. A valid point indeed.
– But—I went on—this being so, it would on the other hand be of use to our cause if the Rusyns threw a bomb.
– Yes, indeed—said Sherlock Holmes—This hadn't occurred to me, but it's an excellent idea.
– But, wait a moment—I said—Imagine what would happen if it became known that our police had organized an attempt on the life of foreign guests merely to compromise the Rusyns!
– I really hadn't thought of that. You're quite right.

By this time Sherlock was showing signs of discomfiture, since he didn't know if I was serious or not. He left without making any further inquiries. And, of course, there were no bombs thrown either on this occasion or the next.

In the spring of 1932, the ministry sent a high official to carry out an inspection of the university, and the Rector asked me to act as his guide, and, in addition to showing him all the physical plant, explain the university's administrative structure. We had been working on the principle that whenever any question arose that could not be settled on the basis of the 1920 Statute—which was really only in the nature

---

[162]That is, the Ruthenians.

of a framework—we would fall back on the old, long lapsed, Austrian statutes and protocols in the collection by Beck and Kelle.[163] Our justification for proceeding in this way was that there was surely no need to consider yet once more questions to which answers had long before been obtained based on the many years of practical experience of the score or more academic institutions of the former monarchy. However, it soon became clear that this apparently exceedingly thorough specialist in the administration of tertiary education had never heard of this collection.

But worse was to come: a delegate of the Ministry of Religious Creeds and Public Enlightenment went to my successor as Dean of the Mathematics and Natural Sciences Division of the university, and asked him to explain the role of the office of the dean. It might be thought that he could easily have done his homework and obtained this information from the 1920 Statute, which was so brief that he could have read it in its entirety while travelling on the train from the Ministry to Krakowskie Przedmieście Avenue,[164] or that in order to satisfy the specialist the dean in question might have consulted the Ministry—as he must surely have done fairly often for other reasons—as to the prerogatives the minister granted deans beyond those clearly specified in the 1920 Statute. However, such a request could occur only to a pedantic Galician lawyer and never to an amateur in the art of wielding authority, whose actual purview owes much to joyful creativity.

* * *

In 1931 my father-in-law died in Kraków, and in summer my wife, my daughter Lidka, and I went to Kosów to see if the cure there would improve Lidka's health. There was some risk involved in travelling there since from Kołomyja[165] one had to travel by taxi, and there was usually just one such to be had at the station, so that its owner charged exorbitant fares. And then the driver kept nodding off because of the heat, to be prodded awake by an assistant whose sole function seemed to be just this. Kosów is oddly located, askew from where one would have thought a better site. It has a quite dry climate, and though there was plenty of fruit available, it was of only middling quality.

Our boarding house *Halina*, popular with Varsovians, was not too bad. Its tone was dictated by the nearby sanatorium run by a Dr. Tarnawski.[166] Those taking his cure had, in particular, to wear prescribed clothing of a sort of white netting. I paid the local specialist tailor 22 złotys for the material and workmanship for my own

---

[163]Beck von Mannagetta, Leo, and Kelle, Karl von (Eds.): *Die österreichischen Universitätsgesetze. Sammlung der für die österreichischen Universitäten gültigen Gesetze, Verordnungen, Erlässe, Studien- und Prüfungsordnungen* (The laws of Austrian Universities. A collection of ordinances, edicts, and regulations relating to education and examinations at Austrian universities), Manz: Wien, 1904.

[164]"Kraków Suburb" Avenue, one of the grandest streets in Warsaw.

[165]A city on the Prut, now in western Ukraine.

[166]Apolinary Tarnawski (1851–1943) established his Kosów sanitorium using natural medicines and procedures in 1891.

costume, and it lasted several seasons. So here is a good example of how the Jews exploit the gentiles—I mean my exploitation of the tailor.[167]

In addition to the usual summer occupations—walks in the woods and leisurely boat rides on the river—there were others such as the consumption of *makagigi*[168] and *gluden*, a strange almond layer cake embedded with various kinds of fruit, in little Jewish sweet shops. There were also stores selling goat meat, an Armenian delicacy. In nearby Kuty, we found an excellent restaurant where one might see patients of Tarnawski secretly consuming steaks, forbidden under his vegetarian regime. In the Kosów market one could buy kerchiefs, shirts, and socks embroidered in many colors, but the most interesting things available there were the beads on sale in the stalls. These were supplied to the vendors by country women in the process of replacing their valuable old-fashioned necklaces by modern trash. The beads were large—up to a centimeter in girth—and made of colored glass or porcelain. There were white ones, ones painted in pastel colors, gilded black ones, some of glass, some carved elaborately, and some of a most beautiful green hue, with prices ranging from 5 to 12 groszy.[169] There were also strings of cherry-colored "Kuty pearls", each the size of a cherry stone, very beautiful, and made of some material unknown to me.[170] It was somewhat of a mystery as to how all these beads—like nothing traded anywhere else in Poland—ended up in Pokucie.[171] It was Vincenz who explained to me that a hundred years earlier, when Venice belonged to Austria,[172] a regiment recruited in the eastern Carpathians was stationed in Venice for many years, whence the soldiers, when visiting their homeland on leave, brought necklaces of these beads with them as gifts for their girlfriends. Since this continued for many years, there were enough such visits to saturate the district with these beautiful little spheres.

I remember a hike to Kuty I took in the company of my wife, a Mrs. Smolińska from Poznań, and another lady and her children, when we ran into a dreadful hailstorm, and the slippery balls of ice densely carpeting the hillside made climbing impossible for the time being. On another trek, this time with Mr. Smoliński, we decided to return through a dense wood, and when night fell we lost our way. We knew we had to climb upwards, and though exhausted, my wife insisted we use sticks to tug each other a few meters higher. As luck would have it, the forest opened up into a glade, and by navigating by the stars for a couple of miles, we reached the

---

[167] At that time of the Great Depression the złoty was worth about ten US cents.

[168] A cake made from caramel, honey, and poppy seeds, sometimes with walnuts or almonds.

[169] A grosz is a hundredth of a złoty.

[170] According to the *Internet Encyclopedia of Ukraine*, the Hutsuls living in Pokuttya are famous for their artistic jewelry with inlays of—among other things—mother-of-pearl and glass beads.

[171] The region around the upper Prut, which changed hands several times over the centuries. It belonged to Poland from 1919 till 1939 (and during certain earlier periods).

[172] Venice was under Austrian rule from 1797, when it was handed over to Austria by Napoleon, till 1866, when Austria was defeated in the Seven Weeks' War with Prussia and had to cede Venice to the newly independent Kingdom of Italy.

highway near Kosów. We also went to Jaworów to buy *liżniki*,[173] shaggy Hutsul blankets, very decorative and warm, and multicolored—although with aniline rather than the former earthen dyes. The blanket I bought there served us well till the war as a cover for the couch.

During our absence from Lwów, dark clouds had been gathering ever more menacingly on the academic horizon. In particular, the Ministry was proposing a new law pertaining to tertiary institutions, whereby it could on its own say-so liquidate departments, divisions, and even whole faculties—and if a professor had his faculty annulled under him in this way, then *eo ipso* he also would cease being a professor. This proposal had the appearance of objectivity, since it was justified in terms of a need to reduce expenditures and touted as aimed not at individuals but at eliminating expensive and redundant institutions. The Jan Kazimierz University organized a meeting of its professors at which it was resolved that everyone was against the proposed legislation, but this resolution had no official status since there existed no official body such as a professors' association. It followed that someone had to put forward the resolution as his own, and when it was suggested that I do this, I agreed. That was in March 1932.

At the end of that year my Father fell seriously ill. When they phoned me with the news from Jasło, I fetched Professor Rencki[174] to accompany me home. My Father's situation was dire: he had caught the flu in early December, and after recuperating had gone to his office on a freezing morning and suffered a heart attack. Then when he had lain down, the resulting ischemia—reduction of the blood supply—in one of his legs had left a blue mark in the middle of his calf. Such symptoms in a man of 78, together with his diabetes and atherosclerosis, represented a formidable syndrome, offering little or no hope. After examining him thoroughly, Rencki took me outside and told me that death was imminent. Then Father called me back in to give me specific instructions as to Professor Rencki's supper: where we were to sit, which wines should be served, and so on. After supper I accompanied Rencki to the railroad station, but he urged me not to wait to see him off since, given Father's condition, every minute was precious. However, I knew Father had unusual reserves of strength, and, sure enough, there began a struggle with death conducted by him in concert with my Mother, my sisters, and my wife, lasting upwards of four months.

The student disturbances and resulting suspensions of lectures meant that I was able to be present in Jasło for most of that time. We arranged for Dr. Ostern[175] to visit to analyse Father's blood, and Professor Parnas arranged for Dr. Baranowski,[176] then

---

[173] A word used by Ukrainians for thick woolen blankets.

[174] Roman Rencki (1867–1941), Head of the Institute of Internal Medicine at Lwów University. Murdered by the Gestapo in the Massacre of the Lwów Professors in 1941.

[175] Paweł Ostern (1902–1943), brilliant Polish biochemist. Committed suicide while being arrested by the Gestapo.

[176] Tadeusz Baranowski (1910–1993) was then completing his medical studies under Parnas in Lwów. Ultimately pursued a highly successful academic career in medicine, at first in Lwów, and later in Wrocław. Rector of the Wrocław Academy of Medicine 1965–1968.

his assistant, to administer prescribed injections to my Father regularly, to which end Baranowski took up residence in our house for a few weeks. We also hired a German nurse. Two other doctors (Dr. Kommehl and Dr. Lanes) came almost every day. Since part of the ritual of care involved keeping his diabetes under control, in addition to giving him injections of insulin the doctors had worked out a precise diet to be adhered to, requiring the careful measurement of the amounts of protein, fat, and carbohydrates in every portion of food fed to him. This prompted him to joke in his wonted fashion: "Just look at those two. They have been calculating the amounts I'm to be fed for supper for an hour, and tomorrow for breakfast I'll be fed the result of their mistakes." And when the doctors arrived late, he would say: "Because of the doctors' unpunctuality I won't be able to die on time."

Father suffered a great deal but didn't show it: he showed no impatience, and kept the nurse by him for hours. I was very grateful to my friend Janek Adamski for visiting him to tell him what was going on in town. We fought a stubborn but hopeless battle with death, which of course we couldn't possibly win. We summoned from Kraków the diabetes specialist Dr. Maksymilian Blassberg. The worst of it was that the gangrenous stain on his leg gradually grew larger, and we had to decide whether to have the leg amputated. Although Dr. Kommehl was for it, we ended up deciding that subjecting a 78-year-old person on his deathbed to amputation of a leg would be to risk causing him unnecessary further agony, and in any case, as Mother said, Father would not consider a life on crutches as worth living.

He died in May 1933. That was the first lightning bolt that struck us. The second was preparing itself: in July 1933 Hitler set fire to the Reichstag as a preliminary to setting fire to the world. In Jasło my Father's funeral procession moved slowly along Kościuszko Street past closed store-fronts and lampposts swathed in black, and in front of the Town Hall, the Savings Bank, the District Council Offices, and the club *Sokół* (The Falcon) there hung black flags. All of Jasło attended the funeral, and many came up to me and said: "You do not and cannot know what your Father was really like. He did a great many people favors that only he and the recipient of the favor ever knew about."

My Father's death meant the end for me of a rather long carefree period of my life. I now had to settle his affairs, which were rather complicated because his property consisted of houses, certain grounds, the old brickworks, a few morgs of arable land, and a wooded plot of some 30 morgs. But there was no liquid money, and the debts incurred during his illness had to be paid. To realize his property, we first had to obtain a certificate of inheritance, in which matter I was greatly helped by my brother-in-law Wiesław,[177] a lawyer. Both he and another brother-in-law, Zgliński, loaned us money to pay the debts, including repayment of an earlier loan from the City Savings Fund in Lwów. Incidentally, at that time the situation with credit was so tight in Poland that even a loan secured by several signatories with substantial financially unencumbered property took a long time to be processed.

---

[177] In fact it was Gustaw Müller, the husband of Felicja, the eldest of the author's sisters, who helped in the matter in question.

The arrogance of the bank directors was commensurate with the exorbitant interest they charged.

In the summer of 1933, Hitler and his cronies managed to assume sovereignty in Germany. As everyone knows, this "revolution"—if one can call it that—began with the fire in the Reichstag, which the Hitlerites blamed on the communists. In fact, once installed in power, the Nazis began court proceedings against a certain van der Lubbe,[178] a Dutch communist, whom they accused of setting the fire. Whoever had a radio could listen to Göring's speech, full of fury directed against van der Lubbe and also Dimitrov,[179] a presumed partner in the arson. Only the most dim-witted of German provincials actually believed the accusations.

Our family doctor was then a Dr. Ziemilski, whose wife became a close friend. She told us the following story. Her brother, Stefan Askenase,[180] knew an eminent Belgian violinist whom Ribbentrop[181] had invited to Berlin about a year before. When, after travelling all night, he arrived at the Berlin railroad station, there was no one to meet him, although Ribbentrop had said he'd be there. So he made his way to Ribbentrop's residence on Wilhelmstrasse, where Ribbentrop received him hospitably and ordered a servant to see to it that a substantial breakfast was prepared for his guest. This was soon brought in on a little wheeled table, but when the hungry guest took up his knife and fork, Ribbentrop, clearly amused by his own cleverness, asked:

– Aren't you afraid of poison?
– Who would want to poison me?—replied the violinist.
– But I would!—exclaimed Ribbentrop.

A little later Ribbentrop took his guest for a ride in his chauffeured car, and when they passed the ruins of the Reichstag, the Foreign Minister of the Reich placed his finger against his lips in a gesture urging silence, afraid of inappropriate questions that might be overheard by the chauffeur. After a short interval he whispered in his ear:

– What you saw there was Hitler's greatest folly.

Well, one can only presume that Ribbentrop later had occasion to see greater ones!

Of course, that deed of the former lance corporal had repercussions for the fate of everyone then living. For example, we at Lwów University were soon subject

---

[178]Marinus van der Lubbe (1909–1934), a Dutch communist, fled from Holland to Germany in 1933, where he worked in the communist underground. Accused of setting the Reichstag fire, he was guillotined in January 1934.

[179]The other four defendants at the trial, Georgi Dimitrov, Ernst Torgler, Blagoi Popov, and Vassili Tanev, were cleared.

[180]Polish-born (in Lwów) Belgian classical pianist and pedagogue. Lived from 1896 to 1985.

[181]Ulrich Friedrich Wilhelm Joachim von Ribbentrop (1893–1946), Foreign Minister of Germany from 1938 to 1945. Hanged for war crimes following the Nuremberg Trials.

to the sympathetic reaction of our Polish youth to the example set by our western neighbor.

It was around that time that we invited Professor Lichtenstein to come from Leipzig for a few months and give a course of lectures on his work. Since it fell to me to find suitable accommodation for him, I had a fresh opportunity of examining Lwów from the perspective of a traveller looking for temporary accommodation. I found it very difficult to find anything suitable, and ended by taking a room less than satisfactory. Nearly all of the boarding houses I looked at had dark stairwells, foul-smelling corridors, and rooms cluttered with furniture acquired according to some unknown principle and serving some unknown purpose. And the bathrooms served as storerooms for discarded suitcases, baskets, old furniture, and, in many cases, jars of preserves ranged on planks over the bathtubs—all going to show that the current tenants aimed at thwarting the builders' intentions that that particular space should be used for washing oneself. After some time living in the tolerably comfortable room I eventually found for him, Lichtenstein related of his neighbors, a young couple, that they continually made and received phone calls always with one and the same content: "What are you doing? We're not doing anything either. So come over. Or should we come over to your place? So who should come, us or you? So come over! Why not? In that case it's better to go to Roma's place. Bye, bye! What? really? Bye, bye." He was much amused.

Lichtenstein's lectures were on a rather specialized topic: integro-differential equations and the various kinds of equilibria of spinning bodies. During one of his lectures he quoted an English work on, I think, mechanics, and recommended it to the audience, saying that it was well worth reading even though "it doesn't contain a word about set theory." Banach absented himself from Lichtenstein's lectures on principle, and when he heard about his ironical remark, he assumed an air of great surprise that Lichtenstein should so "insult Polish mathematics", as it were. Incidentally, Banach so muddled the finances of *Studia Mathematica*, that it was only thanks to a grant from the oil firm *Małopolska* and an earlier subsidy from my Father that the journal was saved from going under in 1933.

For some years from 1933 we spent our summers in Krynica. I remember that in 1933 my Mother stayed at the guest house *Kasztelanka*, but that I preferred the adjoining *Skarbówka* because it was nice and clean, provided tasty food and good service, and was reasonably priced. The two guest houses were set in a sloping garden luxuriant with flowers, splendid marguerites and roses predominating. In this garden I had the following conversation with a young lady studying in Warsaw, clearly oblivious of who I was—or might be. She declared that, in her opinion, university entrance examinations were harmless, since their real purpose was simply to exclude Jews from institutions of higher learning.

– But what—said I—if the Jews are well prepared for the exams?
– Our professors ask the Jews entirely different questions from those asked the rest of us—she answered.

– I assure you—I said—that our professors, that is, the professors of the Mathematics and Natural Sciences Division of the Jan Kazimierz University, ask everyone the same questions.
– O, go on, sir. At the Warsaw Polytechnic, for example, they ask a Jewish applicant to draw from memory the portal of some church, and of course he can't do it because he doesn't go to church.

I replied:

– I doubt that this is possible. After all, the professors swear an oath to apply the laws pertaining to tertiary institutions to all in the same way.
– My dear sir—she said—who pays attention to such things as oaths?

In connection with this, it is relevant to remark that *Skarbówka* was run by a Mr. Zamorski, whose son had acquired a kind of fame by throwing a petard during a student demonstration, and then blaming the deed on the first Jew he laid eyes on. However, the Lwów medical students' association acknowledged him as the perpetrator, and expelled him from their midst.

By this time the new law I mentioned earlier had been passed, and one of our mathematics departments had already succumbed to it, that is, been liquidated. I received a telegram to the effect that I should return to the university in its time of crisis and discuss the issue with my colleagues. The telegram did not say which department it was, and although I had reason to think it might be mine, the idea did not greatly upset me. However, it was not mine but that of Ruziewicz, who was quickly snapped up by the Business Academy. Furthermore, he could keep his retirement pension, along with his title, and was permitted to work as docent so still able to give lectures in the university.

At that time I began collaborating with Stefan Kaczmarz on a book on orthogonal series—I would never have ventured to write such a work by myself. I continued to be interested in applications of mathematics. For example, I had worked out a method for computing demographic concentration and dispersal of settlements: these cannot be expressed by single numbers but only by means of curves of concentration plotted from data. Of course, this idea might also be applied to measure how the properties of the landed gentry break up over time.

Since a firm hand was lacking in Jasło, I had to make frequent trips there to put things in order, each visit having of necessity to be short. My Mother and aunt were too old to be able to defend themselves against masons demanding excessive payments for repairs, against exorbitant local taxes, theft of wood from the woodlot, the tardiness of lodgers in paying their rent—being inclined to want to live rent-free—and against the parish council which was levying an additional fee for the right to erect a monument over Father's grave. It is a curious fact that fully half of our tenants, and the tradesmen and other workingmen that I encountered in those days had spent some time abroad—in Germany, France, America, or even Russia or Syria. Since the brickworks no longer functioned, I decided to fill the clay pits with water and municipal trash with a view to converting the site to fertile land. I also had to see to it that the oil-producing firm *Małopolska* paid my Mother the dividends she was entitled to. On one occasion the firm's representative responsible for such matters told me that the payment would be determined by the company lawyers, to which I retorted that in fact my Mother was not subject to *Małopolska*'s deliberations as to what was or was not an appropriate payment, but that *Małopolska* was subject to being sued in the civil courts. And indeed a situation did arise where I had to bring them to court. The fraud engineered by the directors of *Małopolska* was as follows. A subsidiary of *Małopolska* called *Gazy Ziemne* (Earth Gas) was formed, completely owned by the parent company, which went ahead and bought petroleum rights from landowners "where the former rights to drill for oil had lapsed though non-exploitation of the agreement", that is, because no drilling for oil had actually taken place. In this way the subsidiary firm, masquerading as a new company, acquired petroleum rights to lands from which my Mother would formerly have potentially earned revenue, but no longer since those rights had lapsed, as it were, and which were now owned by a "different" company which had purchased them directly from the respective landowners, thus side-stepping the former obligations to my Mother. Although the judges in the case knew perfectly well they were dealing with a brazen swindle, they had difficulty formulating the breach in strictly legal terms, and my case dragged on, not reaching the "third instance[182]" till 1939, when the war made it moot. Another piece of dishonesty I had to deal with involved a Jew from Ułaszowice who had bought a building site from Father, but had put down only part of the price. I tried to negotiate a deal with him whereby he would take a proportionate part of the plot of land in question, but it was impossible to come to any agreement as he was a most incredible prevaricator. Finally he sold his rights to the land to some peasants and left with his family for South America.

Fortunately, my Kraków sister came to Jasło from time to time to help settle pressing matters. However, the difficulties tended to increase since revenues from my Father's estate were insufficient to pay for necessary repairs, and to cover all debts and taxes and the cost of care for my ailing Mother. Again and again we had to combat dishonesty, which was endemic in business circles in Poland at that time. It was as if cheating and failing to keep a promise were regarded as good business practice.

We loved foregathering in Jasło because of our nostalgic sentiment for the house we were brought up in, and to spend time with our Mother in her last years and to see old friends like Janek Adamski, Ludwik Oberländer, and others, whom we knew we could count on—and also because it represented an inexpensive summer resort. There I learned to prepare certain culinary specialties, particularly certain herring dishes. And there was a place to dance, since Mr. Dunaj, the owner of the restaurant, found that it paid to put on music two evenings a week. Most of the patrons were local youth, three-quarters of them Jewish. Barred from the universities and the civil service, they were making the best of it by learning to foxtrot and squandering the little bit of money that came to them from their parents' businesses—which were also feeling the rigors of the hard financial times. When in the evening we heard

---

[182]The Polish version of a supreme court.

the whistling of an oriole, we knew it was Janek Adamski come to go dancing with us. While the ladies were dressing there was time for a piece of *torte* washed down with the cognac Janek brought. If it had been up to him he'd have brought enough cognac to have us all reeling before we even got to the restaurant. I regretted that my cousin Dyk Schoenborn did not accompany us, since he would have made a very convivial addition to the group at our little table. However, he had become chief of the local branch of the party known as "The Nonpartisan Bloc for Cooperation with the Government" (BBWR),[183] and in the Jasło *powiat* he did everything he could to bring dislike down on himself. For example, he would threaten officials who refused to listen to him with being transferred, which, since in fact he wielded little power, succeeded only in increasing the number of his enemies. Nevertheless we sometimes went to his home, where we danced to the gramophone or dance music from the radio. There we would sometimes encounter Mrs. Juniewicz, daughter of the *starosta* Zoll, and occasionally Count Józef Michałowski and his Irish wife, inevitably accompanied by a Mr. de Grouchy, whose claim to fame was that he was a descendant of the Napoleonic colonel of that name[184] whose portrait hangs in the *Musée Carnavalet* in Paris. Sometimes Kraków acquaintances of Dyk would show up, and from time to time certain exotic oil men would also turn up. This is how we diverted ourselves during those summers we spent in Jasło.

In the meantime the storm clouds gathering in Lwów grew darker and darker. Following Krzemieniewski, Father Gerstmann[185] was appointed temporary Rector, pending the outcome of a search for Krzemieniewski's successor. One of Father Gerstmann's first actions was to attend the funeral of the murdered anti-Semitic student activist Grodkowski, wearing the Rector's chain of office. The Association of Jewish Students considered this an affront to them, and submitted a letter of complaint to the Senate. The letter had to go through me, as official monitor of this association, and, although its tone was harsh, since it contained nothing that was not true I gave it my *imprimatur*, warning them that they were certainly calling down retribution on themselves for such a letter—and their association was indeed suspended for one whole trimester and other penalties imposed. One of the members of Senate in favor of harsh punishment was the law professor Kamil Stefko,[186] who was appointed by the government not quite as Rector, but as a person "fulfilling the duties of a Rector", the list of candidates suggested by the Senate being ignored.

At that time the most powerful figure in the Ministry of Religions and Public Enlightenment was Father Żongołłowicz, a graduate of the St. Petersburg Spiritual Academy, and supporter of Piłsudski; and reporting directly to him was an inter-

---

[183]*Bezpartyjny Blok Współpracy z Rządem* (BBWR), an ostensibly non-political organization existing from 1928 to 1935, affiliated with Józef Piłsudski and his Sanation movement.

[184]Emmanuel de Grouchy, Second Marquis de Grouchy (1766–1847), French general and marshal.

[185]Father Adam Gerstmann (1873–1940), Polish Catholic priest, doctor of theology, professor at Lwów University and its rector in the academic years 1928/29 and 1932/33.

[186]Kamil Stefko (1875–1966), professor at the University of Wrocław after the war. Founder and first Dean of the Faculty of Law at that university.

mediary by the name of Lipski. There was then an atmosphere of suspicion of all by all permeating throughout Poland, best illustrated by the case of the policeman who, when making a deposition in connection with some clandestine affair, said "Suddenly I noticed a suspect individual," and when the judge asked him how he recognized that that individual was suspect, answered: "Most honorable members of the court, for me every civilian is suspect."

Mr. Lipski was similar to this policeman. For example, when Borsuk,[187] one of the most talented graduates of the Warsaw school, applied for his *Habilitation*, Lipski kept the application on his desk instead of immediately sending it on to the Minister, because Borsuk, when a first-year student, had been a member of the extreme nationalist youth movement. I took advantage of the fact that I was on good terms with the then Deputy Minister Konstanty Chyliński, a former Lwów historian, whom I used to occasionally meet in the café *Wisła* (Vistula)—one of the finest cafés I have ever frequented—during the summers we spent in Krynica, and I informed him about Borsuk. Knowing the sort of people I was dealing with, I reminded him of the book *Szczenięce lata* (*Marley & Me: The Puppy Years*) by Wańkowicz,[188] in which the Borsuk family, of which Karol Borsuk was a scion, is cited as genuine Polish borderland nobility. Of course, for Żongołłowicz heraldry was more important than topology, so the confirmation went through smoothly.

During our vacations in Krynica, I also used to get together once in a while with Leon Kozłowski,[189] the then Deputy Minister of the Treasury. I remember him trying to convince me there would be no war because "there are no contentious issues between us and the Germans." There was a rumor that this Mr. Kozłowski had once been seen seated at a small table in a café with a few of his political colleagues and a German envoy showing them a plan for the dividing up of Czechoslovakia, and that by coincidence the Czech envoy to Poland had happened to be standing nearby and overheard him. When I asked Kozłowski whether this had really happened in a café, he answered brusquely: "No, that was elsewhere." He once asserted to me that devaluation in wartime was impossible because the warring countries would want to keep exchange rates secret. It took me about ten minutes to convince him of the impossibility of such secrecy. It struck me as very strange that the Deputy Minister of the Treasury should be so easily out-argued by a layman in such a matter.

---

[187] Karol Borsuk (1905–1982), Polish mathematician, specializing mainly in topology, subsequently professor at Warsaw University.

[188] Melchior Wańkowicz (1892–1974), Polish writer, journalist, and publisher. Best known for his reporting for the Polish Armed Forces in the West during World War II, and for his book on the battle of Monte Cassino, in which those forces participated.

[189] Leon Tadeusz Kozłowski (1892–1944), Polish archaeologist, and politician. Prime Minister of Poland from 1934 to 1935.

In Krynica I also met Samuel Dickstein,[190] who, although 80 years old, was still extremely lively. He lived in a guest house located higher up than the rest, and walked up and down the steep road several times a day.

I also met Julian Tuwim[191] there. He was staying in the house where the spa was located, and he kept his window shades tightly drawn till midday so that the sunlight would not wake him. He was continually besieged by female admirers bringing him flowers. He told me that the extent of his present popularity was due to the fact that Kiepura was absent from Krynica just then, so he had the ladies' undivided attention—Tuwim and Kiepura being at that time public heartthrobs of the same lofty stature. There were even ladies who tried to get to Tuwim through so-called café "fordancers".[192]

I rate Tuwim's poetic power and unerring poetic aim very highly. Pawlikowska's[193] poem about him gets to the bottom of his poetic individuality. Wittlin, who expressed what he had to say so very directly, compared Tuwim to a blazing bush. Tuwim had many of his own poems by heart, as well as many of those of Pushkin. His translation of Andreï Belyĭ's[194] *Letters from Tula* is positively brilliant in the way it transforms Russianness into Polishness.

On the other hand, the cult of Kiepura, which succeeded that of Paderewski,[195] was a measure of the degree to which the average standard of aesthetics had degenerated in Poland over the years since the end of World War I. The Kiepuras, these Cyganiewicz brothers[196] of Polish song, lived for some time in the hotel *Patria* in Krynica, a splendid but very costly structure that he himself, the so-called "Master", had built. The waiters there used to call him "Mister Master", his younger brother "Mister Brother", and his father "Mister Chairman". I actually met "Mister Brother" when we were seated together at a small table in the *Patria*. As the due of a non-singer he gave me an exaggeratedly limp and casual handshake. I once saw

---

[190]Samuel Dickstein (1851–1939), Polish-Jewish mathematician. One of the founders of the Jewish movement *Zjednoczenie* (Unification), advocating the assimilation of Polish Jews. Considered the preserver of Polish mathematics in Warsaw during the nineteenth-century Russian occupation. Killed by a German bomb at the beginning of World War II. All of his immediate family perished in Nazi concentration camps.

[191]Julian Tuwim (1894–1953), considered one of the greatest of modern Polish poets. In 1919 cofounded the "Skamander" group of Polish experimental poets with Antoni Słonimski and Jarosław Iwaszkiewicz.

[192]Men paid to dance with women.

[193]Maria Pawlikowska-Jasnorzewska (1891–1945), Polish poet, known as the Polish Sappho and "queen of lyrical poetry" of interwar Poland.

[194]Pseudonym of Boris Nikolaevich Bugaev (1880–1934), Russian novelist, poet, and literary theorist and critic. Vladimir Nabokov considered his novel *Petersburg* one of the four greatest novels of the twentieth century.

[195]Ignacy Jan Paderewski (1860–1941), Polish pianist, composer, diplomat, and politician. Prime Minister of Poland in 1919.

[196]Well-known Polish strongmen and wrestlers of the interwar period. See Chapter 2.

Kiepura the elder in the company of Marta Eggerth,[197] evidently returning from a game of tennis, with spanking new racquets and snow-white shoes. It was as if they had stepped from the front page of *Światowid*.[198] Witkowski's[199] wonderful jazz band played every day in the *Patria*, and there was dancing every afternoon and evening, the public continually calling for more: "Too little, too little!" This was echoed by the owner Kiepura when Metro-Goldwyn-Mayer or Paramount—I forget which—phoned to offer him half a million dollars to sign a contract for three months to make a singing version of some film or other. The opera star Adamo Didur was his friend and sometime financial adviser. I never met him, although he often came to Jasło to visit his friend Antoni Zoll, and one of his daughters, Mary, Count Załuski's wife, lived in Iwonicz, close to Jasło. Didur's wife, a Mexican lady,[200] also lived there, and in the early twenties the whole of this company would occasionally come to Jasło, causing a sensation—especially if "Mister Master" himself came along.

In Krynica I also got to know the neurologist Maksymilian Rose,[201] a professor of physiology in Wilno specializing in the anatomy of the brain. In 1935, following on the death of Marshal Piłsudski, the government supplied him with a sizable sum of money to study the great man's brain. Of course, anyone with a scientific bent would be sceptical about these investigations' resulting in anything as illuminating as some of the gentlemen from the BBWR were hoping for—in particular, it was unlikely that anything could be inferred from the ganglions of this brain as to how Poland might continue to exist squeezed between Hitler's Germany and Stalin's USSR. Professor Rose told me that a certain professor at the University of Wilno had willed his brain to Rose's department, and that at his demise his brain duly arrived and was carefully anatomized. It was observed that certain parts of the visual centers were poorly developed, so Rose turned to the family of the deceased to ascertain if he had demonstrated any visual handicaps when living. However, as far as they knew he had always had average vision, taking to wearing reading glasses only after the age of forty—also quite normal. The dead man's diaries likewise failed to reveal anything unusual about his vision. But finally, when Rose talked to the man's colleagues, he made an interesting discovery. He was told that when he had been Dean, he could not, even after a year in the post, learn which door led from his office to the corridor, and, once in the corridor, how to exit the building—his secretary had had to show him the way out every day. The conclusion drawn was that the areas of the brain responsible for the ability to orient oneself in space overlap with the visual centers. The celebrated neurologist also told me that the number of connections between

---

[197] Hungarian lyric soprano. Often performed with Kiepura. Fled to the US with Kiepura to escape Nazism. Born in 1912.

[198] A social and cultural illustrated weekly published in Kraków between 1924 and 1939.

[199] Bernie Witkowski, American band leader. Son of a Polish immigrant to the US. Mixed traditional Polish music with American jazz.

[200] Angela Aranda Arellano (1874–1928), Mexican singer. Daughter of Spanish immigrants to Mexico.

[201] Polish neuroanatomist and psychiatrist. Lived from 1883 to 1937.

neurons greatly exceeds the number of thoughts and impulses that one has—to such an extent that even if one assumes that the average person can experience millions of such phenomena every second, this activity represents but a tiny part of what nature has in reserve in the form of possible connections.

* * *

Political events unfolded with sometimes dizzying unpredictability, and without us, the powerless onlookers, being able to discern their sense or tendency.

In Jasło I asked my cousin Dyk, as someone privy to the plans of the ruling coterie, what would happen when Piłsudski died, the man whose name represented a guarantee of the competence of those around him who had so willingly taken on responsibility for the fate of the nation. Without Piłsudski's aura of infallibility, what will they use to reassure the people? But Dyk merely answered with the cynicism typical of his political non-party: "A new legend will be fabricated! We have already chosen the central figure of the new legend: Rydz-Śmigły!"[202]

When Piłsudski died in 1935, I was Dean for a second time, though only for a single trimester, since the incumbent had had to relinquish the post before his term was up. That's how I found myself attending Piłsudski's funeral in Kraków as representative of the Jan Kazimierz University, and was able to observe many of those taking part in the procession. As the open carriage bearing the coffin moved out of the railroad station to the accompaniment of a mournful drum beat, we waited in our academic "togas" for the appropriate moment to join the procession. Ahead of the bier walked Archbishop Sapieha,[203] praying continuously, and at the same time looking as if he regarded the duty of leading the procession as a form of penance for his manifold sins. Alongside the military escort strode Wieniawa-Długoszowski,[204] followed by the family of the deceased Marshal, demonstrating by their modest appearance the Marshal's genuineness in not caring about personal gain. After them came the President of the Republic of Poland, followed by ambassadors and envoys, two of which stood out from the rest: Lord Cavan,[205] representing Great Britain, and Hermann Göring, thick-bodied and clearly pleased to be witnessing this particular funeral. At this point we joined the procession, and I was unable to see much of those coming behind. I was aware of the fact that the members of the *Sejm* and Senate[206] came after both us and the members of the supreme court, which

---

[202]Edward Rydz-Śmigły (1886–1941), Marshal of Poland, Polish political figure, painter, and poet. Hero of World War I and the Polish-Soviet War of 1919–1920. Succeeded Józef Piłsudski as General Inspector of the Polish Armed Forces in 1935 on Piłsudski's death, but lacked the latter's authority.

[203]Prince Adam Stefan Sapieha (1867–1951), Polish Roman Catholic cardinal and archbishop of Kraków.

[204]Bolesław Wieniawa-Długoszowski (1881–1942), Polish general, politician, diplomat, physician, and poet. Piłsudski's *aide-de-camp* during the Polish-Soviet war.

[205]Frederick Lambart, 10th Earl of Cavan (1865–1946), Field Marshal and Chief of the Imperial General Staff 1922–1926.

[206]The lower and upper houses of the Polish parliament.

7 In the University Town Lwów

represented a breach of propriety, and having members of the august Academy of Learning coming after those of the new upstart Academy of Literature was a gross impropriety. As we arrived at the *Wawel*,[207] we saw ranged alongside us various regiments in full regalia. The faces of the young soldiers in their antique-looking steel helmets looked very beautiful, and I couldn't help but gaze on them with admiration. They, on the other hand, looked amused by us in our academic get-up, especially by the corpulent Rector of the Mining Academy outfitted in accordance with his standing as a Knight of Malta.

The funeral of Marshal Piłsudski was to be the last manifestation of Polish state power, as the government then embarked on a continual decline. Much of the blame lay with Mościcki.[208] I recall when he visited Jasło as President on a tour of the Kraków Voivodeship by car. He stayed overnight in the Town Hall, and I watched his departure from a front window. His retinue consisted of some dozen cars in which there sat young men in bowlers. The tempo of the visit was rather too hurried, and it had the air of an escape from somewhere by a well-organized gang of crooks. Of course, this was only my impression, but it was a very powerful one, albeit produced more by Mościcki's hangers-on than by the man himself. It was as if various shady characters had taken advantage of the president's poor acumen to infiltrate his entourage. But I could be wrong, since although Paderewski's intelligence was well above average, I remember my uncle telling me that whenever he entered the premises of the Council of Ministers in Paderewski's time as president, he would instinctively button up the pocket in which he carried his wallet, such being the impression made on him by the secretarial and majordomo types moving through the corridors of power.

As the reader may have noticed, Krynica was a place where one could meet a variety of interesting people. Another, very unusual, such person was an biochemist by the name of Lejwa, whom I looked up there on the recommendation of Dr. Blumenfeld. He came from Kielce,[209] where he had been educated in a Talmudic school, and had begun to learn Polish only when he reached the age of about 12. His reading in Polish then took fire and he became entranced by the beauty of the novels of Sienkiewicz. Having been taught to reflect, he arrived at the conclusion that what he had had drummed into him at the Talmudic school—that all Goyim[210] are intellectually crude and what they write not worth the candle—couldn't after all be true. He went on from this to conclude that if the Talmudists didn't know

---

[207] An architectural complex dating in part from the ninth century, built and rebuilt over many centuries atop a limestone outcrop on the left bank of the Vistula in Kraków. It includes a royal castle, an armoury, and a cathedral, and is of great historical interest to Poles. Site of royal coronations and interment of distinguished Poles.

[208] Ignacy Mościcki (1867–1946), Polish chemist and politician. President of Poland from 1926 to 1939, he entered into a power-sharing agreement with Rydz-Śmigły in 1935, whereby he would continue as president, with Rydz-Śmigły as *de facto* leader of Poland. However, Rydz-Śmigły lacked the moral authority of Piłsudski, and the government degenerated into bitter in-fighting between the factions headed by these two and others.

[209] A city in the southeastern part of central Poland, capital of the Świętokrzyskie Voivodeship.

[210] Non-Jews.

about Sienkiewicz, then there must be many other things worth knowing of which they were ignorant. Acting on these conclusions, he embarked on the study of the learning of the Goyim, acquired a high-school certificate, went to Paris, where he acquired a very good degree in chemistry, and returned to Poland, where he became an acknowledged expert on hormones, employed in the State Institute for Hygiene. I found out from him that the fad for injecting male hormones into married women to increase their libido was the purest quackery since they resulted only in the masculinization of the subject, doubtless not the hoped-for outcome.

His most interesting stories were of his youth in Kielce. He still had difficulty understanding why one might like to go for a walk in the woods, or appreciate the beauties of nature, since during his younger days in Kielce, *bucher*[211] either sat at home or stayed at school reading the Talmud, and were in any case discouraged from spending any free time they might have had walking around the town by stories of fierce dogs and stone-throwing urchins. He related how when he returned from Paris, a friend from the Rabbinical School once visited him at five in the morning and insisted on having him explain "Vos iz dos 'ein-stein'?"[212]. But to appreciate this story one really needed to see the very funny way in which he mimicked, with the appropriate gestures, his friend's innocent posing of the question.

He also brought up an interesting biological question. There are small, whitish fish, rather common in our waters, whose bellies become reddened during their spawning period, and when these fish were injected with a certain substance extracted from the human pituitary gland, the same effect was observed. The natural question to ask then, was what role that substance plays in the human organism. Dr. Lejwa also called my attention to the importance for biology of being able to obtain a quick estimate of areas of superficial regions of an organism: for example, the potency of certain hormones used on roosters was measured, he said, by the increase in area of their combs due to swelling after injection with a unit amount of the hormone, but an accurate method of measuring such irregular areas as that of the surface of a cockscomb was lacking.

Days and even years were flashing by: the book with Kaczmarz had come out, and ever new problems and projects were continually arising, so that the turmoil in the country became ever more visible. In 1937 there was a mass strike by Polish peasants aimed at the ruling Sanation party, protesting the hardships ultimately caused by the Great Depression. The anti-Semitic movement was becoming ever more virulent, led by such idiotic upstart "heroes" as Doboszyński,[213] the same

---

[211]"young men"

[212]Yiddish for "What is 'ein-stein'?"

[213]Adam Władysław Doboszyński (1904–1949), Polish politician. One of the leaders of the National Party (*Stronnictwo Narodowe*). In 1936 he organized a march on Myślenice, a small shtetl near Kraków. His militia occupied the town, disarmed the police, plundered Jewish stores, and attempted to set fire to the synagogue. He was ultimately arrested and sentenced to three and a half years in prison. This was just one of many anti-Semitic incidents and pogroms of the period, which multiplied in spite of organized Jewish resistance.

man who in a youthful work proposed the reintroduction of *ius primae noctis*[214] as a right of all students at institutions of higher learning.[215] Yes, the mental virus known as anti-Semitism had so infected the populace, that the issue of "Ghetto benches"[216] seemed to be the most important in Poland. Begging your pardon, but I had forgotten about the issue of ritual slaughter.[217] It seemed that the most fanatical of the proponents of so-called "humanitarian" slaughter of animals would gladly kill people to spare the suffering of cattle. The arguments of Father Trzeciak,[218] an "expert" on the Talmud, and in this capacity consulted by the *Sejm*, captured attention far more than the newly agreed-upon constitution—although that was in any case of questionable significance. Ruziewicz told me the following story about Trzeciak. Some years ago Lwów University had numbered the eminent orientalist Professor Andrzej Rawita Gawroński[219] among its luminaries, and also a docent Moshe Schorr, who would later be appointed to a position at Warsaw University. Well, Trzeciak became involved in some scholarly dispute with Schorr, who, unlike Trzeciak, was genuinely erudite. Assuming that Gawroński must surely be a passionate anti-Semite like himself, Trzeciak wrote asking him to put Schorr in his place, but Gawroński answered that in scientific matters he was never guided by what, for merely political reasons, people liked or disliked, thus deflating Trzeciak's fustian. However, only a few specialists knew about Gawroński, while everyone knew about Father Trzeciak. And the Jews, who are so easily out-manoeuvered politically, failed to notice that the issue of ritual slaughter was only a pretext for the anti-Semites—who in this matter seemed cleverer than they—so that their reaction of passionately defending their right to this tradition was misjudged, playing as it did into their enemies' hands. In that situation people like me benefitted the most, since the new laws still allowed one to eat the high quality meat of animals slaughtered by Jews—though no longer according to ritual—while practicing Jews could not do so for religious reasons, and Christians likewise, though for statutory reasons. Thus someone like me, a so-called Jew though paying no attention to the Rabbi's strictures, could buy good meat cheap. But there were, of course, few who treated these matters with any degree of humor.

---

[214] The supposed feudal privilege of the lord of the manor to bed his peasants' brides on the night of their wedding.

[215] See: Adam Doboszyński, *Słowo ciężarne* (The Pregnant Word), Warszawa, 1932.

[216] Official segregation in the seating of students, introduced in Poland's universities, beginning with the Lwów Polytechnic in 1935. By 1937 the practise had become conditionally legalized. Under this system Jewish university students were compelled, under threat of expulsion, to sit in designated left-hand side sections of lecture halls.

[217] Polish laws limiting ritual slaughter—forming part of the traditional Jewish religious practice— were passed in 1936, and in final and drastic form in 1939, following the Nazi invasion of Poland.

[218] Stanisław Trzeciak (1873–1944), anti-Semitic Polish priest, and self-proclaimed expert on Jewish lore.

[219] Andrzej Gawroński (1885–1927), Polish orientalist and polyglot. His father was Franciszek Rawita-Gawroński, a Polish historian, writer, and publicist. Andrzej dropped the "Rawita" because it had been used as a pseudonym by his father.

We continued spending our summers in Krynica, and during one such I met a Mrs. Knoff there, staying at the guest house *Skarbówka*. While we were walking in the park, she asked me to explain to her what occupied mathematicians, so I began to make drawings in the dirt with my walking-stick to illustrate my words. It was then that the idea came to me of writing a mathematical "picture book" that would answer the questions of this charming lady. Fortunately, I found someone to help me with this project, namely Marek Kac,[220] who was then a student in the Krzemieniec *Lycée*. I judged him to be gifted, and to test my judgement I gave him the definition of "independent functions" in the probabilistic sense[221]—a concept which, as I later found out, had been introduced earlier by Kolmogorov[222]—and asked him to see what he could do with it. This was just before the 1935 summer vacation, and at the end of the break he was able to show me several interesting results on the concept. With him as assistant, I then embarked on two collaborative projects: writing research papers on independent functions and producing the mathematical "picture book." When Dr. Marcinkiewicz[223] from Wilno presented a paper on Brownian motion in my seminar, we realized that the whole theory of these motions, worked out initially by Einstein and Smoluchowski, should properly be based on the concept of independent random variables, since then all of the statistical assumptions reduce to just one: the *independence* of successive collisions of molecules with each particle. Whereas before this independence had been left rather vague, now it acquired mathematical precision.

---

[220]Marek (Mark) Kac (1914–1984), Polish mathematician. Obtained his doctorate from Lwów University. Worked mainly in probability theory. Went to the US in 1938 on a scholarship from the Parnas Foundation. Later held positions at Cornell, Rockefeller University, and the University of Southern California.

[221]That is, functions considered as random variables.

[222]Andreĭ Nikolaevich Kolmogorov (1903–1987), preeminent Soviet mathematician.

[223]Józef Marcinkiewicz (1910–1940), Polish mathematician. Studied under Antoni Zygmund. Mathematics docent at the Stefan Batory University in Wilno. Murdered by the Soviets in the "Katyń massacre". (He was in fact murdered in Kharkiv, Ukraine, along with thousands of other members of the Polish educated class—on Stalin's instructions. The term "Katyń massacre" originally referred only to the massacre by the NKVD at Katyń Forest, near the villages of Katyń and Gnezdovo (about 19 km west of Smolensk, Russia), of Polish army officers from the Kozielsk prisoner-of-war camp in 1940. However, the term "Katyń massacre" subsequently came to refer to the execution by the NKVD of about 22,000 members of the Polish intelligentsia, including about 14,550 prisoners-of-war (mainly Polish army and police officers) arrested in Soviet-occupied Poland, sent to camps chiefly in Kozielsk, Starobelsk, and Ostashkov, or to Soviet prisons in Kharkiv, Kalinin (Tver) and elsewhere, then shot and buried in mass graves. Soviet officialdom long claimed they were Nazi victims, continuing to deny responsibility for the massacre till 1990.)

Encouraged by this, we tried our approach on the Ehrenfest model,[224] on Peano functions,[225] on certain number-theoretic problems, and so on.

At the same time, we were working on the "picture book". We went to the studio of the photographer Miss Wanda Diamandówna, on Fredro Street, and broached the idea with her. She was very taken with the notion of photographing mathematical objects, and, with tremendous patience, she devoted to its realization a great deal of time as well as space. It might seem on the face of it that it must be easy to photograph a twisted and glued paper strip, but in practise it turned out to be more difficult than photographing a beautiful lady. The lighting had to be arranged so that it would be perfectly clear what the object was illustrating, which required that the part of the strip's shadow that happened to fall on the strip itself should not be as dark as the background, and the whole thing had to look plastic. We had to make and photograph a dozen or so such models, and photographing a single one of them required over two hours of work. On one of the models we wanted to place a paper fly, and I went to the Magic and Souvenir Store in the Mikolasch Passage to see if they had one. After searching minutely among his wares, the storekeeper announced that he had no fly, but did have a paper elephant. I bowed to him and intoned in the language of *The Thousand and One Nights*: "O sir, you are undoubtedly the cleverest of merchants if, without making incantations or even smiling, you can turn a fly into an elephant—a feat that till now has only taken place in fables." We also had to mobilize an army of five hundred lead soldiers: setting them up on a table and photographing them took Marek Kac close to a week. To illustrate the Gaussian probability distribution I used shot allowed to run at random down an inclined board—a fairly awkward experiment to arrange. In the end it turned out that an ordinary board without any impediments to the rolling shot gave a better result than one with uniformly distributed nails or hexagonal obstructions. Getting a good shot of a honeycomb was easy, but arranging four soap bubbles next to one another and photographing them took two weeks of practise at blowing bubbles! I very much wanted to have a real cardboard model of a regular dodecahedron in the book, but the problem was to build such a polyhedron made of two separate halves that would pop up like a top hat when one opened the book. This cost me a sleepless night, but at last I hit on an ideally simple solution: a rubber band around the model's middle. In making anaglyphs[226] I had the help of Mr. Franciszek Otto,[227] assistant

---

[224] A model of diffusion proposed by Paul Ehrenfest to explain the Second Law of Thermodynamics.

[225] Functions similar to that invented by Peano, which maps a line segment continuously onto the unit square. Giuseppe Peano (1858–1932), Italian mathematician and logician. One of the founders of set theory.

[226] Drawings or photographs in two off-set colors which when viewed through glasses each lens of which filters out one of the two colors, trick the brain into perceiving a stereoscopic, that is, three-dimensional, image.

[227] Franciszek Otto (1904–2000), Polish mathematician. Following World War II adjunct professor of descriptive geometry at Gdańsk Polytechnic. His brother Edward was also a professor of descriptive geometry.

to Professor Bartel, who was meticulous to the point of pedantry in his work, and also had the advantage that he was skilled at taking stereoscopic photographs for the anaglyphs.

It was planned that the book would appear simultaneously in Polish and English editions. My contract with the publishing house *Książnica Atlas* specified 4000 copies, with the 2000 Polish copies being retained by the publisher, and the 2000 English copies representing our payment as authors and suppliers of the illustrations. The book would have appeared in 1937 but for a fire on the publisher's premises which delayed publication by a year.

In the meantime the young idiots and the cunning old villains who were goading them became more and more rabid. During the academic year 1935–1936 Jan Czekanowski[228] was appointed Rector of the Jan Kazimierz University. So far was he from advocating university autonomy, that he had submitted his inaugural address beforehand to Senator Kozłowski for censoring. He was followed by Kulczyński,[229] a courageous and decisive person who understood the seriousness of the situation, and saw through subterfuges and quibbles. He seized the bull by the horns and organized a student referendum on the "bench ghetto" question, the result of which showed that only a minority—a clear minority—of Polish students supported it. The fury expressed from among the various ranks of *canaille* knew no bounds, and, after dissolving the refractory student associations, Kulczyński was forced to close the university. However, lacking the support of many of his colleagues and of the minister—at that time Wojciech Świętosławski[230]—he eventually resigned in protest.[231] One can say the same things about Świętosławski as about Kulczyński, but with the opposite sign: he was a coward and a fool, without intelligence enough to formulate a definite political line.

We spent the summer of 1936 again in Krynica, and this time I met a lawyer called Szumański who was involved in the defense of a certain person accused of killing a peasant during the Przytyk pogrom[232] by shooting at him from an upstairs window. Since the line of entry of the bullet into the body had been determined by the autopsy, the case came down to obtaining a professional opinion as to

---

[228]Jan Czekanowski (1882–1965), Polish anthropologist, statistician, and linguist. Introduced numerical taxonomy into comparative linguistics. In 1913 he developed a still much-used index of similarity between two samples of some kind of objects.

[229]Stanisław Kulczyński (1895–1975), Polish botanist and politician. Rector of Lwów University from 1936 to 1938, when he resigned in protest at the "Ghetto Law". A member of the "Polish Secret State" during World War II, he participated in the well-known Polish underground education project. After the war first Rector (1945–1951) of the University and Polytechnic in Wrocław. Deputy chairman of the Polish Council of State 1956–1969.

[230]Polish physicist. Father of modern thermochemistry. Lived from 1881 to 1968.

[231]In 1938.

[232]This occurred in the town of Przytyk in east-central Poland when local peasants, incited by extremist politicians of the right-wing National Democratic Party, attacked Jewish stores in March 1936. The Jewish community fought back, but in the trials following the restoration of order, Jewish participants in the disturbances were unfairly penalized. As a result there was a significant increase in Jewish emigration from Poland at that time.

whether, according to the laws of mechanics, the angle of entry of the bullet was compatible with its having been fired from the window in question. Not being myself a specialist of the right sort, I wrote a letter for the defense stating that in principle the question could be settled, and that Antoni Przeborski, professor of theoretical mechanics at Warsaw University, and Maksymilian Huber,[233] professor of mechanics at the Lwów Polytechnic, would be appropriate expert witnesses. It is significant as characterizing that unstable epoch that writing such a letter and allowing it to be used as court evidence was viewed as an act of incredible civic courage!

* * *

Who was ruling Poland? Good question. The President of the Polish Republic[234] organized a Polish Harvest Festival. During the celebrations, rain washed the dye from what looked like multicolored blotting paper piled on his head down onto his face, and his wife, normally a reasonable woman, wept to see what a laughing stock the eminent chemist was making of himself. Then there was the Prime Minister of the Polish Republic, Leon Kozłowski, author of the decree concerning Bereza Kartuska,[235] which gave the authorities the power to lock up in that prison camp persons who they merely *thought* might be guilty, even if evidence was completely lacking. Justifying the law before the *Sejm*, he said: "The hand of Justice will strike with a vengeance those locked up in that place of isolation. Ruffianly students, Ukrainian arsonists, and Jewish communists will together do useful work there building roads in Polesie." Much later, in January 1939, I had the opportunity of asking Kozłowski how he reconciled the Bereza law with the constitution, to which he responded that, yes, it did indeed represent a form of governmental *lawlessness, but a necessary one* in view of the situation in Poland. Then I asked him why he used the phrase "the hand of justice" in his speech when the law states specifically that guilt is in fact not an indispensable condition for being incarcerated in Bereza Kartuska, to which the former Prime Minister replied: "Yes, that was a mistake on my part."

The conditions at the concentration camp in Bereza Kartuska were extremely harsh, inmates being beaten and starved routinely, and outdid in cruelty its model, the camp at Dachau in Bavaria. Of course, the claim that useful road construction was carried out was completely false; the exclusive object of the legislation was to imprison people by administrative fiat, without right of appeal, for three months at a time—in flagrant violation of the constitution. By way of example, I mention the case of a Kraków alderman who, in order to register disapproval of the unrest

---

[233] Maksymilian Tytus Huber (1872–1950), Polish expert in mechanical engineering.

[234] Ignacy Mościcki—see above.

[235] A law permitting the detention of those "whose activities give reason to believe that they threaten public security, peace, or order" in the prison at Bereza Kartuska in Polesie Province (now in Belarus). The law was in force from 1934 to 1939, and was passed when Kozłowski was both Prime Minister and acting Minister of the Interior.

organized by students, advocated lowering to just one złoty the subsidy called "Brotherly Assistance" extended to students of the Jagiellonian University by the City of Kraków. A few days after making this motion at a meeting of the city council, the alderman found himself unceremoniously abducted and sent to Bereza Kartuska.

At that time the Minister of Education was the above-mentioned Wojciech Świętosławski. When a chair became vacant in the Department of Chemistry, we found ourselves in the position where there was just one serious candidate, namely Kazimierz Fajans[236] of Warsaw, who was then the Head of the Institute of Physical Chemistry in Munich. The recent passing of the Nuremberg Laws[237] meant that he would be well-disposed to coming to Lwów. Although certain parties reproached him for trying to obtain a position in Poland "merely" because of the Nuremberg Laws, those of us who had a premonition of what these laws prefigured, appreciated that this was a completely appropriate step. Thus Fajans came to Lwów to test the waters, staying with his cousin Professor Parnas. At the end of his visit, as he was boarding the train from Lwów to Warsaw, a young man threw an egg at him. In any case, the University Senate rejected the proposal of the Division of Mathematics and Natural Sciences to appoint Fajans, and then the Council of the Department of Chemistry proposed offering the position to one of their own, Roman Małachowski,[238] who had voted against Fajans on the grounds that "Jews are enemies of the Polish nation," and had recommended offering the position to a certain native German from Königsberg. After very little discussion the Senate passed the resolution and sent it on to Minister Świętosławski, who declared, however, that he couldn't pass it on to the President for signing. One might say that Fajans' life was saved by an egg, since he soon afterwards left for Cambridge, and eventually obtained a position at Ann Arbor, in the US.

* * *

In that twilight time I was very busy. One matter that occupied me was the search for a *modus vivendi* for the different branches of the Polish Mathematical Society: the Warsaw Branch's ambition to dominate and the mistrustful stubbornness of the Kraków Branch would have led to a breakup were it not for the mediation of our Lwów Branch. Since I was then chairman of the Lwów Branch, it fell to me to invite delegates from the other branches of the Society and lead the debates—and after a session lasting around 12 hours, I managed to get everyone to agree on a new constitution. The main "holdouts" were Zaremba Sr. and Kazimierz Kuratowski.

In the Fall of 1936, the Polish Mathematical Congress took place in Warsaw. The hubris of the Varsovians was evident in the proportion of plenary lectures they assigned themselves: fully half of them were to be given by Warsaw mathematicians,

---

[236] Kazimierz Fajans (1887–1975), Polish-American physical chemist. Pioneer in the field of radioactivity. Established, together with the English radiochemist Frederick Soddy, the existence of isotopes of certain radioactive elements, a major discovery of physical chemistry.

[237] Anti-Semitic laws passed in Nazi Germany at the 1935 Nuremberg Rally of the Nazi party.

[238] Murdered by Germans in 1944.

and the rest divided up amongst four other branches. I myself was assigned a plenary lecture on independent functions.

At that time I was investigating measurements using X-rays, motivated by a problem of gynecological diagnosis communicated to me by a Dr. Emil Meisels, a radiologist working in Lwów: how can one determine that the head of a foetus will pass through the pelvic opening, the shortest diameter of which—the so-called *conjugata vera*—runs from the joint between the pubic bones to the anterior pelvic promontory? Despite various attempts, I hadn't found a practical solution, but in the process I became interested in another, more general problem: how to make organs or foreign objects inside the human body visible. An X-ray photograph only shows projections of such organs on a screen, and not how they lie relative to one another. Just after the Congress in Warsaw I bumped into Dr. Meisels on the square of St. Mary's Cathedral, and assured him that I would solve this problem, even though all I had was a conviction that I had somehow to exploit the principles of stereoscopic vision. Once again with the help of Bartel's assistant Franciszek Otto, I tried all manner of things. First I attempted making anaglyphs on glass with the actual object placed under the glass, but here I ran into the problem that, in order to view the anaglyph in relief, one has somehow to gain conscious control of the eye muscles responsible for accommodation. Eventually I discontinued this avenue of approach, and began to experiment in Jasło with two wine glasses, trying to arrange their reflections in a window pane so that I saw them as a single image. After many a trial and error I finally hit on a procedure that would produce a stereoscopic X-ray photograph. However, Dr. Meisels, who had just then returned from a conference of radiologists in America, told me that he had seen something similar there, although he couldn't say what exactly. After further prodding, he produced the name of a Professor Hasselwanger from Erlangen. I wrote to this professor and received a sheaf of articles on his work. It turned out that as early as World War I he had made practical use of stereoscopic X-ray photography: the idea had been broached by Ernst Mach just a few years after the original invention of X-ray photography by Röntgen,[239] and had been realized in practice by certain Frenchmen. This discovery shattered me, broke my spirit. When I asked my helper Otto if he thought I should give up, he encouraged me to persist, so I kept thinking and experimenting, constantly returning to the problem of superimposing the reflections of a wine glass in a double window pane on the real wine glass, but was held up by circumstances for some time.

Meanwhile my student Marek Kac completed his doctorate, and to celebrate he invited my daughter and me to a reception at Dr. Hetper's[240] place, organized by him and his aunt Miss Bronia. Dr. Hetper was then working as assistant to Professor

---

[239]Wilhelm Conrad Röntgen (1845–1923), German physicist. Produced and detected the short-wave electromagnetic radiation known today as X-rays in 1895, for which he received the first Nobel Prize in physics in 1901.

[240]Władysław Hetper (1909–1940), Polish mathematician. Studied logic in Kraków, obtained his doctorate in Lwów, and habilitated just before World War II broke out. Was arrested by the Soviets and perished in the Gulag.

Chwistek, and was specializing in logic. He was an ideally righteous person, a devout Roman Catholic, tall and handsome, and an excellent sportsman. He was friends with Herzberg,[241] also a logician, but a devout communist—although one of almost evangelical principles. While he himself lived in what must be called poverty, he never refused help to anyone, and was always willing to share the little he had with those worse off than he. I met him for the first time at this reception, where I was told the story of how he had been summonsed to appear in court because of notes found in his rooms which looked as if they were in code. The case was resolved when Professor Chwistek, called as witness, explained that the putative "ciphers" were merely standard symbols of mathematical logic.

For a few years at the end of the 1930s I acted as the university's representative on the Scholarships Committee, which included a delegate from the Polytechnic, three people representing various petroleum enterprises contributing to the Parnas Foundation, as it was called, and its chair Mr. Hłasko, director of the firm *Małopolska*. Naturally, most of the scholarships went to engineering students, especially those specializing in petroleum technology in Professor Pilat's school, and to medical students. It was much less common for students in theoretical disciplines to be awarded scholarships, and this was a recurring theme in the debates between the oil industry people and myself. I remember that on one occasion I tried to obtain a scholarship for a specialist in Semitology[242] recommended to me by my colleague Father Aleksy Klawek, who had a very high opinion of him because he had been able to decipher an Aramaic text that had defeated other specialists. When the oil industry types on the committee argued against this on the grounds that petroleum specialists were of paramount importance for Poland, I responded that, as someone raised in the oldest of the oil-producing regions of Poland, I was fully aware of the importance of oil, and even liked how it looked and smelt, but that at the same time I couldn't allow myself to forget that its purpose was to provide oil for the lamps by whose light scholars may decipher the texts of past ages.

Dr. Józef Parnas[243] was also a member of the committee, and usually mediated such disputes, so that in 1938 I managed to obtain a scholarship for my student Marek Kac[244] to go to the US. Since I wanted to continue working with him in the meantime, I gladly accepted his invitation to spend that summer in his home town Krzemieniec, taking my daughter along.

Lidka had by this time completed first-year law, but of course not without incident. As might be expected, she refused to comply with the wishes of those trying to enforce the "Ghetto benches" law, and always took her place in the lecture

---

[241] Jan Herzberg (1908–1941?), Polish logician. Educated in Kraków, he came to Lwów in 1938. Went missing in June 1941.

[242] The study of the Semitic languages and the literatures and history of the Semitic peoples.

[243] Appointed first chairman of the Lwów *Judenrat* when the Germans took Lwów in 1941, but then shot for refusing to cooperate with the Nazis.

[244] Though recommended by Steinhaus, Kac was not awarded the Parnas scholarship on his first try in 1937. This was perhaps fortunate since had he gotten it in 1937, he would most likely have returned to Poland in 1938. In 1939 there was, for obvious reasons, no question of returning.

halls wherever it suited her. Since the young university stalwarts were somewhat afraid of calling down disciplinary measures on themselves should they begin manhandling a young woman, they recruited Polytechnic students to do their dirty work, and so it happened that on one occasion four of these valiant warriors took hold of my daughter and removed her from her seat. Even though she was certainly not possessed of unusual strength and in any case offered no resistance, those brave lads evidently felt that a war party of four strong was the necessary minimum. However, there was an occasion where a single such wrong-headed individual entered the hall where a docent was lecturing and without removing his cap stood in the middle of the room demanding that the lecture be halted—perhaps in protest at the non-observance of the law regarding seating of Jews—and threatening that "it will be bad for all" otherwise. The docent asked him his name, but the student refused, demanding instead the docent's name. The docent complied, but later failed to report the incident to the Rector.

Aleksander Rajchman was largely correct in his article written for a certain political weekly at that time in maintaining that the professors were mostly to blame for the students' excesses. For his outspokenness he was subject to a disciplinary hearing by the university, but then it turned out that the university in fact lacked the necessary competence to judge him since the pertinent law did not apply to docents. The presiding authorities asked docent Rajchman if he would agree to be judged on the basis of the statute as it applied to professors, but Rajchman refused, naturally, and the university's humiliation was total.

What poor types of people one sometimes had to deal with in Poland in those troubled times is illustrated by a story told to me by my son-in-law-to-be Jan Kott. One day he was visited by two young people from the radical-nationalist camp and asked to write an "ideological declaration" for them; they offered to pay him well as long as the declaration was sufficiently strident in the radical-nationalist manner—the precise content being of little importance to them. Their aim was to use the declaration to help overthrow the current leaders of the association they belonged to and seize power for themselves. When Jan asked if his racial origins perhaps disqualified him for composing such a document, they replied that the only thing of any interest to them was his ability to produce it. Their task would then be to exploit the inflamed rhetoric of the declaration to sway ruffians among the students to support them.

As planned, I spent the summer of 1938 staying with the Kac family in Krzemieniec. Kac senior had a doctorate from Leipzig University, where he had been a student of Wundt,[245] and supported himself and the four other members of his family by giving private lessons in German and Hebrew, and teaching religion in the Krzemieniec *Lycée*. His wife's sister, Ms. Bronisława R., also lived with them. It was she who showed me the sights of Krzemieniec: the ruins of Queen Bona

---

[245] Wilhelm Maximilian Wundt (1832–1920), German doctor, physiologist, psychologist, and philosopher. A founding figure of modern psychology.

Sforza's castle, built in the sixteenth century, the palace in nearby Wiśniowiec,[246] and the river Ikwa which, *pace* Słowacki, does not flow through Krzemieniec. Last but not least, she led me to a small Turkish café where they served excellent black coffee.

Krzemieniec was indeed beautiful. At the center of the town there stood a group of elegant buildings forming the famous Krzemieniec high school or *lycée*, founded in the early nineteenth century.[247] Its towers, arcades, and squares all rested on elevated foundations, in effect transforming the whole complex into a rather splendid palace. Around this there were smaller houses, formerly residences of the nobility and wealthier burghers, now mostly owned by Jews. Each house was painted a different color, each had its own peculiar little stairways and galleries, and each seemed oriented in a different direction. In those days Krzemieniec was very popular with painters, and I saw the Jasło artist Stanisław Szczepański there, and bumped into Mrs. Cybis in the street, the wife of a Kraków painter and a considerable artist in her own right. I had met these people much earlier at a party in our house at Jasło arranged by my brother-in-law Leon Chwistek. Mr. and Mrs. Cybis had then been living at the residence of Count and Countess Mycielski in Wiśniowa, also both artists. Among the other guests at that party were Mr. and Mrs. Kinga Turno, accompanied by a brother of Mr. Turno. I remember that time as one when there was still some wine remaining in the cellar, when I ate for the first time sandwiches with a filling mainly of tomatoes and mayonnaise, when after dinner we all went to dance in the basement of Dunaj's restaurant, and when Mrs. Turno suddenly fixed her gaze on me and asked me if I hated her very much, a question I found very amusing.... The party was particularly memorable since it was to be the last held at our house. However, the tomato-mayonnaise experiment failed in Krzemieniec because real olive oil was lacking. By the way, since some Russians call olive oil "Provençal" oil, it follows that they would have to call olive oil from Provence "Provençal Provençal oil".

Marek Kac and I had enough time free of distractions to actually do some joint work, spending a few hours each day talking about independent functions. The work we did was not devoted to a specific problem, but more about experimenting with new definitions and concepts—to my mind the most exciting kind of mathematical work. It was also useful as preparation for a lecture I was to give in the Fall in Geneva.

I relaxed by playing tennis with Kac and his brother, who was then a student at the *Lycée*. The *Lycée* had been renovated, a deed worthy of respect carried out under the then minister Poniatowski. The Krzemieniec *Lycée* did not pale in any respect even before the much-lauded English public school: its architecture was marvellous, it had over a hundred years of tradition, spacious grounds, excellent teachers, and succeeded in awakening among both students and teachers the kind of respect for

---

[246]Both Krzemieniec and Wiśniowiec are today in western Ukraine.

[247]By Tadeusz Czacki (1765–1813), Polish historian, pedagogue, and numismatist. Inspector of schools for the province of Volhynia, Podolia, and Ukraine. Worked for the emancipation of Jews in Poland.

genuine culture and continued effort without which a stratum of society deserving of the name "intelligentsia" can never emerge. Although I didn't know Poniatowski personally, I had heard a lot about him from Marek Kac, from whom I gathered that he was not a realist in that he failed to perceive the villainy of the members of the National Unity Camp and the Nonpartisan Bloc for Cooperation with the Government surrounding him. These used him as a sort of model idealist serving as a front for their actions, dictated by anything but idealism.

Once around that time a young lady came to my office in Lwów and handed me a document detailing the abuses then being carried out on the prisoners at Bereza Kartuska, revelations that would arouse the indignation of any dispassionate observer, if not make his hair stand on end. I sent the document to Minister Poniatowski, and got a reply from the chief functionary in his ministerial office claiming that "the facts in the document are invented, and its object is to foster mistrust of the government amongst scholars." Now this reply had the opposite of its intended effect, since it convinced me that the document in question was veridical: the only valid means of rebutting an accusation of abuse is to establish a commission made up of ordinary citizens to examine the prison and question the prisoners. An unsupported denial is tantamount to an admission of the abuse.

Although I have had relatively little to do with ministers and other political authorities, on those occasions when I did come into contact with them I have almost always been surprised—if not shocked—by what emerged. I once conceived the idea of letters whose contents have been officially confirmed as genuine. This would necessitate an arrangement whereby the post office would by some means obtain a copy—for example, by making one itself—of the registered letter and send that in place of the original in an envelope bearing an official stamp confirming that the contents are identical with the original. Thus there would have to be people in the Post Office whose job it was to retype or otherwise make copies of letters whose senders were requesting this service. I wrote to the ministry in charge of postal services setting out my idea in this—admittedly imperfect—form. I received a reply telling me that the idea was a good one and expressing the ministry's intention of putting it into practice. So then I sent a second letter asking them whether I was to be rewarded for my idea. The reply to this gambit contained as its key statements that "in principle, improvements and ideas issuing from the public are not honored with remuneration," and that "your idea will be implemented in a form different from the one presented." Now let's consider this reply closely: since essentially everyone who *can* write writes letters, the first statement implies that the ministry in question arrogates to itself ownership of all innovations of any use to it—except those thought up by illiterates. The second statement essentially implies that in its new form the idea is no longer mine, in apparent reaction to my requesting some sort of payment. A more honest approach would have been to tell me up front, that is, in their first letter, that there would be no recompense, and to suggest that if this did not suit me then they would refrain from using the idea. Thus one perceives here a certain amount of stupidity and dishonesty.

However, this attitude of overweening condescension towards the "little man" was not new to me; I had encountered another instance of it among treasury officials just after my Father's death. The inheritance was to be taxed. However, a portion of it was in the form of a lifelong annuity to be paid to a sister of my Father's living in Hungary. This was to be paid out of rental income, and the amount per year of the annuity was to be deducted from the inheritance and set up as a separate account itemizing this rental income, on which we were also to pay tax. So we paid tax on the annuity—in the form of rental income—and separately on the balance of the inheritance, which meant that the state treasury was receiving the same in tax as it would have if the inheritance had not been encumbered by the annuity. When my Hungarian aunt died, the tax people sent us a letter stating that they would now be levying tax on the amount of the annuity on the grounds that our income from the inheritance had increased by that amount—in effect that we must pay tax a second time on the rental income that had been going to our aunt. On one occasion, furious at the intransigence of an official over this matter, I said to him: "I am in your power because you have executive control backed by the full might of the government, but I also have some power. If you continue to treat me this way, I will make good the damage to my finances in one fell swoop, by withdrawing a sum of several thousand złotys, deposited as a bequest to Jan Kazimierz University, and hence, indirectly, to the state treasury." Although I had no real intention of carrying out this threat, I at least had the satisfaction of seeing the official look taken aback. I abhor the despotism of bureaucracy.

* * *

At that time a proclamation composed by Mr. Adam Koc[248] was posted up on the walls of public buildings of every town and village in Poland. Any competent journalist could have written it in a hour, but presumably Koc worked on it for several months, in effect making it a very expensive document. But why quibble when we could always thank the Almighty that "although things are wretched, still, we've got Koc!" The proclamation heralded the arrival of fascism in Poland, and the attendant consequences were soon in evidence. The shift to the extreme right in Polish politics paralleled the growing accord in Polish foreign policy with that of Hitler's Germany. As evidence of this tendency we had earlier the invitation to Göring, the Great Huntsman of the Reich, to go ahunting in Białowieża Forest,[249] but now, far worse, we had the invitation to Dr. Goebbels to give a lecture, organized "within the framework of a general Polish-German cultural understanding" under the auspices of the Hellenist Professor Tadeusz Zieliński,[250] who had come to Warsaw University from St. Petersburg after World War I. This "cultural understanding" looked more like toadying. For example, in a textbook

---

[248] Adam Ignacy Koc (1891–1969), Polish politician, journalist, and soldier.

[249] An ancient forest straddling the present-day border between Belarus and Poland. A remnant of the immense primeval forest that once spread across the European plain.

[250] Tadeusz Stefan Zieliński (1859–1944), prominent Polish classical philologist and historian. Translated several ancient Greek authors into Polish and Russian. Professor at the University of St. Petersburg from 1890 to 1920, then at Warsaw University from 1920 to 1939.

written by Professor Zygmunt Łempicki[251] for Gymnasium students, and certified by the Ministry for Public Education, Heinrich Heine's[252] *Lorelei* was printed without the poet's name[253] and described as a *Volkslied*,[254] in accordance with the principle *Amica veritas sed maior amicus Goebbels*[255] or, "the company you keep largely determines who you are." This piece of flunkeyism—an example for the ages—was in any case entirely unnecessary since every person with even a minimal knowledge of German would immediately recognize the poem as written in Heine's characteristic lyrical style, or even know it as his. It would have been simpler for the compiler to simply omit the poem, thereby avoiding the test of courage involved in deciding whether or not to give the name of a convert from Judaism prominence in a school textbook.

Shortly after the appearance of Koc's posters, I was visited by the Wards, a very pleasant and well brought up young couple who had come to Poland a few months earlier from Cambridge, where they were students. He was a mathematician, had a grant to travel abroad, and had chosen Stanisław Saks[256] in Warsaw as someone he must visit. After a couple of months there, he had decided to come to Lwów. Since my wife was away just then, I could accommodate them both comfortably in our home. They loved to walk around the town investigating, and they soon knew more about the topography of Lwów than I did. Some aspects of the city surprised them tremendously: the "slavery" of the housemaids, as they called it, the terrible poverty in evidence in the quarter behind the City Theater, and cheese made from sheep's milk (*bryndza*), of which Mrs. Olive Mabel Jessie Ward—to give her her full name—said that it must be good but would I please release her from the obligation to try it. They came to like Lwów very much: the old world architecture, the high level of learning at the university, the neighboring countryside—practically everything, in short. Ward noticed the Koc posters plastered up everywhere and wondered greatly where the National Unity Camp could have obtained sufficient funds for gluing paper over the whole of Poland. He was kind enough to help me a great deal with the English edition of my picture book, entitled *Mathematical*

---

[251] Polish literary theorist, Germanist, philosopher, and cultural historian. Lived from 1886 to 1943, when he perished in Auschwitz.

[252] Heinrich Heine (1797–1856), great German poet and essayist of Jewish background. Converted to protestantism in 1825. His radical political views led to many of his works' being banned in Germany, and he spent the last 25 years of his life in Paris.

[253] Another source maintains that in a Grade Three textbook prepared by Łempicki and Gustaw Elgert and published by the publishing house *Książnica Atlas* in 1936, the poem in question *is* attributed to Heinrich Heine.

[254] "folk song"

[255] "Truth is a friend, but Goebbels is a greater friend", parodying Aristotle's "Plato is dear, but truth is dearer."

[256] Stanisław Saks (1897–1942), Polish mathematician. Member of the "Scottish Café" in Lwów. Following the invasion of Poland in 1939, he joined the Polish resistance. Wrote an extensive monograph *Theory of the Integral*, and co-authored *Analytic Functions* with Antoni Zygmund. Executed by the Gestapo in 1942.

Snapshots, and even found a few substantial errors. I kept in touch with them by mail for a while after they left, and I remember asking him in one letter what he thought of the Munich Agreement.[257] He replied that he remembered all too well the slogans of the Conservatives' election posters of 1935: "We shall stand by our word"—meaning that they would support the League of Nations,[258] and honor their other promises. But these ringing words, which helped the Tories to victory in the elections, had been forgotten by the time of Munich.

One could feel the storm gathering. Just as prior to a storm one sees gusts of wind blowing litter and dry leaves about, so too, in Poland, did strengthening squalls blow nonsense about the country. Here is an example: the slogan "To Madagascar!" was suddenly heard resounding throughout the land. This craze originated with a book about Madagascar written by Piłsudski's former adjutant Lepecki, who had travelled to Madagascar at considerable government expense, supposedly to look into its suitability for settlement. In the book he gave figures for the area of the whole island, and of the part suitable for settlement, but then it was claimed that his figures were out by several percent, and the French government, who had thought Madagascar belonged to France, protested that at no time had it agreed to any proposal to settle citizens of any foreign country there. An apology was then forthcoming from the Polish government, which assured the somewhat miffed French that the whole episode had been blown up out of nothing by the yellow press.... Then a member of the National Unity Camp proposed sending the geographer who had had the temerity to contradict the findings of Lepecki, former adjutant to the great Marshal Piłsudski, to the prison camp at Bereza Kartuska. Somewhere in Pińsk a Jewish storekeeper announced that his store was for sale, advertising the fact with a sign in the shop window reading "Everything must go as we're leaving for Madagascar"—for which piece of levity he was fined by his local *starosta*. It is unknown if he was being fined because he intended to leave or because he intended to stay.

* * *

In the Fall of 1937 I went to Geneva. Each year a so-called "colloquium", that is, a small international conference of specialists in a single area of mathematics, was organized there, the area of specialization varying from year to year. These meetings were financed by a certain millionaire who wished to remain anonymous. On this occasion the special theme was probability, and from Poland Mazurkiewicz and I were invited, but Mazurkiewicz declined. The travel expenses were paid by the conference committee, and I stayed in the home of my wife's cousin Olga Pamm, who had been living in Geneva for a few years, her husband running a business exporting Lusina watches.[259]

---

[257] The agreement of September 29, 1938, signed by Germany, Britain, France, and Italy, ceding to Germany the region of western Czechoslovakia considered in Germany as part of the Sudetenland. This was one of several attempts at appeasement of Hitler by Britain and France prior to World War II.

[258] Which had passed sanctions against Italy for the invasion of Ethiopia.

[259] An exclusive brand of Swiss watches, produced by Pamm Frères Ltd., Geneva.

7 In the University Town Lwów

The trip to Geneva was a lively one. When the train was approaching Basel, but still in German territory, a man in the same compartment suddenly put his face close to mine and whispered conspiratorially: "I won't say a word till we have crossed the border." Just before Basel the train left German territory, and he and I were left alone in the compartment. Now that we were safely inside Swiss territory he began to tell me his story. He was a laborer, with a Jewish grandfather. Friends in the Leipzig police department had warned him to flee Germany, having learned that the German police at the Czech border had already been issued a warrant for his arrest, which explained why he was travelling to Switzerland. We shared a room for the night in Basel, talking till the wee hours about the regime he detested. One of his assertions has stuck in my memory: "I am not so much troubled by the anti-Semitism of those who have always been anti-Semites, as by the professors and scholars, contemptible lackeys that they are, who discovered this propensity within themselves when Hitler came to power." Although an ordinary workingman, he understood these things better than many a member of the intelligentsia. I promised him that on my return through Leipzig I would telephone his parents to let them know that he had reached Switzerland safely.

On the electric train from Basel to Geneva I met a young German woman on her way to a Christian Youth Convention. It was clear from the way she talked, in particular about this convention, that she was not in the least infected with Hitlerism. In fact, no one in that third class compartment was: all studiously avoided giving the Hitler greeting. I still see them before me: a priest, an officer, an old country woman, and a typical Jew, almost a caricature of the type. He wore a medal earned as a soldier in World War I, and waxed ecstatic about the brilliance of the Minister of Finance of the Third Reich. The others listened to him with fellow-feeling tempered with scepticism. I was reminded of Tuwim's little poem:

> O Mensch, gib Acht
> Dumm ist die Macht
> Und dümmer als der Schacht[260] gedacht.[261]

Among those attending the conference was Heisenberg,[262] then at Leipzig. His lecture was about the departure of the behavior of photons from that expected according to classical probability theory. He expressed great interest in the approach to the notion of stochastic independence that I indicated in my lecture. The serious

---

[260] Hjalmar Schacht (1877–1970), German economist, banker, and liberal politician. Served in the Weimar Republic. Became a supporter of Hitler and the Nazi Party, and helped implement Hitler's policies of redevelopment, reindustrialization, and rearmament. Forced out of the government in 1938 through disagreements with Hitler and other prominent Nazis, he became a fringe member of German resistance to Hitler.

[261] "O Man, watch out
Power is stupid
Even more stupid than Schacht had thought."

[262] Werner Heisenberg (1901–1976), German theoretical physicist. Made fundamental contributions to quantum theory. Famous for his "uncertainty principle." Nobel laureate in 1932.

business of the conference was of course interspersed with excursions, on one of which we visited a beautiful house on Lake Geneva belonging to Professor M., a chemist. During afternoon tea I sat next to Heisenberg. Suddenly someone asked about Karl Kraus, and our host gave an answer showing his contempt for Kraus. So I spoke up, defending Kraus in no uncertain terms. Feller, who was then living in Stockholm,[263] reminded me that I had said exactly the same thing at a reception in Courant's home, which had won him over to me. By a curious coincidence, Feller's talk was on a metric approach to measuring probability in the case of infinitely many trials, and he was surprised—and disappointed—when I showed him a paper[264] in Volume IV of *Fundamenta*, which had come out in 1923, setting out just this approach. Among those present at the conference were Cramér[265] from Stockholm, Paul Lévy[266] from Paris, a mathematician brimming with ideas but not always sufficiently critical of them, Eberhard Hopf,[267] Mr. Dodd[268] from Texas, Fréchet,[269] accompanied by his student Döblin[270]—to mention just a few of the many foreign luminaries attending. From Switzerland itself, there was Mr. Fehr,[271] Mr. M., and Mr. W., and, from Zürich, the first-class mathematician Pólya,[272] whom I had met before, and who gave two lectures, one on "going astray in a network of streets", and the other on a mathematical theory dealing with gravel on the beds of streams.

---

[263] He had fled from the University of Kiel to Copenhagen in 1933 to escape the Nazis, and taught there and then in Sweden, before emigrating to the US in 1939.

[264] By Steinhaus.

[265] Harald Cramér (1893–1985), Swedish mathematician, specializing in mathematical statistics and probabilistic number theory.

[266] Paul Pierre Lévy (1886–1971), eminent French probabilist. Professor of Analysis at the *École Polytechnique* from 1920 to 1959, with a gap during World War II dating from his dismissal in 1940 in accordance with the Vichy Statute on Jews.

[267] Eberhard Hopf (1902–1983), Austrian-American mathematician and astronomer. Made significant contributions to the fields of partial differential equations, integral equations, fluid mechanics, and differential geometry.

[268] Edward Lewis Dodd (1875–1943), American mathematician and statistician at the University of Texas at Austin.

[269] Maurice Fréchet (1878–1973), eminent French mathematician. Made major contributions to topology, probability, statistics, and functional analysis.

[270] Wolfgang Döblin (known in France as Vincent Doblin) (1915–1940), German-French mathematician. Elected to the French Academy of Sciences at the age of 21. Took his own life when his company was faced with imminent capture by the Germans in 1940. A notebook of his, which remained unopened and forgotten till the year 2000, showed that his work on Markov processes had been far in advance of its time.

[271] Henri Fehr (1870–1954), professor of algebra and higher geometry at the University of Geneva.

[272] George Pólya (1887–1985), eminent Hungarian mathematician. Professor at ETH Zürich from 1914 to 1940, and thereafter at Stanford. Contributed to many areas of mathematics. Wrote very popular and influential books on what might be called "mathematics education."

I was particularly impressed with the stenotype machine that was shown to us. This looked somewhat like a typewriter, except that the typed text was recorded on a continuous paper strip, and multiple keys were to be pressed simultaneously, like chords on a piano, to spell out whole syllables, words, and phrases at a single stroke, after which the ribbon would shift by a whole line. As demonstrated to us by a trained stenotypist, it was much quicker than a typewriter, and, being also effectively soundproofed, could be used to record conversations without disturbing them.

I had been instructed by the relevant ministry that—as always when a professor travels abroad—I should visit our embassy in Geneva. When I arrived there, Ambassador Komarnicki[273] was away, so I was received by the secretary of the legation Mr. Jan Meysztowicz. Now at that time there was considerable controversy in Poland surrounding the Polish Association of Teachers, which Składkowski[274] had berated for its "defeatism and pacifism". I had the temerity to tell Secretary Meysztowicz, after quoting these words to him, that just because someone does not want war does not mean they are lacking in patriotism, adding by way of example that if, say, a country is too weak to fight then he who would rather there be no war may be helping to prevent a catastrophe. The secretary, however, stoutly maintained that this argument was fallacious.

The ambassador, Tytus Komarnicki, was unpopular in the League of Nations, since he was an exponent of the politics of Józef Beck, whose aim was to put an end to what remained of the authority of that institution. I decided to visit the Palace of Nations in Geneva where the beleaguered League had its headquarters. Since the Palace was located some four kilometers from the town, I needed a lift, and eventually managed to find a Dr. Wasserberg, from Kraków, in Geneva with some official position at the League of Nations, who was prepared to do me this favor. The Palace was splendid, and I was grateful to Dr. Wasserberg, without whom I would probably never have seen it. He shared an office with a Mr. Neyman from Warsaw, from whom I learned that the various Polish officials associated with the League had gotten together to institute a foundation for financing travel abroad for young Polish scholars—and it was from this foundation that I managed to obtain funding for my student Marek Kac to travel to America.

One evening Dr. Wasserberg came to the house I was staying in to take me for a walk. As the evening was balmy, we decided to go to the park on foot, leaving the doctor's car, with his coat and briefcase, parked in the street for the few hours of our stroll—unthinkable in Poland! Even more wonderful for me was the way newspapers were left at night in a basket hanging on a lamppost next to a newsstand, now closed for the day, together with a cardboard box for depositing payment. If you had no change, then you put in a franc and took out your 90 centimes change. At the closing banquet, where several of us made speeches thanking our hosts, in my

---

[273] Tytus Komarnicki (1896–1967), Polish diplomat and historian.

[274] Felicjan Sławoj Składkowski (1885–1962), Polish physician, general, and politician. In 1936 became Poland's last prime minister before the outbreak of World War II.

speech I said that this cardboard box had made a greater impression on me than the view of the Alps, Lake Geneva, the beauty of the town, and the wealth of its inhabitants. I gathered that my cousin's husband, who had been living in Geneva for around a dozen years, periodically needed to have credit available to maintain his business. When I asked him about the conditions for obtaining this credit, he said that he simply wrote a letter to his bank requesting a loan of 10,000 francs, say, and shortly afterwards a bank messenger would bring him the 10,000 francs and a receipt for him to sign.

– How is this possible—I asked in surprise—Without collateral, or some kind of guarantee, even an I.O.U.? This must surely be a very expensive short-term loan.
– No, it's a loan for a year at four percent.

I began to understand that we in Poland have to secure loans with collateral and pay interest at ten to twelve percent per annum because of all those who bounce checks and declare bankruptcy. However, although they readily extended credit to my host, they would not give him Swiss citizenship—apparently because he and his wife were both from Kraków, and persisted in speaking Polish at home. Borrowing money in Switzerland is far easier than actually becoming a Swiss citizen.

For us Poles, the trappings of wealth evident everywhere in Switzerland would have defied our imagination had we not seen them with our own eyes. In Geneva alone at that time there were over 30,000 private cars—milkmen and bakers owned cars and your average butcher had two. The streets were well paved with asphalt, and so too were the main arterial roads between the major cities. The roads were illuminated at night with orange sodium vapor lamps which make it easier to see ahead in foggy conditions.

My hosts had some business to settle in Lausanne concerning furniture they had ordered, and they also wanted to visit their little son, who was in a sanitorium in the mountains not far from Lausanne, and they invited me along. The road followed the lakeshore almost the whole time. We ate a dinner of fried fish washed down with wine in a small restaurant. Mr. Pamm would not drink more than a single glass, explaining that the laws relating to drunk driving are very strict in Switzerland: in case of an accident, however minor, the police immediately take a blood sample from the driver at fault, and if the test reveals alcohol, he is arrested, regardless of his position in society. When we arrived in Montreux, at the other end of Lake Geneva, I had a few hours to myself, waiting for my hosts to return from the sanitorium, so I made an excursion to the *Château de Chillon*, returning on foot along the shore, jam-packed with villas, guest houses, hotels, and roadside vendors selling grape juice squeezed from small winepresses into little glasses. Englishmen came there to relax and play tennis, while their daughters, chaperoned by governesses, explored the countryside. I stopped for an afternoon snack at an exquisite little café with a view of the lake. There was a dance floor there, and expert and non-expert dancers danced casually in shoes with soft felt soles on the perfectly smooth parquet. There were several handsome young men among the dancers, but no very attractive women that I could see. Generally speaking, Switzerland is not very *chic*—it's as if Calvin knocked all licence out of the heads of his compatriots.

My cousin took me on several more expeditions in her car, mostly in and around Geneva. Once we went to a town on the border with France, and ate at an inn on the French side. Again I felt awe at the extraordinary affluence of the Swiss when a few dozen Swiss peasants—at least that's what they looked like—came across the border in their cars to partake of the fine French food, served in the form of a seven-course repast at seven Swiss francs a head.

Swiss railroads were electrified, and if, standing in a station, one looked up and down the track at the network of wires strung on poles hung with insulators and fuses, the enormous cost of such an investment became clear. But the payoff in terms of smooth, soot-free rides without the chug-chugging of the steam locomotive surely made it worthwhile.

I had had a brainwave to do with measuring electricity consumption, which I wanted to discuss with the *Société de Compteurs Électriques* in Geneva. I had been thinking for some time about a rational means of regulating electricity rates, and I wanted to see their reaction to a proposal of mine for constructing an electricity meter. The problem in question is well known to electrical experts. The cost of producing electric current is made up of two parts: costs proportional to the amount of electrical energy produced—the cost of coal, for example—and costs proportional to the upkeep of the necessary infrastructure—maintenance and mortgage or rental payments on buildings, machinery, and the supply network. This explains why it is most profitable for a power station to supply current in the same constant amount day and night, day in, day out, summer and winter,[275] since if the average consumer uses more current at one time than at some other, then during the period when demand is low the power station's installations are idle, while still incurring costs of the second sort. Electric companies compensate for the non-uniformity in consumption by imposing differential tariffs on electricity use. For example, they might break the month up into two-hour periods, choose from the resulting 360 periods the three during which average consumption is highest, then of these three periods take the one where consumption is the least, and multiply the average amount of current consumed during that period by 360 to get a figure— a sort of "virtual" consumption—on the basis of which an appropriate price can be calculated. This is one compensatory means of levying tariffs on the consumer, and there are others—for example, that employed by the "Ferraris meter"[276]—but they all fail to take into account the non-uniformity of the short-duration highs of the consumption curve. My method was totally unrelated to other approaches to imposing fair and efficient electricity tariffs, but required the construction of appropriate meters, to which end I obtained the collaboration of Mr. Rosenzweig, an engineering assistant at the Lwów Polytechnic. We made a few such prototype meters, and wrote a paper describing one of them, published in the journal of

---

[275] That is, 24/7, as they now say. This has to do with the well-known difficulty in storing electricity efficiently.

[276] Developed from the discovery in 1885 by Galileo Ferraris of Turin that two out-of-phase AC fields can make a solid armature rotate.

the *Schweizerische Elektrizitätsvereinigung*.[277] After this article had appeared, I received a letter from the official responsible for tariffs at the Budapest Electrical Works praising our solution to the problem as superior to those hitherto in use.

At the conference in Geneva I was frequently in the company of the elderly Texan Mr. Dodd and his English wife. She was interested in buying a watch, so I thought of getting her a good one at a good price through mine host Mr. Pamm, and indeed he was willing to sell a watch to her at the factory price, which was a third of the asking price of the same watch in New York. I went rather late one evening to Pamm's office to get the watch. He began trying to persuade me to buy a fashionable wrist watch for myself. The one I had was only of nickel; for me a watch was something to be used and I was indifferent to its decorative qualities, and in any case I reasoned that there was no point in spending a lot on a wrist watch since they were too small—the mechanism was too delicate—for them to be very reliable. Mr. Pamm countered this argument by saying that wearing a pocket watch on a fob chain was like wearing a nineteenth-century top hat. But I was worried that it might be getting too late for me to bring the watch to the Dodds at their hotel, so I glanced over at the clock on Pamm's desk, an big old onion-top clock, compared to which my wrist watch might be thought the latest thing. When I put this to him, he blushingly admitted that this was the only really accurate timepiece in his office. Incidentally, the Dodds decided not to buy the watch I had obtained for them because they thought it too cheap. In fact, at that time there were just two brands of watch of high quality: *Patek Philippe* and *Schaffhausen*. Some time ago, a representative of *Patek Philippe* told me that Americans coming to Switzerland always demand "the world's best watch", rejecting all the commonly available ones, so they manufacture for them a watch containing a few more rubies, with a different face, a different pattern etched on the case, and costing twice as much—though not one iota more reliable mechanically.

In Geneva I was introduced to Mr. Spława-Neyman,[278] who was wondering whether he should return to Warsaw or accept the offer of a university position in the US. He had been in England for several years, and wanted to know what sorts of changes were occurring in Polish universities. I put the matter very clearly to him: did he want to be subjected to recurring student riots—such being the outlook for a professor in Warsaw—or did he want to pursue his scientific interests in relative peace and quiet? Naturally, he decided on the US.

It so happened that the Nobel prize-winning physicist Raman[279] was spending a few days in Geneva at that time. He gave a lecture on his discovery that when

---

[277] Swiss Electricity Consortium.

[278] Jerzy Spława-Neyman (1894–1981), prominent Polish-American mathematician and statistician. Born in tsarist Russia to a family of Polish background, he was repatriated to Poland in 1921 after the Polish-Soviet War. In 1924 spent some months in London working with Egon Pearson. In 1934 he left Poland again for an extended visit abroad, chiefly in Britain, and in 1938 obtained a position in Berkeley, where he remained for the rest of his life. Introduced the "confidence interval" in a 1937 paper.

[279] Chandrasekhara Venkata Raman (1888–1970), Indian physicist. Awarded the Nobel Prize in 1930 for his discovery of "Raman scattering" of light rays.

7 In the University Town Lwów

a beam of light traverses a transparent material medium, some of the refracted light changes its wavelength. Raman looked very much like Father Skowron from Siekłówka, except that his eyes were darker. He spoke English very well and lectured clearly. I availed myself of the opportunity to ask him a question about the course of a projectile—like a bullet—that enters into a human body obliquely,[280] which I had also put to Heisenberg. His answer was in effect the opposite of Heisenberg's—an instance of two Nobel laureates having radically different opinions. This confirms that there are problems—like this one—which cannot be resolved without performing an experiment.

* * *

After returning to Poland, I met a very interesting young man by the name of Maurycy Bloch,[281] a fellow student of my daughter's studying law. Although only 19, he was constantly involved in what he called "big deals", sometimes requiring his presence on the property of his relatives, at other times in Warsaw. On one occasion he was involved in selling balsa wood to the French for airplanes, and on another he was off in Vienna trying to prevent the confiscation of a house there that had formerly belonged to his family. He told me things of which otherwise I could have had no inkling: for example, that a certain bomber of 1937 vintage, sold to the Polish government in Warsaw in 1938, would fetch half a million złotys, while the same bomber loaded on a ship at Gdynia[282] and furnished with Chilean or Ecuadoran papers, would fetch a million dollars, and if sold to Chiang Kai-shek through an appropriate Chinese port would go for a million and a half. Not long after I met him he wangled a government commission to go to America to buy two Morgan horses[283] for the army. This is how he explained to me why he was given this commission even though he was no expert on horses: He averred that the general in charge of this matter in the ministry said to him: "We need two mares, and the estimated cost of bringing them over, including transport, is 40,000 złotys. If I send an officer from the Uhlans who speaks English, then he will undoubtedly be an elegant young man from the aristocracy, who will get on very well with his American opposite numbers because he will be able to ride and carouse with them. And, naturally, at the end of the first drinking bout, they will pin their regimental insignia on the breast of his uniform as a symbol of eternal friendship, and then when he gets drunk the next time he will remove this regimental badge, attach it to the tail of a herring, and hang the herring in a toilet. Then an official in the Department

---

[280]This may have been suggested by the incident mentioned earlier in connection with the Przytyk pogrom.

[281]Maurycy Bloch (also called variously Morcio and Moryś) (1917–1995), scion of a wealthy Polish family, immigrated to the US just prior to World War II, where he set up in business.

[282]A city in the Pomeranian Voivodeship, on Gdańsk Bay. An important Polish seaport.

[283]One of the earliest breeds of horse developed in the US. Tracing back to the stallion "Figure", later named after his best-known owner Justin Morgan (1747–1798), the breed excels in many disciplines.

of State in Washington will send an appropriate letter of protest to the American *attaché* in Warsaw, who will come to me for a private conversation. We will then have to dispatch two aides on a mission of apology, and instead of 40,000 złotys, the total cost of acquiring the mares will have risen to 400,000 złotys."

It was he, Moryś, whom I asked to present my idea for a secret letter code to the Ministry of Foreign Affairs. The code had been suggested to me by an idea I had found in James Joyce's *Ulysses*, where the need arises for encoding a telegram containing information on a change of route of the viceroy's carriage so as to avoid an assassination attempt by Irish separatists. Using my code, encoding and decoding were straightforward, the key was readily changed, and without it an encoded message was very difficult to decipher. Its only flaw was that because of the unusual symbols used, it was not telegraphic. Morys brought a sheet of paper to the ministry with the encoded message "Chlorodont[284] is best" on it, and the relevant people there appeared to set to work to decode it. However, what they actually did was to send agents to his hotel in his absence, where they rummaged through his suitcase looking for a description of the code and the key. Naturally, I made no analogous request of him again.

* * *

Matters such as those, and also work connected with the publication of my book *Kalejdoskop matematyczny*,[285] meant that I was absent from Jasło when a strange person showed up there. He was a Czech of ethnic German background whom my niece had met on a trip. He had promised to visit her in Jasło, and one day did show up, taking a hotel room. I should mention that at that time there were no males in our house to afford any protection of the females from intruders. The new arrival claimed to be an "economist", and gave out that he was a foe of National Socialism who had fled Czechoslovakia when the Germans took over. He questioned my niece about the number of employees in the post office and the *starostwo*, and so on, saying that he needed such data in connection with his statistical work in economics. When leaving, he promised to write and give his new address, but was never heard from again. I had a strong feeling that he was one of Hitler's spies.

In Lwów I continued to work on localization,[286] but got nowhere. In February 1938, a young fellow wearing a student cap entered the room where I was lecturing, came up to the table behind which I was standing, and, pointing to a small group of students standing by the window, said: "These Jews are standing." What he meant to convey was that, by standing, these Jews were disobeying the order to *sit* on the left side of the lecture hall. I told him to leave the room, and he answered "I don't feel like it!", at which a cluster of youths armed with sticks rushed into the room and began attacking the standees, knocking one of them to the floor. While this was going on I walked towards the door, but it was being held closed from outside. I

---

[284] The brand name of a toothpaste first marketed in 1907.

[285] The English version was entitled *Mathematical Snapshots*.

[286] That is, the problem of determining the location of an object inside a patient's body.

gave it a good shove, and found myself in the corridor. Later, when I talked this incident over with my students, they all condemned it, but at the same time they exaggerated, it seemed to me, the danger threatening those who tried to stand up to the aggressors. The hooligans involved belonged to an organization headed by one Rojek, known from earlier, when Krzemieniewski was Rector, as an organizer of student rallies. This organization was responsible for murders: youths belonging to it had killed a bookkeeper in Stryjski Park[287] and a veterinary student, and had stabbed another student who had resisted the demands of this gang of bandits.

Despite such distracting events, I kept in mind my promise to Meisels that I would find a way to actually see organs or foreign objects three-dimensionally within the human body. A few days after the above incident, I was walking along St. Nicholas Street around noon, a light snow was falling, and on the sleeves of my black fur coat I could see several snowflakes, shining like stars. There was snow on the ground, so the street was quite brightly lit by the winter light, and, since I had already acquired the habit of playing with reflections in windows, I stopped in front of the show window of a ladies' hat shop called *Jenny* to experiment once more. I could see the reflections of the "stars" on my coat sleeves very clearly, and observed that by moving my arm I could cause one of these stars to descend onto the surface of a ladies' hat displayed in the window, and then by moving it again, cause the star to seemingly move inside the hat and out the other side, all the while remaining visible. As I walked in the direction of home along Zyblikiewicz and Kadecka Streets, my mind was turning over what I had seen, and by the time I reached home I had my invention of an "introvisor" plainly there before my mind's eye.

From this moment on my theorizing gave way to experimentation. I first made a mirror as follows. I took a small pane of glass, lightly silvered on one side, into which a small screw had been inserted with its head sticking up above the glass, and fixed it in position above a base consisting of a piece of board covered with black paper, with a second screw positioned between the base and the pane so that when looking from above the image of the first screw exactly coincided with the second screw, which could be seen through the glass. When I pushed the second screw into a small potato and then viewed the potato with the screw hidden inside it through the glass, it was as if the potato had become transparent since now I could see the image of the first screw inside the potato, as it were. Holding a sharp knife I put my hand around between the pane and the base and set to slicing the potato until there was just a thin layer of it between the blade and the head of the screw embedded in it—this I could judge by looking through the mirror at the image of the first screw. And sure enough, I could feel the screw with the point of the knife just where the image indicated it would be. I carried out several other experiments, but I will spare the reader the details of these. When it came to applying this idea in the operating theater, I used a brightly colored bead positioned with respect to the mirror symmetrically opposite the object of attention inside the patient's body. Two

---

[287] One of the largest parks in Lwów.

X-rays taken from two given positions were used to determine the position of the object and thence that of the bead.

I gave a brief description of my method in a note[288] published in Paris in the spring of 1938, and a more detailed one in the Jubilee Year Book of the Jasło Gymnasium for 1938, marking the 70th year of its founding,[289] which appeared in the summer of that year. I also gave lectures on this topic to the *Lwów Scientific Society* and the *Polish Physical Society*. In the latter lecture I gave a practical demonstration using an X-ray machine. Dr. Meisels had placed his X-ray laboratory at my disposal, and there, with the help of Marek Kac and Franciszek Otto, I carried out an actual localization, namely of lead shot in a bread roll, with the aid of which a Dr. Scharage performed a make-believe operation of removing shot from a live brain. I was visited by Professor Bartel, who evinced great interest in my invention. Later he told me about the optical illusion whereby a wire fence with identical wire apertures may seem to be closer or further away than it actually is—an illusion proving the importance of the role in vision of cognitive identification of images. In the theory of vision as it stood at the time not enough attention was paid to this fundamental aspect of the plasticity of vision.

Actual practical application of my device began when our friend Dr. Ziemilski put me in touch with Major Alfred Bong, a military doctor, Mrs. Frank, and Dr. Grabowski, a docent at the university. These all showed great interest in my "introvisor", and asked me to give a talk on it to the local radiologists, so I gave another lecture, this time in Dr. Grabowski's well-equipped and very modern laboratory. Although I had patented the invention in March 1938, there didn't as yet exist an apparatus for use in serious surgical operations. The first such apparatus was built by Dr. Bong at the Lwów military hospital, and the first operation in which it was put to use consisted in the removal of a fragment of a needle from the hand of a certain Corporal Reguła. First a local anaesthetic was administered to his hand, then the necessary X-rays were taken and the colored bead positioned appropriately in front of the mirror of the introvisor, and finally, in ordinary daylight from the windows of the operating theater, and with the surgeon peering through the mirror, the metal fragment was excised. Each of these steps took but a few minutes. In all, four operations were carried out using the introvisor built by Dr. Bong, two of which were quite serious. I should mention that this method has nothing in common with the so-called "Viennese" method, where the surgery is performed during the X-raying. My introvisor actually shows the foreign body—or rather the bead's image substituting for it—inside the patient's body.

* * *

In the spring of 1938, Henri Lebesgue, the great French mathematician, came to Lwów on an official visit: our department had prevailed in having the title

---

[288]H. Steinhaus, "Sur la localisation d'objets au moyen des rayons X", *Comptes Rendus des Séances de l'Académie des Sciences, Paris* 206 (1938), pp. 1473–1475.

[289]*Księga Pamiątkowa 70-lecia Gimnazjum w Jaśle (1868–1938)*.

of doctor *honoris causa* bestowed on him. He had published a full account of his revolutionary new approach to the integral when he was 27. His measure theory and integral were appreciated sooner in Poland than in France, and quickly led to generalizations of the notion of measure, the work of Banach and Tarski, the theory of linear operators, generalized Fourier series, and applications to probability theory. All of these developments and more were connected in one way or another with Lebesgue's work—especially with his doctoral dissertation and the two so-called "Borel tracts".[290] Before an audience consisting of members of the Senate and the department, I conferred the degree—a handsomely produced diploma made of parchment printed with rotund Latin phrases—and Banach gave a speech about Lebesgue's contributions. In response, Lebesgue talked about Polish mathematicians who had some connection with Paris, having diligently boned up on the history of Polish mathematics, obtaining some of his information from the Polish Library on the *Quai d'Orléans*. He knew more about Polish mathematics than we did since there were no real historians of mathematics amongst us—the docent Mr. Auerbach[291] not counting since, although he had studied the history of the mathematics of ancient Greece, he was primarily a classical philologist. Naturally, we had Lebesgue to visit at our home a few times during his stay. The occasion for one of these visits was a dinner to mark the consecration of the graves of the fallen defenders of Lwów,[292] to which we invited General Musse, a French military *attaché* at the French embassy in Warsaw, and the consul and vice-consul at the French consulate in Lwów. We discussed politics a little, and I assured General Musse that, official policies notwithstanding, Polish society would never countenance war with France. Lebesgue talked mostly about the Parisian scientific milieu, and his background. His father had worked as a typesetter, and his grandparents had been of peasant stock. He told us that it was a mathematician at the Sorbonne in the thirteenth century who had first observed that the two feet of a normal human being are not congruent by means of an orientation-preserving Euclidean transformation,[293] and only then did shoemakers begin making different left and right shoes.

---

[290] Collections of Lebesgue's lectures, the first of which is titled *Leçons sur l'intégration et la recherche des fonctions primitives*.

[291] Marian Auerbach (1882–1941), Polish classical philologist. Murdered by the Gestapo in Lwów.

[292] On May 26, 1938, at the Cemetery of the Defenders of Lwów, a special place of interment memorializing those Poles and their allies who died in Lwów during the Polish-Ukrainian and Polish-Soviet War of 1918–1920, a monument to the French participants in the defence was unveiled. The ceremony was attended by the Polish Marshal Rydz-Śmigły and Major-General Félix-Joseph Musse, French military *attaché* to Poland.

[293] That is, they can only be superimposed by means of a transformation involving a reflection.

In his official speech Lebesgue had mentioned that the Malaxa prize[294] for 1938 had been awarded jointly to Leray[295] and Schauder. This was a great distinction for our docent, since this award was given to the mathematician who had made the greatest contribution of recent times to mathematical analysis. In fact, Leray, a charming Frenchman of great culture and elegance, visited us a year later.

An incidental fact concerning the appointment of Mandelbrojt over Leray as successor to Hadamard at the *Collège de France*: one of our national-radical newspapers reported that I had arranged the doctorate *honoris causa* for Lebesgue as a superior alternative to having a Jew chosen as Lebesgue's assistant in the Sorbonne. However, this is the purest fantasy since, first, professors at the Sorbonne did not have assistants, second, none of us in Lwów would have dreamed of proposing Mandelbrojt instead of Leray, and, third, having Hadamard's support, Mandelbrojt certainly didn't need ours.

* * *

I spent the summer of 1938 in Morszyn[296] for health reasons, staying at the guest house *Orion*. Although the town was surrounded on all sides by forest, Dr. Rencki, who was director of operations there, had so landscaped the immediate surroundings that the patients remained unaware of this: they saw only beautiful meadowland, diversified by copses of oak trees, an extensive park, and a new *Kurhaus*. I met a Dr. Szymanowicz of Kraków there and showed him my introvisor, which seemed to impress him as possibly having important applications in the military. I went walking with others to the nearby town of Bolechów, and in that inhospitable, windy little town I witnessed Morszyn patients furtively wolfing little marinated fish, washed down with plum vodka.

At that time I was conducting an extensive correspondence—in particular concerning the patenting of my introvisor. A professor Suchowiak from Warsaw helped me obtain a Polish patent, but advised against applying for Russian or Japanese patents on the grounds that the patent offices of these countries had a tendency to respond to submissions by foreigners with the claim that the invention in question has already been made. And in the Japanese case there was an added wrinkle: as proof of priority—someone else's—they send a portion of a document in Japanese, which, if one is not literate in that language, can be turned upside down or back to front without making any difference in one's being able to make neither head nor tail of it.

* * *

---

[294] A prize in the mathematical and physical sciences announced by the engineer N. Malaxa of Bucharest, and awarded for the first time in 1938.

[295] Jean Leray (1906–1998), French mathematician. Worked in partial differential equations and algebraic topology. The "Leray spectral sequence" was seminal to the development of spectral sequences and sheaves. Collaborated with Schauder from 1932 to 1935.

[296] A popular spa. Before World War II it was run by the Lwów Medical Association. Now Morshyn, in western Ukraine.

7 In the University Town Lwów

In the Fall of 1938 I attended a conference of radiologists in Warsaw, at which I demonstrated my introvisor using a model built *ad hoc*, with the help of Dr. Knaster. Dr. Bong was there of course, and I met many old acquaintances there, such as Mrs. Rajchel,[297] who was now married and living in Warsaw, and several mathematicians whom I had not seen for some time. Naturally, there were many foreign attendees, especially from Germany and Switzerland. And there was a plethora of receptions. One such party, held at the home of a radiology docent, surpassed by the abundance of elaborate dishes and drinks anything I had seen before—but this was not especially to be wondered at since an X-ray laboratory is a money factory. A certain Professor Ch. from Switzerland was present, and I had wanted to buttonhole him over a claim he had made that most people do not in fact see stereoscopically. Perhaps this holds true for doctors and those working in the biological sciences, since they are always using only one eye to peer through microscopes, but I am completely certain that normal people have stereoscopic vision—not to mention cats, who use it to so unerringly jump up on fences and what not. However, there was no chance of my taking this up with him, as any possibility of serious discussion was drowned out by the copious quantities of alcohol flowing.

Dr. Bong informed me that General Rouppert,[298] head of the military health service, had ordered him to arrange a demonstration of the introvisor before the Minister of Military Affairs. Since every supplier of medical equipment has its spies sitting in cafés eavesdropping and hanging around other appropriate venues to see who has been summoned to that ministry, immediately after my presentation I had representatives of X-ray firms and doctors' offices proposing partnerships with me. However, my goal was rather to obtain a US patent and market the apparatus in America, to which end I thought of making use of those of my former students who had gone to the US, and also Maurycy Bloch, who was just then planning to move there. I had several discussions with him while in Warsaw, learning many interesting things from him—for instance, that currently there was international collusion going on between weapons manufacturers making it impossible for any new firms to compete.

Warsaw itself in late 1938 made the worst possible impression on me. It had the aspect of a roiling morass of speculators, confidence tricksters, gangsters, and spies. Champagne, caviar, and money flowed in abundance. Married women wore furs of all hues clearly beyond the means of their husbands. Everywhere there was a vulgar flaunting of wealth, everywhere the brutal principle "He who pays the piper calls the tune" operated, and, simultaneously with noxious exhibitions of anti-Semitism, there were demonstrations of hubris by rich Jews who in every restaurant occupied the best tables, ordered the most expensive champagne, commanded the attention of the most beautiful *danceuses*, and ignored the Goyim expressing their disapproval by making a display of quitting the premises.

---

[297] A physician from Lwów.

[298] Stanisław Rouppert (1887–1945), head of the health service of the Polish armed forces from 1926 to 1939.

At about that time, in Lwów, when my daughter was ill, one of her doctors, Dr. Ludwik Fleck,[299] who owned a private chemical and bacteriological laboratory where he carried out experiments, sent me a short paper on the statistics of leukocytes.[300] There are different types of leukocytes: lymphocytes, monocytes, macrophages, and so on. The number of leukocytes per unit volume of blood is more or less fixed,[301] so that variation in this number may indicate disease. Dr. Fleck was taking samples, each of a few hundred cells, from the blood of a single patient, and looking at the distribution of the number of leukocytes over the samples. In order to have a basis for comparison, he took a different set of samples using the following procedure: he mixed some of the patient's blood with sodium citrate in a test tube to prevent coagulation, and shook the tube vigorously, before taking samples from drops placed under the microscope. His reasoning was as follows: according to the laws of probability, by comparing the distribution of the number of leukocytes in the samples taken directly from the patient's finger with that of the leukocytes across samples from the homogenized blood, it should be possible to infer whether the distribution of the leukocytes in the blood is subnormal, normal, or above normal. On studying Dr. Fleck's tables I quickly saw that the distributions for both sets of samples were highly subnormal by comparison with the theoretical Bernoulli distribution, which applied here, but of which Dr. Fleck, not knowing much probability theory, was unaware. When I explained the situation to him, he quickly understood the issue. This is just one example of the situation where a natural scientist—a biologist, say, or even an engineer—is rendered very liable to error through ignorance of the appropriate mathematics. The fact that the distributions σ_n, σ_h, and σ_b of numbers of leukocytes over samples taken respectively from normal blood, homogenized blood, and theoretically perfectly uniform blood (for which the distribution will be Bernoulli) came out in that order of departure from the Bernoulli distribution, had to do, in fact, with a tendency of leukocytes to resist dissemination once in static equilibrium. Dr. Fleck's work was of great interest to me, but of course, since I had no desire to work with a microscope or otherwise dirty my hands engaging with experimental method, I remained merely an empathetic observer of the many further trials undertaken by him, his co-worker Mrs. Ewa Altenberg, and others in the laboratory.

At that time there turned up amongst my students a lad from Borysław by the name of Kazimierz Borek with a gift for probability. I set him the problem of formulating the mathematical methodology appropriate for tests for determining the

---

[299]Ludwik Fleck (1896–1961), Polish medical doctor and biologist. Developed in the 1930s the concept of "thought collectives" as a means of explaining how accepted scientific ideas change over time. (Compare Thomas Kuhn's later notion of a "paradigm shift", Foucault's "episteme", and Richard Dawkins' concept of a "meme".)

[300]White blood cells, the cells of the immune system, involved in defending the body against infectious disease and foreign materials. They come in five different types (including lymphocytes and monocytes), all derived from a multipotent cell in the bone marrow called the hematopoietic stem cell.

[301]At between 5000 and 7000 per cubic millimeter.

concentration of pharmaceutical preparations—more specifically, of those preparations obtained from certain animal organs containing only a roughly estimated quantity of the active chemical one is interested in. Borek brought to this assignment a great deal of understanding and thoroughness. The firm *Laokoon* helped him locate English and German articles on this topic, and also provided funds for him to visit the State Institute of Hygiene in Warsaw. My intention was to make him a specialist in applied mathematics.

* * *

As already mentioned, in the summer of 1938 there was a ceremony marking the 70th jubilee of the founding of our Jasło Gymnasium, with speeches, concerts, church services, banquets, balls, and what have you. However, when I compared the level attained by graduating students of some 40 years earlier with that of the current crop, I noticed some falling off. The reminiscences of earlier students contained in the Jubilee Year Book bore a marked similarity to each other, resembling in some ways those of the classical memoirs of, for instance, Prus, Nowakowski, or Zbierzchowski.[302] In this book I had reproduced my article on the introvisor which, although I would have thought it simple enough, I believe no one understood—except perhaps for Dr. Stanisław Kadyj the eye specialist, son of our family doctor.

I must mention here also my article apropos the popularization of science ("Dialog o popularyzacji"), published in *Wiadomości Literackie* (Literary News) in 1936. This took the form of a dialogue between an adherent of popularization and an opponent. I played the role of opponent (in the character of Mr. B) and Dr. Ignacy Blumenfeld, who co-authored the article with me, that of the adherent (Mr. A). Although we felt our dialogue to be scintillating, it did not itself turn out to be very popular, nor was it much liked by those who actually read it. I think the reason for this may be that the dialogue presented only thoughts without definite conclusions, so that readers waiting suspensefully for a witty verdict were disappointed.

* * *

In that busy Fall of 1938, my daughter Lidka enrolled at the Sorbonne, study at Lwów University having become an impossibility for her. Ever growing and changing crowds of protesters participated in marches along Kadecka Street. On one such occasion, I heard the triumphant refrain "We have a ghetto in the 'technic, in the 'technic", prompted by the ruling of the Polytechnic's Rector Joszt[303] that "ghetto benches" would be instituted there. One had the feeling that a wave of the most frenetic madness was sweeping through Poland. My daughter found a flat on

---

[302] Bolesław Prus (born Aleksander Głowacki) (1847–1912), distinguished figure of Polish positivism, novelist, journalist, and short-story writer. Zygmunt Nowakowski (born Zygmunt Tempka) (1891–1963), Polish writer and actor. Henryk Zbierzchowski (pseudonym Nemo) (1881–1942), Polish writer, poet, and playwright, connected with Lwów.

[303] Adolf Joszt (1889–1957), Polish chemist, professor at the Lwów Polytechnic, and Rector from 1936 to 1938.

the *Île de la Cité*, and soon afterwards my brother-in-law Mr. Dittersdorf moved with his family from Bratislava to France, so she had relatives not too far away. My cousin Mrs. Fraenkel and her husband, also living in Paris, looked in on her from time to time, and the Lebesgues also took an interest in her welfare. In the spring of 1939 we received a letter from her announcing her engagement to Jan Kott, a student of Romance languages, studying in Paris on a scholarship from the Polish government. I was unable to attend the June wedding, so my wife went alone. The young couple honeymooned in St. Nazaire in Brittany. I wanted them to stay in France because I thought that the Maginot Line[304] probably represented a reasonable guarantee of safety in the—by now clearly inevitable—event of war.

When my wife returned in July, we went to Jaremcze[305] and in August moved to the Krzemickis' guest house *Nad Prutem* (Above the Prut) in Kamień Dobosza.[306] The Prut did indeed flow right under the windows of the house in a deep ravine. There were tennis courts, pleasant woodland, and a beautiful park close by. Sierpiński and Ruziewicz showed up, and we talked a little mathematics. I remember that I was then interested in the problem of "equipartition of the sequence in a random model". One evening we went to dinner at a neighboring guest house where one of my sisters happened to be staying, and she introduced me to a couple from Warsaw whose 12-year-old daughter had a fragment of needle embedded in her knee. Attempts had been made by surgeons in Warsaw to remove it, but they had been unable to pinpoint the location of the fragment. I very much wanted to take this poor girl to Dr. Sołtysik, a surgeon in the military hospital in Lwów who had used the introvisor successfully four times, but a surprising turn of events prevented me carrying out this plan: my daughter and my new son-in-law arrived unexpectedly from Paris. I was, of course, very happy to see them, even though I would have preferred to have them safe in Paris.

I read a report in the press about the case of a certain Hutsul who was overheard saying in some inn: "You will soon see Russia making a deal with Germany against Poland." He was had up for spreading malicious rumors, but before he was tried, an official bulletin of the Polish Telegraphic Agency announced that indeed a pact had been concluded in Moscow by the German foreign minister Ribbentrop, and the Hutsul was freed. The judge in the case actually congratulated the Hutsul on having greater perspicacity in such diplomatic matters than Beck. Some of the Hutsuls in Kamień Dobosza asked me if there would be a war, "because, sir, if there will be one, we shall all perish."

The guests at the Krzemickis' guest house included Mrs. Tomaszewska, *née* Połaniecka, her Ruthenian cousin Mr. Kiedryn, and Mr. Ostrowercha, also Ruthenian. Mr. Kiedryn was a journalist, and had heard that in the opinion of General

---

[304] A line of concrete fortifications, tank obstacles, artillery and machine-gun posts, and other defenses along France's borders with Belgium, Luxemburg, Germany and Italy.

[305] A resort town in the Eastern Carpathians on the river Prut. Till 1939 in Poland, now in western Ukraine.

[306] A place close to Jaremcze noted for its rocky outcrops.

Ironside,[307] Poland's preparedness for war was nil. Mr. Ostrowercha voiced a preference for the Soviets over Hitler. Mr. Berson, an engineer whom I had met in Lwów, was also staying in Kamień Dobosza with his wife and daughters. In Lwów we had worked together on applications of probability to problems relating to competition among businesses. One such question was the following: under what conditions is it practical for a cartel of firms to set up a new firm financed by the cartel and operated at a loss in order to undercut competing firms outside the cartel by means of artificially low prices and force them into liquidation? Of course, here the crucial question is how to ensure that the newly created firm does not draw more customers away from the cartel than from outside firms. We arrived at a theoretical solution, but the cartel that had originally approached Berson with the problem neglected to implement it, choosing rather to raise and lower prices experimentally several times—which ended up costing them a million and yielding no useful conclusion. At least we had the satisfaction of knowing that our theory was not to blame.

In Kamień Dobosza I asked Berson to make a professional draft of a design for a new introvisor that I had been working on, one that could be used for serious operations. In coming up with the design, I had thought to ask the brain surgeon Dr. Domaszewicz to allow me to be present at an operation to remove a brain tumor, since I had never actually witnessed such an operation. I took Franciszek Otto with me, warning him that the sight of the exposed brain tissue would certainly be too much for some, but he assured me that he would be unmoved. However, the operation was a difficult one, and after half an hour Otto turned pale and had to go and lie down in an adjoining room and sip water till he recovered. I remained: the sight of blood and the instruments attached to the patient's head had a direct visceral effect on my companion, but seemed to act on me only through conscious reflection.

But that was earlier, in Lwów; let's return to Kamień Dobosza. After the arrival of Lidka with her new husband, a company of young folk formed around them. These comprised Józef Nacht,[308] a young poet and a very pleasant fellow, accompanied by some friends, a friend of my son-in-law together with his *fiancée*, and a few others. One of these young people had with him an anthology of Belloc's[309] verses, in which I read the following little poem[310]:

> I was playing golf the day
> That the Germans landed;
> All our troops had run away,

---

[307]Field Marshal William Edmund Ironside, First Baron Ironside (1880–1959), British army officer. Chief of the Imperial General Staff during the first year of World War II. In May 1940 transferred to the post of Commander-in-Chief of the Home Forces.

[308]Józef Prutkowski (theatrical pseudonym Nacht) (1915–1981), Polish actor and writer.

[309]Joseph Hilaire Belloc (1870–1953), Anglo-French writer and historian.

[310]This poem is actually not by Belloc but by Jocelyn Henry Clive "Harry" Graham (1874–1936), English lyricist for operettas and musical comedies and writer of humorous verse. The poem was first published in *Punch* in 1909.

> All our ships were stranded;
> And the thought of England's shame
> Altogether spoilt my game.

My wife and I spent a couple of days on a last marvellous hike. We first travelled by narrow-gauge railroad to the place where it branched off towards Kostrycz, where we were joined by a few friends, including Irena and Ewa Łomnicka, Irena's husband, and Mr. Hubert, among others, all of whom seemed to know that the war was coming. We walked in a group up the mountainside gathering raspberries and reached Kostrycz in daylight. There we spent the night in a splendid chalet managed by the Polish Scouting and Guiding Association, with a beautiful dining room of dark wood warmed by a fireplace lined with old Hutsul tiles. The bedrooms were comfortable, the blankets fluffy, and everything was spotless and functional, without ostentation. There were albums for the guests to record their impressions, containing sketches, improvised poems, and the signatures of past guests. One group, evidently consisting of officer-skiers, had left a sketch of a lady with a very prominent backside, apparently their guide. The sketch bore signatures with arrows pointing at the lady's bottom—all in good fun, perhaps, but even the youngest boy scout could not fail to get the crude gist.

Next day we hiked along the ridge to Żabie, meeting no one apart from a shepherd or two with their flocks. We descended to Żabie, where the Czeremosz flows, a tributary of the Prut, went from there to Worochta in a horse-drawn wagon, and finally by rail back to Kamień Dobosza. This was to be the last hiking trip I would take for many years.

* * *

The weather continued splendid. Each morning I saw a bird, blue all over like a patch of sky, soaring above the Prut. Then the bird disappeared, and mobilization placards appeared in the town. My son-in-law had to go to Warsaw to join the defense effort, and the local Hutsuls were commanded to guard the bridge over the Prut and the tunnel under it near our guest house. German planes flew overhead, apparently on a heading from Slovakia over Hungary towards Lwów. Every day we went to a neighboring house—owned by a Mrs. Sadlińska—or to the local inn to listen to the latest bulletins from Warsaw on the radio. These broadcasts would frequently be interrupted by spasmodic outbursts such as: "Attention, attention, they are coming; kota24[311]...passed"—which were partly encoded warnings about the German bombers attacking Warsaw at that very moment. We drew comfort from the declarations of solidarity with Poland by Britain and France. Once, when a broadcast from London concluded with the playing of "God save the king", the innkeeper ordered the Hutsuls present to stand up. Sometimes, when the bombs fell close to the Warsaw radio station, words such as "This is unbearable!" would escape the commentator's lips.

---

[311] Code for the altitude.

Thinking that my acquaintance Stempowski[312] would be at home in Czerniowce,[313] I wrote him a letter in French asking if he could procure a Romanian visa for the three of us. I addressed the letter to Stanisław Vincenz, at that time living in Słoboda Rungurska,[314] for him to forward to Stempowski, and hired a Hutsul as letter-carrier to take my letter to Vincenz. In fact, Stempowski happened to be at Vincenz' place in Słoboda Rungurska, so you might say that fortune smiled. *En route* to Słoboda Rungurska the Hutsul was stopped by a police patrol, which gave the letter to a doctor in Jaremcze to translate, and then returned it to the Hutsul with the admonition that he tell no one that it had been intercepted. Although the Hutsul said nothing, I heard of the delay from the translator himself. Stempowski's response to my letter was to advise us against going to Romania, being of the opinion that, as he put it, the enormous political pressures that had been building up there would likely soon cause that country to explode.

Cars were continually passing through on the way to the border with Hungary at the *Tatarowska Przełęcz* (Tatarów Pass); others were choosing to take the last trains back to Lwów. I was in no hurry to return to Lwów since I expected it to be bombed. My Mother had gone to Lwów with her sister and two of her grandchildren, and I wrote telling them to stay at our house there. In case I should be needed, I also sent a card with my address to the Lwów military hospital.

Police appeared at our guest house looking for Mr. Ostrowercha, and my wife managed to slip him some bread and cheese before they took him away. After some time I made the decision to move with my wife and daughter to Tatarów[315] to be closer to the pass. However, we had run through our money, and had to borrow 500 złotys from a Mr. Mikuli, a petroleum industrialist and friend to the Krzemickis. In the circumstances this was a tremendous favor, all the more so in that I had known him for only a few days. He also advised me against going to Hungary, saying: "What will you do there? Here you are somebody, there you'll be nobody." However, I stuck to my guns, and after some searching found someone with a carriage who was prepared to take us to Tatarów. My acquaintance Karol Kossak, a painter of the Hutsul lands, and nephew of Wojciech Kossak,[316] ran a guest house there, and the three of us moved in. Among the lodgers was a Mr. Rejchan, son of another well-known artist,[317] and his German wife from Berlin. They had arrived in two cars, and had two purebred Alsatian dogs with them. The wife strutted about in a red blouse and high officer boots, with a riding crop in her hand.

---

[312] Stanisław Stempowski (1870–1952), Polish-Ukrainian writer and politician. Prior to World War II he was Grand Master of the Polish Freemasons.

[313] Now in western Ukraine.

[314] A village now in western Ukraine.

[315] Formerly a mountain resort village in southeastern Poland.

[316] Karol Kossak (1896–1975), Polish painter and illustrator. Wojciech Kossak (1857–1942), the most famous of a celebrated family of Polish painters.

[317] Stanisław Rejchan (1858–1919), Polish painter, illustrator, and pedagogue.

We listened to radio broadcasts from Warsaw, London, and Paris, and once—possibly on September 12—a bulletin from Moscow announcing the mobilization of army units in Leningrad, Kiev, and other locales adjacent to Poland. We were, to put it mildly, stupefied by this news. We discussed this new turn of events also with some other guests at the Kossaks' guest house: the Romaszkans and the Glińskis. Romaszkan, in particular, seemed to be a person of great intelligence, but was, unfortunately, quite ill. However his wife, a vivacious and attractive woman, was determined to cross the border at the pass come what may, and the Glińskis, who had a car, being of the same mind, all four soon decided to set off for Hungary. This appeared to be a general trend, since it was somehow becoming clear that Polish resistance to the Germans, though heroic, was foredoomed.

I went with the Glińskis and Mrs. Romaszkan to assess the situation at the border. In order to see if I would be admitted to Hungary, I stood in the queue to the Hungarian border outpost, only in the end to be refused—through perhaps injudiciously mentioning that I was Jewish. I approached an officer of a Polish detachment of the "defense of the nation" and he offered to "toss" me over the border, that is, smuggle me across.

After my return from this reconnaissance sortie, we watched as the road through Tatarów gradually became clogged with soldiers and civilians on foot, carriages, and motor vehicles, including buses transporting railroad personnel, post-office employees, and other civil-service officials. A dozen or so officers moved into the guest house. One of these, a colonel in the General Staff, and a cousin of the landlord, had been hurt in a traffic accident on the way to Tatarów, and kept to his bed till he recovered enough to proceed. When I queried him about the situation, he said that as long as we had had only the invading Germans from the west to contend with he had felt that all was far from lost, but that when he received official confirmation of a Soviet invasion from the east, he lost all hope and fled.

We decided to do the same. We managed to get the officer-in-charge of a medical unit to agree to transport us and our belongings, but then lost contact with the unit. We ended up travelling to the pass by means of a horse-drawn wagon, taking only as many of our things as we ourselves could carry. At the Tatarów Pass there was a long line of carts, flat-bed wagons, carriages, and motor vehicles of all descriptions queued up, and it was raining and cold. Alongside the road smoking piles of banknotes from regimental safes burned poorly. The mood of the soldiers seemed quite good. When we were near the frontier, we were recognized by a reserve officer from Lwów on duty, who advised us not to cross over. "Hitler's people are there," he said, "The Hungarian commission has an SS observer as one of its members. Many of our soldiers who crossed over yesterday are returning because everything was taken from them." And indeed we could see Polish soldiers straggling back, telling everyone that the Hungarians had stripped them of everything, down to their cigarettes and even shoelaces. "Go back to Lwów," said the officer, "The Bolsheviks idolize professors. They won't harm you." But my daughter had her own idea of the Bolsheviks, at odds with the officer's, and said succinctly: "Father, you won't be able to work under the Bolsheviks!". However, our wagon driver, who had a peasant's distrust of foreign lands, also advised us to go back. Meanwhile my wife's neck glands were swelling up from the cold and rain, and I was feeling wretched as always at the approach of winter. We had had nothing to eat and the serpent of vehicles moved at a snail's pace. Suddenly a car drew up alongside us, and the driver jumped out and called to me: "Professor, you and your folk should get in with me. Hungary's just ahead, and there things are good!" This was Mr. Kukulski from Jasło, whom I knew from the time we organized tennis tournaments there, now transformed into a military driver. Lidka was for going ahead, but, looking at my wife, I saw that she was not so keen on taking a leap into the unknown, and I myself was afraid that we might become separated from one another, since rumors had been circulating that there were separate refugee camps for men and women in Hungary. I was also wondering how my Mother would manage without us, and thinking that driving on would mean separating Lidka from her husband.

So I decided we would stay.

We had learned that the Soviets were definitely invading Poland, with the intention of occupying all of south-eastern Poland up to the river San,[318] so we were returning to what would soon be Soviet-occupied territory. On the way back to Tatarów, a soldier who had changed into civvies climbed onto our wagon. It seems he had formerly been a businessman from Kutno or Koło.[319] Judging from the half-crazed monologue he subjected us to during the ride, he had been a Polish nationalist all his life, and his nationalism had now risen to a new frenetic pitch.

In the guest house there were new guests: medical orderlies, military doctors, and other officers and civilians, among them Jakub Rothfeld,[320] professor of neurology at Lwów University, with his son, and Mr. Smalewski, the father of one of Lidka's friends. They were all on their way to Hungary, but several had doubts as to the wisdom of going there, and this led to quarrels, which sometimes became so heated that occasionally hands would go to holsters—whereupon I would take the opportunity of going outside for a stroll. The highest-ranking officer there was a second colonel, a member of a family that I had known in Jasło. Observing that military order was breaking down, I voiced my opinion to him: "If I were a military man, I would continue to follow orders as far as reasonable." They had been ordered to Hungary, and to Hungary most of them went. I remember one of the guests who stayed, an apothecary who could not bear to abandon his pharmacy, whom I later saw back in Lwów. One heard gunshots in the woods, signalling suicides of officers unable to stomach defeat.

Wave after wave of fleeing military passed through. There were tanks whose crews swore that the Hungarians would not get their tanks, and officers with their wives in military vehicles, some packed to the roof with furs, carpets, potted palms, and even canaries in cages. These drew indignant looks, but the bulk of ordinary soldiers and officers of the lower ranks deplored but one thing: that they were fleeing

---

[318] A tributary of the Vistula in south-eastern Poland. This occupation had been agreed upon in the Molotov–Ribbentrop pact.

[319] Towns in central Poland.

[320] Jakub Rothfeld-Rostowski (1884–1971).

rather than fighting. They were like the grenadiers in Heine's poem,[321] and their moral attitude elicited my great admiration.

There were many who changed their mind at the border and returned home. The Kossaks provided all they could, including a welcoming ambience. There was little or no lawlessness, neither excessive demands nor thievery. It was rumored that there were nationalist Ukrainian snipers hiding in the woods. On the other hand, the Bolshevik soldiers seemed to be in no particular hurry, maintaining, so it was said, a distance of at least a mile between themselves and the last wagon of the fleeing Polish army. We knew that they were already in Kołomyja, as someone had phoned the railroad station there.

When the last police detachments had passed through, Tatarów was left in a temporary political vacuum. The peasants suddenly began decking themselves out in the colors of the Ukrainian nationalist movement, the local Jews pronounced the community communist, and a struggle for power erupted. One Jew, a pauper, more a boot-patcher than a real shoemaker, who had spent several years in total behind bars for professing communism between the wars, exclaimed at a meeting: "I have been waiting ten years for this day!" They hung a red flag outside the Tatarów Community Administrative Center, and declared it communist territory, occupying it day and night. They were unbudgeable. But when these self-described atheists all repaired to the synagogue on the sabbath, Ruthenian peasants moved in and took over the building. However, having no idea as to what working in an office as an administrator might entail, they had to compromise by admitting a few Jews back in as "office workers". Some military personage or other arrived from Kołomyja and ordered the peasants to divest themselves of the nationalist Ukrainian colors in the name of the Soviet Union. A local militia was formed with a view to preserving order. Mr. Kossak, our landlord, was a member. Downstairs in the dining room, armed with guns and axes, we took turns at guard duty against attacks by bandits. But there were no bandits, just the occasional wanderer passing through wearing decent shoes and with a satisfied air and a bulging knapsack slung over his back. These were mostly people crossing the so-called "green border" at the last minute, so to speak,[322] and their main concern was to avoid peasant militiamen and sentries, which for the most part they did. Our Polish nationalist from Kutno, a former merchant, got himself ready to receive our new bosses, to which end he plucked a red flower from the garden, and with this at the ready in his hand waited at the railroad station for a train from Nadwórna.[323]

It was becoming steadily colder and wetter, cloudier and gloomier. The boot-patcher came to the guest house several times wearing a red armband and with a rifle on his arm, to tell us to surrender our weapons. Our landlord Mr. Kossak was

---

[321] *The Two Grenadiers*, written by Heinrich Heine in 1822, about two grenadiers who on their way home to France from capture in Russia voice their sorrow at the defeat of France and the fall of Napoleon.

[322] That is, before it was closed by the Soviets in late September 1939.

[323] Now called Nadvirna, in western Ukraine.

dismissed from the militia and Mr. Rejchan was told to place his car at the disposal of the local "Soviet"[324]. However, since no one on said council knew how to drive, they then appointed him the local Soviet's official driver. It then transpired that the real object of the expropriation of Rejchan's car was to transport the members of the boot-patcher's family to Tatarów from Kosów or Worochta.

The fields edging the road were strewn with debris left by the decamping army: sheets of tent canvas, cans of conserves, bits and pieces of motor vehicles, and all manner of other rubbish from their encampments. The more valuable objects quickly found new owners. Our merchant from Koło or Kutno was especially in his element in this regard: he appropriated a suitcase left with Mrs. Kossak by a departed officer, among other things. When at last the armored train he had been waiting for arrived, he went to the station with a freshly picked red flower to greet our new masters. He told us later that he was met by a Soviet officer armed with a machine gun, who asked him:

– Are there any Polish officers here?
– No, there are none. They've all gone to Hungary.
– Why didn't you stop them? We would have shot the lot!

Later, after the Soviets had taken over Lwów, on more than one occasion I came across blatant propaganda in the Soviet press to the effect that only the officers of the Polish army had opposed them when they invaded—implying, of course, that the regular soldiers refused to fight their Soviet comrades of the same underclass—but that the brave Red Army nonetheless prevailed against the positions occupied by the Polish officer class. The inane illogicality of this braggadocio, according to which it was a greater achievement to win against an army reduced to five percent of its usual numbers—assuming for the sake of argument this represented the proportion of officers in the Polish army—than against the full complement, could surely be swallowed only by minds very carefully prepared beforehand! What happened in fact was that in Pokucie the invading Soviet detachments strove to avoid conflict with the retreating Polish divisions to such an extent that when their paths did cross, they offered the Polish officers cigarettes and continued on their way.

There came a time when we felt that it made no sense for us to remain in Tatarów. I went to the Community Hall to consult with the interim "Soviet" there, and spoke to a member, a Jewish "warrior"—I can't think of a better term for this half-uniformed, Russian-speaking son of a Tatarów weaver. He asked what the purpose of my trip was, and when I answered that I had to return to the university, he said: "From now on there will be no more scuffling and scrapping at Lwów University," implying by his tone that I was responsible for the "scuffling and scrapping".

Our train trip back to Lwów was interesting. At each station Rusyn youths jumped onto the running boards of the carriages, partly to have a free ride, but also, more significantly, to see what they could manage to steal from the luggage piled up at the end of each carriage. Naturally, the carriages had no

---

[324] "Council"

windowpanes—a circumstance associated with every change of regime in eastern Europe. Since the main Lwów station had been destroyed, we had to get off the train at Persenkówka.[325] While I guarded our luggage alongside the track, my wife and daughter went to look for a horse-drawn wagon to take us into the town. Eventually they found someone to take us, and we drove into the city center along Stryjska Street. On the way we saw signs of destruction and change: telegraph posts knocked over by tanks, dead horses, tangled power lines, bullet holes in the plaster walls of houses, broken windows, and lots of people in high boots and vizored caps, with sacks slung over their backs. These were just the townspeople of Lwów dressed up in what they thought was the obligatory Soviet style. And we saw also the genuine article, the so-called Soviets, mainly soldiers and military railwaymen. They all wore brown greatcoats, had vacant faces, and paid no attention to anyone. The Soviets were on us like a swarm of locusts. We had been told by people on the train who had been in contact with the invaders for as much as a week, that a great number of them were more like beggars than soldiers, without boots or adequate clothing, eating whatever they could scrounge. At last we turned into Kadecka Street, bracing ourselves for the sight of our house: would it be still standing or not?

Our house was still standing!

---

[325]Formerly a suburb of Lwów, incorporated into the town in 1930.

Chapter 8
The First Occupation

In our house we found my Mother, one of my sisters with her two daughters and the *fiancé* of the younger one, my wife's uncle L. G.,[1] our maid, and our third-floor tenant Mrs. Kwaśniewska. We were all to be dining together for some time to come. Next day I went to the university, where chaos reigned. The gates were under guard by young men wearing red armbands and armed with rifles. The majority of them were from the National Radical Camp,[2] and appeared to consider the red armband the most practical disguise—despite which many of them were later unmasked.

In the Rectorate I found Roman Longchamps the Rector, together with a dozen or so professors, from whom I learned what had happened in Lwów between September 1 and September 26. German tanks had already reached the suburbs when General Langer,[3] the designated commandant, called a few of the more prominent Lvovian citizens together to discuss the surrender of the city to the Soviets. While some citizens—Bartel among them—were in favor of this, members of the episcopate voiced their opposition. I was also told that on August 31, Rojek[4] had stood on the steps of the memorial to Mickiewicz, apologized to the members of national minorities on behalf of the so-called national youth movement for its mistaken political views, and called for the defense by all of the threatened

---

[1] Leon Grünwald, alias Leon Zielony.

[2] In Polish: *Obóz Narodowo-Radykalny* (ONR), a Polish extreme right-wing, anti-Semitic, anti-communist, nationalist party formed in 1934 mostly by young radicals from the National party or the Movement for National Democracy. Its members instigated pogroms and other anti-Jewish actions before it was officially suppressed.

[3] This was actually the Polish general Władysław Langner (1896–1972). Over the period September 12–22, 1939, he commanded Polish armed forces in opposing both the Wehrmacht and the Red Army in and around Lwów, an episode later called the "Battle of Lwów". The defense of the city lasted 10 days, until, on September 22, Langner surrendered the city to the Soviets, in order, as he said, "not to further endanger the lives of the people and risk the destruction of the city".

[4] Organizer of student rallies, leader of an organization responsible for several murders. See Chapter 7.

Fig. 8.1 Map showing the invasions of Poland by Germany and the USSR in September 1939 (By permission of the U.S. Holocaust Memorial Museum.)

Motherland. Thus it was that in the space of a few weeks these young people went, chameleon-like, through a succession of political stripes—from national radicalism, through democracy, to Bolshevism. But not all young people were so fickle. On the streetcar I saw one young man wearing the cap of a certain student fraternity, whose countenance was set in a grim look of obstinate intransigence—which I found appealing as evidence of integrity. I liked him.

Professor Kulczyński related how certain Polish officers and NCOs had found themselves in the *Sejm* building besieged by Soviet forces. When the NCOs were advised to strip off their insignia, one sergeant, known to be Jewish, had said: "I refuse to remove the stripes of a sergeant of the Polish army!" However, the mass of poor people living in the district behind the theater had turned out in force, decked out in red stars and knotted red kerchiefs, to greet the Bolsheviks, causing some mirth among the Soviet officers. Others had disarmed Polish officers in the street, and when Soviet tanks arrived, had actually kissed their armored hulls and reached up to caress their cannon. Clearly, such episodes were more than anything else mere expressions of joy and relief at what seemed like a miraculous salvation from Hitler, and the overthrow of an interwar regime that had been rapidly becoming more fascist.

Before the *Sejm* building, coming from the direction of Kościuszko Street, I saw Father Gerstmann threading his way breezily through the crowd. The Rector had informed us that he knew of the impending meeting there, and had urged us to attend and bring with us as many students as possible. So we gathered in the great hall (*auditorium maximum*) and sat down at the benches, while up on the dais were ranged uniformed representatives of the Red Army, members of the Komsomol,[5] Mr. Kornijczuk,[6] as a member of the intelligentsia of the Ukrainian SSR, our Rector, and others unknown to me. The Soviet officers were in full uniform, and with their peaked caps they reminded us of Germans. Professors Krzemieniewski, Chwistek, and a few others were invited to join those on the dais. Mr. Kornijczuk spoke first, as follows: "On my way into Lwów I could see how deeply attached its citizens are to their town. I also saw the damage to the streets caused by the bombing. However, neither you nor we are to blame. Those bearing responsibility for this are the ones who defected to Romania." This was a well-chosen tack since we were in fact indignant at the officers who, with their wives, mistresses, dogs, and carpets had gone bag and baggage away to Romania, leaving us to fend for ourselves. And we were most thoroughly angered by that fellow Rydz [Rydz-Śmigły], who had promised not to yield up even so much as a button, yet had given the whole coat away, by that fellow Beck with his empty talk of a Poland from sea to sea, by that fellow Składkowski who as late as September 10 was imposing penalties on those with unpainted fences, and who had arranged for a false air-raid alarm to be sounded in Kosów in order to have a pretext for fleeing with his retinue and stocks of brandy, and, finally, this President who was no president. Our university especially might well feel disgust at such betrayals since not long before (following the events in Zaolzie[7]) at the initiative of ardent supporters of the National Unity Camp, it had conferred doctorates *honoris causa* on Messrs. Mościcki, Beck, and Rydz. Thus Kornijczuk's words touched a sore spot with us. He went on to assure us that the university would continue to be a Polish one, and in case of need, another, Ukrainian one would be founded. Furthermore, he hoped for the support of the Polish intelligentsia. However—he continued in a certain tone—anyone wishing to continue with nationalistic and reactionary political activity would learn that *my ne vegetariantsy*[8]. It emerged that the main item on the meeting's agenda was to have the text of a dispatch to Stalin[9] ratified. Some of our professors also spoke. In particular, Professor Krzemieniewski turned his gaze (in exactly this order) from the Rector to the Red Army officers, and then to the rest of us, while expressing the hope that the values represented by the university would be preserved, and calmly argued that it was no nest of reaction or oppression. After him various students,

---

[5] An abbreviation for *Kommunisticheskiĭ Soyuz Molodezhi* (Communist Union of Youth), the official Soviet youth organization.

[6] Oleksandr Kornijczuk (1905–1972), Ukrainian playwright and socio-political activist. In 1943 appointed Deputy Commissioner for Foreign Affairs of the USSR.

[7] Zaolzie is the Polish name for the lands beyond the Olza River, now comprising the eastern part of the Czech portion of Cieszyn Silesia. The region was formed in 1920, when Cieszyn Silesia was divided between Czechoslovakia and Poland. The division was unsatisfactory to all parties and the resulting conflict led ultimately to the annexation of the region by Poland in 1938.

[8] "We are not vegetarians."

[9] Joseph Stalin, Marshal of the Soviet Union (1943–1945), and Generalissimus (1945–1953).

Poles, Rusyns, and Jews (the most numerous), took turns to address the meeting. They all agitated for work, work, and more work. The quality of their speeches was poor, full of commonplaces about the proletariat and the nation—meaning the "people", a popular word with the new regime—, justice, and the unvanquished Red Army—unvanquished indeed, since it had as yet not seen any fighting.[10]

The leaders of the Soviet force occupying Lwów completely ignored the speeches and posturings of the students; for them the professors' attitude was paramount. Kornijczuk later said that he'd been unable to sleep the following night because he had been so tremendously irritated by the fact that none of the Polish professors had spoken in a conciliatory manner to the Ukrainians. What had been most interesting for me at this meeting was the easygoing and super-enlightened attitude of these Soviet notables towards matters we would have considered difficult and sensitive. Thus they said that the language of instruction could be whatever was convenient: Polish, Russian, Ukrainian, or even Yiddish, this representing no real problem for them. They expected scientific articles and books to be published literally by the wagon-load, now that all administrative and other difficulties had been swept aside—at one blow! Everybody was now to be admitted to university, tuition would be free, and if space be lacking, then a second, third, or even fourth university would quickly be established. Cost would be no object. Following on these heady pronouncements, these Soviet bigwigs left us, never to be seen again—but then others came who didn't have any idea as to what the first lot might have said to us.

There were many stories and anecdotes circulating at that time concerning the first encounters between Lvovians and the Soviet soldiery. For example, once a captain accompanied by four privates with fixed bayonets came to our house looking for billets for the soldiers, but when he heard that I was a professor he called a retreat. As they tramped back down the stairs, the captain began a discussion with me in the presence of the maid and our janitor. "Do you know who Marx was?" he asked, and when I said I did, he continued: "I myself have read Marx's *Das Kapital* till my hair turned grey. It's very difficult reading." Then he turned to the attack, saying: "You are a mathematician, but mathematics is also a class science."[11]

Quite soon I perceived that we were dealing with liars and people of little— say a quarter of the average—intelligence whom the liars had turned into zombies. They made invariably empty promises, big fibs like trump cards to be dealt at every opportunity: "You had a pension equivalent to a thousand rubles; now you will get two thousand." They promised to double everyone's income. At the same time, placards appeared everywhere denigrating the Polish army, depicting our officers dressed in rags, in old-fashioned shakos and with swords, camping out in the bush alongside broken-down wagons like Gypsies. These caricatures were intended, of course, to degrade an army which, in addition to having to contend with overwhelming German forces, had also to deal with a stab in the back by the

---

[10]Except in the civil war of 1917–1922, when it defeated the White Army and their allies—but that was a good generation earlier, so a different incarnation, one might say.

[11]Meaning perhaps, a product of the class system.

Soviet Union, and they filled me with loathing and contempt for these perpetrators of lies. There is nothing more incompatible with the notion of honor than bringing a valorous army into disrepute through lies—an army which in fact fought very bravely against impossible odds; some units even managed to escape to continue the struggle from abroad. And whence came this undeserved mockery? From a power which had violated its non-aggression pact with Poland,[12] and had come to an agreement to partition Poland with an enemy of democracy, of all things Slavic, and of communism. But as if this were not enough, the official Soviet line was that "the rabbit had started it all", that is, that Poland had attacked Germany in the first place. And to bolster this blatant untruth, the war was called the "Polish-German war" in their newspapers, while Germans and Soviets were touted as people from nations wanting only peace, as against the war-mongering French and, especially, British. The Soviets didn't look too kindly on Italy either, which shows that they were perhaps getting themselves tied up in some sort of knot with all their fibbing. However, the young Polish Jews who had become infected by Marxism were so blinded by the apparent fulfillment of their cherished socialist dream, that there was no lie issuing from the Soviets, no matter how nonsensical, that they could not swallow. People like Herzberg and Wojdysławski[13] believed everything the Soviets told them—so even believed in the Molotov–Ribbentrop Pact!

* * *

Soon two plagues descended on us: the plague of meetings and the plague of reorganizations. Meetings were called without much warning: a clerk would frequently bring an unsigned note announcing a meeting in half an hour to deal with a very important matter. The "very important matter" might be anything from a resolution that everyone must vote in all elections, or that we should demand that Western Ukraine[14] be united with Soviet Ukraine. I remember discussing this issue with some non-academic Ruthenians who were participating, all of whom were in favor of the annexation. When I explained that voting on the resolution would not guarantee that it would carry, they responded by saying that it didn't matter since Ukrainian affairs were not our concern. When I then asked them why, if the issue didn't concern us, we were forced to vote, they were at a loss to answer me. There were meetings at which elections were held to choose members of a committee for eliminating private houses, land, woods, coal mines, and factories, and other meetings to pass resolutions aimed at fulfilling other clauses in the Stalinist constitution.

The administration of the university went through radical changes. Divisional Council meetings were disallowed. A notice suddenly appeared stuck to the wall

---

[12] Signed in 1932. Unilaterally broken by the USSR on September 17, 1939, in accordance with the Molotov–Ribbentrop Pact.

[13] H. Menachem Wojdysławski (1919–1942), Warsaw topologist. Wrote an influential paper titled "Rétractes absolus et hyperespaces des continus", *Fund. Math.* 32 (1939), pp. 184–192.

[14] Now understood to include occupied south-eastern Poland.

in the main entrance announcing that on account of his unpunctuality Professor Zierhoffer[15] was herewith relieved of the deanship of our division. He was replaced by the Ruthenian Biskupski, an adjunct in the department of mineralogy. Such lack of ceremony was typical of the Bolsheviks, and would have been amusing if their behavior had not been so unpredictable in general. Ruziewicz, who prior to the war had become Rector of the Business Academy, said that a Soviet official wearing the standard peaked cap had come to his office and announced without any preliminaries: "You are no longer Rector. I am now Rector." He then asked to speak with the Academy's librarian, who happened to be a little late arriving to work that day and so not immediately available. On hearing this, the Leninist commanded that he be relieved of his post. The fact that this man had worked for 20 years as librarian to the complete satisfaction of the Business Academy was a matter of total indifference to this Soviet official.

Soviet officials generally behaved like ill-educated, intransigent children, trained to a sort of ideological ruthlessness. I called them ...,[16] and it turned out that by chance Dr. Vincenz, although far from Lwów at the time, had hit on a similar name for them—although of course only for private use. He had been arrested when he returned with his son to Słoboda Rungurska from Hungary for illegally breaching a border. When he came before a Soviet court in Lwów, the public prosecutor asked him for his *curriculum vitae*, from which he learned that Vincenz had written his doctoral thesis on the philosophy of Ernst Mach. He said that this made the case much more serious since Mach's philosophy was in disagreement with Marxian materialism. Dr. Vincenz then explained contritely that he had written this work in 1915, when the epochal discoveries of Lenin and Stalin, the greatest philosophers of all time, were not widely known. Although this seemed to cut little ice, he was in any case released—probably because of his sympathetic attitude to the Hutsuls.

In the Soviet Union Joseph Stalin was regarded not only as a great philosopher, but also as the all-time greatest writer, leader, politician, historian, and friend of all peoples, an inspiration to discoverers of new knowledge, and patron of subarctic seamen. His most important work, albeit largely unattributed, was supposed to be *The History of the Communist Party of the Soviet Union (Bolsheviks). A Short Course*.[17] This work falls naturally into two parts.

---

[15] August Zierhoffer (1893–1969), Polish geographer and geologist, specializing in geomorphology. Studied at the Jan Kazimierz University in Lwów, and held a professorship there from 1932 to 1939.

[16] The word in question is illegible in Steinhaus' manuscript.

[17] This was an abridged version of the collectively written *History of the All-Union Communist Party of the Soviet Union*, published in 1938. Stalin is believed to have been the author of the section of Chapter 4 entitled "Dialectical and Historical Materialism," an exposition of the foundations of Marxist philosophy, and possibly of other parts of the tract.

8 The First Occupation

The first of these is devoted mostly to "theology", one might say, since it is concerned with the struggle to subdue Satan as personified by Trotsky,[18] who was to be blamed for all the evil that had befallen the world in recent times: from defeat in the Polish-Soviet War in 1920 up through all the catastrophes of the twenties and thirties afflicting the USSR, to the period 1934–1937 of the Moscow show trials justifying the slaughter of the party's former intelligentsia.[19] In large part the *Short Course* was merely an attempt to lend these murders legitimacy: since it was extremely difficult for independent-thinking, reasonable people to swallow the confessions of the accused in the Moscow trials, it had been decided to publish a book justifying them and make it required reading for all workers, soldiers, secondary and tertiary students, professors, and so on—in short, for everybody—who were to imbibe its teachings as if it were gospel. The richness of the book's invective against the so-called Trotskyites—Zinoviev, Bukharin, Radek, etc.—is revealing of a fury that feared its own impotence. Having buried hundreds of thousands of real and imagined opponents, it demanded the covering over of their graves with mountains of rock so that not a single grass shoot of the truth might ever emerge.

I knew of a case where a person was tried in Lwów in 1940 for allegedly having said in a private conversation in Wilno in 1937 that Trotsky was not a traitor. The sole argument of the prosecution, representing our local Stalinists, consisted of the unsupported assertion that if Trotsky had ruled Russia he would have disarmed her just as the leftist French had disarmed France,[20] so that, *eo ipso*, Russia would have gone as booty to Hitler.

One thing that should be kept in mind as distinguishing our intelligentsia from the Soviet brand is that unlike the Russian intellectuals ours had not experienced Soviet rule and knew only of the early period of the Soviet Union, that is, of a culture defined by the Cheka,[21] "commissars," and the associated propaganda—such as "down with the bourgeois educated of the tsarist regime"—and so forth. From 1937 this culture became overshadowed somewhat, I would say, by the objectives of turning the USSR into a major industrial and military power—objectives to a

---

[18] Leon Davidovich Trotskiĭ (1879–1940), Russian-Jewish Bolshevik revolutionary and Marxian theorist. Second only to Lenin among the leaders of the October Revolution. Played a major role in the Bolshevik victory in the Russian Civil War following the revolution. Removed from power, deported, and eventually assassinated through the machinations of Stalin.

[19] Mostly "Old Bolsheviks", the leadership of the secret police (NKVD), and others who might potentially have challenged Stalin's authority. For similar reasons Stalin also purged almost the whole of the officer class of the Red Army in the late 1930s.

[20] In fact, it is now known that in the spring of 1940 the French armed forces were superior to those of Germany. The 6-week defeat of France is now attributed rather to French military incompetence and psychological unpreparedness. See, for example, Chapter XI of: Tony Judt, *Reappraisals: Reflections on the Forgotten 20th Century*, The Penguin Press, 2008.

[21] Russian acronym for "Extraordinary Commission", the first Soviet State security organization, decreed by Lenin in December 1917, and headed by Feliks Dzerzhinsky. Under its auspices many thousands of abductions and executions were carried out. Forerunner of the NKVD.

large extent independent of the underlying communist ethos. I shall return to this theme below.

I now turn to the second part of the *Short Course*, represented by Chapter 4, which expounds the philosophy behind Soviet communism. The word "materialism" is used in two senses in the book. When used in the phrase "historical materialism", I had no great objection: this is the name of a well-known doctrine according to which historical processes are at root economic ones, an idea that makes some sort of sense and therefore lends itself to rational discussion. But when it appeared in the phrase "dialectical materialism", I was at a loss to understand what was meant, even though I tried hard to figure it out. In the *Short Course*, where the philosophical material is allegedly based on Stalin's words, the concept is not defined, and likewise none of the learned Fathers and Doctors of the Communist Church seems to have been tempted to formulate a definition of the concept. It was as if Soviet communism, like a dualistic religion with Stalin partaking both of Ormuzd and—to a lesser extent—Ahriman,[22] needed to have a mystery at its heart to lend it mythical value. Every apologist of "Leninism" falls back in the last resort on "dialectical materialism" as the only means of avoiding being squeezed in the pincers of the principle of contradiction.[23] This materialistic dialectics requires the new principle *Tertium datur*, which was justified by falling back on Hegel's trichotomy of thesis-antithesis-synthesis made material.[24] One had the feeling that the creators of this philosophical doctrine had interrupted their philosophical studies 40 years before, when they had gotten as far as Hegel, and then, too busy with their revolutionary activities to get any further, passed on their lecture notes to their heirs as the full philosophical picture. Stalin was undoubtedly capable in some ways, but as a philosopher he is a complete megalomaniacal dilettante: his reduction of qualitative changes to quantitative ones cannot withstand criticism,[25] and his solutions of putative philosophical problems betray a complete ignorance of the philosophical stance.

Similar phenomena could be observed in other domains. To bolster the myth of Soviet superiority, the Soviet powers-that-be had to keep pushing the propaganda that real progress is achieved only through revolutionary Bolshevik thought. This applied also to physics and mathematics. But when you read the papers and books of Soviet scientists, you found nothing that might be designated "Bolshevik scientific thought"; there were a great many excellent Soviet mathematicians, physicists, and

---

[22] The modern Persian versions of the bad and good spirits, respectively, of the Zoroastrian religion of the ancient Iranians and the modern Parsees.

[23] Law of the Excluded Middle or *Tertium non datur* ("no third (possibility) is given"), meaning that every meaningful statement is either true or false.

[24] For example, the progression monarchy—capitalist democracy—communism of a society, resulting from the struggle between classes, with communism as the final utopian synthesis. The idea, but not the name, was first proposed by Marx and Engels. Although one might object to this on various grounds, it doesn't seem to have anything to do with the negation of the Law of the Excluded Middle, except by loose analogy.

[25] But perhaps modern materialists' reductions are more arguable.

biologists, but their works did not show in the least any difference in approach to their subjects over that of western science. Except for the language they were written in—Russian—there was absolutely no evidence that any special "Marxian" mode of thought had gone into creating them. But I must beg your pardon! I forgot that sometimes the papers and books of Soviet scientists did indeed contain prefaces and introductions in which platitudes about "revolutionary Marxian scientific thought" were reproduced, after which they got on with the serious scientific account of say, vacuum tubes, or the vernalization of wheat. It is as if the Latin tag *Duo cum faciunt idem, non est idem*[26] served as one of the favorite mottos of the class of Bolshevik scientific supervisors. At a mathematics conference held in Zürich, one of these, by the name of Kocman,[27] gave a lecture on "Marxian mathematics", at the end of which Bernays[28] asked him how one differentiates "marxistically". The speaker was completely stumped. But if we go along with the Bolsheviks and attempt to get them to deny some of the main pillars of "bourgeois" mathematics—by making it strictly empirical, say, or finitistic, or intuitionistic, or by using three-valued logic—then they will emphatically not go along with us, because they need the technological fruits of traditional mathematical science—they want the real thing. In the depths of their psyches they know they must cultivate traditional European science, but as perpetrators of the big lie, they must deprecate that science before the tribes of Kazakhstan, Kurdistan, and Outer Mongolia, to name but a few. In their industry the boast is similar: they are quiet about the fact that Soviet industry makes essential use of technology developed over centuries in "bourgeois" Europe and America, while trumpeting their industrial might.

A big Soviet lie was their contention that true science had been purloined by bourgeois societies, and that henceforth the initiative lay with Soviet science—and that the Americans were in fact already patenting Soviet inventions illegally. To maintain this untruth, it had been necessary to close the borders: thus from 1937 it was extremely difficult for the average—or even non-average—Soviet citizen to obtain permission to travel outside the USSR, and genuine scientists—as opposed to party hacks pretending to do science—were for the most part no longer allowed to attend foreign scientific conferences.

In Lwów many young Soviet citizens could now encounter the West for the first time, and, *vice versa*, many of us experienced Asia for the first time. However, the circumstances of these encounters were special. Indeed, the people of our

---

[26] "When two do the same, it isn't the same."

[27] This is probably Ernst Kolman (1892–1979), Marxist philosopher, born in Prague, notable for his activities as chief ideological watchdog of Soviet science. One of his publications is titled "Karl Marx and Mathematics". In September 1932, he and Bernays both gave talks in Section VII (Philosophy and History) at the International Congress of Mathematicians in Zürich. The title of Kolman's talk was "Eine neue Grundlegung der Differentialrechnung durch Marx" (A new basis for the differential calculus by Marx).

[28] Paul Isaac Bernays (1888–1977), Swiss mathematician. Contributed to mathematical logic, axiomatic set theory, and the philosophy of mathematics. Sometime assistant and close collaborator of Hilbert.

region of Poland had endured a quick succession of blows descending on them relentlessly: mobilization, war, bombardment, defeat, the exodus of the army, the unanticipated entry of a new enemy, and annexation. Most of the occupying army—both officers and soldiery—had been recruited from Ukraine, and their aim was to establish good relations with the indigenous people of south-eastern Poland. However, poorly fed, ill-clad Soviet soldiers—and there were many such to be seen—could hardly be expected to impress our peasants. They were trying to counteract this unfortunate impression by means of a show of tanks and especially trucks, which were constantly rushing about the streets, often transporting nothing more than a few brooms, half a can of gasoline, and a few tires. However, in Lwów the great majority of Soviet soldiers behaved better than might have been expected of an occupying force: there was no drunkenness evident, they did not accost pedestrians or demand priority in being served in the stores, and in those rare cases where an officer was rebuked by a civilian for improper behavior, he would merely turn red in the face and apologize, or else vanish. One sensed that they had a feeling of cultural inferiority and wanted to be seen as Europeans like us. Their psyche was split: on the one hand they had had it drummed into their consciousness that they were in the vanguard of civilization, but on the other hand their deeper, unavowed, feeling was one of inferiority before the West.

The Lwów populace caught on to this very quickly, and poked fun at them mercilessly. The general feeling was that they were so ridiculous that one had to laugh at them even when they were making one's life difficult. This explained why there was no hatred directed towards the Soviet soldiers: they were considered mere pitiful victims of a deranged system.

There was a small apartment in our house that had been let to the Polish government before the war as offices for their Border Guard, a paramilitary unit whose main task was to prevent smuggling. Naturally, they had escaped with the army, leaving piles of documents behind, including albums containing snapshots of known smugglers and photographs showing a variety of techniques used for smuggling goods on buses, trains, and wagons. This vast quantity of paper was now proving useful to us for burning in our stoves, since fuel was becoming impossible to obtain. Well—to get to the point—this apartment had now been taken over by the families of two officers from an armored NKVD unit. After living there without my permission for two months, suddenly they rang the doorbell and asked for the owner. Combed and perfumed, the officers sat down in our armchairs and asked me to determine rental charges. They haggled a little, as Europeans like to do, noting that the apartment had been in a state of disrepair, and then, having determined the rental costs, departed, never to be seen again. I felt I understood what this had been about: a matter of estimating, purely for theoretical purposes, what the expenses associated with such a stay might have been—as a sort of childishly empty gesture towards taking responsibility for paying for oneself. This childlike quality was general among the occupying forces. None of them behaved like grownups, the only real adults being their Kremlin bosses. Stefan Banach was wrong when he said to me "This is a system run by the intelligentsia." It was a system where the intelligentsia was tightly controlled by the Kremlin.

The people of our middle class were amazed by the behavior of the Soviets: for example, by the unheard-of indignity of an army officer actually carrying his own children in his arms, while their wives, fat, grubby, and slovenly, walked alongside in garish dresses like nightshirts, brightly colored socks, and yellow berets. When a bevy of telephone operators arrived from Kiev, and took up shared quarters with our operators, the Soviet women were astonished to see the elegant and fashionably feminine underwear and nightdresses of the Polish women, and, conversely, our young ladies could hardly believe their eyes when they saw that the Soviet girls' panties were just ill-made men's drawers held up with string. The Russian women were also surprised to see that we had automatic, that is, dial, telephones, since they had been led to believe that this was a Soviet invention unknown in Europe. The Soviet propaganda machine was sorely tried in trying to counteract the culture shock to which its people were exposed in their encounter with a European city such as Lwów, a city which, although battered and depleted, still managed to turn the heads of the poor Kazakhs and Turkestanis sent to show us the best, Marxian, way to live. Thus it was explained that the fact that one could buy a steak for one złoty fifty groszy in Lwów was attributable to the gross exploitation of African labor by Polish capitalists. They also put about the lie that Poland had remained essentially the province of the male, with women having no rights—in particular the right to vote—and no access to higher education or any of the traditionally male professions. They even went to the extreme of producing movies purporting to show the grim economic situation in prewar Lwów. In one such film we were shown a long line-up of ragged people at a grocery store. Suddenly an officer in Polish dress is driven up, alights, goes into the store, and re-emerges carrying several bags of sugar. The store now closes because there's no sugar left for the poor. Another clip depicted a woman who, after lining up for a considerable time, has finally arrived at the counter where they sell cloth. But the cheapest sort costs 100 złotys a meter, and she earns only 50 złotys a month, proving once again that in capitalist Poland the rich took everything for themselves. Not surprisingly, our working class types laughed their heads off at this kind of shoddy agitprop. I am quite sure that no Christian organizations, no anti-communist propaganda even of the extreme type put out by the IKC,[29] no sermons preaching directly against it, could turn our working class more effectively against Bolshevism than did Bolshevist propaganda itself. However, the Bolsheviks had two powerful arguments for the superiority of their system. The first, more general, argument was based on the claim that proud capitalist Europe apparently needed to descend into war every 25 years in order to keep its economic and social system going. The second, more special, but particularly painful, argument, is more-or-less just the contention that might is right: "We, the Soviets, ill-clad, ill-fed, and ill-housed, lack the luxuries you took for granted, but we do have tanks and artillery;

---

[29] *Ilustrowany Kurier Codzienny* (Illustrated Daily Courier), a Polish daily as well as a publishing house, founded by Marian Dąbrowski in Kraków in 1910. Poland's most popular daily in the 1920s. The last issue appeared on October 26, 1939.

all you did was eat, drink, and make merry, and when war came you were caught defenseless, and forfeited your independence."

* * *

At least the latter argument left us with no illusions. The first one, however, relied on official propaganda to the effect that two great powers, peacefully disposed towards one another, had by means of their mutual pact succeeding in extinguishing the possibility of future wars in eastern Europe. More paining than this blather was the acceptance of such Asiatic lies by a certain stratum of our intelligentsia. But in connection with this, it must be taken into account that the composition of the Lwów populace soon changed considerably as a result of the influx of refugees—three quarters of whom were Jews—from Warsaw, Kraków, and the area around Poznań, from all of western Poland, now occupied by the Germans. Lwów's population doubled. After the establishing of the German-Soviet demarcation line[30] in late September 1939, people began crossing from west to east both legally and illegally, especially in the vicinity of Białystok and Małkinia.

One of these was my son-in-law Jan Kott. When last we saw him he had been recruited to the defense of Warsaw, and he did indeed participate in the defense of the capital. He was captured, caught pneumonia, and found himself in a hospital in Dęblin.[31] An uncle came by car from Warsaw, "abducted" him from the hospital, and took him to his home to have him cared for. When he recovered, he caught a train heading through Małkinia to Białystok. When it stopped in Małkinia, German gendarmes dragged those Jews that could be identified as such out of the coaches and beat them with rubber truncheons, but then let them proceed. The local Polish youth helped the Germans by pointing out the Jews, since for the most part the Germans were unable to distinguish them from Poles. At the demarcation line the train halted, and the passengers descended from the train to cross over on foot. They found their way blocked by Soviet soldiers shouting "Davaǐ nazad!",[32] but the crowd of refugees responded with shouts of "Long live the Red Army!", and surged over the line. In this way many thousands—including my son-in-law—penetrated the demarcation line.

Janek also gave us a detailed description of the defeat. He said there had been a complete collapse of liaison, and that defeatist psychology had played a major part, together with the atmosphere of suspicion whereby every civilian was suspected of spying for the Germans. In Białystok he had met Władysław Broniewski,[33] who had earlier held poetic recitals as the "red poet", and they came to Lwów together.

---

[30] As ultimately agreed upon, this followed the rivers San, Vistula, and Narew.

[31] Polish city on the Vistula, considerably to the south-east of Warsaw.

[32] "Go back!"

[33] Władysław Broniewski (1897–1962), Polish poet and soldier. Decorated in the 1919–1921 Polish-Soviet war.

A few weeks later Stanisław Saks, and then Bronisław Knaster, arrived in Lwów. From these, and also from a cousin of my wife, a young engineer who negotiated the greater part of the road from Kraków to Lwów by bicycle, I was able to piece together a good picture of the first few weeks of the German invasion and of the invaders' behavior. I heard how they had bombed passenger trains, how from heights as low as fifty meters their pilots had strafed refugee columns, including children, and even shot at herdsmen and their cattle in the fields. I was told that German partisans had been parachuted into Poland disguised as nuns or Polish policemen, and even—if it can be believed—of one dropped in in motorcyclist's gear, together with a motorcycle carrying a baby in its sidecar.

Then there were the reports of barbarities committed against Jews. From Dr. Knaster I heard of the execution of the whole of the Jewish population of Ostrów Mazowiecka.[34] The Jews of the town, numbering several hundred, women, children, and babes-in-arms included, had been rounded up and herded into the marketplace where they had been machine-gunned on order of an officer. Reports had come in of even greater atrocities perpetrated against Jews—and especially rabbis—in other areas of German-occupied Poland. Incredible scenes of villainy had been witnessed in Warsaw. For example, at the entrance to a house on Marszałkowska Street an agent of the Gestapo had been seen waylaying Jews, dragging them into the entrance and beating them with a rubber truncheon. Other such Germans had been seen unceremoniously stripping the fur coats from passing Jewish women and handing them to non-Jewish passers-by, taking care that no one approached closely enough to the woman for her to communicate her address to them. Sometimes the Polish woman who had been given a coat would, with tears in her eyes, run after the woman from whom it had been stolen and, out of sight around a corner, restore it to her. "How can you do such a thing to a lady?" said a lady to a German officer who had burst into her bedroom without knocking. "*Eine Jüdin ist keine Dame*,"[35] came the answer.

However by the way, this does remind me of the surprise the former Polish prime minister Leon Kozłowski registered once on learning that there were distinct social classes among the Jews. He had naively thought that Jews formed a sort of monastic order where everyone was at precisely the same social level as everyone else, but that, in order to deceive the gentiles, they had adopted various disguises—one as a seller of herring, another as a piano virtuoso, and so on.

But the gruesome reality of the German-occupied zone was for the moment of less concern to us than our own Soviet version. And that proceeded to become frightful enough fairly soon—as if the Soviets wanted to show that they also were red of tooth and claw. They began by arresting all those who had been reserve officers of the Polish army. One of those arrested was Bartel, but he was quickly released. He told me subsequently that during the search of his home, he was asked why he had so many books. He answered in a suitably deprecating manner

---

[34] A town in northeastern Poland.

[35] "A Jewess is no lady".

and they appeared mollified. But his gold watch went missing, so he went to the commanding officer of the Lwów branch of the NKVD to complain, "officer to officer". He was received with stony-faced politeness, but without result. The polite front was maintained while the arrests went on, sometimes as many as a dozen a night. This political action seemed to originate with the convocation of a meeting of trade union delegates—presumed to have been elected by the respective trade unions, but in fact nominated by the party leaders. At this meeting it was decreed that private ownership of factories, of arable plots of area exceeding 25 hectares, of apartment houses, woods, and mines would henceforth be abolished, and that "Western Ukraine", which included Lwów and the surrounding Polish territory, would be annexed to the Ukrainian Republic of the USSR. Halina Górska[36] voted against this decision, demanding that Lwów be made an autonomous Polish region. However, this first phase of Soviet politics as it applied to Poland was very strongly pro-Ukrainian—to such an extent that the names of the streets in Lwów were ukrainianized—and her objections made no impression whatsoever. In addition to Polish reserve officers and their families, the Soviets arrested Polish judges, former state attorneys and other high officials of the former Polish state, bank directors, prominent merchants, and so on. In many or even most cases their families were deported to the USSR rather than being arrested. All this was carried out sporadically, without any apparent system, which made it all the more frightening for being apparently haphazard. Announcements were constantly being made in the streets over loudspeakers of a new round of registrations of Polish reserve officers from among the refugees arriving in Lwów. The cleverer among the latter soon understood that they had best ignore these registrations. The way the Soviets went about making their lists of registrants was incredibly primitive. For the most part they simply recorded the names of those being registered on rough packing paper in the order in which they came, so that it became virtually impossible to verify whether someone was already registered or not. Then we were subject to another bureaucratic nightmare in the form of trade union membership. In the USSR to be considered a human being one had to be a member of a trade union. Membership was decided by a special commission led by delegates from the Communist Party of the USSR, and these effectively decided the admission or not of an applicant since the vote by the commission was a farce: anyone opposing the recommendation of a *politruk*[37] risked being reported to the NKVD as an *agent provocateur* and covert Trotskyite, and was liable to be arrested. This business of registering people in trade

---

[36]Halina Górska (1898–1942), Polish writer and communist activist. After Lwów was occupied by the Soviets, she collaborated actively with them, naively believing their propaganda. Fled with the Soviets from the advancing German army but returned to Lwów soon after. Arrested in June 1942 by the Nazis and executed.

[37]Political Commissar. (An abbreviation of the Russian for political leader.)

unions began in the university in late 1939 and continued to the end, that is, till mid-1941.

* * *

There were ever new bombshells: the Winter War between Finland and the USSR[38] and the absorption of the Baltic countries into the Soviet Union. The Finnish war lasted a little over a hundred days, but it was a revelation to us; even according to Soloviev, a member of the Supreme Soviet of the Russian Federative Socialist Republic, the Soviets suffered over 250,000 casualties. The Soviet Union was considered to have been humiliated, and Finland was essentially forced into the German camp. Prices in Moscow rose 30%. Among Soviet soldiers rumors spread of the frightful ferocity with which the Finns had defended their country: "You don't return from there," they said, and there were stories circulating of the Finns' sophisticated techniques for laying mines in the forest, even of mines camouflaged as cradles containing new-borns—though such tales were doubtless apocryphal. Soviet propaganda became almost hysterical in its denunciation of the Finns and their temerity in revoking the privileges and freedoms long bestowed on them by the tsars.[39] That the real object of the Soviet leadership had nothing to do with improving the lot of the Finns through the imposition of communism and everything to do with a desire to subjugate them and acquire their territory, is so obvious it hardly needs to be argued. There was then nowhere in existence a more genuine democracy than Finland, whose citizens enjoyed the fruits of civilization and a high standard of living despite limitations of territory and the lack of natural resources—the most conclusive proof of the absence of corruption or unreasonable exploitation. The Soviets, for whom Lwów was like a microcosm of the American paradise, were emphatically the least qualified to improve the condition of the Finns. The length of the Finns' heroic defense showed their strength and integrity and the Soviets' relative lack thereof. Nevertheless, there were people in Lwów who were heard to say: "The skin of the Scandinavian bourgeois is creeping in fear," thus giving vent to a sort of envious exultation at the anticipated annexation of the Scandinavian countries.

Meanwhile most—but by no means all—of us were more and more coming to grips with the ins and outs of the Soviet system. Many of the refugees from Warsaw, Łódź, and Kraków seemed to regard their stay as only of relatively short duration, and spent a great deal of time promenading or sitting in cafés. Jews could be seen on Academicka Street and in Fredro Square trading their gold, diamonds, and other possessions. Every now and then the authorities would swoop down and

---

[38] The Finno-Soviet War began on November 30, 1939, with a Soviet offensive, and ended March 13, 1940. Although their army was smaller and they had much less *matériel* than the Soviets, the Finns were able to resist the invasion far longer than expected. In the end Finland ceded only 11% of its prewar territory to the USSR. On the other hand, Estonia, Latvia, and Lithuania were incorporated into the USSR in June 1940.

[39] Finland had been part of the Russian empire from 1809 to 1917.

arrest scores of people, but this had no discernible effect. These same authorities kept promising that *vsyo budet*,[40] but the shortages persisted: among much else, soap, sugar, kerosene, and fuel for heating were in extremely short supply—and it was a harsh winter. Professor Fuliński[41] said to me when we met one day: "Next winter will be worse, and after that a yet harsher one."

* * *

Soon the Soviets added the collection of taxes to their bureaucratic endeavors. They considered that everyone was in arrears with their Polish taxes and pressed for payment. Some among us soon realized that if you changed your place of residence, you could avoid paying these. Thus I was actually quite pleased when a stout Soviet official, accompanied by a former court official from Jarosław who used to deal with housing issues, announced that I was henceforth relieved of the ownership of my house. However, they allowed us to retain ownership of the apartment we were occupying. I discovered that, as a professor, I could actually have retained ownership of the house, as, for instance, Professor Cieszyński[42] did, but he lost rather than gained as a result as he had to pay much higher taxes than those who were just apartment owners.

Not many of us professors were arrested; Leon Kozłowski, Roman Rencki, and Stanisław Grabski were exceptions. Rencki was charged with having hidden substantial sums of money representing revenue from a resort in Morszyn that he had managed on behalf of the Lwów Medical Society. Of course, the Soviets wanted to get their grubby hands on this money. When Grabski was arrested a group of us went as a delegation, headed by Kulczyński, to the Political Commissar at the university, one Captain Lefchenko. Lefchenko duly phoned the local NKVD bureau, and was informed that the matter was *plokho*, that is, bad. We began to argue that Grabski was an old man who would not be able to endure prison conditions, but Lefchenko responded with "He is old, but he did young things." What he was referring to so obliquely was Grabski's periodical *Grody Czerwieńskie*[43] presumably dealing with matters affecting the Rusyns in a Polish nationalistic manner.

I had occasion to visit Lefchenko again, this time on business of my own. The hospital on Łyczakowska Street had been taken over by the Soviet military, and I wanted to find out what had happened to my introvisor. Lefchenko said that they

---

[40]"Everything will [soon] come."

[41]Benedykt Fuliński (1881–1942), Polish biologist. Professor at the Lwów Polytechnic 1919–1941. Died in 1942 during a severe depression caused by the conditions of Soviet and then German occupation.

[42]Antoni Cieszyński (1882–1941), Polish physician, world-renowned dentist, and surgeon. Head of the Institute of Stomatology at Lwów University. Murdered by the Germans during the "Massacre of Lwów professors".

[43]Perhaps Lefchenko was referring to the essay *Ziemia Czerwieńska odwieczna, nierozerwalna część Polski* published by Grabski in Lwów in 1939. *Grody Czerwieńskie* ("Red Cities") was the ancient name for the borderlands between Rus' and Poland, probably taking its name from the fortress of Czerwień.

would admit me into the hospital without a pass, but then he winked at a man in uniform sitting in the corner, who seemed to be completely unoccupied. This man then turned to me and began asking questions—just to get acquainted, he said. He took down my first and last names and queried me about my Father. When I told him he had been a businessman who ran a store, he insisted on knowing how many people he had had working for him. I answered "One or two." However, it soon became clear that a great deal depended on whether it was one or two, since if he had just one salesman working for him then he was a storekeeper, but if he had two he was a capitalist exploiter, and I, as the son of a bloodsucker, would have had the wrong social background. Thus all my years of work at the university, and all my publications and inventions were of secondary importance compared with the question whether my Father employed one or two salesmen in his store. At that moment I was seized by a powerful, implacable, well-nigh physical aversion for all Soviet officialdom, politicians, commissars, and whatnot. For me they were tedious, dull-witted, lying barbarians who had gotten hold of us like the gigantic monkey that seized Gulliver in Brobdingnag and carried him up onto a rooftop. One was almost completely dependent on their idiotic criteria for this, that, and the other, which would have made one laugh if it weren't for the fact that the whole insidious weight of the NKVD apparatus was behind them.

I envied those who had escaped abroad in time. I knew, for example, that Ulam had managed to get to New York through Gdynia in August 1939, and Dr. Knaster told me that when in mid-1940 Soviet troops invaded Lithuania, Professor Zygmund[44] had managed to obtain a Swedish visa and escaped, eventually to the US, where he obtained a position at one of the smaller universities.[45] Zaremba Jr.[46] went in the opposite direction, obtaining a position in Stalinabad in Tajikistan. In a letter he wrote that he went to the tops of mountains so high that he could see the English positions[47].

Lwów was undergoing a radical transformation. More and more officers—in particular NKVD personnel—arrived, with their wives and children, and, in parallel as it were, more and more people were arrested and sent off to who knows where. Sometimes whole families disappeared. Jaworski Jr., a poet of the same age as my daughter, was arrested merely because he had a friend who was a Trotskyite, and his father and mother were also arrested and sent away because before the war the father had worked in the Ministry of the Treasury and the Soviets had confused his function there with that of public prosecutor. His grandmother died of grief as a result of this catastrophic bereavement, and for some time their house on Janowski

---

[44]Antoni Zygmund (1900–1992), Polish-American mathematician. A major twentieth century mathematician. Professor at the University of Chicago from 1947. (Dedicated his book *Trigonometric Series* to A. Rajchman and J. Marcinkiewicz, respectively his teacher and pupil.)

[45]Mount Holyoke College in South Hadley, Massachusetts.

[46]Stanisław Krystyn Zaremba (1903–1990), Polish mountaineer and mathematician. Son of Stanisław Zaremba Sr.

[47]In India?

Street remained empty. Such trivial mistakes with shattering consequences occurred frequently. Another example concerns the wife of Kaczmarz. Kaczmarz himself went missing during the war,[48] and when his wife was questioned subsequently, she said that her husband had been an adjunct at the Polytechnic.[49] The Soviets misheard the word as "adjutant", and she was sent away with her children. However, supporters managed eventually to explain the mistake, and she was brought back. Then there was my young acquaintance Dr. Wachlowski, who had been a lecturer in international law; he was accused of hatching an anti-Soviet plot and arrested, when all he had done was to get together with a few colleagues to discuss the political situation.

So now our town was packed with Soviet soldiers, guests of the regime, and refugees from all over Poland. There were Jews from Zdołbunów, Winnica, and Żytomierz,[50] and guests of the Soviets from Korea and Ussuriland,[51] come to gaze at this most exotic of the world's wonders, a capitalist European city. It happened that a Soviet soldier visiting the home of one of our working men, say a bricklayer or the driver of a streetcar, on seeing the good quality furniture, the chests of drawers stuffed with items of clothing and fine bed and table linen, pictures on the walls, and attractive curtains in the windows, was brought to say admiringly: *U vas rabotniki tozhe kapitalisty!*[52].

But then suddenly we had to revise our opinion of the Soviets a little, for one day, out of the blue, food appeared in abundance! In the many state food stores that by now had replaced the private ones, there appeared flour, bread, sugar, dried and frozen fish, canned fruit, "butter-honey", and even caviar. There was a flood of food, and the line-ups of late 1939 and early 1940 vanished.

Another surprise: on one occasion a certain theater director invited various Polish literary figures for dinner at a restaurant on Sykstuska Street. Many came, including Broniewski, Peiper, Stern, and Wat,[53] to name but a few. When they were all seated at small tables, the actress Oranowska entered on the arm of a Soviet officer, sat down with him at Broniewski's table and began to carry on an obvious flirtation with him, Broniewski. After a few moments of this the Soviet officer gave her a good slap, whereupon Broniewski pounced on him. As if on signal, a gang of NKVD agents irrupted into the room and began beating the literary guests with fists and rubber truncheons. They were all arrested and taken away in the black cars that

---

[48] According to one version of his fate, he was executed by the Soviets at Katyń in 1940, although his name was never found on any list of those executed there.

[49] One source has him instead a docent at Lwów University.

[50] Formerly towns in eastern Poland, now in western Ukraine.

[51] Far south-eastern Siberia.

[52] "Your workers are also capitalists!"

[53] Tadeusz Peiper (1891–1969), Polish poet, writer, and literary critic. Anatol Stern (1899–1968), Polish poet, writer, and art critic. Arrested in 1940 by the NKVD and sent to the Gulag. Aleksander Wat (1900–1967), Polish poet, writer, co-founder of Polish futurism. Sympathized with communism from the late 1920s. Arrested and deported by the NKVD in 1940.

had been parked for an hour down the street from the restaurant. Next day an article appeared in the Soviet paper *Czerwony Sztandar* (The Red Banner) demanding that the heads be lopped off the "hydra of social fascism", and that the "lackeys of the Polish bourgeoisie", who, while pretending to be friends of the proletariat, were secretly preparing a counterrevolution financed by Britain, be unmasked. The article could have served as a lexicon of the standard Soviet invective. Although it named the arrested *littérateurs*, it contained no word of their arrest, presenting itself as a spontaneous outburst of the tender proletarian consciousness, shocked at the perfidy of the bourgeoisie. At a meeting of the Writers' Union convened soon afterwards, one such proletarian made a proposal that the poets and writers mentioned by name in the article be excluded from the Union, but the chairperson, Wasilewska,[54] showed courage in not submitting the proposal to a vote. All of the arrested were sent off to imprisonment in the USSR, as were many literary personages who had not gone to the restaurant, among them Mrs. Naglerowa and Parnicki.[55] (It seems that Stern was later released.) It thus became clear that the episode in the restaurant was the obscene work of *agents provocateurs*.

To thinking people such arrests were revealing of Soviet psychology. There were many prominent writers and poets—Wasylewski[56] and Boy-Żeleński are just two who come to mind—who were not arrested although they were most certainly not communist sympathizers, and sometimes even distinguished themselves by "reactionary" behavior. This was not the result of pure chance but rather of a fundamental attitude among the Stalinists, for whom nationalists or conservatives or even those without any political affiliation at all were not considered dangerous *per se*: the dangerous ones were those who could not be turned into zombies, or at least into those prepared to keep their heads down and grin and bear it. Thus Broniewski was considered dangerous not because he had been a Polish officer awarded the Order of Virtuti Militari for valor in World War I, or because he had fought with the Polish army against the USSR in 1919–1920, or again because he had written an anti-German poem *Warszawa, która się nie mści* ("A Warsaw that exacts no vengeance") in 1939, when Moscow wanted to avoid irritating Berlin. No, he was considered dangerous because he would simply not be reined in, and the fact that between the wars he had been imprisoned for his pro-communist sympathies was absolutely immaterial.

What the Soviet ruling bureaucracy detested most were those who acted on the naive belief that now, finally, the promised paradise was come, they might say and write whatever they wished. These represented just the active few of the

---

[54] Wanda Wasilewska (1905–1964), Polish-Soviet novelist and communist political activist. Fled from Warsaw to Lwów in 1939, and thence to the Soviet Union. Played a significant role in the formation of a Polish division of the Red Army during World War II. Wrote novels in Polish in the style of "Soviet realism".

[55] Herminia Naglerowa (pseudonym Jan Stycz) (1890–1957), Polish writer and publicist. Teodor Parnicki (1908–1988), Polish writer, notably of historical novels.

[56] Stanisław Wasylewski (1885–1953), Polish writer, journalist, literary critic, and translator.

bulk of ridiculously simple-minded types, many from the Polish intelligentsia, who actually believed what they were told: that there was total freedom of speech and no censorship in the Soviet Union, that policemen were no longer needed, and that the democratic Stalinist constitution was binding on Stalin and his henchmen.

When an election was held, placards proclaimed that "Never before were there democratic elections such as those held in the USSR." And this was true vacuously, since their elections were totally undemocratic. One was presented with a card with a list of candidates' names printed on it, and one was free, indeed, to cross out any or all names—it didn't matter since all candidates were then reported as being chosen by 99% of the electorate. Yet there was pre-election campaigning of a sort, consisting of the posting up of thousands of placards, and the publishing of hundreds of hysterical articles in the Soviet papers, caricaturing or simply lambasting the Polish Socialist Party,[57] the Zionist party, and others, as if any active member of any of these parties could even live in Lwów under his own name. It was as if the Soviet rulers wanted to put on a show for people to see what a true democratic election might look like, much as the NKVD played at the game of landlord and lodger with me.

As for censorship, well, absolutely everything submitted for publication was subject to censorship—even mathematical papers. Those responsible for anything that was printed that disagreed even mildly with the Stalinist catechism included the author, the editor (in the case of journals and newspapers), the type-setter, and the copy editor. Even the layout was censored: for instance Stalin's name should never be split across two lines. People working in the press strove to ingratiate themselves with the great leader; thus certain Jewish zealots publishing papers with tiny circulations in Białystok or Brody[58] decided at some point that the Polish dative form *Stalinowi* was bourgeois, and replaced it by *Stalinu*.[59] And other editors decided that *Żyd* was anti-Semitic and recommended using *Jewrej*[60] instead, not knowing there is also a Polish word *Hebrajczyk* with the same meaning.

Such facts, petty in themselves, in the mass show how very few people are capable of thinking for themselves. Education doesn't seem to have much to do with it: a great many highly educated people—such as university professors—showed a complete lack of civil courage, combined with a slavish readiness to toe the line. Some, even former nationalists, recalled that they had Ruthenian mothers, and began lecturing in Russian or Ukrainian from October 1939, and others began signing their names on official documents using the Russian alphabet. When Saks, Schauder, and I indicated our nationality as Polish on the forms that everyone had to fill in, some of our colleagues said—*sotto voce* and behind closed doors—that we did so out of fear of reprisals by the Polish resistance.

---

[57] *Polska Partia Socjalistyczna* (PPS), active from 1892 to 1948.

[58] A town not far from Lwów.

[59] *Stalinu* is the dative form of *Stalin* in Russian.

[60] *Jewrej* is the Polish transliteration of the Russian word *Yevreĭ*, meaning Jew—*Żyd* in Polish. *Hebrajczyk*, meaning a Hebrew, is rarely used in modern Polish.

8 The First Occupation

Naturally, the university had its own communist party "cell", the ultimate authority in university matters, made up of certain lecturers, assistants, and professors. But don't for a moment think that surely our assistants Marceli Stark[61] and Jan Herzberg, who between the wars had spent time behind bars for their pro-communist leanings, were chosen to belong to this cell! Of course not! The views purveyed by such publications as *Ilustrowany Kurier Codzienny* (The Illustrated Daily Courier) before the war were also wide of the mark. People who have genuine convictions, communist or otherwise, are *bêtes noires* for the Soviet leadership. As I mentioned earlier, they need people who believe only what the party tells them to believe—what it tells them to believe *today* since it might be expedient for them to forget what they were told yesterday. Here is an example: in 1938, in the *History of the Communist Party of the Soviet Union. A Short Course*, Stalin demonstrated uncharacteristic perspicacity in predicting a world war initiated by Germany, Italy, and Japan. This prediction was eliminated from the edition of 1939. Although the prediction had been fulfilled, the Party now ordered that it be forgotten—since it was allegedly Poland, egged on by Britain, that had started the war. I don't know if there was a further edition in 1941 with the prediction restored. If someone writing an exposition of stereoscopy had wanted to mention that Mach invented the X-ray stereoscope, his work wouldn't have gotten past the Soviet censors because Mach's philosophy was forbidden, *ergo* he couldn't possibly have invented anything.

In the Soviet Union there were hordes of people busily twisting facts and falsifying history—down to what one might have thought were banal details. The more intelligent—and therefore possibly more intransigent—members of trade unions were continually being urged to attend re-education lectures put on by their trade union for their re-enlightenment as to the new truth. One of the library's officials, Franciszek Smolka,[62] was asked to give such a lecture, and chose as topic the history of Lenin's sojourn in Poland. Now of course, before such a lecture could be given, the trade union's *politruk* had to vet it. He queried the sentence "Lenin lived [for a certain time] in the village Poronin near Zakopane," since in the *Short Course* it was stated that Poronin is a small town near Kraków. The librarian's objection that he had been to Poronin, and knew for certain that it was a village near Zakopane, was of no avail since the *Short Course* was gospel.

* * *

The four evangelists of the Soviet religion were Marx, Engels, Lenin, and Stalin. Gigantic portraits of these four were hung in various formats everywhere in Lwów: Marx with his big white beard, Engels with the little liberal beard favored in the mid-nineteenth century, Lenin with his goatee, and Stalin, smiling his crocodile smile,

---

[61] Marceli Stark (1908–1974), Polish mathematician and editor. During World War II survived the Warsaw Ghetto and various concentration camps. Connected with Wrocław University postwar, then, in 1951, moved to Warsaw, where he held several responsible positions at the Mathematics Institute of the Polish Academy of Sciences. Published several textbooks which were very popular in Poland and translated into other languages.

[62] Polish papyrologist, librarian, and historian of philosophy. Lived from 1883 to 1947.

and pink, as befitted "the friend of nations and the sun," with just a moustache. Some people made effigies of Lenin and stood them on their balconies for all to see and pay homage to. Most of these were extremely badly made, and sometimes resembled small corpses, doubtless evoking in passers-by a pitying smirk.

Soviet art—social realism, so-called—was more draftsmanship than true art. It was carried out by tessellating a photograph by small squares and the canvas by larger ones, copying the contours of what was in each small square of the photograph onto the corresponding square of the canvas, and then coloring in. This was the technique they used to produce the vast cartoon-like propaganda posters covering the walls of public buildings lit up with spotlights at night.

The Soviets evidently liked electric light. Whenever I passed the NKVD headquarters on Pełczyńska Street in the evening, the lights were turned on in practically every room, and one could glimpse through the windows massive portraits of Lenin and Stalin, and sometimes of Voroshilov, Molotov, or Timoshenko. If you had a portrait of Lenin hanging in your home, but none of Stalin, you ran the risk of being labelled a Trotskyite—like a Christian believing in God the Father but not God the Son. In the vestibule of the main entrance to the university they placed a huge plaster sculpture of Lenin, seated and wearing a smile of fathomless benignity, in the act of initiating Stalin, standing next to him, into the darkest secrets of Marxian wisdom. Next to this was another statue, showing the two of them standing engrossed in conversation. This arrangement of a quartet of statues of two subjects was somehow characteristic of Soviet aesthetics, which tended to invite ridicule from Lvovians. The saying "The difference between us and them is that we know the difference whereas they don't," was known to all Lvovians. However, it was not quite accurate. One of the Soviets who early on in the occupation came to the chemical plant *Laokoon* to take it over, said to its manager Izydor Blumenfeld "Now you'll be coming down in the world," and another said "Don't complain; at least you have seen something of the world and have been able to experience a variety of things. We have been enduring this monotone existence all our lives."

It seems that the year 1937[63] marked a tremendous change in the conditions of Soviet life. From that time on Russian poetry and fiction ceased, and only certain circumscribed forms of music continued to be composed, although the exact sciences largely continued to flourish.[64] Certain Soviet scientists paid visits to Lwów. Of the mathematicians among these, there were Aleksandrov, Lavrentiev,[65] Soloviev, Bermant, and a few younger mathematicians from Kiev and Odessa. All of the lectures they gave were first-class; perhaps I may conjecture that the innate acting ability of the Russians contributes to their high standards when it comes to delivering lectures.

---

[63] Probably closer to 1930, when the first of Stalin's purges began.

[64] Except for the biological sciences, which came under the leadership of the incompetent agronomist Lysenko.

[65] Pavel Sergeevich Aleksandrov (1896–1982), eminent Soviet topologist. Mikhail Alekseevich Lavrentiev (1900–1980), outstanding Soviet applied mathematician.

8 The First Occupation

When it came to education, the Soviets were unstinting with funds and much more liberal than in other areas. They set up Jewish, Polish, and Ukrainian schools. They also added a department of education, devoted to teacher training, to each of Lwów's three tertiary institutions. Problems did sometimes arise, probably as a result of the somewhat capricious way in which orders as to the management of the schools descended from above. For instance, their locations seemed to be forever being changed, and this relocation usually involved moving their libraries and other equipment. They reduced the school week from the former seven days to six, but then changed it back to seven days. The constant attempts at ukrainianization of the schools and tertiary educational institutions were a nagging source of irritation to Lvovians, in their hearts opposed to the general imposition of Ukrainian as the chief language of instruction and everyday dealings, then still ongoing. Courses in Ukrainian were organized to facilitate the transition, but I did not attend any since I detest all coercion. I did peruse an elementary grammar on my own, and later even read two works in Ukrainian: Gogol's[66] *Taras Bulba* and Irina Wilde's *Povnolytni diti* (Children who have come of age).[67] I also read a few novels of Ivan Franko,[68] an indubitably gifted author. Our university was renamed after him in an effort to win over the Ukrainian intelligentsia to the Bolshevik cause. However grateful the Ukrainian intellectuals may have been for such concessions, many of them continued to look in anticipation to Germany to liberate them from the Bolshevik yoke.

A language is not just a code for communicating between people, but provides a well rounded picture of the very soul of the people whose native tongue it is. The Ukrainian language is a peasant language; whenever anybody speaks Ukrainian he feels in his heart that he is playing at being a peasant. My reluctance to deliver my lectures in Ukrainian did not at all derive from a dislike of Ukrainians; had I been told to lecture in the Masurian[69] dialect or that of Polish uplanders, I would still have refused. Many assert that mathematics is independent of the language in which it happens to be written, but for me this is not so; this is the view of the narrow specialist, and I am not one of those. There were no Ukrainian mathematicians—or if there were, they were doing mathematics in Russian or German. Many mathematical concepts invented during the interwar period were initially given Polish names coined by Puzyna, Dickstein, and others of their generation,[70] and

---

[66] Nikolaĭ Vasilievich Gogol' (1809–1852), great Ukrainian-born Russian satirical novelist and dramatist.

[67] Irina Wilde (1907–1982), Ukrainian writer. The novel of hers read by Steinhaus was published in Ukrainian in 1938.

[68] Ivan Yakovych Franko (1856–1916), Ukrainian scholar from Galicia, novelist, poet, journalist, and political activist. He was the first to write novels and modern verse in Ukrainian. On January 8, 1940, the Jan Kazimierz University of Lwów was renamed "Ivan Franko L'viv State University".

[69] The Mazurs are a Slavic ethnic group with historical origins in Mazovia, a region of cultural and historical importance of east-central Poland.

[70] Indeed, Polish mathematical terminology is very specific to Polish. For example, the word for "integral" in most western languages, including Russian, is similar to the English variant, but in

these terms contained within them a suggestibility lacking in their foreign variants; they spoke to us much as the terms invented by the great German scientists spoke to them. Thus I do not regard mathematics as merely a collection of theorems to be communicated in any language one wishes. Some of those among my colleagues who, without objecting overmuch and sometimes even enthusiastically, went over to giving their lectures in Ukrainian, were later to acknowledge that I was right in the following: The language of the people born east of the San river in a certain period is for us Poles not just "yet another language" but an essential form of communication. If such people decide to be Poles—ultimately the decision is up to them—then they must abjure the easy speech, the lazy manner of speaking of the vast billowing grassy open steppe and wheat fields, and the primitive Ukrainian *selo*[71], and shape their speech to the versatile, precise, and stately architecture of a literary language. To speak Ukrainian in Lwów was to give up a certain cultural attitude. The faculty members who decided to deliver their lectures in Ukrainian—perhaps to avoid an imputation of Polish nationalism—made spectacles of themselves. Every sentence they uttered sounded false since constructed in a pedantic professorial manner rendering it foreign even—or especially—to Ukrainian ears. A Polish speaker who doesn't habitually speak in peasant argot will never learn to speak Ukrainian. But please do not think that I am here giving voice to a mere nationalistic bias. In many ways the relation of Ukrainian to Polish is analogous to that of *Wasserpolnisch*[72] to German in Silesia, or Slovak to Czech, or Breton to French, or Yiddish to Hebrew.[73]

* * *

As was to be expected, the Soviets gave positions in the university to a fair number of Ukrainians, most of whom were not of very high academic caliber. The Ukrainian Kiryło Studyński, who had been a professor before, and had always jibbed at swearing loyalty to the Polish state, was now in favor with the Soviets. However, contrary to expectations, they did not appoint him Rector. He was an elderly gentleman of the sort who likes to attend dinners given by bishops. He lacked the feeling for what was good form in a given situation, so was liable to tactlessness. After his audience with Stalin, he told everyone who would listen that Stalin was a more impressive ruler than any he had seen, including Kaiser Wilhelm II and Emperor Franz Joseph. This did not go down well with the Soviet apparatchiks because there was a rigidly prescribed set of titles that could be used of Stalin, and

---

Polish it is the Polish word *całka* for something like "wholeness". Steinhaus's opinion of Polish mathematical terminology was that "It is not our strong point". (Compare S. Hartman, "Uwagi o *Słowniku Racjonalnym* Hugona Steinhausa", *Wiadomości Matematyczne* 26 (1985), p. 229.)

[71] "settlement, rural town"

[72] "Water Polish", a mixture of German, Polish, and Czech used by the Poles of regions of Upper and Lower Silesia from the seventeenth century, now extinct.

[73] In the modern world certain former dialects, so-called, have recently become national languages, such as Slovak (since 1993) and Ukrainian (since 1991), and now need to be developed so as to serve as media for scientific communication—just like the creole English in days gone by.

"ruler" was not among them. So a certain Marczenko, a cowardly boor, was made Rector instead.

Although our Soviet "guests" needed a lot of coaching before they acquired some manners—such as removing their caps at appropriate times—it must be admitted that they got the university machine up and running. They made Banach Dean of the Division of Mathematics and Natural Sciences, and thanks to his belief that cooperation was optimal in our situation, he managed to get Saks and Knaster appointed professors—not an easy thing to do since the Soviets seemed to have an innate dislike for Varsovians. The heads of departments in the Division were Banach, Schauder, Żyliński, Zarzycki,[74] Mazur, and I, while the extraordinary professors—that is, without departments under them—were Saks, Knaster, Chwistek, Jacob,[75] Auerbach, and Orlicz. Our docents were then Eidelheit, Szpilrajn,[76] and Wojdysławski. This represented a powerful collection of mathematicians, and in normal times such a team would have achieved a great deal. Of course, the Polish Mathematical Society was no more, but the mathematicians of the division continued the scientific meetings informally. We also managed to bring out Volume IX of *Studia Mathematica*, with the slight modification that each paper now had to include a Ukrainian summary. Following Lavrentiev's visit, a number of us—Banach, Zarzycki, Schauder, Mazur, Saks, and I—became corresponding members of the Kiev Academy of Sciences, a smart move, in particular since it entitled each of us to a payment of a few hundred rubles a month, which, supplemented with our university salaries, enabled us to live tolerably well.

From Warsaw and Kraków news of what had been happening to the professors there filtered through to us. We heard that some professors formerly at the Jagiellonian University in Kraków had been sent to a concentration camp in Sachsenhausen,[77] and also of arrests in Warsaw. We heard that Aleksander Rajchman had perished, Koźniewski[78] had committed suicide, and, after escaping from German custody, Hetper had been arrested—as an officer—by the Soviets

---

[74] Miron Zarzycki (1889–1961), Ukrainian mathematician.

[75] Marian Mojżesz Jacob (1900–1944), Warsaw actuary. When the Germans entered Lwów, he began working in their *Organisation Todt*. In 1944 he returned to Warsaw, where he died.

[76] Meier Eidelheit (1911–1943), Polish mathematician. Obtained his doctorate in 1938. Murdered by the Nazis. Edward Szpilrajn-Marczewski (1907–1976), Polish mathematician originally from Warsaw. Worked in measure theory, set theory, general topology, probability theory, and universal algebra. When the Germans entered Lwów, he returned to Warsaw, changing his name to Marczewski, a name he subsequently retained. Played an important role in establishing a center for mathematics in Wrocław after the war.

[77] Now a suburb of Oranienburg, a town in Brandenburg, Germany.

[78] Andrzej Koźniewski (1907–1939), Warsaw specialist in actuarial science.

while attempting to cross the border into Soviet-occupied Poland at Przemyśl,[79] and sent to Starobielsk.[80]

Although we in the exact sciences appeared to be relatively better off than our Warsaw colleagues, there were many professors at Lwów University who had become redundant under the Soviets, such as the theologians and the philosophers, philosophy having been usurped by dialectical materialism. We came to an agreement amongst ourselves to donate ten percent of our salaries to our defrocked colleagues, but there were some who, despite in some cases having incomes from more than one source, failed to meet this obligation.

As I mentioned before, apparent inconsistencies were the order of the day. Certain of the professors known for their right-wing views before the invasion seemed to get on swimmingly with the Soviets, while those with democratic views—such as Ganszyniec[81]—were sometimes in opposition. Certain others whom one would have expected the Soviets to embrace were at best relegated to minor positions. Such a one was Stefan Rudniański,[82] a Marxist Jew from Warsaw, and possibly a member of the Comintern,[83] who wore a little pointed mephistophelean beard. He looked like a surety to become head of the department of economics, but the most they would do for him was to appoint him lecturer in some specialized subject of little significance. This was a prime illustration of the fact that communists who had not been trained in the Soviet school were not considered to have absorbed the Soviet orthodoxy, and were therefore of little use to the Soviets, while non-party—even rightist—professors who were accomplished in an appropriately useful field might be given special treatment. Thus the preferred instructors in Leninism and Marxism were those who knew the catechism off by heart and could be completely relied on not to deviate from it by as much as a hairsbreadth. For any other than these parroters of received wisdom to give instruction in Marxism and Leninism—to allow laymen to make critical assessments of the prevailing theology, as had been the case before the takeover—was out of the question.

I recall a debate between Professor Kleiner and an emissary from Moscow named Gagarin who had been sent to inform the professoriat of the details of a significant reorganization of the university that had been ordered from Moscow. The discussion took place at a professorial meeting with the mandate of carrying out the indicated restructuring. The delegate of the relevant ministry Gagarin sported a red star on his tunic and was very sure of himself. What happened was that Kleiner proposed

---

[79]Town in south-eastern Poland on the San river.

[80]Town in present-day western Ukraine. Site of a prison camp for Polish prisoners of war, especially officers. Some died in the camp, but the majority were murdered in the Katyń massacre.

[81]Ryszard Ganszyniec (Gansiniec) (1888–1958), Polish philologist, professor of cultural history.

[82]Stefan Rudniański (1887–1941), Polish philosopher and pedagogue. Had been active in the Polish Communist Party.

[83]"The Communist International", an organization founded in Moscow in 1919 dedicated to fighting for "the overthrow of international bourgeoisie and the creation of an international Soviet republic as a transition stage to the complete abolition of the state."

that in addition to the Department of Marxism and Leninism, we should institute a department of the history of philosophy so that the students could find out what the political and philosophical currents had been in Europe prior to the creation by Marx and Lenin of the final philosophical system. But Gagarin responded: "Our students now learn the one true philosophy, and what Comrade Kleiner is proposing is what we call fetid liberalism." Some time later, in some other connection, someone quoted Gagarin's words to the Rector, who responded with *"Gagarin durak!"*[84] and a few months later Gagarin's star was extinguished for reasons unknown. This was a common occurrence in the Soviet firmament: someone would proclaim some program of reform or other, the program would fall into disfavor for some—mostly unfathomable—reason, whereupon responsibility was dumped on the proponent and he was removed to some remote region of the universe.

However, in at least one respect the Soviet university was superior to ours as it had been run before they took over, namely in the matter of textbooks. Not only did Soviet scientists produce excellent textbooks, but they were also very inexpensive. I myself acquired several extremely good texts for practically nothing. But more than that, they instituted the system whereby for each of its courses, each department acquired a dozen or so copies of the prescribed text which were then loaned out free to the students. I should emphasize that here I am speaking of scientific books.

* * *

The Soviets also ended up managing the homes of us, the dispossessed, reasonably competently. Tenants paid their rent and charges for electricity and water at the bank. Although repairs carried out by Soviet workmen tended to be unprofessional, one could hope for improvement with time. I was pleasantly surprised when my request for the installation of a gas line to the house at 14 Kadecka Street was actually fulfilled. This had been an improvement I hadn't been able to afford prewar. The Bolshevik regime showed itself capable in a few other areas as well: they completed buildings that had been left half-finished in September 1939, laid a second gas line from Daszawa to Lwów, started on the construction of a highway from Lwów to Kiev, replaced our Polish train tracks by ones of wider gauge, and so on.

But I must be permitted to observe that they carried out all these projects in a haphazard way, and employed too many workers, whom they paid inadequately. In the USSR all work done by laborers was piece-work; every job performed by a laborer was paid for according to a predetermined scale, so that the calculation of the earnings of builder's laborers, who might have to do many different kinds of work in a single day, became a terribly complicated business. As a result of this complexity, and because in any case the piece-rates were continually changing, these calculations were carried out by specially trained persons called *tabelshchiki*[85]. However, in practice the *tabelshchik* would generally just take a stab, and write

---

[84] "Gagarin's a fool!"

[85] "tabulation experts"

down some fictitious figures in his notebook, ensuring only that the workers received enough for them not to complain too much. In fact, the lot of the Soviet worker was deplorable: they were paid the bare minimum for food, with very little extra for clothing, and allocated third-rate living quarters. Then there were the much-touted Stakhanovites[86] who exceeded the norms of output by 20 %, then 50 %, then 100 %, only to have the management of the factory, say, reclassify as a 100 % norm what a few bare months ago had been 150 % of a norm, so that they didn't get to enjoy their enhanced earnings for very long—nor their Stakhanovite status, since now they were back to being ordinary workers. At Soviet workers' meetings it was difficult to get worker demands discussed since the official line was that in the USSR they work for themselves—the fruit of their labor is all their own—so that exploitation of them is a logical impossibility. He who dared to give voice to complaints that the Soviet laborer was poorly recompensed for his labor, lived in a wretched room in a communal dwelling, and had no medical care, would soon find himself deported to Archangelsk or Kazakhstan. At workers' meetings they voted for resolutions prepared beforehand by *politruki*, which might recommend, for instance, increasing the number of hours in the workday, or the elimination of certain holidays, and so on. Or there might be a resolution that the workers express their indignation—or horror—at Britain's or Germany's or Finland's treatment of their workers, or at their political infamy, depending on the flavor of the day as determined by Comrade Molotov. And all resolutions always passed unanimously.

However, the working men of Lwów found themselves in an especially bad situation—not only materially, but also psychologically. In prewar times they could at least curse the capitalists, attributing their indigence to the fact that the lion's share of the profit generated by their labor went perhaps to their greedy capitalist bosses. They knew—and sometimes even saw with their own eyes—how those capitalists dined in fancy restaurants, travelled by car, and wintered on the French Riviera. But now this picture was no more, and although there was now no capitalist to blame, yet they lived in damp basements on cabbage and potatoes, and did not have enough money to buy shoes for their children. The explanation of this paradox provided by the Soviet newspapers and the local *politruk* was logically unassailable: there were peculating commissars exploiting the workers by means of corrupt practices such as bribery to secure well paid positions, which they then used for their own ends—such as the consumption of champagne in large quantities. However, our vigilant rulers were in the process of rooting out these rotten apples, on whom the righteous wrath of the people would then fall: they would either languish in prison—in the officially unacknowledged Gulag—or be executed.

But who or what exactly benefited from the Soviet worker's work if not he himself? What parasitical phantoms exploited the "excess value" created by his labor? I discerned three such phantoms: the phantom of industrialization, the

---

[86]Those workers who, following the example of the miner Aleksei̇̆ Grigorievich Stakhanov (1906–1977), worked extra hard in order to exceed production targets and demonstrate the superiority of the socialist economic system.

phantom of imperialism, and the phantom of Russian incompetence. The first of these commanded the Soviets to continue building power plants, railroads, iron and steel works, and factories producing industrial plant far beyond the current needs of the present—or even the next—generation. You might say that in this regard the next generation was exploiting the current one. The second phantom commanded them to produce field guns, rocket launchers, tanks, and airplanes in quantities that were simply astronomical; this was the phantom of Soviet imperialism or megalomania, dreaming of flying the Hammer and Sickle from a mountain top in the Pyrenees, or from the top of Mont Blanc.[87] The third phantom was the one insisting that the workers work without appropriate tools and proper organization, resulting in at least half of their effort being wasted.[88] Only a country as rich as the USSR could continue to function amid such wastage. For example, they built huge warehouses in Lwów for storing fruit and vegetables, most of which went rotten before it was brought to market: tons of tomatoes rotted away to slush in barrels, and potatoes froze in the railroad wagons delivering them. On the other hand, the produce that did somehow get through to peoples' tables was often of excellent quality: good tomatoes and cucumbers, from time to time various kinds of fruits and cheeses, potable Caucasian and Crimean wines, and even chocolate—although this was not pure, but admixed with milk and other ingredients, since the cocoa bean had to be imported to the USSR at that time. But in fact only the four cities Moscow, Leningrad, Kiev, and Lwów were so well supplied with food and other consumables; those living in the countryside had to somehow make do or go on shopping trips to the nearest of these cities. The Soviets did produce luxury wares: their own brand of "champagne", canned lobsters, caviar, some quite good perfumes, and so on. In theory these products were for the workers, but for most they were too expensive, the Soviets adhering in this to the capitalist principle of having large mark-ups on luxury goods.[89] One visible effect of all this was that our university janitors looked pale and undernourished; their hours of work had gone up significantly, and their monthly pay—now in rubles—was really enough for not much more than a week.

The inadequacy of the wages and salaries paid by the Soviets meant that a black market soon sprang up and thrived; almost everyone engaged in some sort of underground economic activity.[90] Following on ten hours of work, workers were

---

[87] This second phantom would seem to have come into its own only following the invasion of the USSR by Germany and during the Cold War.

[88] Many would say the waste was caused mainly by the lack of incentive and the inevitable failings of a centrally planned economy.

[89] They also established special shops full of luxury and otherwise unobtainable goods—such as the chain called *Beriozka*—for foreign visitors to spend their valuable foreign currency in, and for the ruling elite of the Party, from which the common folk were excluded.

[90] A significant part of the Soviet underground economy in the former USSR consisted of so-called *blat*, the doing of a favor—such as the gift of a joint of good quality meat stolen from the abattoir or a table in a restaurant purporting to be full—in return for payment or a favor in return, the whole system depending on extensive networking for its effectiveness.

often expected to spend another hour listening to improving lectures by *politruki* or simply by experts on some uplifting topic or other.

On one such occasion such an after-work talk was announced in the factory *Chervenyj Lampovshchyk* (The Red Lamp Manufactory), but to the factory workers' great relief the advertised speaker failed to show. Alas, the over-zealous local NKVD commissar stepped into the breach, and the exhausted workers had to stay. The subject of the lecture was the United States of America, and the speaker gave its area, its population, listed its main cities, its chief rivers, and so on—in other words, just related the basic information contained in every schoolbook of world geography. After some minutes of this tedious stuff the commissar interrupted himself to put in a request for volunteers to act as intelligence agents—actually spies, or, more accurately, snitches. Someone objected: "But we have no time. Much of our free time is already spent waiting in line-ups at the stores." Replied the commissar: "That's just the point. It's because of the speculators that much of what is produced doesn't reach the stores, as a result of which there are shortages and you have to stand in line. If you point them out to us, we will arrest them, and there will be no more line-ups." In fact the so-called "speculators" were for the most part suburban bandits in cahoots with the police who through bribery managed to acquire goods in substantial quantities—vodka, for example—by by-passing the line-ups or before they reached the stores, which they then sold on the street at 100% profit.

Another lecture I was told about actually dealt with a real issue, namely the deportation of people to far-flung regions of the USSR that was going on all the time. It was not only former members of the bourgeoisie who were liable to be deported, but also anyone who had escaped to Lwów from Silesia, or for that matter from anywhere else in German-occupied Poland. The *politruk* giving the lecture explained to his worker audience that the deportations represented an anti-espionage measure, and told them, by way of illustration, a long, tedious story about a Russian lieutenant who got married and then discovered his wife was a spy. Our working men plied him with questions such as: "Was she pretty?" "Was she young?" and so on. The commissar did not understand that they were pulling his leg and answered these frivolous queries as if they had been seriously put.

In fact, everywhere they encountered them—in the street, on the streetcars, or in the stores—the Lvovians poked fun at the "chubariks", as they called the Ukrainian and Russian invaders, as at unsophisticated bumpkins. But although the butts of their banter did indeed feel their inferiority—in terms of naivety and gaucheness—and were ashamed of it, they had the secret satisfaction of knowing what was happening to the arrested in the prisons or in banishment, so knew that thousands of those jaunty Lvovians who were now mocking them were destined to meet with the harshest of fates. They were sometimes heard to warn them openly: "You don't know what the NKVD is capable of! How can you dare to talk so loudly and carelessly about such sensitive things?"

There didn't seem to be any particular prohibition against people listening to short-wave radio broadcasts from the West, and I would sometimes go to my cousin Dyk's flat to listen to news bulletins from Paris and London. Soviet officers' wives who lived in the same flat would sometimes pass by our room and must have known that we were listening to foreign broadcasts, but never seemed to be concerned. In April 1940 we heard of the German occupation of Norway, and then in June the BBC announced the surrender of France to Germany. We both wept bitterly at the news, and did not sleep at all that night. For us the fall of France sounded the death knell to our hopes of getting out from under the Soviet behemoth.

From then on when I woke up in the morning, with the return to consciousness would come the dread of the prisoner who awakens in jail. I did indeed feel like a prisoner, and looked with hatred upon my jailers—as represented, for instance, by the Soviet troops marching each day to and from their exercises by way of our street. They seemed to be always marching to the fields not far from our house where they performed their military exercises both in winter cold as low as $-10°$ and in the noontime heat of summer. They seemed reasonably well accoutered and very well armed, at least judging by the number of pistols and machine guns they carried. Is there any army in the world where in a battalion of five or six companies in compact marching order each company sings a different marching song? The resulting cacophony didn't seem to trouble them at all, although since their outbound march sometimes went by our house at three in the morning, it most certainly disturbed the people living along their route. They sang *"Za natsiei, za partiei, za Stalinom poĭdyom!"*[91]

From month to month of the Soviet army's 22-month long occupation of Lwów, its behavior, and indeed the way in which the whole system functioned, kept changing. For example, greater formality was introduced into the army, so that one no longer saw officers carrying their children, and now they were always saluting one another. Terminological changes came thick and fast; thus the political commissars were now to be called "educational officers." In parallel with these changes there were similarly frequent shifts in political fashion—sometimes so rapid that the supporters of the new regime amongst us were sometimes caught off guard. For example, on the front page of an issue of *Czerwony Sztandar* (The Red Banner) I saw an article written by a certain Lwów Jewess expressing her great admiration of the Soviets for their institution of free education for all, and reproaching the former "Poland of the upper classes" for imposing fees making it impossible for the sons of workers and peasants to obtain a tertiary or even high school education. But then, on turning the page, I saw an official announcement of a new ruling according to which henceforth fees *would* be levied for secondary and tertiary study—fees so high, in fact, that a worker could not possibly afford them. The accompanying justification for the introduction of fees was typical: the welfare of the Soviet worker had by now improved to such an extent—so it was alleged— that he or she could now shoulder this relatively trifling burden.

* * *

The arrests continued with increasing zeal. Having mopped up all the former Polish judges and army officers, they now turned to former leaders of political

---

[91]"We go behind our nation, our party, [our leader] Stalin."

parties such as the Polish Socialist Party, the Zionists, and the Polish People's Party "Piast",[92] then the heads of social organizations such as the Red Cross, the Folk School Society (*Towarzystwo Szkoły Ludowej*), and so on. Without warning we were subject to a new ruling concerning passports: every refugee from German-occupied Poland would be issued not with a regular passport[93] but with one qualifying its possessor as subject to a certain legal clause called "Paragraph 11", which had formerly applied only to vagrants, the unemployed, members of the former bourgeoisie, and so on. In practice, this meant that the holder of such a passport reporting to the authorities of a parish to register his presence, might be sent away as an undesirable, and if he then continued to be excluded in this way might end his vagabondage in Siberia. The bearers of such passports were also forbidden to enter any town less than 300 km from the border.[94] In any case the bulk of the intelligentsia among the refugees from the west—Jewish and Polish alike—were feeling chary of the idea of acquiring a Soviet passport, fearing that it might cause them to be branded as pro-Soviet and close to them forever the option of returning home. Hence there arose the paradoxical situation whereby a clamoring thousands-strong crowd gathered in front of the German consular offices to apply to their Repatriation Commission to return to the German-occupied zone. Incidentally, this affords yet another counterexample to the continuing assertions by the yellow press that the terms "Bolshevik" and "Jew" are synonymous. Just as, after the conclusion of the Treaty of Riga[95] in 1921, several hundred thousand Jews—very few of whom knew any Polish—immigrated to Poland from Bolshevist Russia, taking advantage of Stanisław Grabski's[96] broad definition of the concept of a Polish citizen, so now once more thousands wanted to leave Soviet-occupied Poland for the West.

On the other hand, information continued to arrive of the restrictions placed on Jewish activity in Germany, and worse—of persecution, abasement, and murder of Jews, of their ghettoization, and the systematic escalation of the atrocities perpetrated on them. The Germans published photographs of Jews massed before the building of the Repatriation Commission in Lwów to counter what they called

---

[92]Name taken from the Piast dynasty, the first ruling dynasty of Poland, lasting from the tenth to fourteenth centuries, and as late as the seventeenth in Masovia and Silesia.

[93]In the former USSR everyone was issued with an (internal) passport as a means of identification.

[94]According to other sources 100 km from the border—see, e.g., B. Gleichgewicht, *Widziane z oddali*, Wydawnictwo Dolnośląskie, Wrocław 1993, Chapter IV.

[95]Treaty between Poland, Russia, and Ukraine ending the Soviet-Polish War of 1919–1920, and ceding to Poland parts of Belorussia and Ukraine. (The ceded regions had in fact belonged to Poland prior to the third partition, and in part prior to the second partition.)

[96]Władysław (1874–1938) and Stanisław Grabski (see Chapter 3) were brothers serving in the *Sejm* in the early 1920s, the former intermittently as prime minister. The Treaty of Riga entitled Jews from the Ukraine, where they had been subject to a wave of pogroms, to choose the country they preferred as their domicile, and several hundred thousand chose Poland. The Grabski brothers favored the assimilationist policy called "Polonization", whereas Piłsudski advocated a more tolerant approach.

the *Greuelpropaganda*[97] of the British about the cruelties of the Hitler regime. Among those trying to return to the German zone were former enthusiasts of Soviet socialism, former hard-core communists, leftist literati, and ordinary folk who might have been tailors from Zgierz or Pabianice just wanting to return home. When any of the former extreme leftists among these happened to run into a former judge who had perhaps sentenced them to a few years in prison for their radicalism, they now blamed him for not sending them to the Soviet Union to cure them once and for all of communism. One of these escapees through the "green border" was heard to say before he left: "There [under Hitler] I am threatened by death of the body and here by death of the soul."

The strange Soviet world of contradictions was taking over Lwów. Early on they had begun recruiting Polish literary figures, historians, and scientists to compose textbooks on literature, received history, mathematics, physics, and so on—work which in turn led to a demand for proof readers, copy editors, librarians, museum custodians, and archivists. The number of textbooks grew rapidly, and, at least as far as the mathematics and physics texts were concerned, they were of very high quality. Libraries and museums or depositories were continually being moved from one building to another, in order either to combine them or subdivide them into smaller units, and in the process many books were destroyed or lost. Each library had its associated archive, often with much long-accumulated valuable material, but here the Soviets made depredations, seemingly more interested in the unused sides than the printed sides of the sheets of paper making up much of the archival materials: records of government activity at many levels and over many years. Thus as a result of the shortage of writing paper, the Soviets erased from the record a great number of acts passed by our municipal and provincial authorities, as well as registrations of mortgages, marriages, births, deaths, and so on. One cannot really blame them for what were essentially unplanned actions based on expediency; for them it was so much scrap paper recording facts outside their ken and in any case of absolutely no interest since relating to a pre-Soviet time.

After the Ukrainian branch of the Soviet Writers' Union had been purged of its politically unwashed element, there remained: Wasilewska, Borejsza, Szemplińska, Kosko, Rudnicki, Jastrun, Szenwald, Halina Fryde, Górska, Kurek, Parecki, Pasternak, Boruchowicz, Winawer, Ważyk, Kott, Przyboś, Putrament, Nacht, Gil, Wasylewski, Boy, and Ginczanka,[98] as well as a few other Ukrainian and Jewish

---

[97]"atrocity propaganda"

[98]Jerzy Borejsza (born Benjamin Goldberg) (1905–1952), Polish-Jewish communist activist and writer. Chief of the communist press and the publishing house *Spółdzielnia Wydawnicza "Czytelnik"* during the Stalinist period of the People's Republic of Poland. Elżbieta Szemplińska (1910–1991), Polish poet and prose writer. Allan Kosko (1907–1986), Polish poet, writer, and translator. Adolf Rudnicki (1912–1990), Polish-Jewish author and essayist, best known for his works on the Holocaust and Jewish resistance in Poland during World War II. Mieczysław Jastrun (1903–1983), Polish-Jewish essayist and poet. Lucjan Szenwald (1909–1944), Polish poet and communist activist. An enthusiastic supporter of the Bolsheviks in Lwów 1939–1941, he joined the Red Army in 1941 and took part in the formation of the Soviet-affiliated Polish 1st Tadeusz

writers. The Writers' Union had its headquarters in the mansion on Kopernik Street that used to belong to Count Bielski. Of course, here as everywhere else, the rule that one had to formally belong to a "trade union" was solemnly observed. Writers were especially liable to censure and had to answer any charges of political incorrectness publicly. It was a matter of great relief to me that my son-in-law Jan Kott seemed to know instinctively what double-speak was required of him in this business. Asked if he was a communist, he answered that he had a catholic view of the world, so belonged to no particular political party. He was then asked what he thought of the arrests and deportations, to which he replied simply that he didn't understand them. In spite of what looked to me like prevarications, he was accepted into the union. Borejsza was installed in the National Ossoliński Institute,[99] now responsible for the production of school textbooks. The museum had been turned into an outpost of the Kiev Academy. Now photographs were exhibited there in glass cases, showing numbers of children hanged from boughs, with captions informing the viewer that the children had been Ukrainian and that the hangings had been carried out by the Polish rulers of Western Ukraine in the interwar years. Naturally, no precise dates or locations were specified. Another photograph depicted a company of Warsaw policemen charging a defenseless crowd of workers. But the Varsovians among us recognized the picture: the "defenseless workers" were actually youths from the extreme right-wing radical-national movement involved in an anti-Jewish demonstration, whom the Polish police were attempting to disperse—so that the police were actually carrying out an *anti-fascist* action rather than the fascist one the Soviets would have had us believe was the case. Thus here was the august Kiev Academy engaged in the noble pursuit of trying to perpetrate—essentially incompetently—transparent falsehoods.

Kościuszko Infantry Division. Halina Fryde was a Polish poet; her name appears in Polish literary journals, e.g. *Skawa. Czasopismo Literackie*, Nos. 6 and 7 (1939). Jalu Kurek (1904–1983), Polish poet and prose writer. Stanisław Franciszek Parecki (1913–1941), Polish poet, translator, and cartoonist. Drew satirical cartoons for *Czerwony Sztandar* (The Red Banner) in 1940. Leon Pasternak (1910–1969), Polish poet and satirist. Related to the Russian poet Boris Pasternak. Michał Maksymilian Boruchowicz (1911–1987), Polish-Jewish writer. Bruno Winawer (1883–1944), Polish writer of popular social comedies and science fiction, journalist, and physicist. Adam Ważyk (1905–1982), noted Polish writer and poet. From 1955, well known for his criticism of the postwar Polish leadership. Jan Kott (Steinhaus's son-in-law) (1914–2001), well-known Polish writer, literary critic and theoretician of the theater. Julian Przyboś (1901–1970), Polish poet, essayist, and translator. One of the most important poets of the Kraków avant-garde. Arrested by the Gestapo in 1941. Jerzy Putrament (1910–1986), Polish writer, poet, editor, and politician. Franciszek Gil (1917–1960), Polish writer and journalist. Zuzanna Ginczanka (1917–1944), Jewish-Polish poet.

[99] *Zakład Narodowy im. Ossolińskich* or the Ossolineum, one of the most important centers of Polish science and culture, until 1939 combining a library, publishing house and the Lubomirski Museum. Founded for the Polish Nation in 1817 by Józef Maksymilian Ossoliński, scion of a Polish noble dynasty dating from the fourteenth century. A branch was opened in Lwów in 1827. In 1947, the Ossolineum was partly resuscitated in Wrocław.

8 The First Occupation

Somewhat surprisingly, all of Lwów's monuments were left standing: the statue of Mickiewicz, and that of Kiliński[100] in Stryjski Park, and those of King Sobieski[101] and Hetman Jabłonowski.[102] The statue of Mary Mother of God on Mariacki Square was also left. Once when passing by I saw that seven white pigeons had perched on her halo, and some of the people gathered around assured me it was a good omen.

Soviet atheism is of a kind peculiar only to them. Illustrative of the Soviet style of scepticism was the Soviet person who stood before the Basilica of the Virgin Mary[103] asking passers-by if she, Mary, lived on the ground floor or the first floor. Then there was the Soviet pilot who attended a lesson at a school in the Notre Dame Monastery, and afterwards told the schoolgirls that he had flown higher than 5000 meters yet seen no sign of God—to which one of them responded with: "Had you crashed you would have seen him." Initially the Soviets had laid great stress, via propaganda, on converting the Lvovians to atheism—to such an extent that on more than one occasion Catholic school children organized protests, prompting the NKVD to send a couple of tanks to fire over their heads. However, with the passing of time their efforts in this direction seemed to flag.

Fear and loathing of the Soviets increased very significantly after the infamous St. Julius' night[104] of April 12/13, 1940, when an unprecedentedly large number of arrests were made.[105] Hundreds of cars and trucks arrived escorted by NKVD agents, and soldiers entered apartments and arrested those whose names were on lists they had with them. The arrestees were given an hour or so to gather together a few belongings and then taken to the railroad station, where they were packed into freight cars, several hundred of which had been especially brought up. Some of the freight cars were left standing in the rail yards for as long as three days before departing, and, while many of those crammed into them managed to escape, many others, mainly children, froze to death in the unheated wagons. During that ghastly night from April 12 to April 13 many of my acquaintances and friends were taken away—and not just from Lwów but throughout Soviet-occupied Poland. My good friend Stanisław Adamski of Jasło was taken, together with his wife, daughters, and granddaughters, the ostensible reason being that his daughters' husbands had been Polish officers. They were deported to Kazakhstan, where they settled on the steppe, but they managed to survive there and were allowed to write letters back home and receive letters and parcels from relatives and friends. They were just one

---

[100] Jan Kiliński (1760–1819), a commander in the Kościuszko uprising in the late eighteenth century against the Russian rulers of Poland.

[101] Jan III Sobieski (1629–1696), a notable monarch of the Polish-Lithuanian Commonwealth.

[102] Stanisław Jan Jabłonowski (1634–1702), Polish nobleman, magnate, and outstanding military commander. He halted the Tatars at Lwów in 1695.

[103] To give it its full name, The Archcathedral Basilica of the Assumption of the Blessed Virgin Mary.

[104] The feast day of St. Julius, pope from 337 to 352, is April 12.

[105] One estimate has the Soviets deporting around 1,200,000 Poles to various parts of the USSR, notably Kazakhstan, in 1940 and the first half of 1941.

of hundreds of families deported from Lwów, thousands from our province. Anyone with a connection with the former Polish military, no matter how remote, was liable to deportation. Many of the Polish peasants who had settled on large manorial holdings subdivided by the Soviets were now also arrested because, as an arresting officer was heard to say, "The Communist Party has information that they are spying for Poland!" One of Lidka's friends, Miss Smalewska, was arrested at that time together with her mother, the wife of a former Polish medical officer. The letters we received from them told of life among primitive tribespeople waiting for the death of the deportees unceremoniously dumped among them to acquire their few humble possessions. The settlers had to walk some tens of kilometers to the post office to collect the packages of food and clothing sent by their relatives, and sometimes on the way back they would be robbed and occasionally even killed. Hundreds of our peasants and workers were deported north to the shores of the White Sea, the vast majority of whom did not survive the extreme conditions. Przyboś read me the letters from some of these people, full of expressions of terror, of suffering from the extreme cold, and descriptions of almost complete destitution—but also of the wonder of the Northern Lights. The deportees ranged from Jewish merchants from Gródecka Street and Żółkiewska Street to prostitutes singing "Poland Has Not Yet Perished"[106] as they were herded into the freight cars.

Now we felt we had at last gotten to the bottom of the Soviets, but in fact we hadn't. Fully half of the Lwów intelligentsia were missing, among them Sierpiński Jr. and the lawyer N. Łomnicki, who had made a futile attempt to escape with his *fiancée* to Romania. People began to look for ways of escaping from this "frying pan": some tried Hungary or Romania but were picked up by the NKVD, which had these borders heavily guarded, while others headed for Lithuania, now part of the USSR, where they were for the most part caught and deported; such, for example, was the fate of my wife's young cousin Dr. J. Grünhut and his wife. There were some who had invitations from relatives living in the US, and, with the help of US consular officials, were able to obtain passage through Japan to America. Most of the Jewish contingent of the former Jasło intelligentsia was deported, including Dr. Karpf, Schochet, and H. Kramer and his wife, among many others. The myth that these were the sort of people favored by the Soviet regime persisted in the face of evidence that the conditions under which they were deported were particularly harsh. They were beset on three sides: by the Soviets, who accused them of spying for Germany, by the local Poles, who suspected them of being Bolshevik agents, and by the Germans, who would undoubtedly have killed them simply for being Jewish. A good number of former Lwów judges, lawyers, and government officials acquired new identities with accompanying documentation, moved to new quarters, and started earning a living by working in factories or construction, keeping mum, of course, about their former more exalted social standing. One of my neighbors, Mrs. Tomczycka, *née* Petold, was deported, along with her children, as the wife

---

[106]*Jeszcze Polska nie zginęła*, the first line of the Polish national anthem, a lively mazurka composed by Józef Wybicki in 1797, two years after the Third Partition of Poland.

of a Polish officer in the reserve (*sic!*), while said husband stayed undetected in the countryside under the assumed name of Jastrzębiec, and several people from Jasło who had formerly held relatively prominent positions in the town, such as Dyk Schoenborn, Ignacy Weiss, and Ludwik Oberländer, after coming to Lwów went to work in Soviet-run chemical and medical cooperatives.

If a refugee could prove that he had been in Lwów since before September 1, 1939, then he was considered a native. My wife's uncle, Mr. Zielony, decided to change his birth certificate for one certifying that he had been born in Lwów, and as a Roman Catholic rather than a Jew. He found a forger, who produced a beautiful document, and I found a rubber stamp of the type used perhaps at the time of his birth and saw to it that the Latin was flawless and the color of the ink was appropriate. In spite of all this, he was still afraid the NKVD would discover the forgery, but managed to get through the requisite succession of petty formalities. All that remained was to pick up the passport, which he did—only to notice that in it his nationality[107] was given as *Jewrej*[108]! Did they mistake his new Catholic birth certificate for a Jewish one? Who knows!?

The arrests and deportations continued inexorably: Irena Wachlowska, *née* Łomnicka, Mrs. Aleksandrowicz, Wertenstein Jr., and the Jaworski family, and so on, and so on. Every now and then determined young men from Warsaw, say, would sneak across the demarcation line to Lwów, resolved to get to Hungary—and a few actually made it. One such was Kazimierz Kott, a cousin of my son-in-law, whom I met for the first time only in this time of troubles. He was young—not yet 25—and enterprising. Before the war he had left home, joined the British merchant navy as an ordinary seaman, obtained a helmsman's diploma, and then completed a program at a business school, returning home to Warsaw with more money than he had when he left. After the German invasion he had joined the Polish resistance, but had been arrested. He asked to go to the toilet, and the soldier escorting him undid one of his handcuffs. Seeing that the toilet window was open, he had climbed through and fallen onto the heads of three German soldiers standing just below. By luck, a horse and cart laden with straw happened to be passing out through the prison gate, so Kott ran out in its lee, and, before the Germans had realized what had happened, he had leaped into a passing horse-drawn cab and was well away. Although a reward was posted for his capture, he managed to get through to the Soviet side of the demarcation line. So that's how he came to be in Lwów, where I met him. He was a handsome, self-possessed young man, of obviously strong nerves. Naturally, the Polish underground in Lwów supplied him with a Soviet passport, and when the Polish government-in-exile in Britain heard of his exploits, he was made a lieutenant in their army. In Warsaw, however, the Nazis exacted retribution by executing his family and *fiancée*. As for him, he became disillusioned with the Lwów branch of

---

[107] Internal Soviet passports always indicated the nationality of the holder: Russian, Jewish, Georgian, American,....

[108] From the Russian for Jew.

the underground, considering it not sufficiently serious about fighting the invaders, and decided to return to Warsaw. He was killed trying to cross the "green border".

* * *

It would have been very sad in our apartment if it were not for the visits of the Knasters, Oberländers, and various refugees and writers who dropped in to visit Lidka and my son-in-law Janek. Janek had kept on changing his profession: from copy editor to translator to accountant's assistant to archivist to librarian, and, on top of all that, to candidate for the Chair of Romanistics. Life in the USSR was easier if you had many irons in the fire. I observed strange—almost uncanny—changes in some of these acquaintances and friends of Janek and Lidka. For example, Lidka's friend Ginczanka broke off relations with Lidka after "conversion" to communism, and Przyboś suddenly stopped talking about certain sensitive topics, stammering and turning pale when they were broached. This was doubtless to be explained by the fact that he was an editor of the literary magazine *Nowe Widnokręgi* (New Horizons),[109] and therefore bound to be in close contact with the NKVD. Among the other editors were Boy-Żeleński, Wanda Wasilewska, and Usiejewiczowa in Moscow. Boy-Żeleński was one of those people who, without being a believer, had nonetheless decided to take on protective coloration and pretend to toe the communist line. Gil also compromised somewhat by writing anti-religious columns. However, there were others who managed to maintain their integrity, such as Nacht, who continued to produce his little cabaret poems, and Tadeusz Hollender,[110] who refused to join the Writers' Union because he did not want to play the role of Stalin's fool and write poems to order on the theme of, say, "Colorless red ideas sleep humbly arrogant."

Hollender was the only person I met who predicted that the war would last more than five years. It seemed to me then that the young people had a better grasp of the situation than us. For example, young Zbyszek Oberländer, who used to weave kilims[111] on the loom of Mrs. Maria Bujakowa, *née* Łomnicka, seemed to be able to read between the lines of the war news we heard over the radio. He told his father that he would definitely not be returning to Jasło while it remained in German hands. Others did go back—with varying results—to what was now German-occupied Poland. For instance, Janek Adamski returned to Jasło, and Ms. Węgrzyńska attempted to take advantage of the fact that she had been born in Vienna in 1915, so was now "legally" German. But she didn't get back to her home in Jasło because the Germans redirected her to Saxony to work. There she met a Pole, became his wife, and escaped with him to Switzerland. Her brother tried to cross the demarcation line near Sanok, but mistimed the crossing, was

---

[109] A socialist literary monthly, founded by Wanda Wasilewska in Lwów and published under the auspices of the Soviet Writers' Union.

[110] Polish poet, translator, and humorist. Born 1910. Wrote underground satirical articles and poems during World War II. Arrested by the Gestapo in 1943 and executed in the infamous Pawiak prison.

[111] Carpets of woven tapestry.

arrested by the Soviets, and deported. Professor Jan Hirschler[112] turned himself into a *Volksdeutscher*, and went "legally" to Posen in the German zone. One can only presume[113] that there the Germans confronted him with the patriotic speeches he had been in the habit of making, many of which had been printed in government reports of Polish national festivals. In Lwów he was viewed as a traitor, anxious to ingratiate himself with the German invader. I remember that, as one of the examiners for my *Habilitation*, he had queried the nationality of my as yet unborn children.

Thanks to Count Schulenburg,[114] the qualifications for entering the German zone were somewhat relaxed, and another segment of those in the Soviet zone could go west, including Professor Ehrlich, Mr. Emil Bader and his wife, Ms. Lanckorońska, a docent specializing in art history, and Mrs. Rzewuska, who used to teach my Lidka French, and who left in safe-keeping with me a collection of a few dozen of the most exquisitely beautiful porcelain figures.

Thus Lwów was being emptied of its people, and even more so with the second lot of massive deportations of the night of June 28/29, 1940. This time the majority of the deportees were refugees, most of whom were of course Jews, and I was in a position to observe the arrests from close quarters. Late at night a column of cars appeared on the street, and soon our doorbell rang. I opened the door to two NKVD officers asking for my son-in-law. Just the day before, after overcoming many obstacles, Janek had acquired a Soviet passport, which he had thought would allow him to stay in Lwów, so he protested vehemently against his arrest. But the soldier said simply: "So you have a passport, so get ready to leave." My wife, daughter and I tried to help Janek pack his things—not an easy task because a soldier's bayonet was always poking at us as we moved from room to room. It seemed the soldiers were afraid we might try something. They looked under Janek's pillow for a revolver, and then one of them went into my room and began a search, ignoring my protests. He turned everything upside down roughly, but took neither money nor my watch. We were especially tense since there were five people hiding in a sort of attic above the kitchen to avoid deportation: Uncle Zielony and his wife, our maid, and two others. Although the soldiers accompanying the NKVD officers went into the kitchen, they did not suspect there was a hiding place above their heads. They were free with advice for Janek and Lidka as to what he should pack; they even recommended taking a toothbrush and toothpaste in order to show off their familiarity with the finer points of western civilization. As they were leaving

---

[112] Jan Hirschler (1883–1951), Polish biologist, professor of zoology and comparative anatomy at Lwów University. Member of the Polish Academy of Sciences between the wars. In 1940 he went voluntarily to Germany, where he worked *inter alia* at the Kaiser Wilhelm Institute for Biology in Berlin, renouncing his Polish citizenship in order to do so. Returned to Poland in 1948, where he was imprisoned till 1949 for collaborating with the Nazis. From 1949 he resumed his research work in Gdańsk.

[113] An incorrect presumption, apparently. See the preceding footnote.

[114] Friedrich-Werner Graf von der Schulenburg (1875–1944), German diplomat. Last German ambassador to the USSR before Operation Barbarossa. After the failed plot of July 20, 1944 to assassinate Hitler, he was accused of being a co-conspirator and executed.

with Janek, I saw that my niece Muta, whose room was on a higher floor, had also been arrested, and was also descending the stairs with the soldiers, carrying a bundle of her possessions. For some reason the Soviets were arresting her but not her sister and brother-in-law. We quickly shoved a blanket into Muta's bundle, and then the whole group—Janek, Muta, and the soldiers—went off down Kadecka Street. They were taken to the school attached to the Mary Magdalene Church, and when Janek showed his passport he was freed immediately, and was back among us within the space of 15 minutes. We knew from him where Muta was being held, and we sent our housekeeper to see if she could get a few other necessaries to her, which she managed to do. We were unable to find out which railroad station she left from, or when she left, but after some weeks we received a letter from her: she had been transported to Sverdlovsk,[115] on the eastern side of the Ural Mountains. There she was put to work mowing hay, and each day had to walk eight kilometers each way to the fields where she worked. Our parcels did get to her, although sometimes they took over a month. She met friendly people who looked after her to a considerable extent. Then she broke a hand, and had to spend a few weeks in hospital in Sverdlovsk—a splendid hospital, she wrote, with facilities and equipment such as only luxurious western European sanitoria have. She was well cared for there, in particular well fed, and when her hand had healed sent somewhere else to work.

The dispatching of parcels of food and other items to exiles was hampered by the fact that the Lwów Post Office was forbidden to accept them. The reason for this seemed to be that Lwów had been designated a "model town" for showing off to foreigners sceptical of the Soviet "miracle", so they needed to prevent food from being shipped out. However, we got round this by sending our parcels to Zdołbunów via a freight handler, from where they could be sent on by train. It must be admitted that by and large the exiles did receive the parcels sent to them, although of course there must have been many cases where there was nobody left behind willing or able to send anything. Correspondence also usually got through, provided there was no specific prohibition, such as that applying to Dr. Józef Grünhut, who was in punitive exile,[116] and was forbidden to contact his wife—who had also been deported—or receive anything from her or anyone else. Although we received letters from him intended for his parents, the answering letters sent through us—his parents being afraid to write openly to him lest they be identified and also exiled—never got to him. His parents finally decided to move to Truskawiec,[117] where his father, old Dr. Bernard Grünhut, had managed to obtain a position as doctor at what was now a Soviet style *kurort*.[118]

Many people from Ukraine, Russia, and further afield came to the Truskawiec sanitorium for the rest-cure. I was told of one of them, a lecturer on Marxism and

---

[115]Formerly Ekaterinburg, and since the collapse of the USSR once again called by that name.

[116]Perhaps in a camp of the Gulag?

[117]Resort town some 90 km from Lwów, located at the foot of the Gorgany mountain range in the eastern Carpathians.

[118]Resort or sanitorium.

Leninism, that he took advantage of the rich diet to such an extent that he made himself sick, and then claimed that the Polish staff of the sanitorium were trying to poison him. Despite this, he stayed on and continued to give free rein to his gluttony. Spending several weeks each year at a sanitorium formed an integral part of the Soviet system, and was one of their chief pleasures. Another perquisite of the system was the *komandirovka* of a worker, that is, a few weeks' work assignment involving travel to somewhere more interesting than where he lived—to Tbilisi,[119] say, or Feodosiya[120]—ostensibly to do something related to his work, where he could spend four weeks free of committee meetings, the intrusions of the NKVD, and largely without fear of being exiled or sent to the Gulag for five years for coming late to work.

However, there is much that was negative in the worker's life to set against these "perks" and other socialist benefits. For example, it was not very difficult to get oneself exiled or imprisoned for shorter or longer terms. Sentences to deportation were determined by a people's court consisting partly of fellow workers. The accuser would propose a motion and ask for those opposed, and since no one dared oppose it, the sentence of five years' deportation for arriving ten minutes late to work would pass unanimously. As I mentioned earlier, one of the trials endured by Soviet workers—by no means the least—had to do with the inflicting of punishments on themselves out of a need to demonstrate their solidarity with the communist program—such as increasing the hours of work, or contributing from their already meager funds to some cause or other, or hamstringing themselves in some other way—albeit mostly only in theory. Some of them must surely have wondered at the magic trick whereby they had somehow acquiesced in being deprived of their individual freedom.

An acquaintance of my cousin who was a believer in communism had at some point gone to work as doctor somewhere in Russia. She returned in a state of nervous paralysis, terrified to the core by what she had encountered there, and almost afraid to speak. All we could get out of her was that she had seen huge factories there where the workers endured conditions so bad as to be fatal to many of them. She said, for instance, that they had no running water and were reduced to quenching their thirst from rain puddles, when it would have been a relatively simple matter to dig wells. When she had demanded certain improvements in hygiene, the NKVD had so struck terror into her that she had descended into her present state of nervous disorder. My assistant Mr. Stark was another former communist whose disillusionment involved some psychical derangement, since he had been forcing himself for too long not to see things that existed and *vice versa*.

For a certain time there was constant agitation in factories and offices for people to join the International Organization for Aid to Revolutionaries, whose aim was to help revolutionary organizations in other—principally European—countries. However, since Germany was officially a friendly nation, France had fallen to

---

[119]Capital of Georgia.

[120]Resort town on the Crimean shore of the Black Sea.

Germany, and Britain did not imprison its communists, the political commissars were in the rather embarrassing situation of not being able to say what the pressing concerns of the IOAR might be. But at last they discovered a worthy project, namely that of freeing Thälmann[121]! However, it began to seem incongruous, even comically tortuous, that an organization with a membership of millions should concentrate all of its resources on the freeing of just one man from imprisonment by a fascist state with which the USSR had friendly relations, and soon, following an order passed down discreetly from on high, the IOAR muted its agitation and ceased bothering us. Although I myself managed to avoid becoming a member (as well as avoiding having to lecture in Ukrainian), others among the professoriat who evinced a reluctance to join were forced to attend evening classes in Marxism and Leninism.

Another fad of this period of our first taste of Sovietism was that of "socialist struggle", whereby each department in the university was told to challenge other departments to socialist competition, to which end each department had to come up with a program of scientific and social projects to be completed during the coming year. Naturally, this was largely a rather hollow undertaking most of which remained at the paper stage: it's obvious that you cannot oblige yourself to solve scientific problems. After some time we learned how to cope with "socialist struggle" by padding out our work with things we had done in the past but had left unpublished. But the net effect of all this tomfoolery was to waste our time and distract us from real research work. Much the same situation obtained in factories. I was told of an engineer of no particular accomplishments whose supervisor told him he had three months in which to produce an invention or innovation. Then the supervisor, in an excess of zeal, made the three-month plan a six-week one, and was disappointed when the poor engineer was unable to come up with anything. But the secret so well learned by just about everyone was to cope with this sort of thing by putting on a show without substance—for example, by taking some detail from an obscure technical journal and presenting it as a new discovery. Since no one bothered checking for originality—one's immediate superiors themselves being concerned only with show—this would often do the trick, leading to the heaping of spurious praise on the "discoverer" and the publication of his picture in the newspapers.

Like magpies, the Soviets liked pretty objects: a colored pencil, a fountain pen, a watch—all such things fascinated them.[122] One couldn't leave even a pencil on an office table without it being stolen. This explains to some extent why toy stores were the second most numerous kind of retail establishment (after food outlets) in the Soviet Union. These stores, which seemed more like markets than sophisticated urban shops, sold fairly primitive wooden toys, toy drums, trumpets, fiddles, and

---

[121] Ernst Thälmann (1886–1944), leader of the Communist party of Germany during much of the era of the Weimar Republic. He was arrested by the Gestapo in 1933 and held in solitary confinement for 11 years before being shot on Hitler's orders in 1944.

[122] Perhaps because the production of consumer goods was sacrificed to large-scale industrial production for much of the first 40 years of Soviet socialism.

so on. Our market was now cluttered with stalls, above which there had been erected vast canvasses illustrated with idyllic pastoral scenes of the paradise of the Ukrainian Soviet Republic: buxom smiling wenches and plucky farm hands in traditional peasant costume against the background of a landscape of milk and honey. However, such propaganda failed to sway even the simplest of our Lvovians, who evaluated Soviet prosperity more directly on the basis of the poor quality of the underwear of the Red Army soldiers and their women.

A toy with which the Soviets were very taken was electric light. The aforementioned factory *Chervenyj Lampovshchyk* (The Red Lamp Manufactory) had been set up to supply our province with halogen light bulbs, but was unable to induce the new provincial administrators to pay the many thousands of rubles due them for the supply of the bulbs. Over and above its fundamental lack of incentive, the Soviet centralized economy suffered from all the woes of the capitalist system: if the director of a factory was not paid for what his factory had delivered, then he in turn could not meet payment for his raw materials. And added to these difficulties was the requirement that on no account should the prescribed budget be exceeded—in order to satisfy the overriding demand for zero inflation. Although I myself did not make a study of the basic causes of the inefficiencies of the Soviet economic system, I knew from experience that they were routine—and as often as not due to mere human caprice. At last, after having exhausted all indirect means of demanding payment—by phone and the post—the director of "The Red Lamp Manufactory" went in person to the main provincial administrative office to collect the debt, only to be put off by some dignitary with the usual "zavtra"[123], so that the thing seemed hopeless. But then a light went on in the dignitary's head, so to speak:

– Do you have colored bulbs?—he asked.
– Indeed we do.
– Can you supply me with a red one and a blue one?
– Most certainly.

Well, two bulbs were appropriately painted and sent, and the tens of thousands of rubles owed the factory were paid.

It may not be too harsh to say that there was something immature about Soviet man—or at least that his natural instincts had been largely suppressed. The average Soviet officer would pass by beautiful girls and elegant ladies without turning around to follow them with his eyes—as if castrated. Their laughter was never unguarded, never given full rein. Prudery prevailed, and the students who before the occupation had been demanding that the Senate remove so-called "indecent" advertisements posted up along the streets of Lwów, now got what they had wished for. It was not only such *risqué* billboards that were not contemplated in the USSR, but also any sculpture and painting to which the most censorious of censors might object, as well as operettas and comedies of any degree of frivolity. And neither was there any tragedy produced—just as there was none in Aldous Huxley's *Brave New*

---

[123]"tomorrow"

World. Irina Wilde, whom I mentioned earlier, wrote a novel about a Soviet scientist who is confronted with the choice of either marrying his beloved or pursuing, all alone, investigations which might lead to a discovery guaranteeing a decisive enhancement of Soviet production—I forget whether it was military production or some other kind. The young scientist sacrifices his personal happiness for the sake of the Motherland. This novel—which is, not to put too fine a point on it, dead uninteresting—came in for harsh criticism from the Soviet establishment, which charged the author with being fundamentally ignorant of the realities of Soviet life. It seems that in the USSR, as opposed to capitalist countries, no such conflict between personal happiness and the welfare of the state was possible: it was open to everybody to be a great inventor and marry whomever he wished into the bargain since the Soviet state provided both for his work and for his hearth and home, every sort of eventuality having been foreseen and ideally accommodated for.

The Soviets had hardly any social life. Not only did they have no leisure to pursue it, but they also had little inclination to mix socially with people who might at some later time find it expedient to inform on them for some incautious comment. Dr. Kontny, an acquaintance of mine, told me how when he was working in an archive, he applied successfully for a transfer to work in forestry. When he went to the archive to tell his co-workers he was leaving, two of them, who till then had behaved as collegial friends, began openly to discuss whether or not they should submit a report to the public prosecutor concerning his willful abandonment of his office employment—a crime carrying a penalty of five years' exile.

The Soviets kept us abreast of current political events—at least as they wanted us to see them. Special lectures were given by a certain Comrade Trofimov that were clearly aimed at the intelligentsia amongst us, being subtly worked out versions of official Soviet opinion—or rather what passed for official opinion since, although officially the USSR was promoting friendship with Germany, there was an undercurrent of opinion—especially among Soviet officers—predicting a "march on the Germans." Comrade Trofimov's lectures sometimes struck an intermediate note—for instance, when he claimed that France had been defeated because it had not admitted communists to the government, while England remained undefeated precisely because it had admitted them. He called de Gaulle a "great French patriot", and bestowed praise on Greece, as militarily no weaker than Italy. From these lectures it became clear to me that the Soviet leadership was biding its time, keeping its political options open, while it skulked and waited for the prophesied mutual destruction of the European capitalist powers. Despite the bluster of some of the Soviet officers, it was clear that that leadership did not yet want war with Germany. Wagonloads of wheat and gasoline continued to pass through Lwów from east to west, and when our railroad workers attempted sabotage, they were rounded up and deported *en masse*.

We settled down for a long occupation. Soviet bureaucratic practices had by now penetrated into all administrative areas. All officials were forced to waste some of their meager salaries on life insurance policies and government bonds. We became schooled in the art of getting by. Thanks to my colleague Knaster, who had translated my *Kalejdoskop matematyczny* (Mathematical Kaleidoscope) into Russian, in the process making numerous corrections to the drawings, I signed a contract with *Gosgortekhizdat*[124] for its publication in Russian. I had to change some of the examples, such as the one involving proportional representation of electorates, which I modified to proportional representation of sporting groups: the system of proportional representation being unknown in Russia, it was better not to mention it. I also added a theorem I had recently proved about dividing three arbitrary regions[125] of a plane in half by means of a single circle, and three little pictures to do with physics. There were also a couple of new figures drawn by Mrs. Lesia Sierpińska, namely of wine being poured into three vessels and a billiard ball on a rhomboidal billiard table. The manuscript was sent to Moscow at the beginning of March 1940, and I eventually received a payment of 60% of the revenue from sales. Thus I now had one item on my scientific work program.

We were forever filling out forms—if not in connection with the university then our Union, and if not the Union, then the Academy.[126] One of the most important questions on such forms asked what the form-filler had been doing at the time of the October Revolution. My son-in-law always answered that indeed he had had the wrong attitude *vis-à-vis* that great event back then, but that it shouldn't be held against him since he had been only three years old. However, one's age was of no relevance, since one's parents may have had grievously faulty attitudes. The second most significant question asked if one had relatives abroad. And there was also a question about travel abroad, to which I answered that in 1939 everyone had come from abroad to visit me so that I had had no occasion myself to go abroad. Unofficial departures somewhither and arrivals somewhence were equally anathema to the Soviet authorities, since a new arrival may well be a spy and a departee a malcontent or spy. When at that time I happened to read the memoirs of the Marquis de Custine[127] about his travels in Russia a hundred years earlier, I saw that many of the things that struck us as not just foreign but strange, were the same as those that had had the same effect on this French aristocrat back in the time of Tsar Nicholas I. My friend Ludwik Oberländer had read Mickiewicz's *Lectures on Slavic Literature*,[128] and he found in them another confirmation of the peculiarly Russian nature of the procedures and customs being forced on us. It was not so much "communism" but "russianism" that felt so alien to us.

The official line was that the Soviet system was not yet "communist" but still at a preliminary stage of "socialism." The difference between the two had to do

---

[124] A Soviet state publishing firm specializing in technical works.

[125] Presumably of finite two-dimensional measure.

[126] The Academy of Sciences of the Ukrainian SSR.

[127] Astolphe-Louis-Léonor, Marquis de Custine (1790–1857), French aristocrat and writer, best known for his travel writing, in particular his account of his visit to Russia in 1839 titled *Empire of the Tsar: A Journey through Eternal Russia*.

[128] Composed beginning in 1840, when Mickiewicz was appointed Professor of Slavonic Studies at the Collège de France. In these lectures he indicates, in particular, the importance of Byron and Napoleon for the formation of the poetic ideal in Poland and other Slavic countries.

with the impossibly rarefied ideal of the withering away of the state and with it all social injustice: "Now everybody is rewarded according to his work, but then he will receive according to his needs." Such is the official formulation of the economic difference between the socialism obtaining now and the true communism of the future.

The young people around me tended to assess the political situation differently from me. For example, my son-in-law was of the opinion that if not central Europe, then certainly eastern Europe was ultimately faced with the stark choice between rule by Germany or by Russia, since nations further afield had neither the resources to spare or strength, nor a genuine desire to get involved in the affairs of smaller nations such as Poland, far from their field of influence. According to Janek, although such nations might use their political weight to sway the balance a little one way or the other, none of them possessed sufficient means for imposing on the citizens of Poland, say, their own social and political system. So even if we found a way of rejecting Sovietism, we would then be stuck with Hitlerism.

Parnas was one of our professors who came in for much praise from the Soviets. He had a dozen or so people at the professorial level working in his Institute for Medical Chemistry, as well as several assistants, was generously funded, and was invited to Kiev, Kharkov, and Moscow to expound his institute's findings and attend scientific meetings. I too was invited a few times to attend meetings of the Academy in Kiev, but I always declined on the grounds of poor health. The journey lasted a dozen or more hours, which I would have found very wearing, and, in addition, it would have been hateful to me to have to look at the other passengers and listen to their speech. But most others took advantage of such opportunities. Sometimes, despite official invitations, no one travelled because the relevant official in the university's *Spetsotdel*[129] had failed to formulate and get approved in time the documents authorizing *komandirovki*[130], without which train tickets would not be issued. Professor Parnas told me how, when he was in Kharkov, he had visited an X-ray Institute and had started telling the professor showing him around about my introvisor: "We have a person in Lwów...," he began, and the other finished the sentence: "...Professor Steinhaus, who has made it possible to see objects hidden from direct view." However, this flattery had no effect on my aversion for travelling to meetings of the Academy.

My friend Dr. Ziemilski was at this time working in a tuberculosis clinic, and, thinking to do me a favor, gave my name as that of an expert in statistics to the docent from Kiev in charge of the clinic, a Dr. Plushch, who asked me if I might carry out statistical analyses of their data. But I failed to complete the—on the face of it, reasonable—tasks set me by the group of doctors working in the clinic since the data were inadequate and furthermore the terminology—relating to "pneumothorax"[131]

---

[129] Special Section (abbreviated).

[130] "trips on official business"

[131] A quantity of air or other gas in the pleural cavity between the lungs and the chest wall, associated with collapse of a lung.

and similar chest complaints—was foreign to me. Nevertheless, at some point Plushch sat me down in an office to act out a scene to be entitled "A Famous Researcher in Consultation with a Mathematical Expert." We were recorded, but thankfully there was no photographer present. Although all of his questions were very general, and he cannot have gleaned anything he didn't already know from our conversation, he insisted that I be paid a hundred rubles for the interview. I'm sure that in the clinic's record of scientific activities, a note would have been entered something like: "From 10:00 to 12:00 the director of the clinic was engaged in a scientific conference with Professor Steinhaus." The Soviets waste half their time on such comedies. For example, at the end of one of my lectures, a photographer from *Vil'na Ukraina* (Free Ukraine) asked if he might take pictures of a consultation with one of my female students. He was not discouraged by my telling him that there was nothing worth photographing about a mathematical consultation, and asked me to adopt a suitably scholarly pose. I told him to go ahead and snap what he wished, but that I would not pose for him. As a result, in *Vil'na Ukraina* there appeared a shot showing Miss B. at the blackboard, and some obscure unrecognizable figure that was me.

Naturally, that photographer had asked me for the spelling of my name, so I handed him my card. However, he needed to know how to write my name in the Cyrillic alphabet, since the newspaper was in Ukrainian. I then told him that I didn't know how to write my name in Cyrillic, and that, in any case, the transcription was part of his job. The poor man didn't seem to be able to grasp that the Roman alphabet suffices for us. Many of these ignorant people thought that we wrote with "Polish" letters, and held it against Saks and myself that we signed official documents in our usual way. I heard that once when a student who had come to us from Kiev saw an English book among the books of a fellow student, he had asked him how he could tell that it was written in English if the alphabet in which it was printed was Polish! In their simplicity and naivety they were fair game for us.

One could not really say that the newspapers were censored, since there were more censors than genuine journalists working on them, so that what was published hardly needed to go through an intermediate stage of censorship. Moreover, the articles were attributed, so their authors had to be very sure that they were toeing the official line: an incorrect ordering of the names Stalin, Lenin, and Molotov might incur exile to Siberia. A word would come from above, and every newspaper would publish articles on the harmfulness to crops of the mole cricket—effectively a reminder to the collective farms that it was time for spring plowing. This need to remind the workers to take appropriate actions was a typical feature of the Soviet economy: the lack of incentive meant that one had to prod the collective farmer into doing what one would expect a normal farmer to have been thinking about well beforehand. One searched the Soviet press in vain for articles of reasonably high quality on popular science or literature. On the other hand, international agreements made by the USSR were laid out in great detail in every newspaper. For example, an agreement with Japan about the lease of fishing rights in the Pacific Ocean was given much space, with details as to the size of catches permitted and the indemnity paid.

Many of our great national figures, literary and otherwise—such as Mickiewicz and Kościuszko[132]—were assimilated into the Soviet canon, but, conversely, Peter the Great and Suvorov[133] were now to be admitted as "ours." In the same breath as Mickiewicz one now had to mention Pushkin, as well as various minor poets of national significance, such as Lesya Ukrainka[134] or the Kazakh poet Jambyl Jabayev,[135] close to a hundred years old, who wrote rather naive little poems. All such representatives of local culture had to be lumped together and extravagantly praised, so as to make it clear that nationality doesn't matter, that poets of all stripes form a single cooperative, reciting their verses in the manner of a massed choir singing *The Internationale*[136] in Russian, to the deafening accompaniment of a gigantic orchestra. However, on one occasion an evening performance honoring Mickiewicz alone was permitted, and since it was organized and executed exclusively by our writers, stage-managers, producers, and actors, it was splendid. However, the Soviets don't understand the finer things, and what they don't understand they don't like.

I must relate the story of what happened to a certain Lwów doctor in the spring of 1941. The Soviets took great pride in their medical services. In Lwów they had organized an ambulance service which had at its disposal an airplane for delivering medical assistance in emergency cases at some distance from large urban centers. A call came in from near Jaworów[137] requesting gynecological aid for a peasant woman in childbirth, and the doctor on call, who had been playing tennis, was ordered to go immediately, in his tennis flannels, to the airport. The pilot lent him his leather jacket so he would not freeze, but then lost his way, and the plane blundered into German-occupied airspace before finally landing in a fallow field a mile from the village where the patient was waiting. The pilot took his jacket back and flew off, leaving the doctor to flounder in the mud. When, mud to the thighs, and almost frozen stiff, he finally reached his patient, she took one look at him and refused to let him near her. Somehow or other he managed to get back to Lwów towards evening. Some days later he was summoned by the NKVD to explain why he had flown over German-occupied territory, why he had dropped leaflets there, and what had been written on them. Happily, after a few hours of intense interrogation, what

---

[132] Andrzej Tadeusz Bonawentura Kościuszko (1746–1817), Polish-Lithuanian military leader during the uprising of 1794 (the "Kościuszko Uprising") against Imperial Russia and the Kingdom of Prussia. He had fought earlier in the American Revolutionary War as a colonel in the Continental Army.

[133] Aleksandr Vasilyevich Suvorov (1729–1800), last generalissimo of the Russian Empire. Fought against the Ottoman Empire, against the Poles in the Kościuszko Uprising, and against the French Revolutionary Army in Italy, never losing a battle.

[134] Originally Larysa Kvitka-Kosach (1871–1913), Ukrainian lyric poet and dramatist.

[135] Kazakh traditional folk singer. For some time deputy to the Supreme Soviet of the Kazakh SSR. Lived from 1846 to 1945.

[136] A famous socialist, communist, social-democratic, and anarchist anthem. Words written by Eugène Pottier (1816–1887) in 1871, and set to music by Pierre de Geyter (1848–1932) in 1888.

[137] A small town about 50 km west of Lwów, now in Ukraine.

had actually occurred got through to them and they let him go with just a reprimand. So much for Soviet emergency aid.

The fateful summer of 1941 was close when I heard that Dr. Fleck would be giving a lecture on leukocytes. From the very beginning of the Soviet occupation I had had little opportunity, and zero desire, for contact with biologists because I had become used to the idea that true scientific work in that field had been given up.[138] But in Gröer's Institute, where Fleck had his laboratory, this was certainly not the case, and I shared their enthusiasm for the good work they were doing. The fact that the five types of leukocytes occur in normal blood in approximately stable proportions suggested to me the idea that they might produce substances reacting differently with each type. At any rate, this was the source of our idea of manufacturing a chemical that acted differently on the different types of leukocyte. And in fact Fleck was able to synthesize such chemicals—for instance, one that destroyed lymphocytes but did not harm monocytes. I actually saw this demonstrated under a microscope in the Institute of Health. From this to the artificial synthesis of blood, or at least to the cure of diseases due to disturbances in the normal composition of leukocytes in the blood, would not seem to be such a big step.

I remember that it was Saturday, June 21, 1941, when I left the Institute of Health at midday, went to nearby Wólka, and lay down on a grassy slope to worry over Fleck's last words to me, namely that the war was like a wound in a state of delayed sepsis. Earlier, Professor Tarnawski had told me that he had heard that among the Jews it was widely believed there would be no war between Russia and Germany, to which I had retorted that this would seem to point to war, the Jews being such notoriously bad politicians. And my cousin Dyk had declared on a recent visit that he had heard that Hitler had delivered a 48-hour ultimatum to Russia.[139] There were also rumors about that anti-aircraft batteries were being set up outside Lwów with orders to fire on German planes.

At three in the morning on Sunday, I was awoken by the droning of a great fleet of planes. I looked out of the window and saw in the grey light of early dawn a pink sky over the whole of which white puffs of smoke were bursting intermittently. These were exploding shells from Soviet batteries. Down in the street I saw a squadron of soldiers running by in formation—and not singing! All this was more than enough to convince me that war had indeed broken out. I woke my wife to tell her the ghastly news. Of course, we didn't doubt for an instant that the Germans would take Lwów. Later in the morning we bought a local newspaper, but there was nothing whatsoever in it to indicate that anyone had anticipated war. What it did contain was a long article on the various Sunday distractions available to the populace outside the city, purportedly organized by the caring Soviet authorities for the Lvovian proletariat—

---

[138] Perhaps because the renegade peasant pseudo-scientist Trofim Lysenko was then director of Soviet biology. Lysenko's work was formally discredited in the USSR only in 1964.

[139] This must have been merely rumor, since Germany's attack on Russia came as a surprise—despite many indirect warnings—to Stalin.

but these were, of course, just the public baths, parks, and woods that we had always enjoyed. We now had leisure to sum up for ourselves the past 22 months of Soviet occupation—the enforced leisure of an air-raid shelter.

The war had begun some three hours before its declaration, which took the form of a note handed by the German ambassador to Molotov at six in the morning. It was interesting that the Germans did not submit their list of demands to the press nor did the USSR publicize them. Thus the two sides kept their citizens in the dark as to what was being hatched. Some of the last prewar issues of Soviet newspapers had poked fun at Stafford Cripps[140] for rumor-mongering about a German-Soviet conflict. They had also contained news items about the landing of German army units in Finland, but without comment. But now the tenuous front of friendship was riven: next day's papers were full of raging invective against Hitler. Suddenly it transpired that it was he who had masterminded the invasion of Poland. We Poles ceased to be suspect, and were subject to appeals to volunteer for the Red Army. There now occurred some sporadic exchange of fire between Ukrainian squads and Soviet units, and pedestrians who were found to have Ukrainian nationality according to their passports, were liable to be arrested. Soviet tanks rumbled ceaselessly through the streets and volleys of shots rang out continually from various parts of the city, added to which there was a German bombardment every few hours.

Now the insincerity of Soviet pronouncements to the effect that Western Ukraine—that is, the part of Poland they had occupied—was now an integral part of the Ukrainian Republic, and that we—that is, Poles, Jews, and whatnot—who had been living there before the Soviet occupation were full Soviet citizens with all the attendant rights, just like the Soviet occupiers, our "guests", became crystal clear. The NKVD immediately began to organize the removal from the city of all wives and children of Soviet officers, officials, professors, etc.—that is, of the occupiers—without apparently giving a thought to the welfare of the longtime citizens of Lwów. They did offer transport east to a few of our professors, but only Professor Parnas took up the offer. Professor Chwistek found a spot for himself on a Soviet lorry at the last minute. Many, preponderantly Jews, began leaving the city on foot. Joanna Guzówna,[141] a friend of my daughter, went with the departing Soviet soldiers in a light summer coat and without even a suitcase. Professor Gładyński drove off in his car, only to be waylaid and killed outside Lwów by persons unknown.

There were, in fact, many ways one could perish: from German bombs, artillery fire, from the bullets of nationalist Ukrainians taking it upon themselves to murder fleeing communists and Jews, and from the bullets of Soviet officers shooting at refugees whom they considered deserters shirking military duty. At a hastily convened meeting of the Writers' Union, Wanda Wasilewska delivered herself of a passionate appeal to remain, guaranteeing that Lwów would be defended, and

---

[140] Sir Richard Stafford Cripps (1889–1952), British labor politician. British ambassador to the Soviet Union 1940–1942. Served in a number of positions under the National Coalition led by Churchill during World War II. Served in the Attlee government postwar.

[141] Joanna Guze (1917–2009), Polish translator, critic, and art historian.

then departed for the east by car. Szemplińska rode away on a motorcycle, and I think Józef Nacht left with her. Soviet womenfolk stood wailing in the street with their children outside their dwellings, forced to wait hours for the promised lorries to transport them away from danger. But in fact the evacuation of their own folk was well organized, and all were eventually picked up. They rode in the backs of open trucks cradling the smaller children in their arms, with small bundles of possessions, and some provisions in buckets or other containers—much as when they had arrived in Lwów.

We, the glorious cultural *élite* of Lwów, had to stay in our houses, which were now officially declared "closed" by the Soviets, as if they would be coming back to declare them officially "open" again. Many among us felt some satisfaction at the removal of the removers. Their troops departed in an orderly fashion, without much noise or confusion, and without undue haste, it would seem—from which we concluded that they had in fact been prepared for such an eventuality. The German bombardments met with relatively little opposition. There was a great deal of anti-aircraft artillery in place, but its effectiveness seemed limited—although Mr. Jan Dobrzański said he saw three planes shot down by a single battery. Planes tumbled from the sky every day, but the batteries were slowly being withdrawn from the city, and day by day there were fewer of them. Following every air raid, Soviet planes could be seen circling over the city, presumably intending to pursue the German bombers, but giving the impression of having arrived too late. Soon German artillery found its range, and commenced firing on the Russian positions at Wólka and the old toll gate. Once a shell fell in our garden, and on another occasion on the street outside our house. On yet another occasion, I found myself on Łyczakowska Street when 24 bombers arrived overhead. I ran with a crowd of people to the nearest air raid shelter, the basement of an old house on that street. I had never known that these old dwellings had such large and deep basements; they must surely have served as warehouses at some distant time.

Interlude: Flashes of Memory

One day prior to the onset of this war, I was sitting on a bench in Stryjski Park, when I noticed a swarm of small flies flying about in disorderly fashion. As I continued to observe them, they suddenly moved *en masse* about ten meters to the right and began flying about in pairs. Then they returned to the place where I had first noticed them and resumed their apparently random flitting. This behavior repeated itself every minute or so. Looking more closely at the swarm, I saw in the passage from one locale to the other what at first I took to be strings of bright, minuscule beads, but then recognized as tiny bright flashes of light, occurring in sequences, with the flashes in each sequence separated by about a centimeter. I soon realized that each succession of flickers was made up of successive flashes of sunlight reflected by the wings of an individual fly. Since the speed with which they flew from the place where I first saw them to the other one could be estimated to be at least

60 km per hour, it followed that the frequency with which the wings of an individual were beating was approximately 1670 per second, since

$$60 \text{ km/h} \equiv 6{,}000{,}000 \text{ cm/h} \equiv 6{,}000{,}000/3600 \approx 1670 \text{ cm/s}.$$

Thus a very conservative estimate would put the frequency with which an individual fly's wings beat at more than 1000 beats per second—a conclusion arrived at without instrumentation of any kind.

* * *

A clerk by the name of Alter who had been working before the war in the Philips Company, once told me the following story about the ceremony marking the opening of a Philips plant in Warsaw. His boss had told him to go to the offices of the Polish President to request official participation in the ceremony, but then had himself immediately phoned the president's office to let them know the clerk was on his way and to insist, in the name of Philips, that the president's representative guarantee the participation in the opening ceremony of a presidential cavalry column as well as a mounted band of "buglers with silver bugles." Mr. Alter had then been greatly surprised at how his requests in the name of Philips were graciously acceded to and how accurately and smoothly the requested arrangements were carried out at the castle.[1] And it was much the same with the Warsaw episcopate, which had also been invited.

* * *

During the period when the Bolsheviks were in charge in Lwów, Mr. Olczak paid a visit to Kharkov, where he consulted with certain Soviet experts in geodesy. He was invited to the home of one of these gentlemen for dinner, consisting of an *entrée* of bread and green tomatoes, a main course of bread and ripe red tomatoes, ending up with tea. There was no butter.

* * *

Not long after the end of World War I—in 1920, I think—the case of Dietrich and Sołoneńko caught public attention. One of these two was a railroad machinist and the other a tradesman of some sort, and they had obtained a bomb from Russia with instructions to place it under a railroad track at a suitable spot. The one who had been assigned the task of actually placing the bomb under the track took it straight to the police. The police told him to place the bomb according to his instructions, and when this was done they went to the spot with a bomb specialist to defuse the bomb. They then arrested the two would-be terrorists. At the trial the one who placed the bomb under the track on instructions from the police was used as a witness against the

---

[1] The Royal Castle in Warsaw, traditional residence of Polish monarchs, and from 1926 to 1939 the seat of the Polish President, Ignacy Mościcki. On December 2, 1930, President Mościcki, together with Anton Philips, officially declared open a new Philips plant in Warsaw. In addition to lightbulbs and lamps, it produced the first radios.

other one, who was sentenced to death. The judge asked the policeman responsible why he had ordered the placing of the bomb as instructed, but received no answer. Clearly, this had been a planned provocation on the part of the police and the public prosecutor—a disgraceful attempt at subterfuge, preventing a proper evaluation of the guilt or otherwise of the parties concerned.

* * *

It might be said of Stalin that he was a "conventionalist." For him it was as if all of the bourgeois social traditions—their external manifestations, such as bourgeois manners (snobbishness, for instance), legal systems, literature, and so forth—were merely the results of certain basic conventions imposed by the wealthy so as to optimally promote their own interests, whereas he, Stalin, ruled according to a different set of conventions, ones leading to strict censorship, total control of the press and of what counted as literature and scholarship, isolation from foreign bourgeois nations to prevent contamination, and constant, inescapable propaganda. The implementation of the underlying conventions was by no means limited to the present: the historical record was modified to fit the system, and it might have been expected that any day geography or even mathematics would need to be appropriately refashioned. During the Soviet occupation of Lwów from September 1939 to June 1941, genetics, the general theory of relativity, and Schrödinger matrices[2] were just a few of the topics on which it was forbidden to lecture. These and other prohibited topics were considered "eccentric forces" acting to disturb the Soviet system of conventions, and therefore to be condemned. Even "materialism"—at least in any consistent variant—was on the index of prohibitions.

In this system scientists were respected only hypocritically, in the sense that they were accorded respect the way sophisticated, complex, and essential assemblages of machinery were appreciated in the Soviet Union—but handled roughly nonetheless. The Soviets, including some members of their intelligentsia, liked the prefix *spets*,[3] since in a system where lip service was paid to the principle that everyone counts the same, this prefix carries with it a cachet of exclusivity.

Formulating a new way of looking at the world—the proper task of the scholar—was of course prohibited in the USSR, so that genuine philosophers, writers, poets, and even scientists of a certain bent, found themselves severely constrained—to the point where many of them gave up the unequal struggle. It was given out that Mayakovskiĭ[4] committed suicide, but the fate of a great number of other novelists,

---

[2] Perhaps the author is referring here to Heisenberg's matrix mechanics, the highly successful quantum mechanical formalism for which Heisenberg was awarded the Nobel Prize in Physics in 1932.

[3] An abbreviation of the Russian word for "special" or "specialist", as in *spetssluzhba* meaning "special service".

[4] Vladimir Vladimirovich Mayakovskiĭ (1893–1930), Russian poet of the revolution, representative of early Russian futurism. He committed suicide in 1930 by shooting himself, whether out of unrequited love or disillusionment with the course the Soviet Union was taking under Stalin is not clear. Some conjecture that he was purged.

poets, scholars, and stage and film directors purged in the 1930s remains swathed in mist: were they liquidated, did they die from the rigors of life in the camps, or did they take their own lives? We do know that the mathematician Shnirelman[5] committed suicide. Sholokhov[6] lost the will to write, and Alekseĭ Tolstoĭ[7] was accompanied everywhere by a snoop with instructions to keep his person under surveillance at all times. There were no Soviet painters, for who would paint according to Soviet strictures? There was music, but it had mostly degenerated into mere virtuosity.

The Soviet intelligentsia was very different from our Polish variety. The doctors, engineers, and teachers who belonged to the party worked behind the scenes at night, helping transport people to the Gulag. Having had it drummed into them to follow their leader's commands blindly, the Soviets lost all sense of honor and responsibility to oneself, and were without individual concerns such as "fashion". Stealing someone's private property was considered a minor transgression, and respect for age had fallen by the wayside.

From time to time we were vouchsafed an exhibition of the kind of mind-training the Soviet rulers had a predilection for: suddenly there would come a pronouncement from on high directly contradicting the received word of the day before. For example, in early 1941 a speech by Malenkov[8] was published in a leading daily condemning the waste, incompetence, and mindlessness of the management of Soviet factories, including a great many concrete examples. He took aim in particular at those directors of plants who, so he averred, continued to take pride in their higher social origins and to keep capable workers of humbler backgrounds from promotion to higher positions. This speech—composed of course by Stalin—became the new dogma. It was the subject of lectures at meetings at every level, and "criticism and self-criticism" in the style of Malenkov became *de rigueur* everywhere. There would be nothing very surprising about all this if not for the fact that the hapless directors, so rudely removed from their positions, arrested, and transported to the Gulag, had, up to just a little while ago, long been enjoying apparently secure, comfortable jobs, while in the yards behind their factories dozens of tractors and other kinds of machinery lay rusting under mounds of snow, or in their warehouses lay tons of metal for the supply of which a nearby workshop had put in multiple requests over many months without response—to mention just two specific sorts of inefficiency and dereliction. Prior to Malenkov's speech optimism

---

[5]Lev Genrikhovich Shnirelman (1905–1938), Soviet mathematician. With L. Lyusternik invented the "Lyusternik–Shnirelman category", a global invariant of a topological space.

[6]Mikhail Aleksandrovich Sholokhov (1905–1984), Soviet novelist. Best known for his 1934 novel *Quiet Flows the Don* describing the lives of the Don Cossacks prior to and during World War I and the revolution. Nobel laureate for literature in 1965.

[7]Alekseĭ Nikolaevich Tolstoĭ (1883–1945), Russian and Soviet writer, especially of historical novels.

[8]Georgiĭ Maksimilianovich Malenkov (1902–1988), close collaborator of Stalin. Leader of the Communist Party. Overshadowed by Khrushchov in 1953.

was compulsory, since to publicize faults with the system was to deny the dogma of its perfection—and in fact those who dared point them out had been liable to arrest and imprisonment. Yet now the scandals were suddenly being trumpeted to the skies, and what had been forbidden just a little while ago now became the thing to do, with the result that thousands of little Malenkovs crawled out of their holes and began vociferously criticizing and tale-bearing.

The newspapers took up the refrain, publishing stories of unmasked miscreants sabotaging the Soviet enterprise. I remember one article reporting on the passing of severe sentences against members, corresponding members, secretaries, and so on, of the Kiev Academy of Sciences. It seemed that for a few years a group of a dozen or so of these people had been denouncing the directors of various institutes, and certain officials of the Academy—among them some of their own coworkers—accusing them of sabotage, counter-revolution, and Trotskyism. The newspaper listed the names of these poor wretches, who had then been sentenced either to death or exile, and pronounced them resuscitated, that is, officially whitewashed. Their former accusers, however, would now reap the whirlwind. Such were life's excitements for the Soviet intelligentsia.

* * *

In the last week of the ascendancy of the Soviets, that is, the week of June 22–29, 1941, they proclaimed the mobilization of all able-bodied men, but no Pole reported. Nevertheless every day hundreds of males from neighboring villages were impressed into service and marched to Lwów. When one group of such recruits dispersed to evade bombardment, three of them turned up at the gate of our house at 14 Kadecka Street seeking shelter. I let them in and struck up a conversation with them. They were Ukrainian peasants, one of whom claimed to be something of a biblical scholar. I advised them to wait for a while just inside the gate, and then, instead of proceeding to the Cadet School as ordered, to go in the opposite direction and, by way of the woods, get back home. They said they were afraid of meeting other peasants on the way, to which I expostulated "How's that? You're afraid of your own Ukrainian people?" The biblical scholar answered: "It is said that after this war every man will fear every other." At last they began to see the correctness of my advice, and, as they were leaving, the scholar turned and said: "A blessing on this house and its inhabitants." I was much moved by this innocent demonstration of faith.

There were other incidents at the gates of houses as a result of the insistence of those in charge of anti-aircraft defense that the inhabitants of each house arrange guard-duty at the gates. Unaware of this measure, other soldiers sometimes took those standing at the gates for enemy diversionaries who might open fire on them. Thus, when the young mathematician Skarżeński was standing before the gate to his residence, a Soviet soldier came up to him and demanded to see his passport. However, Skarżeński didn't happen to have his passport on him, and when he turned to go inside and fetch it, the soldier shot him in the back of the head. Thus did a young mathematical adept perish pointlessly—like a certain Ukrainian who died when a bomb scored a direct hit on him at the corner of Słowacki and Sykstuska Streets.

* * *

Sunday, June 29, 1941, was an interregnum, free of both Soviets and Germans, the last such for many years. Dr. Kontny and I walked to Stryjski Park through the Stryjski Cemetery. Kontny had his camera with him and found some interesting things to snap—such as the crowd gathered around a corner kiosk, breaking down the doors, and robbing it of candy and bottles of raspberry juice and lemonade. Looting was in fact going on throughout the town: while the fires set by the departing Soviets in some of the larger apartment houses on the Hetman Ramparts[9] and at the corner of Copernicus Street burned, crowds looted the stores. It was as if during the whole of the Soviet occupation the inhabitants of Lwów had had a single frustrated *idée fixe*, to which, now that the Soviet yoke was lifted, they unhesitatingly gave full rein. In our part of the city the main attraction was the barracks of the Cadet School, where the Soviet soldiers had abandoned supplies of flour, rice, coffee, and other wares. Although German patrols had doubtless already arrived at the other end of town, it was not till Monday that the roar of their motorcycles was heard in our district.

* * *

Once during the Soviet occupation, a young NKVD officer came to our house asking for me. As I happened to be out, he said he would be back at seven in the evening but absolutely refused to say what he wanted, so all were concerned that he would arrest me. Thus we made our preparations, giving some of our things to our friends. When he arrived as promised, he mentioned Mr. Łukawski, a former lieutenant-colonel in the Polish reserves who had lived above us until arrested by the Soviets in December 1939. Waving a sheaf of papers relevant to the case, the NKVD agent began interrogating me about the lieutenant-colonel's political convictions— in particular, his views on the October revolution. I said to him:

– Mr. Łukawski was first an Austrian officer, then became a Polish officer, and now, if you should so desire it, he would become a Soviet officer.
– I understand—said the young agent—He is a professional soldier.

And indeed that's what he had been. Łukawski had been Austrianized, then settled in postwar Poland—even to the extent of heading some sort of association of postwar Polish settlers—and when the Soviets took over he had taken down the portraits of Rydz-Śmigły and company in his apartment and hung up one of

---

[9] In Polish *Wały Hetmańskie*, one of Lwów's main promenades before World War II. Named after Hetman Stanisław Jabłonowski, defender of Lwów against the Tatars in 1695. A monument to him used to stand at this site, but was destroyed by the Soviets after the war.

Lenin instead. After his disappearance his wife had herself avoided the same fate by reporting to the German Commission set up in Lwów as being *Volksdeutsch*, and had gone to Vienna. I conjectured that she had been making representations there to get her husband returned to her, and that was why his case was being investigated in such detail by the NKVD.

Chapter 9
The Second Occupation

On June 30, 1941, the Germans occupied Lwów. They were preceded by a small advance party of Ukrainian volunteers—fewer than a battalion[1]—who marched through the city singing. A curfew was immediately imposed, and already on that first evening one could hear the heavy tread of two helmeted German soldiers patrolling a deserted Kadecka Street in the moonlight. The curfew was enforced to the utmost: persons found at large after 9 pm were immediately shot, no questions asked, and little calling cards attached to their corpses showing the time of their demise: 9:10, 9:25, and so on. The Ukrainians pasted up yellow posters announcing the liberation of Ukraine from the dominion of the USSR and forecasting its purification from "outsiders",[2] but without any details about whom they had in mind. Although the posters carried a facsimile of Bandera's[3] signature, it did not seem to have received the *imprimatur* of the Germans, who simply ignored it. At the university there were bands of "lesser heroes" than Bandera, busy smashing up the statues of Lenin and Stalin that the Soviets had erected. There wasn't much other university business to occupy them since the Germans appeared to be in no rush to reopen that institution. In fact, the Ukrainian professors, who had immediately assumed control of the university's institutes and departments, treating their Polish colleagues with a crude lack of ceremony, soon found out that the Germans had no intention of handing the university infrastructure over to them for their own use, since for them it was just so much war loot. I heard indirectly of the opinion voiced by one of the Germans assigned to take control of the chemical plant in

---

[1] So a few hundred at most.

[2] Now termed "ethnic cleansing".

[3] Stepan Andriyovich Bandera (1909–1959), Ukrainian politician and leader of the Organization of Ukrainian Nationalists in Western Ukraine. Proclaimed an Independent Ukrainian State in Lwów on June 30, 1941, allied with Nazi Germany. However, ultimately the Germans refused them recognition, and in September 1941 Bandera and his followers were arrested. A controversial figure in modern Ukraine because of his cooperation with Nazi Germany 1939–1941.

Lwów, namely that since Germany had for a considerable time officially recognized Soviet rule, and since, therefore, all buildings, factories, forests and lands that had been nationalized by the USSR were in German eyes the legal property of the USSR, it followed by an article of international law—unwritten but generally acknowledged—that that property could now be considered the German victors' legitimate spoils of war.

The first German troops I saw were part of a mountaineering division. They marched along followed by mules laden with machine guns, boxes of ammunition, and other equipment. They were quite young, and many of them wore sprigs of edelweiss in their caps, but it was clear that they were very weary. Passers-by gazed curiously at them, and some women even lifted their hands in greeting. I also saw uniformed motorcyclists with flowers adorning their handle-bars—and it was not only Ukrainian and German women who were welcoming the new set of conquerors with bouquets. These were followed by dozens upon dozens of military trucks and other vehicles, some as large as railroad wagons, endlessly rumbling through the streets of Lwów.

It was not long before the Germans began investigating the prisons and other places of Soviet incarceration, where they found grisly evidence of Soviet brutality. In the basement of the *Brygidki* building,[4] and in places of detainment such as those on Łącki Street, among others, they found hundreds of bodies of people murdered by the NKVD. Not all of these were political victims. It transpired that all those arrested in the last few days had been summarily executed before the Soviets' departure. There were deeper layers of corpses in a much more advanced state of decay, but it seems that the scale of the murders had risen precipitously from the time the Ukrainians had begun their sporadic attacks on Soviet troops. Evidently these murders were not intended to deter the Ukrainian *francs-tireurs* directly, since they had been carried out in an Asiatic manner, that is, mysteriously, in strict secret. It was as if the NKVD did not wish to publicize their heinous deeds, perhaps because they were cognizant of the fact that at some future time they might have to answer for them. Everyone was surprised when the Germans admitted journalists, photographers, and anyone else who so desired to view the grisly spectacle in the basements. There were many bodies there of people one had thought deported, for example, Roman Aleksandrowicz and Mietek Bader, to name but two. The Germans immediately issued a press release stating that there were no Jews among the victims, and the murderers had been Jews serving in the NKVD. It hardly needs saying that both these statements were utter falsehoods. In any case the percentage of Jews in the NKVD would more than likely not have exceeded ten percent, and even these would have been recruited from beyond the

---

[4] *Brygidki* is the building of the former Bridgettine order of nuns founded in Lwów in 1614. After the convent was secularized in 1784, it was used as a prison. It was one of several sites of mass murder of political prisoners by the NKVD in Lwów in June 1941, when approximately 7000 prisoners—mainly Poles and Ukrainians—were executed.

Zbrucz,[5] since locals had not been hired except as minor informers. Since the people could not take revenge on the guilty, who had withdrawn with the Soviet army after destroying all documents, they turned to the Jews as objects of vengeance—which was precisely what the Germans intended. Several thousand Ukrainian peasants armed with clubs entered the city and set to beating the Jews they encountered in the street, dragging others out of their houses, frog-marching them to Łącki Street and commanding them to exhume the corpses in the basements. With my own eyes I saw how such thugs treated a family consisting of a grandmother, her daughter, and grandchildren, holding hands in an attempt to slow the pace. Within the space of a few days some 3000 Jews had perished in and around Lwów. It seems that Dr. Eidelheit[6] and Mosler[7] were killed in this uprising. The Germans took no part in the massacre, considering that the spontaneity of the treatment meted out to the Jews by ordinary people served to confirm before the world the nationalist-socialist thesis of Jewish guilt for unspeakable crimes. Not a word was uttered concerning the fact that the murders committed by the NKVD had been in progress even while Molotov was breakfasting with Hitler and his entourage in Berlin.[8] Once I saw a Ukrainian chasing a Jew along Kadecka Street. He caught him and ordered him to march ahead with his hands over his head, kicking him constantly. A German officer who happened to be passing by in a car, stopped, got out, and began asking the Ukrainian what was going on. His answer must have been unsatisfactory to the officer, since I could see how he ordered the Ukrainian to go one way and the Jew to go in the opposite direction, thus putting an end to the latter's torment. German officers were frequently known to act in such humane fashion—but, unbeknownst to us at the time, their highest authorities were otherwise inclined.

Within three or four days of their arrival, the Germans arrested the following professors[9]: Antoni Łomnicki, Włodzimierz Stożek with his two sons Eustachy and

---

[5] In Ukrainian, "Zbruch", a tributary of the Dniester, considerably to the east of Lwów. From 1921 to 1939 it marked the border between the Second Polish Republic and the USSR.

[6] Meier (Max) Eidelheit, Lwów mathematician. Another source gives the year of his death as 1943. See W. Żelazko, *A Short History of Polish Mathematics*. Warszawa 2007.

[7] S. Mosler was a mathematics docent at Lwów University 1940–1941.

[8] Since this may have suggested that the Germans connived at those murders?

[9] This was the beginning of the tragic event known as the "Massacre of the Lwów Professors" of July 1941. The author mentions by name about 22 prominent Lvovians, mostly professors, arrested together with some of their family members or guests, and then murdered by members of the German Nazi occupation force. According to contemporary sources the number of Polish academics from Lwów killed in July 1941 was 25, with altogether 45 identified victims. From Steinhaus's list above, the following persons killed on the Wóleckie Hills (*Wzgórza Wóleckie*) were omitted: Dr. Jerzy Grzędzielski (Head of the Institute of Ophthalmology at the University of Lwów), Professor Edward Hamerski (Chief of Internal Medicine at the Academy of Veterinary Sciences in Lwów), Henryk Hilarowicz (surgeon, professor at Lwów University), Father Dr. Władysław Komornicki (theologian, a relative of the Ostrowski family), Professor Witold Nowicki (Dean of the Faculty of Anatomy and Pathology at the University of Lwów) with his son Jerzy, Professor Włodzimierz Sieradzki (Dean of the Faculty of Legal Medicine at the University of Lwów), Professor Adam Sołowij (former Head of the Department of Gynaecology and Obstetrics of the

Fig. 9.1 Monument erected in 2011 on Wóleckie Hills outside Lwów in memory of the victims of the "Massacre of Lwów professors" by the Nazis in July 1941 (Courtesy of Stanisław Kosiedowski, public domain.)

Emanuel, Stanisław Progulski[10] with his son Andrzej, Roman Longchamps with his three sons Bronisław, Zygmunt, and Kazimierz, Stanisław Ruziewicz,[11] Stanisław Pilat,[12] Antoni Cieszyński, Władysław Dobrzaniecki,[13] Tadeusz Ostrowski,[14] Tadeusz Boy-Żeleński, Roman Rencki, Dr. Stanisław Ruff[15] with his wife Anna and son Adam, Kasper Weigel,[16] Jan Grek,[17] Roman Witkiewicz,[18] Kazimierz Bartel,[19] Włodzimierz Krukowski,[20] Franciszek Gröer, and Henryk Korowicz.[21] Altogether 22 professors were taken, mainly from the Medical Faculty and the Polytechnic. Among them was Professor Rencki who by a sort of miracle had evaded death in a Soviet prison, only to be re-arrested and murdered by the Germans a few days after being freed. When the Gestapo went to the home of Jan Grek to arrest him, they also arrested Tadeusz Boy-Żeleński who had been staying with Grek as his guest, being in fact his brother-in-law. This was extremely bad luck, since if he had not happened to be there they would perhaps not have thought of arresting him. When they went to the home of Professor Ciechanowski to arrest him, he was out, and they did not bother to go looking for him. They sought to arrest Professor

---

National Public Hospital in Lwów) and his grandson Adam Mięsowicz, Dr. Tadeusz Tapkowski (lawyer), and Professor Kazimierz Vetulani (Dean of the Faculty of Theoretical Mechanics at the Lwów Polytechnic). There were a further four people murdered in the courtyard of *Bursa Abrahamowiczów* (a school in Lwów): Katarzyna Demko (a teacher of English), Dr. Stanisław Mączewski (Head of the Department of Gynaecology and Obstetrics of the National Public Hospital in Lwów), Maria Reymanowa (a nurse), and Wolisch (a merchant; full name unknown).

[10] Stanisław Michał Progulski (1874–1941), Polish pediatrician and professor at the University of Lwów. Celebrated member, with his son, of the Lwów Photographic Society.

[11] Stanisław Ruziewicz was arrested on July 11, 1941, and killed on July 12.

[12] Stanisław Pilat (1881–1941), professor of petroleum technology, Chief of the Institute of Technology of Petroleum and Natural Gases at the Lwów Polytechnic.

[13] Professor Władysław Dobrzaniecki (1897–1941), Head of the Surgery Faculty of the National Public Hospital in Lwów. Arrested together with Eugeniusz Kostecki, husband of his servant.

[14] Tadeusz Ostrowski (1881–1941), Head of the Institute of Surgery at the University of Lwów. Murdered with his wife Jadwiga.

[15] Stanisław Ruff (1872–1941), Head of the Department of Surgery of the Lwów Jewish Municipal Hospital.

[16] Kasper Weigel (1880–1941), Chief of the Institute of Measures at the Lwów Polytechnic. Murdered with his son Józef.

[17] Jan Grek (1875–1941), professor at Lwów University, pathologist, internist, and well-known art collector. Killed together with his wife Maria.

[18] Roman Witkiewicz (1886–1941), Head of the Institute of Machinery at the Lwów Polytechnic.

[19] Kazimierz Bartel was arrested a day earlier than the others, that is, on July 2, 1941, and murdered on July 26 in Brygidki Prison after refusing to collaborate with the Nazis.

[20] Włodzimierz Krukowski (1887–1941), Polish scientist and electrical engineer. Head of the Institute of Electrical Measurement at the Lwów Polytechnic.

[21] Henryk Korowicz (1888–1941), Polish economist. Rector of the Academy of Foreign Trade in Lwów (changed to the Lwów State Institute of Soviet Trade during the Soviet occupation), Head of the Institute of Economics at this Academy. Arrested by the Gestapo on July 11, 1941, and murdered on July 12.

Grzymalski,[22] not knowing that he had been dead for several years. Some of the professors arrested—such as Progulski—were old people unconnected with politics. While they were placing Professor Progulski under arrest in his home, Progulski's son happened to wander in, whereupon one of the Germans said: *Der kann auch mitkommen.*[23] When Ostrowski's wife saw that they had come for her husband, she cried out: "And we had been so happy that the German troops had arrived at last!" *Halt's Maul!*[24] responded an officer. When an *Einsatzgruppe* came for Bartel, one of them asked him why he hadn't gone off with the Bolsheviks. "Because I'm not in the habit of running away", he replied. They expelled his wife from their home, allowing her to take two suitcases of belongings, and telling her that anyone giving her shelter would be bringing woe upon themselves. Professor Gröer gave his name as "Freiherr Franz von Gröer" and was freed after a few hours' detention,[25] the only one of those arrested to survive. Professor Łomnicki was refused permission to take his hat and coat. Professor Pilat's request of the German arresting officer to light a cigarette received the rebuff "Can't you see that I don't smoke!", and when he started to say something to his wife, he was stopped with the words *Keine Dialoge!*[26]. When they came to Dr. Ruff's home to arrest him, his son Adam had an epileptic seizure and was shot in the presence of his parents and other arrested professors, who were then ordered to carry out the body. While this was going on, one of the members of this particular *Einsatzgruppe* was intermittently whistling a merry tune. I believe that all those arrested during the night of July 3/4, 1941, were shot. Mrs. Bartel was informed of her husband's death, and when asked who had ordered it, was told that it was Hitler.

Following these murders Germans turned their attention to the Jews. A dozen or so Jewish doctors, lawyers, and other professionals vanished without anyone knowing what had happened to them. A compulsory contribution of 20 million rubles was levied on the Lwów Jewish community.

During the first five days of the German occupation I hardly left the house. We packed up the more valuable of our household items, and I buried Dr. Grünhut's money in the garden. From our window we could sometimes see small groups gathered before certain houses, apparently engaged in conversation. Soon such a group, made up of Szydłowski, the landlord of the house at 12 Kadecka Street and a few of his friends, appeared in front of our house. Janek said of them: "This is one

---

[22] Possibly Wiesław Grzymalski, Polish engineer and architect.

[23] "He can come along too."

[24] "Shut your mouth!"

[25] That Gröer tried to pass himself off as German is denied by Zygmunt Albert—see: Z. Albert, "Mord profesorów lwowskich w lipcu 1941 roku" [The massacre of the Lwów professors in July 1941], in *Kaźń profesorów lwowskich. Lipiec 1941. Studia oraz relacje i dokumenty zebrane i opracowane przez Zygmunta Alberta* [Execution of the Lwów professors. July 1941. Studies and relations, and documents gathered and edited by Zygmunt Albert], Wydawnictwo Uniwersytetu Wrocławskiego, Wrocław 1989, p. 44.

[26] "No conversations"

of the Hitlerite foxholes of Kadecka Street." On the opposite side of the street units of the SS and other formations had their offices, and we could see some of them standing for days on end at the windows, with sheer curtains fully drawn, observing the dwellings opposite as if trying to determine what was going on in them.

In the meantime, Pazia, our maid, was in the process of losing her mind. She had been running from prison to prison inspecting the bodies of those murdered by the NKVD to see if any of her relatives were among them. After the departure of the Soviets, in her home town of Przemyślany[27] her cousins—promising young students and the pride of the family—had been found murdered in the local jail, and, seized by an ungovernable fury, her uncle had exacted revenge by single-handedly killing eight Jews. The pernicious lies spread by the Nazis to the effect that the Jews had perpetrated the murders were the more easily accepted precisely because the Soviets had left, whereas the defenseless Jews were to hand as ready scapegoats.

On Thursday, July 3, we finally understood why the SS men had been staring through their windows at our house. One of them, wearing sideburns and looking for all the world like an Austrian sergeant, came to enquire about my profession and religion, and also the names of our neighbors, thanked me, and left. Next day, July 4, my son-in-law and daughter hastened to move to the Ziemilskis' at 3 Czarniecki Street, and were gone by early afternoon. Then around four we heard repeated ringing and kicking at the door, and when I opened it I saw the same SS man, this time accompanied by a tall dark-haired man with a small moustache, clearly an SS officer. I asked them to enter before me, but the newcomer pushed me with his boot so that I stumbled in ahead of them. He then set to asking those present what they were doing in the apartment, paying particular attention to my cousin Dyk. After a glance at Dyk's birth certificate, the SS officer said: *Ach, du bist Graf Schoenborn von diesen Jüdischen Aristokraten.*[28] Dyk was so naive as to then admit that his mother had been Jewish, at which the lout slapped him, shouting *Du bist Jude!*[29], and told him to get out. I was thus left with my wife, my sister, and my niece.[30] Next the thug said to me: *Schau, der sitzt ganz gemütlich hier*[31]—I was seated in an armchair—and told me to stand up. He then drew on his gloves and dealt me two hard blows to the face; at the third blow I fell down. Patting the holster of his pistol, he began asking questions: why my daughter and son-in-law were absent, where we kept our clothes, what my pension had been under the Soviets, why my passport had me as a Pole, and so on. They pulled out a navy-blue pair of my pajamas with red piping and began insisting that it was an NKVD uniform. This was a sort of game; it was completely obvious that they themselves didn't for a moment believe this nonsense. The officer pulled my wife's hair, laughed at her dismay, and then

---

[27] A town about 45 km southeast of Lwów.

[28] "So you are Count Schoenborn of the Jewish aristocracy."

[29] "You are a Jew!"

[30] Muta's sister.

[31] "Look, he's sitting so comfortably here."

suddenly sat down at the piano and banged out a jolly melody. This whole display lasted only a very short time; clearly this was an unofficial escapade of their own, the whole *mise-en-scène* aimed at maximal intimidation. They asked for gold, coffee, and soap, packed the soap they found together with some of our bed-linen in one of our trunks, and told me that I personally should bring the trunk to them at eight o'clock next morning: *Der könnte bei uns Arbeit haben.*[32] But when I asked for the address, they consulted quietly together and gave vague directions "to the house nearby."

This visit served to clarify my mind as to our situation, and I told my wife to pack a small suitcase with the most indispensable things, and at seven that evening we left through the garden at the rear of the house. On the way my wife picked a last rose. We had prepared an opening in the wire netting of our back fence, and we passed through it into the yard of our neighbor Mr. Tomanek, and thence through the garden of a house adjoining his into the street, thronging with SS men and lined with their black cars. Seeing only a lady with a rose walking slowly and nonchalantly with her middle-aged gentlemanly companion, they did not realize that we were not residents of that house. That was Friday, July 4, 1941, the last day we spent in our house in Lwów.

---

[32]"He could work for us."

Chapter 10
Homeless Wandering

During the first few weeks the Germans let the Ukrainians make fools of themselves, and the Ukrainians took full advantage of the opportunity. Ukrainian "warriors" could be seen standing on the running boards of German vehicles showing the drivers the way. They also took it upon themselves to organize their own militias and police departments, and in their self-appointed role as enforcers might be seen harrying Jews on their way to work. Initially, the Germans made Mr. Polański, a naturalist, mayor of Lwów, and seemed for some time disposed to be conciliatory to the Ukrainians, even permitting the publication of the Ukrainian language newspaper called *Vil'na Ukraina*.[1] However, it didn't take long for the Germans to change the name to something less tendentious like *L'vivskyj Distrikt*[2]—I don't recall the exact name. From that rag one could discover that, although the Nordic people are racially the most superior, the next best is the Dinaric race,[3] of which the purest, most undiluted specimens happen to be the Ukrainians. Poles and Hungarians all too often have Jewish blood in their background somewhere, and are therefore to be counted inferior—or so at least this paper averred. If at some future date the Ukrainians should demand equality in the name of democracy, it would then be opportune to remind them of their racial superiority. Once when my sister turned to a Ukrainian acquaintance for help, he said: "We can do nothing for you since we must carry out German orders." When the time for reckoning comes we must quote such words back at them, and it will then be appropriate to remind them that the Ukrainian police, also known as the Ukrainian militia, helped Germans in Kraków and Warsaw carry out raids and arrests, thereby extending their field of activity to regions to which no Ukrainian politician had ever made claims. During

---

[1] "Free Ukraine"

[2] "L'viv District"

[3] Also called the Adriatic race. A sub-category of the Europid (or white Caucasian) race, so termed by physical anthropologists in the early twentieth century. In more recent times this and similar typologies have come under criticism, especially since the rise of molecular anthropology.

the Soviet occupation of eastern Poland, the Ukrainian committees appointed by the Soviets actually consulted with Poles in political matters, but once the Soviets had departed the mass of Ukrainians broke off all allegiance to their own intelligentsia as having authority over them, turning into a leaderless horde, at first submissive to the Germans, and then totally capricious.

On quitting the house on Kadecka Street in the evening of July 4, 1941, we went to stay with the Indruch family in their home on Snopkowska Street, Mrs. Indruch being an aunt of Danka Smalewska, a friend of Lidka. They received us gladly, but it was clear that we couldn't stay for more than a few days since their apartment was a small two-room affair, and furthermore their landlord was a Ukrainian, a stupid woman, disagreeable to her tenants, who would not agree to our staying without being officially registered.

The streets of Lwów were jam-packed with columns of military vehicles heading east through Lwów. Among them there were massive trucks with double wheels on which white figures of a fox, or lion, or seal, etc., had been stencilled, and bearing white stripes on their rear ends to prevent collisions. It all looked very impressive, very well organized, but Mrs. Stasia Blumenfeld was of the opinion that this was no army for the conquest of Russia, being, as she said, too orderly, too preplanned, even too beautiful, for the vast, trackless forests and bogs that still lay between the Germans and Moscow. At the Indruchs' we were introduced to Mr. Witkowski, a former Polish officer, and his wife. He had not registered with the Soviets and in this way escaped the worst. Both the Indruchs and the Witkowskis knew very well what most Jews failed to understand, namely that in order to survive the war one should not blindly follow the bidding of the authorities of the moment, and that one should alter one's name, occupation, and corresponding documents if the current situation was such as to justify such changes. Many Jews had an exaggerated idea of proverbial German sagacity and organizational ability. They seemed to regard them as more-or-less omniscient, as if every German gendarme was an expert on seals and stamps—and therefore able to detect falsified documents—and the German authorities so painstaking as to write to every parish and congregation to gather information about every Jew or neophyte. This belief in German efficiency and thoroughness, together with the conviction that obedience to German commands would enable them to survive the worst times, albeit in squalor and degradation, were the chief reasons for the deaths of many thousands of people.

The "economic exploitation of Lwów" by the conqueror was on view everywhere. For instance, members of the Gestapo, military policemen, and officers occupied the most comfortable and convenient apartments, having thrown out the owners. If the latter were not Jews they were allowed to take their furniture with them, and this furniture was replaced with that of Jewish homes. Thus there was going on what seemed like a continual moving of furniture in trucks from one house to another, with Jews forced to carry and arrange the furniture running alongside. This period of "installation" lasted several months because the Germans were wont to change their apartments at a moment's notice—after all, the apartments, furniture, and labor associated with moving cost them absolutely nothing. The supply of Jewish labor was organized by the Jewish community itself, based in the former

Jewish Community Center. Among other things, it supplied the Germans with several thousand laborers each day, and to organize this—in particular to keep a roster of those available—a few dozen were employed as officials and even policemen. Every now and then a car filled with armed SS men would drive up to the Jewish community building, bringing a demand for 24 workers, say, or bed-linen for a hundred beds, or furniture for a few rooms. And whatever they supplied had to be first-rate or else. On one occasion they requested a number of cameras, and on another dozens of pairs of galoshes—and, noticing the galoshes of the Jewish officials left in the entrance-way, they took them as a first instalment.

Since we no longer had incomes of any sort, Janek went into business. He began by selling German soldiers clothing, including socks, which he bought in bulk in various stores. He then expanded into sugar, fruit, and, in short, whatever he could get his hands on. He would consult German military magazines, where commissaries advertised wares sold in bulk at a discount. These paid army personnel to transport the wares in military vehicles illegally to designated places for pick-up; the only risk was that the trucks might be stopped *en route* by the SS or the military police, suspicions aroused by the direction in which they were travelling.

From the Indruchs' I moved for a few days to the Blumenfelds'. My wife spent a couple of days in the apartment belonging to Miss Parandziejówna, a retired treasury official. German military police and SS men were prowling everywhere. The Blumenfelds told us how they had once come to their house and demanded that Stasia—Mrs. Blumenfeld—come to their barracks to scour the floors. *Ich bin keine Jüdin,*[4] she objected, to which one of the Germans responded with: *Wenn du lügst, erschiesse ich dich wie einen Hund*[5]—showing perhaps an extreme regard for the truth. But many persisted in the failure to grasp that whoever tells the Germans the truth, pays attention to their cant, follows their instructions, and is honest in their dealings with them—in particular, does not falsify one's documents—goes to an undoing more horrible than might be predicted even on the basis of *Mein Kampf*. That manifesto contains a chapter describing the way one deals a nation a succession of calculated blows in order to destroy it, together with the statement *Ich werde die Judenfrage lösen.*[6] It was not till the war was well advanced that anyone understood the full implication of these words. The example of the treatment of the Jews in the Warsaw ghetto was, at least in the initial stages, not clear enough, since for some time life was allowed to persist there in some remote semblance of normality—albeit with the steady imposition of greater and greater restrictions. One might conjecture that over the 22-month period of the pact with the USSR mass

---

[4] "I am not a Jew."

[5] "If you're lying I'll shoot you like a dog."

[6] "I will settle the Jewish question."

murder had still not yet emerged as part of German policy[7]; however thereafter it was soon adopted and implemented.[8]

To make Lwów's inhabitants happy, the Germans decided to divide the city into four quarters: German, Polish, Ukrainian, and Jewish. Although monumentally absurd, this idea didn't strike people as such, so used had they become since the times of mutual cultural understanding between Poland and Germany—involving visits to Poland of such luminaries as Goebbels—to accepting the ideas of the brilliant Adolf. However, in German military circles it was well understood that forming a German quarter meant exposing Germans to selective bombing, as well as having to give up the possibility of installing German soldiers in buildings on streets of tactical or logistical significance. Thus for practical reasons the measure was reduced to the delineation of a Jewish ghetto, followed by a proclamation of a timetable setting out the order in which Jews were to leave "Aryan" districts and enter the ghetto. A definition of "Jew" was announced: a Jew was anyone having among their four grandparents at least three who were Jewish.[9] The genius who dreamed up this definition apparently failed to notice that it contains a circularity.[10]

The Witkowskis took it upon themselves to arrange for our luggage to be brought over from our neighbors'—that is, the Witkowickis'—apartment on Kadecka Street. Janek also went over there, sneaked into the garden and dug up some of the money that old Dr. Grünhut had accidently left before leaving with his wife for Truskawiec—enough at a pinch for us to live on for a few years. We often changed domiciles, occasionally separating; for instance, at one point I stayed for a few days in the home of Mr. Jan Dobrzański in Łyczaków, while my wife stayed with Mrs. G., the sister of Mr. Mozołowski.[11] Though many people helped us willingly, there were those who refused. After a week of such wanderings—on July 13, to be precise—

---

[7] Many concentration camps were built in Germany between 1933 and 1939. Between 1939 and 1942 the number of camps—now including those on occupied territory—increased fourfold. In the early spring of 1941 the SS, together with doctors and officials of the "T-4 Euthanasia Program", began systematically killing selected concentration camp prisoners in "Action 14f13".

[8] At the infamous Wannsee Conference of January 20, 1942, Reinhard Heydrich, appointed by Hitler as chief executor of the "Final solution of the Jewish question", presented a plan for the deportation of the Jewish population of Europe and French North Africa to German-occupied areas of eastern Europe, and the exploitation of those fit for labor to work on road-building projects, the surviving remnant to be annihilated after completion of these projects.

[9] The "three grandparents" definition had been adopted as early as the ratification of the Nuremberg racial laws of 1935. By the protocols of the Wannsee Conference (see the preceding footnote) "half-Jews" (those having just one Jewish parent) were to be considered—regardless of their religious affiliation—as Jews (with some exceptions) while "quarter-Jews" were to be classified with those of German blood (again with certain exceptions).

[10] In the sense that it involves the grandparents' own status as Jews and thus leads to a regress over the generations. However, the definition makes more sense when one recalls that the concept of a "Jewish nation" had at some time official standing in several Middle European countries, including the Habsburg monarchy and possibly Poland.

[11] Possibly Włodzimierz Mozołowski (1895–1975), professor under J. K. Parnas in the Lwów University medical faculty.

10 Homeless Wandering

we moved in with my sister who had all the while been living with our Mother in Professor Fuliński's house at 82 Tarnowski Street. He himself lived on the ground floor with his wife, daughter, and four sons, while my sister and Mother had rooms on the first floor. Some of Fuliński's cousins were also living in the house. At that time our apartment in the house on Kadecka Street was still empty. Alone of the tenants there, Halina and her husband on the floor above knew where we were and brought us news of what had been happening. On the morning of July 5, SS men had come for the trunk we'd left behind, containing bed-linen and soap, insisting that I be found to bring it to their quarters myself. They had then invited Halina and her husband to drink vodka with them, and queried them as to my whereabouts. Not long afterwards the apartment was taken over by German NCOs.[12] Hearing this, I made a point of going to what was then still functioning as a Housing Office to cancel my lease on the apartment taken out under the Soviets, a step that most of my acquaintances thought pointless—as indeed it doubtless was. Halina and her husband presented themselves to the new residents of the apartment as service personnel—he as cook and she as maid. In this a neighbor, Ms. Tola from Warsaw, helped them; indeed, this lady showed both them and us a great deal of kindness in those difficult times. The German NCOs held parties lasting till morning to which they invited the young girls from the third floor—one of whom had belonged in 1939 to a group of resistance fighters engaged in disabling German tanks by means of gasoline bombs—and girls from other parts of town. They ransacked our library, looking in particular for letters and photographs of possible use against us, but could not have found anything of any significance since I had taken care to burn all important papers before leaving. I had not wanted to take any of them with me because they would be in conflict with the change of identity I was already planning at that time, and thus liable to incriminate us.

The day we moved in with my sister was also the last day before white armbands with a blue Star of David sewn onto them became obligatory for Jews.[13] Professor Fuliński, who received us into his house very cordially, advised us not to register with the authorities, but his sons, fearing German sanctions, had registered us on our behalf, so to speak. Given this *fait accompli*, we—myself, my wife, and my sister—decided we had better obey the order to report to the office on Zielona Street for work assignment at 4 pm that day. There we encountered hundreds of people crowded outside the office waiting to be interviewed, with German military policemen, aided by Ukrainians, lining them up two by two. When I told one of the Ukrainians—a commissar, by the look of him—my age and profession, he said I could go, and likewise my wife and my sister when they showed him medical certificates. From this time on we resolved never again to register or report to the German authorities—despite the existence of various official lists of names most probably including ours, new variants of which were promulgated daily, with sanctions threatened against those not obeying summonses. One felt one had to

---

[12] In German, *Unteroffiziere*.

[13] According to other sources, this happened on July 8.

defend oneself even from the officials of the Jewish community, who had compiled their own lists, and before the housing administrators, some of whom were Jewish yet had somehow managed to hold on to their positions. However, many of these officials were very helpful to their Jewish confreres. The first chairman of the Lwów *Judenrat* was the lawyer Józef Parnas, who was arrested and murdered after demonstrating great courage in refusing to obey certain of the Nazis' orders. The deputy chairman, Dr. Rothfeld, tried to help us as much as he could. Another official of the Jewish community, a Dr. Jacob, managed without apparent difficulty to get my wife, my sister, and myself exempted from queueing up for forced labor impressments. However, he, like a great many others, failed to take the full measure of the Nazis, and so prepared, in the absence of all precedent, a list of Lwów Jews by suburb. The head of the forced labor unit in Lwów was of course a German from the *Arbeitsamt*,[14] with instructions as to the "rational utilization of labor", meaning that he should provide just enough rations for the forced laborers to complete their assigned tasks, in other words to treat the Jewish workers like beasts of burden. But what neither he nor his "workhorses" knew was that all this was a farce, a front to mislead the populace and even their own soldiers as to Hitler's real intentions *vis-à-vis* the Jews.

Bronisław Knaster, whom we sometimes visited of an evening, understood this all too well, and in order to acquire some sort of insurance he tried to obtain a position as "lice feeder" in Professor Weigl's[15] Institute, which, being *kriegswichtig*,[16] was retained and enlarged by the *Reich*—although they renamed it after the German physiologist Behring.[17] Prompted by the same motives, the Blumenfelds went to live in the house of Weigl's sister, near the restaurant *George*.

It had been expected that the Germans would arrest communists and anyone else who had been favorably disposed towards the Soviets, but there seemed to be no such concerted effort on their part. Although Halina Górska was arrested, many others, such as Professor Mazur, whose name, as that of someone approved by the Communist Party of the Soviet Union, appeared on a poster in the university's vestibule in plain view of the Germans coming and going, remained unmolested in his apartment. The former Rector Byczenko was also able to stroll around town with apparent impunity.

The Ukrainians did not gauge very well the correct way to behave *vis-à-vis* the German overlords. The Nazis soon put an end to the comedy starring Bandera

---

[14](German) Office of Work, or Labor Exchange

[15]Rudolf Stefan Weigl (1883–1957), famous Polish biologist. Discovered the first effective vaccine against epidemic typhus. In 1941 the Nazis ordered him to set up a vaccine production unit in his Institute at Lwów University. He employed Polish intellectuals, Jews, and members of the Polish underground, thus affording them protection from the Nazis, and his vaccines were smuggled into the ghettos in Lwów and Warsaw, thus saving many lives. He used human blood to feed lice, the vector for epidemic typhus, but took pains to protect his human "injectors" from infection.

[16]"important for purposes relating to war"

[17]Emil Adolf von Behring (1854–1917), German physiologist. Nobel laureate for medicine in 1901.

by interning his so-called "government" of Western Ukraine in Krynica, part of it disappearing *en route* from Kraków. The Germans began to arrest the more zealous amongst the Ukrainians, and forbade them to write about their racial superiority. In the German language newspaper *Die Lemberger Zeitung*[18] an article was published to the effect that the intelligentsia of the Ukrainian nation is not numerous enough to rule Ukraine effectively, and that in any case those regions of Ukraine acquired at the cost of the blood of German soldiers would naturally be retained by the Germans as *Lebensraum* for the German *Volk*—although the Ukrainian peasantry would be appropriately employed ploughing, planting, and bestowing the fruits of its labor on a Germanized Europe. Thus for the second time in the space of a few decades were the Ukrainian people exhorted to supply foreign usurpers with foodstuffs: to keep up the supply of manure, to surrender their produce in appropriate amounts and in good time—in return for being allowed to celebrate the harvest by singing traditional folk songs while they worked and afterwards around the campfire. Their frustration reached the boiling point when the Germans reintroduced Polish złotys as the means of exchange in Galicia. However, the Germans being too formidable to vent their anger on, they turned on the Poles, and we began to hear of murders of Polish village elders, foresters, those who had settled land after the acquisition of territory by the Polish-Soviet Treaty of Riga of 1921, and especially Poles living in Ukrainian villages. Only occasionally did the Germans make any attempt to prevent such outbreaks.

So the Jews were forced to put on the white armbands with the blue Star of David. Some had armbands made of celluloid, and I saw ladies wearing matching outfits. For the first little while the young people among us endured the continual humiliation with courage and even a sense of humor. But it was not easy. Whoever wore an armband could expect any passing officer or policeman to sneer at him, or whistle, or even call after him *Du Dicker!*,[19] and perhaps order him to carry his briefcase for him. Occasionally they would pick people up in the street and drive them off to do some kind of work beyond the town. I myself walked around town more and more seldom, contenting myself with strolling in the Fulińskis' garden whenever the weather permitted, as if under house arrest. Fuliński himself spent a great deal of time in the garden with his sons. They kept a goose there which became so attached to us that if we hid for a moment behind a shrub she would become quite frantic looking for us.

Professor Fuliński and I enjoyed cordial relations and always used the informal "*ty*"[20] in addressing each other. We discussed politics a great deal together. He had a very extensive knowledge of history. Even at that early stage he was predicting that the Germans would not get beyond the "Tartar trench",[21] which forms a sort

---

[18]"The Lwów Times". Lemberg is the German name for Lwów.

[19]"You fatty!"

[20]Rather than the form of address used in formal conversation, namely "*Pan*" (Sir).

[21]An ancient trench running between the Don and the Volga near where these rivers approach most closely. The Volga end is near the former Stalingrad. The Germans actually went beyond the

of natural moat before the Caucasus some distance to the south—and he was right, as we now know. He also predicted correctly that I would eventually become a professor in Wrocław. He had a rather eccentric take on the "Jewish question", assigning to the Jews the region called "Khersonia"[22] between the Dniester and the Dnieper as an independent Polish "dominion." He tried to persuade me to work seriously on realizing this admittedly long-range project by publicly promoting it once the war was over.

One of the houses adjoining the Fulińskis' belonged to a German woman. During the Bolshevik occupation she and others who were *Volksdeutsch* had escaped to German-occupied Poland, but now she had returned to reclaim her house—only to have Fuliński greet her with "Your people have already lost the war," by which he meant that he was certain the Americans would fight in Europe with a fleet of warplanes a hundred thousand strong.

I would usually descend to the Fulińskis' flat early in the evening for our political discussion, and only much later, around 10 pm, return to our rooms above for supper. All the cooking was done using gas, of which there was very little available during the day, so that sometimes our dinner preparations lasted till 2 am. This, together with the ever-present fear that someone would come for us, robbed us of peace. I sometimes slept in the attic on piles of old books when the fear that the Germans might come searching for someone—such as Leon Chwistek, so-called "communist fellow-traveller"—became especially acute. Thankfully these fears proved groundless, but nonetheless we sometimes also resorted to staying at other people's places for a night or even longer, occasionally separately.

Thus I spent one night at Professor Małachowski's and nearly a week at the Weisses', on Bernardyński Square. Ignacy Weiss was an old schoolmate of mine. In addition to his German wife and three sons—strapping, clever boys—his sister and brother-in-law—the lawyer Stein from Jasło—were then living in his house. Like my cousin Dyk, both Weiss and Stein were working in the Sanitary Cooperative organized in particular to disinfect or delouse clothing as a preventative measure against typhus. One evening during my stay two *Schutzpolizisten*[23] came to the door while we were eating dinner. They came in without removing their caps, and seeing us around the dinner table one of them remarked: *Sie essen noch ganz gut hier,*[24] and then, addressing Ignacy Weiss, *Sind Sie Jude?*[25] When Weiss replied that he was a member of the Consistory of the Evangelical Church, one of them said: *Mit*

---

trench and down through the isthmus leading to the Crimea, where they took Sevastopol′ after a long siege. But on the other hand further north they got no further than Stalingrad.

[22] Kherson is a port at the mouth of the Dnieper River where it debouches into the Black Sea. It is the capital of the region of Ukraine with the same name.

[23] The *Schutzpolizei (Schupo)* was the defense branch of the *Landespolizei*, the state police of Germany. During the Nazi regime, it was made part of the *Ordnungspolizei*, an organ of the Nazi state.

[24] "They still eat quite well here."

[25] "Are you a Jew?"

den Kirchen haben wir auch Schluss gemacht,[26] and then turned to his companion to tell him an anecdote about a Jew who became a Christian when on the point of death, and when asked why, answered that he was destitute and did not want to burden the Jewish community with his funeral expenses. They then turned to me and asked me what my business was there, to which I replied that I came there for meals. Then, while they interrogated Stein, I took advantage of the confusion, took my hat, left hurriedly and went over to Dr. Ziemilski's, nearby. When several hours later I returned, I was told that the visit had been prompted by a denunciation of Weiss and Stein by a member of the Sanitary Cooperative claiming that they were trading in gold. When a search of the premises produced no gold, the Germans had left, seizing only a painting of the wall that had taken their fancy, and warning Dr. Weiss that he must wear his armband in future because if the *Schutzpolizei* caught him without it, it would be bad for him. They had also asked for my address, which was given to them incorrectly as that of our former house on Kadecka Street.

I now think I tended to underestimate the danger inherent in such visits by German military personnel and policemen. One such visit to our temporary abode on Tarnowski Street was a close call, I would say. Two uniformed hoodlums appeared at the door, and, since I happened to be in the garden, my sister opened the door to them. One had a bandaged jaw and the other, speaking German with a Jewish accent, said that they knew the apartment was inhabited by Jews. When my sister denied this, offering to show them the birth certificate of her husband, who—she said—owned the apartment, they answered that they had no right to inspect documents, but would like to know if she knew of any Jews living nearby—to which she replied that as this was of no interest to her she had no idea. Thus, but for her presence of mind, we might very well have been robbed of the few possessions we still had and ejected from the apartment—or worse. Robbery was now the Germans' main activity. In the evening, one could see officers walking with their wives along Jagiellońska Street in the direction of the Jewish district with the aim of taking from the places of abode of the Jews whatever they might take a fancy to. The one comfort left the Jews subject to this arbitrary treatment was that they had not taken their lives.

While living on Tarnowski Street we often had visitors. One such was Mrs. Borkowa, the mother of a former student of mine with whom I had kept in touch by mail up to the time of his arrest by the Soviets in 1941. He had been in the Polish reserve, and his mother told me that his former orderly, a Jew, had pointed him out to the NKVD as a former Polish reserve officer, and that he had suddenly been taken from his flat and taken to Lwów, never to be heard from again. In all likelihood he shared the fate of the Katyń victims. I liked his mother very much. She gave no thought to revenge, and was ready and even eager to provide shelter for us in her home in Borysław.[27] At that time we were indeed thinking more and more seriously about leaving Lwów, and former colleagues such as Kulczyński, Banach, and Rogala encouraged us in this. We were also visited by Schauder, who was then

---

[26]"We've also dealt with the churches."

[27]Town in the Lwów district, now in western Ukraine.

living in a room of a neighboring house. One day, on returning to the room he had been staying in earlier, he had found it taken over by the Germans, and when he objected an SS man present had slapped him. We were also told that Auerbach was subletting a room in the basement of a house rented by a laundress.

The Germans continued with their depredations. They now announced that all objects of artistic value must be surrendered to them—purportedly in order to protect them from damage through wartime activity of one kind or another. While I do not know if anyone obeyed this order, what is certain is that the Fulińskis paid no attention to it. They were survivors of the Bolshevik school, where they had been at one point accused of conspiring against Germany, and their serving-man Staszek had been arrested and tortured by having the soles of his feet held over a fire, but had given nothing away. However, the Germans inevitably found their way to the Ossolineum. An official of the German Ministry of Education demanded that Gębarowicz[28] hand over the Ossolineum's prized Dürer prints, but Gębarowicz is said to have objected and asked the official to show him a written order. To this the German replied: "Be careful! You probably heard what happened to the Kraków professors!" Gębarowicz is then supposed to have retorted "Yes, I heard, but I had thought such stories to be anti-German *Greuelpropaganda*[29]!" Although this apparently caused the German to become more polite, the Dürer prints, among much else, were taken anyway.

It was quite common for a German soldier to kill with his rifle butt a Jew he happened to encounter in the street. Yet the newspapers carried periodic reports of Jews being run over by careless truck drivers, as if this was held by the German authorities to be reprehensible. However, such reports were actually aimed at creating in their readers the false impression of German respect for the law, even as it related to the Jew. The sorts of things that were occurring in fact were, for example, the periodic roundups—from the streets and their homes—of a few hundred Jews, to be taken to the courtyard of the SS building on Pełczyńska Street, and made to lie face down on the ground from morning to evening, with machine guns trained on them, subject to the threat of being shot should they so much as lift their heads. Every now and then an SS *Verteilungskommando*[30] would snatch a few dozen people, take them out to the *Góra Piaskowa*[31] or to Lesienice[32] and shoot them.

The seeming randomness of the German blows is illustrated by what happened to Lila Holzer, a cousin of my wife's, and her husband. One evening two German thugs entered their flat to arrest Mr. Holzer. Apparently swayed by her frantic

---

[28] Mieczysław Gębarowicz (1893–1984), outstanding Polish historian and art historian. From 1920 to 1939 associated with both Lwów University and Polytechnic. Appointed professor at Lwów University in 1936. As director of the Ossolineum during World War II, he endeavoured to save the museum's exhibits and book collection from destruction.

[29] "atrocity propaganda"

[30] "redistribution detachment". Perhaps *Einsatzgruppe* is intended here.

[31] "Sandy Hill", a hill in Lwów's Zniesienie Park.

[32] A suburb of Lwów.

entreaties and the wailing of their little child, they promised her that Holzer would be returned to her. However, when she asked them to give their word of honor, one of them said *Einer Jüdin gibt man kein Ehrenwort*.[33] Holzer was taken to a nearby wood with several others, where they were lined up and, except for Holzer and a Mr. Lindenfeld, a Kraków friend of Holzer's for whose life he had pleaded, shot. Holzer and Lindenfeld were set free, and Holzer afterwards learned that one of the others, although shot through the throat, had managed to drag himself to a nearby farmhouse. I met Lindenfeld a few times, and, although I found him intelligent and with a fair share of common sense, he seemed incapable of fully grasping Nazi perfidy and criminality to anything like its full extent. He thought that by treading carefully—for example, by finding an optimally situated apartment in the ghetto and the safest kind of occupation, and by keeping his head down—he would survive the Nazis. But this was precisely the illusion the Germans strove to engineer. They delivered their blows haphazardly—although with gradually increasing force—now here, now there, so that no one could guess their ultimate, to most people unimaginable, goal. Lila Holzer told us how once when she was waiting at a streetcar stop, she was approached by a German soldier in field uniform, who, on noticing her armband, said: "*Gnädige Frau*, this unspeakable rabble could not possibly win the war." And then he told her that at the citadel[34] he had witnessed the execution of a few hundred young Jews—*prächtige Kerle*[35], he called them—including scientists, doctors, musicians, and so on. He then added proudly "I am a communist, and not afraid to speak out."

On our recommendation, Mr. Fuliński let a room on the first floor of his house to a certain Mr. Z., who was working with Dyk and Stein in the Sanitary Cooperative. Mr. Z. went quite often to the citadel to disinfect the prisoners' barracks, where he saw first-hand the conditions under which the Soviet prisoners of war were being held. They were being systematically starved to the point where a few hundred died every day—an obscene demonstration of German cruelty. Somewhat later, when after the first push into the USSR they captured Soviet soldiers in such enormous numbers that they were unable to find sufficient prison space to accommodate them, they released a great number of them and sent them off walking along roads leading to occupied Ukrainian towns. They died by the hundreds on the roads and those who reached towns begged in the streets. Although the Germans forbade the populace to offer succor, they nonetheless very often did, but, given the huge numbers of soldiers involved, their help was far from adequate. Thus, already in the first few months of their stay, did the Nazis show themselves to be cold-blooded mass murderers both of the civilian populace and prisoners of war.

---

[33] "One does not give one's word of honor to a Jewess."

[34] A military fortress built by the Austrians in the second half of the nineteenth century. Several thousand prisoners of war were killed there by the Nazis during World War II, when it was used as a concentration camp.

[35] "splendid fellows"

Naturally enough, we were preoccupied with the question as to what we should do to survive. Some thought it best to go to the ghetto, while others believed that baptism would protect them—although why they believed this in the face of the definition of Jewishness promulgated by the Nazis was hard to fathom. Baptism into the Christian church had been resorted to by many even during the Soviet occupation, especially, it seems, by Jews deported along with Poles to Kazakhstan and Siberia. It was as if they felt that their ancient God had forsaken them, or even forsaken the world.

Ignacy Weiss was one of those who urged people to take this route to salvation from the Nazis. He had a cousin who knew a priest living near Przemyśl who would perform baptisms, although his charge for the operation was exorbitant; in fact, a significant portion of the fee found its way into said cousin's pocket. I took Weiss' advice, and even entered into an agreement with this priest to take up quarters in his rectory, but he backed out of our agreement and took in someone else, hiring as a housekeeper his young sister-in-law, who also happened to be the mistress of the cousin, the intermediary. This left us with but one trump card, namely the birth certificate of a peasant from near Przemyśl acquired for us by Mr. Tadeusz Hollender for a not unreasonable price. However, we had nowhere to stay. I wrote to the former Rector of the university, Professor Bulanda, asking for help, and he very kindly took a warm interest in our plight and came to visit us to see what might be done. He soon arranged for us to go and live in a house in Rudno[36] belonging to a Mr. Otto, a former official in the university bursary,[37] who held him in great esteem—as indeed did all the former administrative staff of the university—, informing us of these arrangements through the university caretaker Gawlik. Mrs. Knaster[38] offered to help us, and went on foot to Rudno to settle the matter for us, a journey neither easy nor safe. We thus had somewhere to go—a refuge—but found ourselves unable to make up our minds to leave the apartment at the Fulińskis'. The main reason for our reluctance was the worry that we would lose the few possessions we had left, which though indeed meager would require some sort of vehicle to get them to Rudno, and we would then be running the risk of having them confiscated *en route* since the Ukrainian militia suspected everyone travelling with any baggage beyond a bare minimum of transporting Jewish property, which they had instructions from their German overlords to confiscate. The Nazis viewed the hiding of anything belonging to Jews as the greatest of crimes, considering it to be German property. Thus it was that the general populace, which had initially hoped to profit from the extermination of the Jews, lost a great deal of their enthusiasm

---

[36]There are many towns of this name in Poland. This one was one quite close to Lwów.

[37]That is, office of the bursar.

[38]Maria Morska (born Anna Frenkiel) (1895–1945), actress, journalist, feminist, and member of the Warsaw theatrical demi-monde; in particular, was wont to give poetry recitals in the café *Pod Picadorem*. Attended a boarding school in Broadstairs, near Dover, before returning to Poland and tasking up a theatrical career.

when they realized that any purloined goods were not for them but their conqueror. Eventually we decided to stay.

By this time my sister had already left by car to stay with a married daughter in Borysław, prompted by the behavior of a female Ukrainian official of the Housing Office responsible for the region of Lwów where the Fulińskis lived. On July 13 this person had seen my sister in the street wearing an armband, and was persistent about following the matter up. In fact, she had indeed put on the armband; the letters from my sister Irena in Kraków and from my brother-in-law in Stróże, warning against registering for "courses" and wearing "jewellery," had arrived too late. Meanwhile, Professor Loria's[39] daughter and son-in-law arrived by car from Warsaw to fetch him away from Lwów.

The Germans did not neglect to make life miserable also for the non-Jewish Poles. Every so often they would surround a whole block in a residential area of town, drag out all the young people they could find, and ship them off to work in Germany. Unpredictability became the norm. A tremendous trade in goods by road and rail was carried on between Warsaw and Lwów, requiring payment of appropriate bribes to various ranks of German soldiers, and with whole cargoes periodically confiscated by the Germans. Military truck drivers especially demanded extortionate payments for transporting people hither and thither. And of course there was an extensive commerce in documents: passports, visas, and so on. In the city center of an evening, one would find hundreds of ordinary soldiers and many officers strolling about in the gloom accosting any woman who happened to be passing by—and some who weren't. Many of them were drunk, as could be seen from the way they staggered about, but they weren't inclined to be jovially tipsy, seeming rather to be in a befuddled sort of depressed state. Many new second-hand stores and pawnshops had opened in the town, selling goods people had had to pawn—furniture, porcelain and silver items, paintings, gowns, shoes, and so on—to acquire the wherewithal of daily living. At first the Germans frequented these stores, but later, whenever they considered the prices a store demanded to be too high, they would simply close it down officially and take the merchandise without compensation.

It is perhaps extraordinary that when the Germans arranged the furniture they had stolen in the apartments from which they had expelled the rightful owners, they generally paid close attention to the aesthetics of their quarters, even to the extent of choosing the finest paintings stolen from galleries where the Polish school of art was represented. Thus they had for nothing works by Chełmoński, Gierymski, Hofman, and Axentowicz[40] adorning the walls of their beautifully furnished apartments, which would have cost a fortune to acquire legitimately. What subtle souls these

---

[39] Stanisław Loria (1883–1958), Polish physicist. Professor of physics at Lwów University from 1917 to 1941. After the war participated in the organization of a Polish university in Wrocław. Moved to Poznań in 1951.

[40] Józef Marian Chełmoński (1849–1914), Ignacy Aleksander Gierymski (1850–1901), Wlastimil Hofman (1881–1970), and Teodor Axentowicz (1859–1938) were all highly gifted Polish artists.

people possessed! It was not enough for them to sleep in beds not long before warmed by the bodies of their owners and to eat off their dishes after murdering them, they must also have the walls decorated by finely rendered landscapes harmonizing with the countryside they had usurped. But to read their newspapers was to understand them as a nation of dreamers and poets, forced by hard necessity to engage in war to defend honor, hearth and home, and freedom. However, the only good piece of poetry produced by this nation of delicate sensibilities was the two-liner one saw chalked on the sides of the railroad wagons taking them into the depths of Russia:

Wir fahren hin, wir fahren her,
Wir haben keine Heimat mehr![41]

The style of the German-language *Lemberger Zeitung* was, of course, the same as that of the German press generally: thus all were bound to perish before the German juggernaut, which moved inexorably eastwards. When they reached Krzemieniec, I thought of the Kac family: they must surely have been murdered[42] along with the other many thousands of Jews of Zdołbunów, Winnica, and other towns and villages of Ukraine.

It was around this time that I met Ms. Maria Dąbrowska,[43] who during the Soviet occupation had preserved her independence, refusing to join the Union of Soviet Writers. She also maintained her independence of the Germans, refusing to accept payment for translations she had done. She had lived in the Blumenfelds' apartment during the Soviet occupation. Dr. Blumenfeld returned to work in some professional capacity at the plant "Laokoon" which he had formerly managed, and his wife began a small business. She hired trucks to take her to Ponikwa, where she bought fruit, flour, and other foodstuffs in bulk, and had them transported back in the truck to Lwów. It was only two and a half years later that I discovered that the money she earned in this way went to support the underground Polish Home Army.[44]

---

[41] "We travel here, we travel there/Our homeland is no more!"

[42] As noted earlier, Mark Kac had gone to the US in 1938. His parents and brother were indeed murdered by the Nazis.

[43] Maria Dąbrowska (1889–1965), Polish writer, essayist, journalist, and translator. Formerly a member of the impoverished landed gentry. Considered one of the greatest of twentieth century Polish novelists. Active also in politics. Human rights activist in Poland from around 1927. She lived mainly in Warsaw during the war, working in the Polish underground at keeping Polish culture alive.

[44] In Polish, *Armia Krajowa* (AK). Polish resistance in World War II began already at the defense of Warsaw in September 1939. Shortly thereafter, the underground organization "Service for Poland's Victory" (*Służba Zwycięstwu Polski*) was formed, renamed the "Union for Armed Struggle" (*Związek Walki Zbrojnej* (ZWZ)) in November 1939. Then in February 1942, the "Home Army" (*Armia Krajowa* (AK)) was formed from the ZWZ, absorbing most other Polish underground movements over the following 2 years. At its peak, the AK numbered some 400,000 soldiers, as well as a greater number of sympathizers, and became the main Polish underground force. Its most important fields of activity involved sabotage, diversionary actions, intelligence, and propaganda. Its major operation was the tragically unsuccessful Warsaw Uprising of August 1–October 2, 1944.

We were all living under extremely stressful conditions. A single false step, a glance in the wrong direction in the street, might trigger one's demise. Once, when I was staying at the Weisses', my wife came in weeping. She had just seen a Gestapo gangster accost Lidka and Mrs. Ziemilska after she had said goodbye to them and they had emerged into the street. Half an hour later, Mrs. Ziemilska told my wife what had happened. She said that Lidka had avoided detention and its predictably awful consequences by responding to the racist's question *Aber haben Sie keine jüdischen Eltern gehabt?*[45] with an apparently imperturbable *Nein, bloß eine Großmutter.*[46] Thankfully, the devil was satisfied and left. There were quite a few people who chose not to live under such conditions, among them the poet Henryk Balk,[47] who committed suicide, as did Poratyński,[48] an outspoken Polish patriot his whole life through who now found himself forced to wear a Jewish armband!

After the departure of my youngest sister, the eldest moved into the Fulińskis' house. From time to time my wife's niece brought us news concerning the Germans living in our former apartment in our former house on Kadecka Street. They quarrelled constantly. They expressed their hatred of the Gestapo, and were of the opinion that the Wehrmacht[49] should take over the country. She even heard some of them go so far as to call Adolf an idiot! They thought the war would be over in six weeks. They were constantly travelling to Kraków and even further for spare parts for the army vehicles, which were always breaking down; this seemed to be the chief employment of the particular group living in our apartment. When drunk—which was often—they would take to breaking the furniture, whose French polish was in any case ruined from spilled alcohol. When they left it was found that Mrs. Rzewuska's porcelain was all either broken or stolen, while they had made a present of our grand piano to one of the young girls living above them. For Halina and her husband their departure meant the end of a period of gainful employment, and they "improved" their documents and moved to another suburb.

Ms. Żukotyńska,[50] a sister-in-law of my wife's brother, found a source which furnished us with "temporary passports" of the kind issued by the Town Council and the Kraków *starosta's* office back just before September 1, 1939, when there had been a sudden shortfall in the supply of booklets of the prescribed form for regular passports. Thus our new name was duly inscribed in these new passports,

---

The AK was disbanded on January 20, 1945, by which time Polish territory had been cleared of German forces by the advancing Red Army.

[45] "Did you not have Jewish parents?"

[46] "No, just one grandmother."

[47] Henryk Balk (1901–1941), Polish-Jewish scholar and literary critic.

[48] Possibly Jan Poratyński (1876–1941), pharmacist and social activist in Lwów.

[49] The traditional unified armed forces of Germany.

[50] Possibly Małgorzata Wanda Żukotyńska, whose portrait was painted in 1929 by the artist Stanisław Ignacy Witkiewicz (1885–1939) of the Polish "Formist" group.

and I had the sensation of enduring a *Graecus ritus*[51] like having one's name and degree inscribed in the graduate record. We were worried by our sense that to an expert the flaws in this document must surely be obvious, but in the event it was only ever examined once by the half-witted Ukrainians serving for the time being as district authorities.

Mr. Zielony, my wife's uncle, visited us at the Fulińskis' a few times. He had endured some frightful experiences: he had been conscripted to a forced-labor contingent, thrown out of his apartment, slapped in the face in the street, among other indignities—and all because of his persistence in believing in the infallibility and omniscience of the Nazi clerical machine, and thus in the inevitability of his falling victim to it. Young Ludwik Sternbach[52] and his wife decided to go and live in the ghetto after, I think, they had found a room to share there with Mr. Auerbach. As for the ghetto, one should keep in mind that the Germans had there set aside for some 200,000 people[53] a district of Lwów normally accommodating only about 20,000 people—thus representing a tenfold increase over the normal population density. Leaving the city was punishable by death, yet because there were no checkpoints on the roads leading out of Lwów—a fact I verified in the course of several walks towards the edge of town, in particular to the Zielona Tollgate—thousands of people did in fact manage to quit Lwów by a great variety of means. But I myself was still having difficulties over the decision to assume a new identity and leave the city. In the meantime I grew a moustache such as a government clerk might sport, or perhaps a church organist.

I spent a considerable amount of time rooting around in my brother-in-law's library in their flat in Fuliński's house. I found a 100-year-old collection of English jokes, an edition of Saint-Simon's memoirs, a priceless book describing French customs in the time of Louis XV written by one of his courtiers, and a copy of the Marquis de Custine's excellent account of his travels through Russia a hundred years ago, of great perspicacity, proving to me once more how a great many of the ways of Tsarist Russia survived under the Bolshevik regime. There was a book containing a vivid description of an expedition into the Tatra Mountains[54] written by a Warsaw doctor of evident literary talent, a nature book about monkeys with fine hand-colored illustrations attesting to the taste and truly Chinese diligence of the illustrator, and books on Polish history going back to the eleventh century, and the Saxon period.[55]

---

[51] As opposed to a *ritus Romanus*—thus a rite imported from a foreign land.

[52] Former assistant—together with Marceli Stark, Jan Herzberg, and others—of Banach.

[53] A contemporary estimate put the number of Jews remaining at this time in Lwów at between 120,000 and 150,000.

[54] The highest mountain range of the Carpathians, forming a natural border between Poland and Slovakia.

[55] The so-called Saxon period of Polish history dates from 1697 to 1763, when the kings of the Polish-Lithuanian Commonwealth were also Electors of Saxony: August II the Strong and August III. At that time Saxon military forces were stationed in Poland, laying waste to the land

10 Homeless Wandering

I found a description of the conduct of the Saxon army at the time of August II that differed not a jot from that of the Nazis' occupation of Poland from mid-1941.

Although browsing among the books in my brother-in-law's library served to distract me from the reality of our situation, such was the horror of our case that it hardly provided relief from the everpresent feeling of dread. It was as if I was shut in a grim prison with but one ghastly thought on my mind. But I was determined: I simply would not bow to their demands, would not go where they were telling us to go. My wife continued to venture out now and then, but I now resolved to stay indoors or at most to go into the garden. The constant transporting of suitcases, now to the Indruchs', now back to the Fulińskis', the constant hiding when the doorbell sounded, the need to make absolutely sure the entrance door was securely locked, the monotonous walking for hours on end around the enclosed space of the garden, fruitless peering into the street and into neighboring yards to look for safe hiding places or pathways for making a quick exit, the continual view from the window over the front balcony of black sedans filled with Gestapo agents and military police, the noise of the incessant manoeuvering on Snopkowska Street of heavy military trucks travelling east or returning thence, the depressing effect of the continual bombardment by Nazi propaganda—the German press now representing our only source of information about the outside world since they had confiscated my short-wave radio—the unceasing news of arrests of friends and acquaintances, and on top of all this the rainy season setting in, when we would normally have spent weeks relaxing somewhere in the countryside. Thus for us our memories of Tarnowski Street are of the darkest period of the war.

We decided to seek advice from Mr. Bocheński, who had arrived in Lwów in the summer of 1941 as a German collaborator, but had soon reverted to calling the Germans "common bandits," having discovered this from his experience as chief administrator under them of his native town of Ponikwa, near Brody. He was a frequent guest at the Kamińskis' apartment on Linde Street, and through him we got to know the Kamińskis. They were goodhearted people ready to render any assistance they could in our present difficult circumstances. Mr. Kamiński, a forester by profession, was of the opinion that it was best to remain in or near Lwów. His view on the length of the war was, however, far too optimistic, not to say wildly wrong: he thought it would be over in about six months, by the spring of 1942.

The conquering heroes provided Lvovians[56] with ever more opportunities to appreciate their culture. On some streetcars there appeared signs reading *Für Juden zugelassen*[57], and others reading *Nur für Deutsche*[58]. The Ukrainians were learning the hard way that they were not exactly *de rigueur* racially: German police were observed slapping and kicking them—men and women indifferently—for getting

---

during domestic rebellions and the Great Northern War (1700–1721) in which August II embroiled Poland.

[56] Sometimes called "Leopolitans" after the Latin name "Leopolis" for Lwów.

[57] "May be used by Jews"

[58] "For Germans only"

on Germans-only streetcars. A simple-minded peasant without the presence of mind to get himself out of the way of a unit of German soldiers running along the street was killed by one of them with a blow of a rifle-butt to the head, and his body was then kicked out of the way into the gutter. Lvovians summed the situation up in the sentence: "Typhoid fever has gone but cholera has taken its place."

We heard via the grapevine something of what had been going on in Warsaw and Kraków: for example, that a group of professors from the Jagiellonian University of Kraków had been invited to a lecture on national-socialism, only to be arrested and deported after a mere three words had been uttered. From Warsaw came news of the death of Aleksander Rajchman, and a false alarm concerning Białobrzeski.[59] It was also rumored that the Nazis had shot Langevin and Borel in Paris, and that Lebesgue had died in America.[60] We also heard that my friend Jan Adamski had gone to his uncle's place in Wolica[61] to warn him of an impending sweep by the Germans searching for hidden weapons. He was arrested at the station when he returned and sent to Oświęcim,[62] where he died after a few months, unable to endure the torment.

As I mentioned earlier, before leaving our apartment in our house on Kadecka Street I had taken the precaution of preparing myself for a change of identity by destroying all personal documents and letters. Mr. Ciechanowski, the son-in-law of my neighbor Tomanek, agreed to store other documents, but the bulk of my papers— mostly notebooks and books, as well as many copies of *Die Fackel*—went to the Witkowskis'. Some issues of *Die Fackel* remained stored in the space above the kitchen, where they no doubt stayed undisturbed, but other treasured works by Karl Kraus were left in our library on Kadecka Street—not, as far as I know, that any of the Germans living in our flat ever took the slightest interest in them. To the apartment on Tarnowski Street I took only a few photographs, a complete set of *Studia Mathematica*—except, alas, for Volume V—and a typescript draft of a paper I had written in French during the Soviet occupation, where I gave a necessary and sufficient condition for it to be possible to cut three circular arcs in the plane in prescribed proportions by means of a single circle. But while we were living on Tarnowski Street I did no work whatsoever.

On November 23, 1941, one of Professor Fuliński's sons showed us a letter addressed to Fuliński that had been found on the step in front of the gate, out on the street, reading as follows: "We know that you are hiding the family of the former

---

[59]Czesław Białobrzeski (1878–1953), Polish physicist and philosopher.

[60]None of these rumors was true. The prominent French physicist Paul Langevin died in 1946, Émile Borel in 1956, and Henri Lebesgue died in Paris in 1941, but not at the hands of the Germans.

[61]A village near Jasło.

[62]In German, Auschwitz. A town in southern Poland, about 50 km west of Kraków. During World War II the German Nazis built there the largest of their concentration camps—a network of concentration and extermination camps consisting of Auschwitz I, Auschwitz II-Birkenau, and Auschwitz III-Monowitz, together with 45 satellite camps. Here people from all over German-occupied Europe were incarcerated. Various sources estimate the number of victims—gassed and then incinerated, on an industrial scale—at Auschwitz as between 2.5 and 4 million, 90 % of whom were Jews.

university professor St[einhaus], who has the German authorities on his tracks, and that you are using one of your son's names to hide the fact that Prof. Chw[istek], who also has the German authorities on his tracks, is also hiding in one of your apartments, and if Prof. Fuliński does not regularize this matter within one week, then it will be taken up by the 'Polish Committee', which will hold him responsible," signed "Polish Youth for Independence." It was interesting that whoever wrote this letter was unaware that "Prof. Chw." had left Lwów five months earlier. And the turn of phrase "has the German authorities on his tracks" and the reference to the "Polish Committee" were rather odd. It wasn't difficult to deduce that the author of the letter was the lad living in the Fulińskis' basement. This young man felt he had some sort of claim on Fuliński, and also felt resentment towards us because we did not buy food items from him, for the simple reason that when we had initially bought something he had shortchanged us—although Mrs. Fulińska stood up for him when we mentioned this to her.

Luckily we still had a place waiting for us in the Ottos' house in Rudno, and the passports organized for us by Wanda Żukotyńska. Since we did not wish to be at the mercy of a blackmailer even for a moment, we decided to leave. Thus we embraced our metamorphosis. I gave Fuliński my Soviet passport for him to hide up in the attic together with my paper on arcs and circles, we quickly packed the most necessary things, each put on two coats, and, after a heartfelt goodbye to the members of the Fuliński family, set off on foot in the direction of Zimna Woda,[63] each carrying a small suitcase. We happened to bump into Professor Kuryłowicz in the street, who gave us the news that Russia had recovered after the initial debacle, and that the deeper the Germans penetrated into Russia the more difficult it was becoming for them. At the streetcar stop we encountered Ms. Lusia O. who was working in the office of Mr. Walter, former commissioner of gendarmes for Jasło County, and now liaison officer between the Polish and German police in Lwów. Walter had given Ms. Lusia employment in his office under the assumed name Masłowska, to hide her Jewish background. Of course, we didn't tell her where we were going. From the terminal streetcar stop we continued on foot, with a peasant—a Pole—who agreed for a small sum to help us carry our things as long as we stayed on the main road. As I said, our goal was Zimna Woda, however not the well-known summer resort lying to the right of the road to Przemyśl, but the region to the left known as Rudno, and in Rudno the farthest hamlet, called Osiczyna. Since neither we nor our bearer knew the way, we spent a good half-hour or more *en route* going round in circles. During the long walk he conversed with us, mentioning in particular the German murders of Jews. When I observed that some people were happy about those murders, he responded: "Only the stupid are happy. If they do this now to others, you can be sure they'll do the same to you later."

---

[63]"Cold Water", a village near Lwów, about 15 km from the city's center.

Chapter 11
Osiczyna

The Ottos had not prepared for our arrival. Mrs. Otto was in bed. In the sitting room we encountered a tenant by the name of Halina Szpondrowska, whose husband was on active duty in a Canadian armored detachment. Before the war he had been chief of military intelligence in Silesia, had then been taken prisoner by the Soviets, escaped, and somehow reached Canada. They had two sons, the elder of whom, Staś, had graduated from high school during the Soviet occupation, while the younger, Andrzej, was just eight years old. She herself had fled to Osiczyna with her sons during the Soviet occupation because, as the wife of a former officer of the Polish army, they would certainly have deported her. Little Andrzej was so well trained that he never spoke about his father, and when asked at school about his father's occupation, said he was a "railroad man".

The Ottos had a 15-year-old daughter Elza. As I mentioned earlier, Witold Otto had been a clerk in the Lwów University bursar's office during the Soviet occupation, and now worked as caretaker and coach driver in a nearby sanitorium for children with tuberculosis in Brzuchowice, where his brother was both director and chief physician. His father, who had worked in the warehousing end of the railroad enterprise, was retired and living in Kraków. He had been born in Vienna, where he officially became a *Reichsdeutscher*, but his son Witold refused to partake of the privileges of a *Volksdeutscher*, adhering determinedly to his Polishness, which had been reinforced by his participation in the defense of Lwów in the Ukrainian-Polish and Soviet-Polish wars of 1918–1921, and by the death of another brother, a Polish army officer, in the Battle of Westerplatte in 1939.[1] Mrs. Otto had spent much of

---

[1] Westerplatte is a peninsula in Gdańsk (in 1939 outside the city proper), at the mouth of an estuary of the Vistula delta on the Baltic Coast. The Battle of Westerplatte was the first European battle of World War II. The defense of the peninsula from September 1 to September 7, 1939, by fewer than 200 Polish soldiers of the Military Transit Depot against a much stronger German force became a symbol of Polish resistance to the German invasion.

her life in Częstochowa, like Mrs. Szpondrowska, whom she knew because they happened to have both worked in the branch of the Bank of Poland there.

We were shown to a room where two single beds could hardly fit. The window sill served as a larder, and to this end the bottom window panes were papered over. Our standard of living, which had fallen during the Soviet occupation, and then dropped further rather precipitously when we quit our apartment on Kadecka Street, now, on November 24, 1941, fell yet again, bringing us to the brink of destitution.

On the very day we left Lwów, my daughter and her husband Janek had come to the Fulińskis' to see us, only to learn of our exodus. Thus, without so much as a goodbye, on November 23, 1941 we were separated from our daughter and my Mother for what was to turn out to be several years. A few days later, Janek came on foot from Lwów to see us. He brought the sad news that my wife's uncle Zielony had perished.[2] He had gone to live in the Lwów ghetto, and there, as someone over fifty, had been selected to be shot along with several thousand other elderly ghetto-dwellers. My friend Ludwik Oberländer had also been arrested—taken just as his wife was returning home from some errand. Our former Rector Beck[3] had decided to go into the ghetto, accompanied by his son and his son's wife, who, as a non-Jewess, need not have gone with them. From this news I understood that the young man who had written an anonymous letter against us had actually done us a favor in the sense that that act had been the final straw that decided us to quit Lwów. There was still the question of registering with the local authorities. This was settled on December 16 by Witold Otto thanks to his good relations with the district administrator, and also—to some extent at least—to the fact that my fake birth certificate identified me as being born into the Greek Orthodox religion, lending me considerable cachet, especially with the Ukrainians. Among other helpful things he declared that my wife was his cousin from Kraków.

We could now think of bringing our things from the apartment at the Fulińskis'. In this matter Mrs. Szpondrowska was very helpful, instructing a neighbor with horse and cart to go and fetch "her" things. Janek waited at Fuliński's house to hand our belongings over to the driver. Although the latter was stopped several times by Ukrainian militiamen, he managed to get our things to us intact. Mrs. Szpondrowska helped us a great deal in other ways also, since we couldn't go out to such places as restaurants, etc. for fear of running into people who knew us. We also refrained from obtaining ration cards so as to avoid having our names put on the lists on the basis of which people were recruited to do forced labor.

All of those crammed in the Ottos' house had to use the rather pokey kitchen, and this led to constant disagreements mainly because the mistress of the house skimped on fuel. She wanted enough rent from us to make a decent profit, and to

---

[2]Leon Zielony had been an officer in Piłsudski's Legions, and, refusing to heed the advice to hide given by a fellow Legionnaire, or to assume a false name and origins, perished with a group of people of advanced age liquidated in the Lwów ghetto.

[3]Adolf Beck (1863–1942), Polish physiologist. Professor at Lwów University 1895–1935, and Rector in the academic year 1912/13.

this end refused to specify a definite sum so as to keep the tenants indebted for an indeterminate amount that, apparently, she hoped to collect in stable currency once the war ended. She was a youngish, still attractive woman. She was constantly going to Lwów, either on foot or by horse and cart, to settle some matter or other, and was sometimes away for several days. When at home her main concern seemed to be that she not die of starvation; when there was food to be had, she ate enough for two or three, and when it was lacking she stayed in bed with her daughter all morning to conserve energy and fuel. There they passed the time singing litanies and plainchant. When not so engaged, her daughter, a very attractive girl, would spend the time leafing through periodicals such as *As* (Ace) or *Kino* (Cinema), etc., sometimes in the company of her friend Miss Paliwodzianka, who had similar interests, and often dropped in.

We soon got to know our neighbors on Piaskowa Street: Mrs. Żywiecka, a Ukrainian, who owned the peasant cottage opposite, Mr. Helon, who had some sort of job in Lwów, Mr. Leśko, a fitter and turner with the railroad, Mr. Mastelko, a retiree, and several others. There were no Germans of any kind—in particular soldiers or military police—in evidence anywhere in Osiczyna, for the simple reason that no road of any importance passed through it. The road from Lwów to Przemyśl was about a kilometer and a half away at its nearest approach. Military trucks travelled that road, as well as cars packed to their roofs with people and possessions transporting entire families to somewhere in the *Generalgouvernement*.[4] I tried to enter into my new non-intellectual identity, especially when a member of the intelligentsia hove into view. Over the period of seven months or so that we spent in Osiczyna I ran across no one I knew from our former lives, even though there were some of our Lwów acquaintance living not far from us—for instance Professor Szymkiewicz,[5] who was at that time staying in Zimna Woda. The probability of an untoward meeting was small since my sphere of acquaintance was, fortunately, quite narrowly defined.

The problem of fuel for cooking and heating was a rather acute one, because the Germans had taken over the administration of the forests, and had banned the sale of firewood to the populace. There was a similar ban on the selling of coal. Thus the only option left to us was that of foraging illegally in the forest for firewood. I undertook this task with pleasure since it meant going for walks in the woods; one soon developed skill at detecting the presence of the foresters and avoiding them. And the only tools needed for this occupation—in pursuit of which I spent at least an hour every day—were a pole and a rope.

* * *

Since Zimna Woda lay to the west of Lwów, the prevailing westerlies did not bring pollution from the city, with the result that the air was exceptionally pure.

---

[4]"General Government", the German name for the Polish territory under German rule during World War II. The districts of eastern Galicia were added in August 1941.

[5]Possibly Dezydery Szymkiewicz (1885–1948), Polish botanist and forestry expert.

Furthermore, it was sheltered from those winds by the forests lying just to the west, so that the air was also calm. The ground being of a sandy type, it didn't turn to mud after rain. The name "Rudno" derived from the reddish hue of the grasses covering the open steppe there. To the north the steppe descends, so that quite close to where we were staying there were bogs separating Osiczyna from a place they called Połonki. There was a sandy road to the west of Osiczyna which ran between two woods, which, near Brzuchowice, opened out into steppe. I could walk there for hours without encountering a soul, except occasionally perhaps a country woman engaged, like myself, in stealing wood. Sometimes little Andrzejek, the son of Halina Szpondrowska, would accompany me. He was very concerned to help his mother in every way as much as he could, and sometimes dragged a huge branch for quite a distance before crying in frustration when he became exhausted. But he never gave up.

Our meals consisted of morning coffee with milk, or sometimes soup made with fermented rye flour, at noon soup with peas and noodles, and in the evening a slice of bread with pork fat washed down with tea. My wife quickly learned to gather pine cones from the woods for heating, to bake bread in the wood stove, prepare potato soup, and even bake cakes if some special occasion justified a banquet, however frugal.

On Sundays Mr. Otto would come by means of a cart drawn by two horses from the sanitorium where he worked, in order to do what was necessary around the house. He was an active young man highly skilled in the use of axe, hammer, pliers, and awl, and set to dealing with the many things needing doing that had accumulated during the week. He chopped wood, fixed doors and windows, reattached the bucket to the pulley rope over the well, shovelled snow, and so on. I myself learned to chop wood, an art in which I got my first lessons from Andrzejek, who wielded the axe like a master. I was also the water bearer for the whole house, sometimes bringing ten buckets or more in a single day.

Staś Szpondrowski had worked in a railroad workshop under the Bolsheviks, and was forced to continue working there under the Germans. However, this did at least serve to protect him from being deported to Germany, and guaranteed him food rations, albeit meager. He left for work early each morning and returned late in the evening, so that during the day we kept his mother company—in part to afford her some protection from Tosia Otto, who continually pressured her into doing any hard or unpleasant work about the house that cropped up. Mrs. Otto was easily able to persuade Halina that in return for providing a roof over her head, she—Halina—should pay extra in the form of various of her possessions, such as clothing and suitcases, so that over the two years or so that Halina and her sons had been tenants of the Ottos, much of what she had possessed was now in Tosia's acquisitive hands. She also demanded of her and her sons additional payment in the form of various services such as housework, delivery of parcels, work on the grounds, etc. Staś rebelled against this, but Halina had lost all ability to resist, and acquiesced to everything. Given these circumstances it was surprising that she stayed on there, especially since her mother and sister, as well as many of her husband's former friends, were living in Częstochowa and apparently not doing so badly, so that she would have been certain of a warm welcome had she decided to look for refuge there. However, Tosia knew how to play on her fears, telling her that the Germans would eventually find her and arrest her and her sons, and then wreak vengeance on them for having a father who had been head of the Silesian Intelligence Service before the war. Another difficulty had to do with the fact that it was virtually impossible for Staś to quit his work. The treatment of the workers there by the German overseers was in the spirit we had become accustomed to: young and old alike were beaten with rubber truncheons for every mistake or misdemeanor. Staś endured the situation stoically for the sake of his mother and brother.

The janitor of the house in Lwów where Halina had lived before taking refuge in Osiczyna turned out to be an extraordinarily decent fellow. He had arranged that Staś maintain a sort of *pied-à-terre* there, which was also convenient for us since Janek could leave things there for us. Thus Mrs. Knaster left food items for us there several times, and all of these reached us intact on Staś's back. Halina's former apartment there was now occupied by some acquaintances of hers, and all of her furniture and other household items had been appropriated by a former servant, who now refused to surrender them, threatening to denounce her to the Germans should she persist in trying to reclaim them.

* * *

Soon Christmas would be upon us. The Ottos were planning a celebration to include, in addition to the tenants, two of their friends from Brzuchowice: a mechanic originally from Silesia and his son, both of whom worked in connection with the maintenance of central heating systems and taking care of horses.

We were occasionally visited by Mrs. Knaster, an extremely courageous—and optimistic—lady. Among other good deeds, she acted as letter carrier between her husband and myself. Before Christmas Janek and Lidka hired a truck to transport themselves and their things—including some of ours—to Warsaw, and my sister Irena, still in Kraków, found an apartment for our Mother and my eldest sister in Krzeszowice. That our Mother was spared we owe to Professor Fuliński for giving her the birth certificate of his deceased mother, Julia Widajewicz. Before leaving Lwów, Janek handed Mrs. Stanisława Noga, a former servant of the Blumenfelds, a few dozen "four-ruble" coins[6] to be buried. Our confidence in her was fully justified: despite difficulties and downright danger, she gave the prescribed sums to the appropriate people, always adhering to the letter of our instructions.

Janek continued with his transport business, travelling several times between Warsaw and Lwów, without, however, always having time to visit us. That winter was severe, the roads buried in snow and temperatures dropping as low as minus 40°. The Germans ordered all Jews to hand in their fur coats, and all and sundry to deliver up their skis and ski boots.

Information was hard to come by. One of our sources was a man by the name of Dziadura, formerly a prominent national Polish activist in Karwina[7] and the

---

[6]Possibly Austrian four-ducat gold coins.

[7]A Czech town on the river Olza, which now forms a segment of the border with Poland.

Zaolzie area,[8] where he had worked as foreman, but was now making ends meet by delivering sausage and cheese from house to house. He told me some things about the interwar politics conducted by Poland concerning the disputed Zaolzie region. He said that when things got too quiet, the Polish government would organize provocations against the Czechs living in the Polish part of the region by arranging for some youths to break a few windows in a local *Sokół*[9] or in a Polish Gymnasium, and then publishing outraged articles in the Polish *Ilustrowany Kurier Codzienny* about Czech barbarism. Mr. Dziadura took a limited view of the war: for him the Zaolzie issue was still of the utmost importance, and he saw himself as having a future role as a member of a government commissioned by Germany to run Zaolzie.

After a while I began to make mathematical notes—nothing too demanding since I had difficulty making the effort of concentration needed for serious work—jotting down questions and remarks that occurred to me. The winter being a very severe one, famine raged throughout the land. Many people broke up their furniture for kindling, not realizing that it was possible to scrounge for it in the woods, or else fearing to be observed on the way from Zimna Woda to the woods. Food prices rose continually, and transportation of foodstuffs between parishes was forbidden. One peasant who went in the direction of Tarnopol to buy wheat was caught by the Germans on the way back, had his load of grain confiscated, and was beaten and subjected to forced labor for several days. Thus it was not surprising that deaths from starvation or cold were common, especially among the elderly. Some survived by miracle, as it were.

On Christmas Eve little Andrzej pleaded with his mother to let him attend the nativity play put on at the school in Zimna Woda, and eventually she yielded. However, later in the evening a blizzard set in, and it became bitterly cold. It was getting late and there was still no sign of the child. The Ottos' house was three kilometers from the school, and he had to cover that distance in the dark without a light through snow up to an adult's armpits—deaths from exposure had indeed already occurred in such conditions—and we feared for his life. But suddenly there he was, dragging a huge fir branch "for heating for mother."

In Lwów the Germans continued with their arrests and murders. Mrs. Helonowa, wife of one of our neighbors, had started an enterprise whereby she bought quantities of milk, bread and other items locally, and transported them by train to Lwów where she sold them at a profit—a little extra that might make all the difference. Since her children were still quite little, she locked them in the house while she was away. Her husband spent all day at his job in a workshop in Lwów, and the Germans had several times confiscated foodstuffs found on such workers, so that it hardly made sense for him to help her in transporting her wares. However, one

---

[8] A region of Silesia now in the Czech Republic, in dispute between Poland and Czechoslovakia during the interwar period. "Zaolzie" means in Polish "land beyond the Olza".

[9] The Sokol (in Polish, *Sokół*, meaning "Falcon") movement is an international youth sports and gymnastics organization aimed at providing moral and intellectual, as well as physical, training for youth. Founded in Prague in 1862.

day the German military police, assisted by the Ukrainian militia and local Polish police, surrounded the main Lwów railroad station, and began hustling passengers and others who happened to be there, including Mrs. Helonowa, onto trucks, to be taken first to a camp and thence to Germany as forced labor. Happily, fortune smiled on Mrs. Helonowa: a Polish policeman simply took her to the exit and let her go. Many others were not so lucky, even those living in villages well away from urban areas; the Germans, assisted by the Ukrainian militia, now began a sweep of the villages, abducting young girls from their homes at night for deportation to Germany. Thus did the Ukrainians establish a reputation as villains and worse amongst the Poles and even many of their own kind. Although they had ceased forming voluntary military formations auxiliary to the German ones, they persisted in their play at militiamen in the countryside, since there the pickings were easier. The Ukrainian clergy seemed genuinely horrified by the murders of Poles, and launched appeals from their churches for their flock to put a stop to them. On the other hand, they made no mention of Jews, and their silence on this score had the effect of making them seem to condone the terrible massacres of Jews that the Ukrainians had perpetrated when the Germans had first come on the scene.

In addition to their Ukrainian counterparts, Polish police were also collaborating with the Germans, though mostly in the western parts of Poland. The Germans themselves had instituted a wide range of police forces: the *Schutzpolizei (Schupo), Sicherheitspolizei (Sipo), Kriminalpolizei (Kripo),* as well as *Feldgendarmerie, Schutzstaffel (SS) Polizei, Sonderdienst, Geheime Staatspolizei (Gestapo),* and *Wehrschutzpolizei.*[10] A German policeman could even aspire to the rank of general. It seemed they intended to arrest everyone in the world. Soon in or near Lwów they set up camps for the arrestees, shortly to be crammed with Jews, Poles, and Ukrainians, including women and children, and representatives of all stations in life: blackmarketeers, thieves, workers, members of the intelligentsia, and what have you—to put it briefly, they strove to get everybody they could behind barbed wire, where they subjected them to general privation and violence the ultimate purpose of which was to remain obscure for some time.

Meanwhile the German press rhapsodized over the orderliness, security, and cleanliness brought *in diesen Raum*[11]—that is, to our benighted land—at the end of a German bayonet. According to the *Lemberger Zeitung,* everyone had work, proper nourishment, medical care, and even recreation guaranteed them, and anyone unable to find work locally might apply for employment in Germany at wages more than sufficient for the basic necessities of life. At the same time

---

[10]Respectively, Defense Police (with the widest range of duties), Security Police, Criminal Police (who took over in, for example, a murder case after initial action had been taken by the *Schutzpolizei*), Military Police, SS (Protection Squadron) Police (specially formed to implement Nazi ideology, this branch was, under Heinrich Himmler's command, responsible for a great many of the "crimes against humanity" of World War II), Special Police (set up by Hans Frank and operating in the *Generalgouvernement* from 1940 to 1944), Secret State Police, military equivalent of the *Schutzpolizei,* and so on.

[11]"to this area"

the *Völkischer Beobachter*—the official paper of the Nazi party, available also in Lwów—contained stories such as the one about how a German peasant had befriended a Polish prisoner-of-war, invited him to sup at his table, and even tolerated his daughter's forming an intimate relationship with him—as a result of which the impertinent Pole had been sentenced to death, with his name and the verdict announced in the newspapers!

Once the severest frost had passed, we began tentatively to explore our surroundings. Spring in the countryside is different from the urban variety in particular in that there is always something fresh to attract one's interest. It began with the gradual melting of the snow lying on the marshes in Połonki. At sunset the snow sometimes took on green, blue, and pink tints, on the marshes glowing with the yellow and orange hues of embers. I had never before seen such beautiful colors. And the birds were beginning to sing songs of welcome to the spring. However, this did not prevent Halina from plunging into a profound depression, brought on by the realization that she was coming to the end of her means, and exacerbated by the examples of people who had died of starvation or frozen to death that our landlady Tosia Otto was constantly bringing before her.

From Mrs. Żywiecka, in the house across the street, we learned the significance of the deep hole filled with lime on the Ottos' property. It seems that sometime before the war Mrs. Żywiecka had been living in her house with a man by the name of Karolin, considerably older than her, who did odds and ends around her house and yard. He had left his wife, but she had eventually discovered his whereabouts, and when she turned up in the neighborhood, Tosia hired her as a servant purely to irritate Mrs. Żywiecka. Then, begrudging the woman her board and lodging—since she considered the work done for her of little value—she began to agitate for Karolin to pay his former wife alimony from the money that Mrs. Żywiecka must surely be paying him for the work he did for her, so that the wife could in turn reimburse her—Tosia—for the food she consumed at Tosia's table. So taken was she with the legalistic structure she had erected that she went so far as to sue Mrs. Żywiecka for a sum calculated to force the latter—who had no money to speak of—to hand over her little cottage. Her idea was that, in the event her scheme worked out, she would use the cottage as a laundry. The lime in the pit was there for making the whitewash for painting the future laundry. I tell this story—perhaps worthy of Maupassant—to flesh out my character portrait of Tosia Otto.

* * *

Yes, we read the German newspapers. The Polish papers issued under German oversight were impossible to stomach since German propaganda in the Polish language had an especially disgusting taste to it. German propaganda relied on a precise separation of things officially recognized to exist and things whose existence it was forbidden to even hint at. By adhering strictly to this method, that is, by imposing a strict censorship of the press, they could make all seem rosy while at the same time telling the truth—although only in part. Thus one read about the horse races going on in Lwów, a football match between the teams *Ost-Eisenbahner*[12] and *Deutsche Post Osten*[13], a concert to be given in Lwów by a violinist from Hamburg, Tiso's[14] visit to Ribbentrop, and many other such matters. Photographs of the Governor-General of Poland[15] opening exhibitions, receiving peasant delegations, and presiding over the inauguration of party celebrations took up almost as much space in these papers as shots of German tanks, of *Ritterkreuzträger*,[16] of German soldiers aiming point-blank at the reader or staring grimly at the camera from under steel helmets, of fleets of planes, and of great multitudes of bedraggled prisoners-of-war on the march, escorted by well-equipped German infantry. What one neither read about nor saw pictures of in these papers were the millions of starving prisoners-of-war and other camp inmates, the binding of wrists with barbed wire, the Governor-General's ordering of prisoners to dig their own graves, the murders of thousands of innocent women, children, and the elderly, how people, many clad only in their nightshirts, were transported packed side by side like sardines in boxcars, how the arrested were forced to lie for hours with any lifted head likely to be struck with a rifle butt....

Yes, there were pages and pages on National Socialism—its ideals, history and cultural mission—articles about German honesty, respect for the law, natural hygiene, and courage, pages charging the British with barbarism and the French with sybaritism, hundreds of truths issuing from the pens of columnists, among them poets, philosophers, and historians. But all of these noisome truths were mere beguiling, served but one single purpose—that of drowning out the deafening reality that was all about for everyone to hear and see and verify—if, indeed, he himself and his family were not already engulfed by it. One might find in those German papers sonnets composed by tender lyricists, aphorisms of deep thinkers, anecdotes of subtle *raconteurs*, articles by Dr. Goebbels and his disciples, popular science written by well-known biologists and physicists, disquisitions by economists on the current state of the cotton market—some concise and to the point, others prolix and stodgy—, prognoses of the world production of nickel and aluminum after the war, photographs of laughing lads and lasses, ever so strapping in their bathers, and of glamorous actresses. But it was all just like so much fluff manufactured to smother the groans of the millions being murdered in Kraków, Auschwitz, Majdanek

---

[12]"Eastern railroad workers"

[13]"German Post East"

[14]Jozef Tiso (1887–1947), Slovak priest and clerofascist leader of the Slovak State from 1939 to 1945. After the war he was hanged for his treasonable activities in support of Nazism.

[15]Hans Michael Frank (1900–1946), German lawyer, from 1939 to 1945 Governor-General of that part of occupied Poland not directly incorporated into the German Reich. One of his first operations as Governor-General was the so-called *AB Action* aimed at destroying Polish culture, during which more than 30,000 Polish intellectuals and members of the upper classes were arrested, over 7000 of whom were subsequently massacred. He also oversaw the segregation of Jews into ghettos and the use of Polish civilians as forced labor.

[16]Those who had won the Knight's Cross for bravery.

Lubelski,[17] in Lesienice, and a dozen other such sites. The German newspapers were perpetrating a massive swindle, a snow job, whereby they sought to convince by their coverage of such minor and mundane matters as, say, how hemp production was faring this year in the Philippines, or the hidden intricacies of Argentine politics, that they published everything worth reporting, that all was right with the world, that nothing as hideous as *that* could possibly be going on, like putrescence beneath a smooth rouged and powdered skin.

* * *

The snow was too deep that winter for any visitors to come to Osiczyna, but in the spring of 1942 Tosia began receiving visits by a youngish man dressed in rags and badly undernourished, and styling himself a "painter". Halina recognized him as a high-ranking Polish officer from the former Second Detachment. Then one day he suddenly turned up wearing the uniform of an officer of the SS, and began bragging that he personally had murdered many Jews. He said he was on his way to the eastern front, and asked the ladies to help him in a tobacco-smuggling operation. Not wanting to have anything to do with him, we never found out exactly what he was up to.

That spring news reached us of the mass deportations of Jews to Bełżec concentration camp.[18] Old people and the infirm, women and children, were being bundled into boxcars like so much baggage, and sent to the gas chambers there. The population at large was not much perturbed by this, first no doubt because they had troubles enough of their own, and second because they were nursing the resentment generated earlier by the pro-Soviet attitude of many Jews.

Around this time the Germans contracted with the private firm "Hoffman" of Lwów to build an airport at Skniłów, just outside Lwów. On the basis of the lists of names and addresses prepared earlier by the various community agencies, the *Arbeitsamt* had cards sent out to people ordering them to report immediately to the offices of the Hoffman firm on Romanowicz Street. Our neighbor Helon received such a card, and from that time on essentially became a slave of that firm. We, on the other hand, obtained no "invitation" to work for the Hoffman firm—probably because we had no ration cards, and in any case, we had been very careful to cover our tracks, and leave false leads. Those press-ganged into working for the firm were treated much as the Israelites building the pyramids were said to have been abused by their Egyptian overseers, and, indeed, foremen did walk about amongst the laborers striking anyone who showed the slightest sign of slacking off for an

---

[17]Called the "Lublin Concentration Camp" in Nazi documents, operational 1941–1944. It was situated in a quarter of Lublin called *Majdan Tatarski*, whence the name *Majdanek*. Although intended as a forced labor camp, over 78,000 people died there, of which some 59,000 were Jews.

[18]Extermination camp south of Bełżec, a village just off the Lublin–Lwów railroad, operational from March 1942 to late June 1943. It is estimated that over this period about 600,000 people, mainly Jews, were murdered there. The number of Poles killed there for hiding Jews is estimated at 1500.

instant, or had trouble lifting a stone or length of rail. Furthermore, the workday lasted ten hours and most had to trudge the four kilometers each way from Lwów. The mildest attempt at organized protest was met with threats to report the protesters to the police, who would almost certainly then arrange for the liquidation of the trouble-makers. Mrs. Helonowa told us that when her husband informed the bosses that he was suffering from a hernia, they told him to either get it fixed or shut up—and of course he was in no position to afford such an operation. The several thousand workers so employed were in fact doing unnecessary work, since the airport built by the Soviets would have served just as well. In any case, soon the work had progressed to the point where what they had achieved in the way of an airport could be put to use, and then from morning till evening heavy German transports flew in and out. Many flew over the Ottos' house, and at night we could see in the distance the red flares put up to show late arrivals the way.

However, such things hardly impinged on us since the pastoral spring we were experiencing was so full of gentle alarums, transformations, and wonders generally, that every day was packed with fresh delights. With each passing week the sky and the remnants of snow took on new colors, and new kinds of flowers sprouted from the dank earth: first anemones, then primroses, followed by the blue liverwort. Fresh species of birds put in an appearance, and every morning filled the woods with their variegated song. We were lucky to have time on our hands to experience such a spring.

On May 16, 1942, our son-in-law Janek paid us a visit; he had come to Lwów on business for a few days. Capricious Tosia refused to let him spend the night in an empty room on the second floor, so he had to bed down at a neighbor's. The upstairs rooms were now free because Halina Szpondrowska and her younger son had at last gone to stay with her sister M. Laskowska in Częstochowa. The sister had come for her. By some miracle, Staś had been dismissed from his job at the workshop, and was also gone within a month. We were sorry to lose our friends, especially little Andrzejek, but at least we had spent a pleasant Easter with them. Incidentally, at Easter I had learned a new art, namely that of making Easter eggs with pen and ink. You must first ever so patiently draw a necklace of little circles around the middle of the hard-boiled egg, then above and below this a chain of parentheses forming a sort of fishbone design, and so on and so on. After two hours of Chinese labor the Easter egg is ready.

* * *

I heard the cuckoo calling every day from dawn on. Then at last I spied it close at hand and observed its habit of flying soundlessly from one perch to another, and reproducing its call only when it alighted anew. Peewits appeared high above the mud flats, emitting their joyful note before beginning a spinning dive. They were adept at chasing off flocks of crows, and even threatened any person unwittingly trespassing near their nest by dive-bombing them from on high as if they meant to pierce them with their beaks. In the woods for the first time I saw yellow orioles—rather large, canary-colored, looking as if they had wandered in by mistake from the tropics. Their whistle reminded me of that of my old friend Jan Adamski, tormented to death in Auschwitz, who always used to announce his arrival at our house with just such a whistle, standing below our window in the Jasło marketplace. Back then I had learnt how to gather sorrel,[19] which the children in Osiczyna called "kwasok". There was plenty of it in Osiczyna, and it was easy to gather enough of it for soup for the two of us. A little later wild strawberries appeared, growing in dense low green shrubbery and glistening like raspberries, so we sometimes enjoyed the luxury of strawberries and cream.

The Grünhuts, in Truskawiec, never received the items of our winter things that we had dispatched to them, and we subsequently lost contact with them. When we were still in our house in Lwów they had wanted to come and live with us, not understanding that we had become *personae non gratae*—people of absolutely no importance—to such an extent, indeed, that we had been forced to become itinerant refugees from authority.

Janek informed us that he and Lidka were being forced to leave the apartment they first rented in Warsaw because the landlady had begun making a fuss about Lidka's identity, querying the authenticity of her passport—even though it contained a dozen or more official stamps and visas—and saying that she considered her a Jewess. In Warsaw Lidka's type of beauty—resembling that of May Wong[20]—stood out against the standard Mazurian sculpted noses and broad chins more than in Lwów, where miscegenation with Rusyns and Armenians had produced over time a greater variety of facial types, with the result that Lvovians were less sensitive to departures from the Mazurian norm.

A new guest arrived from Kraków, causing us some trepidation. This was Mr. Otto's—Witold's—father, who was, as I mentioned earlier, a *Reichsdeutscher*. Bergson[21] would have included him in the category of *Homo faber*[22]: although I greatly admired his son Witold for his skill at repairing such things as door latches, hinges, wooden fences, etc., he himself was incomparably more skilful, even despite his advanced age. He needed to be active, and immediately set to creating a vegetable garden, digging up the weeds, and carting good soil for it from dawn till dark. He made tools of the highest quality: he attached the head of a hammer to a haft, and the result was superior to any hammer one could buy, and the axe he sharpened was razor sharp. Only after he had taken on the job of fixing the gate did it finally close properly.

I did not talk politics with him, but, indirectly, through his son, I learned that he had tried to talk the latter into officially registering himself as *Volksdeutscher*.

---

[19] A green leaf vegetable resembling spinach.

[20] Anna May Wong (1905–1961), Chinese-American actress and film star. First Asian American to become an international star.

[21] Henri-Louis Bergson (1859–1941), French philosopher. In particular, he viewed immediate experience and intuition as more significant that rationalism and science for understanding reality. His most widely known work was *Creative Evolution*, in which he examined evolution from a general philosophical point of view. Nobel laureate for literature in 1927.

[22] "Man the maker (of things)"

Witold told me that his father refused to believe in the stories of German murders, dismissing them as Polish *Greuelpropaganda*. Since by this time the Germans had already sent hundreds of thousands to untimely graves, I could not but feel that in this case at least *Homo faber* was not *Homo sapiens*.[23] However, from conversations with him I understood that he had doubts as to an ultimate German victory.

And in fact already by the spring of 1942 it was becoming clear that the Soviet Union was proving a more formidable foe than predicted, and that the Germans were becoming the more bogged down the further they penetrated into Russia. The dreadful winter of 1941–1942, long and harsh beyond all expectations, had terrified the average German soldier and weakened his spirit. And then the entry of America into the war in December 1941 must have considerably dampened the hopes of victory of any German able to reason, since Japan—still neutral *vis-à-vis* the Soviet Union—could hardly be considered a counterweight to American military might. Yet the Ukrainian militia—or at least the half-educated country bumpkins making up the bulk of it—were saying: "America! What's the fuss about?! Those few little Jews!" Such outbursts might be taken to demonstrate their unfathomable stupidity especially clearly if one recalled that Ukrainian emigration to the US had been substantial, and many of them had seen the manifestations of American greatness with their own eyes.

In any case, Witold withstood his father's suasions and did not become a *Volksdeutscher*—one of many eligible Poles who resisted the temptation represented in particular by the ration card that came with the appellation. In Rudno there was even such a one who spoke Polish poorly—with a strong German accent—but still lined up bravely with the locals for the lean ration of someone not of the *Volk*. Incidentally, one might have thought that the fact that the Ukrainians had to make do with that same meager food allowance would have brought them to their senses.

Ms. Kamila Speidlówna—Kama—was another person who, although of German origin, adhered to her acquired Polishness. Her family and that of the Ottos were linked by a great many years' friendship. During the years of Polish independence between the wars, she had worked in the Polish post office, and now subsisted on her pension. She was lame in one leg, but hobbled about everywhere by herself, stubbornly refusing help. She visited us quite often in Tosia's absence, and it was she who greatly facilitated our eventual departure from Osiczyna.

Also that spring, some peasants who had travelled to the regions around Rzeszów and Tarnopol[24] looking for corn to buy told us of encounters on the way with armed bands which sometimes took their shoes and other items, and sometimes let them continue on their way unmolested. These were gangs living in the forests, apparently made up of Ukrainians and sometimes also Soviets and Germans who had deserted from their respective Soviet and German armies, although no one seemed to know their exact composition or their aims.

---

[23] "Man the wise"

[24] Rzeszów and Tarnopol lie respectively west and east of Lwów at about the same distance ($\approx$ 130 km) from it. Tarnopol—Ternopil in Ukrainian—is now in Ukraine.

I myself walked in the woods every day, and of course *my* aims were totally peaceful. Once when searching in a grove for wild strawberries, I frightened a young goshawk, which got caught in a tangle of branches of a bush in its hurry to get away. On another occasion I was frightened by what sounded like a rifle bullet whistling close above my head. But it was just a little bird diving at me and emitting its piercing whistle just as it flew over my head. It then alighted on a branch a few meters away and glared at me. When I shifted my gaze he launched himself at me again, aiming right between my eyes, but passing overhead at the very last moment. Here was a bird that could cause harm, so determined was it to defend its nest—and much better prepared for the dangers of life, it would seem, than the feckless goshawk, which I later saw again in the same place as before. In groves such as that one, one could lie undisturbed for hours staring up at the sky. No one ever came that way—not even the foresters—and the sandy road a few hundred meters away was used only by local peasants. No motorized vehicles at all—let alone military ones—ever seemed to use that road.

Over the seven and a half months of our sojourn in Osiczyna, we never saw gendarmes or police of any stripe. We would have remained there but for the zeal of an official in the district administration by the name of Bułat, whose job it was to take a census of the local populace, and who insisted on including us on his list. He told us—and perhaps even believed it himself—that those officially registered as inhabitants would be the beneficiaries of various kinds of largesse, in particular fantastic "allotments" of property, under the rational system imposed by the Germans. In vain did I try to avoid registration by telling him that we were about to leave the district. This Pole lived in fear of being denounced to the Germans by the Ukrainians. His wife called him "old muzzle-head" in my presence, infuriated by his stupidity and cowardice. However, I am grateful to him because he caused us to make up our minds about leaving more speedily. We now had to make sure we left before I received an order to report for work, since this would mean going with him to Romanowicz Street in Lwów. Through their connections, our friends Ms. Kama Speidlówna and [name illegible] put us in touch with a certain Staś Sokólski who worked in the railroad workshops and was prepared to help us make the trip to Stróże, in particular buy us train tickets in Lwów—since they were not sold anywhere in Zimna Woda—and enlist the aid of certain of his coworkers to accompany us and ensure we had places on the train. Such measures were necessary because the trains were always clogged with passengers, many of them peasants—mostly old peasant women—and young ladies from the suburbs of Lwów, all travelling from Lwów to Dębica, where one could purchase goods much more cheaply than in Lwów, and back to Lwów to take their profit. Thus did these "profiteers", as they were called, earn a living.

Tosia was away at the time of our leave-taking. As usual she had suddenly decided to go off on one of her jaunts to Lwów, leaving the house in my care. She was against our leaving because we looked after the house so reliably during her absences in Lwów or Brzuchowice. Through the efforts of her husband Witold, their daughter had by this time obtained a position of some kind in Weigl's Institute for anti-typhus vaccine production in Lwów, exempting her from recruitment into forced labor contingents.

Feeling some urgency about leaving, we did not want to wait for Tosia's return. We carried the few items of furniture that we had borrowed from Halina over to Mrs. Żywiecka's for safekeeping, and Mr. Helon took over to Kama the things we had accumulated during our time in Osiczyna over and above what we could take with us in our suitcases. Thus on the morning of July 11, 1942, after leaving the various household keys together with a farewell note in our room and the keys to our room with a neighbor friend of the Ottos, we went to the station. It was fortunate for us that we did leave thus precipitately, for, as we were later informed by Kama, the order to report for work came on July 13. We were aiming for Stróże, where my wife's mother and brother had been living since early March 1941 on a farm a kilometer from the station. We had been able to keep in touch with them, so they knew we were coming, and we had even been able to agree beforehand on new names. We were to be one family, this being perhaps best in view of the issue of *Generalgouvernement* identity cards called *Kennkarten*, which since June 1941 had been a source of concern to almost all those with reasons for avoiding contact with the German administration.

At the station Staś pushed us into a carriage, and found a place for himself in a compartment adjacent to ours. It was a case of standing room only: the press of passengers was quite extraordinary, the more so for us who had been living in relative isolation. It would seem that most of our fellow passengers were either "profiteers" or workers of one kind or another. The press of people was such that tempers were frayed, and soon we heard one of the workers shouting at someone he considered a "profiteer" that soon there would be a ruling prohibiting train travel to those without work cards issued by the *Arbeitsamt*. The object of this outburst retaliated with a crushing argument to the effect that things had come to a pretty pass when now people objected to Poles travelling by train, whereas before the war even Jews did so without anyone objecting. At one stop we were joined by a well-dressed, clean-shaven young man, with hair neatly trimmed, and sporting a celluloid armband with the blue Star of David. It appeared that he had certain documents permitting travel despite being Jewish. He began a discussion about how many Jews the Germans had killed so far in Poland, claiming, in particular, that this number exceeded the number of Poles murdered. The same argumentative Polish worker opined that the Polish victims were more numerous. At another stop a fearless young Pole boarded the train and began singing a patriotic Polish lament about the devastation of Warsaw by the Germans. Buoyed by his bold show of patriotism, many of his fellow travellers gladly gave him money for his performance.

There was a man standing not far off in our compartment who seemed to be possessed of "St. Vitus' Dance": he continually wriggled and jerked and between spasms handed the two children with him tidbits from a paper bag. It needed no expert to determine that he belonged to the biological order of Primates. He was on his way to Warsaw. In one corner there sat a blond woman with a small child. When her face was in repose she looked like a fairly typical peasant woman, but then at frequent intervals she shouted "Good heavens!", and her evident distress—

or derangement—made one want to avoid catching her eye. As the train was moving out of the station at Przemyśl, a woman jumped on the steps of our carriage. At first nobody wanted to let her in because of the crowding, but then suddenly one of the passengers said "She has to travel," as if he understood something important about her, whereupon comprehension seemed to dawn also on the rest, and they made way for her. There was even a Gypsy in our compartment, abused by those around him as a born thief like all of his kind. I worked my way over to him and started a conversation with him, from which I learned that by long tradition Gypsies may lead a settled life for at most three years. I found it interesting that this nation that recognizes no laws but their own, leading a nomadic existence eked out by thievery, beggary, prostitution, and fraud, still abides by its age-old laws.

On the platform at Przemyśl I saw two German officers, elegantly uniformed, the younger of the two very handsome indeed, pulling at their gloves, smiling ironically as they gazed at the scene of carriages crammed with shabbily dressed, undernourished, and exhausted people. They oozed a sense of security amidst turmoil and poverty—like Englishmen lording it over the natives in the colonies. I've no doubt that in the contrast between their serene, well cared-for miens and the harried, hang-dog appearance of the travellers, they saw but further evidence of the superiority of the *Herrenvolk* over the natives.

In Dębica, we had hoped to be able to remain on the platform and wait for the train to Stróże. However, we were forced instead to pass through a wicket into a yard where a search of passengers' belongings was being conducted. There a uniformed *Bahnschutz*[25] insisted on opening my suitcase by himself, which was impossible since the lock was damaged and I was the only one who knew how to manage it. But when I approached him as he wrestled with it, he shouted *Scher dich weg!*,[26] thus addressing me in the familiar "*du*" form despite my age and the fact that he must certainly have been aware that he was not talking to someone used to being so addressed. Happily, in the end he became convinced we were not transporting forbidden materials and let us proceed. Around me I saw many who had some or even all of their belongings taken from them, though without manifesting great distress at the losses. Evidently a certain resignation had set in, from which I inferred that the German custom of robbing people at railroad stations and along the roads had been standard practice west of the San River since 1939.[27] (At one station we had seen the badly mutilated body of a woman lying across the track—evidently she had tried to run from the *Bahnschutz* and had fallen under a train.) Next we were told to form a line as if for inspection, and as a German officer walked slowly down the line, a couple of Polish policemen demonstrated their zeal to please their superiors by hitting some of those lined up with sticks, apparently at random. This

---

[25] Member of the railroad security personnel.

[26] "Get away!"

[27] That is, in that region of Poland occupied by the Germans since September 1939. Przemyśl lies on the San River, which formed part of the agreed boundary between the Soviet- and German-controlled regions of Poland from September 1939 to June 1941.

was the first time I saw Polish policemen in the role of German stooges. I recalled the official pronouncement of a former mayor of Lwów on the proper role of the police: they exist to provide a bulwark of the state. Thus it was that we, the non-uniformed citizens, were suspect unless cleared by the police, and thus at their mercy both in and out of court.

We spent a few hours waiting in Tarnów, where we parted with our guide Staś Sokólski. We arranged for him to take care of the things we had left behind with Kama Speidlówna in Zimna Woda, but he didn't do it, and later we heard that he had complained that it had not paid him to facilitate our trip and act as guide.

Interlude: Flashes of Memory

The Soviets seemed greatly surprised by the practice they observed in Lwów of keeping dogs as pets—perhaps because they themselves had never been in a situation where there was a surplus of means sufficient to allow for the feeding of a pet. When they appropriated cattle for meat for their troops, they didn't bother to feed them, reasoning that they would in any case be slaughtered in a couple of days.

* * *

When the Polish troops fought for possession of Narvik[1] in April and May 1940, our Lwów communists said that it was reminiscent of the Polish escapade in Santo Domingo.[2] This idiotically inapt comparison represents one of the best pieces of evidence for the destructive effect of dialectical materialism on human brain tissue. At Narvik the Polish army was fighting the same enemy it had fought near Warsaw, an enemy that had proclaimed the annihilation of the Polish State and the extermination of its peoples, whereas in Santo Domingo the Polish legions fought to

---

[1] The battles of Narvik, Norway, were fought from April 9 to June 8, 1940, over control of Narvik's ice-free port. The naval battle was fought in the Ofotfjord between the British Royal Navy and the German Kriegsmarine, and the two-month land campaign by Norwegian, French, British, and Polish troops against German and Austrian ones.

[2] After the Third Partition of Poland in 1795, many Polish officers, soldiers, and volunteers left Poland and formed "Polish legions" in support of Napoleon, assuming that he would come to Poland's aid against the usurping nations. During the Napoleonic wars they saw combat in the West Indies, Italy, and Egypt, and fought alongside the French army in the invasion of Russia in 1812. In 1801 Napoleon sent a large French and Polish contingent to Santo Domingo to topple the rebel black leader Toussaint L'Ouverture. By 1803 the French and Polish troops had been defeated—most dying of dengue fever—and an independent state of Haiti established.

defend France's colonial interests against former black slaves who had never heard of Poland, in the vain hope that France would someday show its gratitude.

* * *

I have personal memories of the arrests of the following by the Bolsheviks: Wacław Grubiński, Wojciech Skuza, Paweł Hertz, Leopold Lewin, Teodor Parnicki, Aleksander Wat, and Tadeusz Peiper.

* * *

I first met Mrs. M. Rem. in 1929, and after we had talked for ten minutes, I informed her that I simply had to go on a solitary walk with her, to which she instantly agreed. Thus some days later we set off walking together in the direction of Ż. Once we had reached the open countryside, Mrs. Rem. turned to me and said "I must ask you for one thing." I interrupted her: "No need to say anything more because I know what you have in mind—a rather strange and unusual thing to say, I know." And that's how it was.

* * *

One morning the Soviets announced that since Polish złotys had become valueless, anyone in possession of these should exchange them for rubles in the State Bank, with that same evening as the deadline. But since hundreds of thousands of Lvovians had złotys, and moreover the announcement had come on a workday, it was utterly impossible for more than a few scores of people to exchange their złotys in the six hours allotted. Just the day before many workers had been paid by their employers in złotys, and now these were declared valueless. When these workers went back to their bosses demanding that the pay clerks do the exchanging, the following typical exchange—so to speak—is reported to have taken place between them and the *politruki*:

– Why do you want to exchange this money?
– Because it has been declared valueless as of tomorrow.
– If that is so, how can you expect Soviet State enterprises to accept valueless currency into their cash reserves?

* * *

The Bolsheviks left Lwów in a hurry, so didn't succeed in destroying all of their documents. Among those left intact were several photograph albums compiled by the NKVD. For the most part these contained photographs of people in the street, or in a streetcar, or sitting in a café, apparently unaware of being snapped. In some cases the same person had been photographed several times—in profile, *en face*, alone at a café table, or in company. There were a variety of subjects: for example, young ladies pushing prams or accompanied by children, Soviet officers, and, strangest of all, NKVD officers in uniform. Perhaps the NKVD suspected some of its own members of not adhering to the party line—or was it merely a question of monitoring the company they kept?

* * *

During the Soviet occupation of Lwów I wrote to M. K.[3] in Baltimore[4] a few times. In order to convey a better idea of that regime, I adopted various disguises in my letters. In one I pretended to be a provincial Jewess telling him about misbehaving children, and in another an insane philologist writing in Latin about the things he had found in the diary of a medieval Spanish saint. It is certain that the Soviets didn't catch on, and that both M. K. and M. B. passed on what I wrote to the appropriate people. Thus I may in this way have contributed in some small degree to the shaping of Uncle Sam's opinion of Uncle Joe.

* * *

In September 1939, while listening to the radio at Karol Kossak's, an alto female voice, trembling with emotion, came on the air. The speaker was a Frenchwoman, and spoke as superbly as only good Parisian actresses can. She gave out that she was appealing to her own circle of acquaintance. Her words were as follows:

— *Je te vois, Roger, tu reconnais ma voix, et toi, Paul, et toi, Maurice. Vous savez bien que je dis la vérité. Les Allemands sont des hommes comme vous, ils ont des soeurs et des femmes, des mères et des fils comme vous; pourquoi se laisser tuer et les tuer, pourquoi?*[5]

Then, as she was speaking from a German radio station, she added:

— *On ne m'a pas appelée ici. Je suis venue seule.*[6]

All one can say about this is that it is the case of a ferocious beast of prey engaging a French whore to batten on French pacifism.

---

[3] Marek Kac.

[4] Kac was actually at Cornell University from 1939 to 1961.

[5] "I see you, Roger, you recognize my voice, and you, Paul, and you, Maurice. You know well that I speak the truth. The Germans are men like you, they have sisters and wives, mothers and sons; why get yourself killed and why kill them, why?"

[6] "No one summoned me here. I came by myself."

Chapter 12
Stróże

We arrived in Stróże on the same day we left Osiczyna, that is, on July 11, 1942. At the station a lad offered to take our luggage to Berdechów[1] in a hand cart. On the way we met a couple walking in the opposite direction, towards Stróże, who turned out to be Tadeusz Cieluch and his wife, the son and daughter-in-law of the owners of the farm where we were to stay. We asked them about Mrs. Szmoszowa,[2] they indicated the way, and we were soon in Berdechów. Fortunately the room adjacent to the little flat occupied by my mother-in-law had become vacant just the day before, so right away we had a place to ourselves, which only needed sweeping and painting white. Neither we nor our landlords the Cieluchs imagined for a moment that we were destined to stay there several years. Cieluch senior, owner of the farm, had been a founder of the Polish People's Movement when Austria ruled southern Poland, and then after World War I, when Poland became independent, he had become vice-governor of the local administrative district, and a deputy to the *Sejm* during Witos'[3] rule. He had three sons: Jan, Tadeusz, and Artur. Jan was in Kraków still working in some capacity in the treasury, while Tadeusz and Artur helped their father on the farm. These two had been investigated for allegedly helping Polish officers escape to Hungary in the first days of the war in the autumn of 1939, were imprisoned for a time in a prison in Nowy Sącz, but were ultimately released for lack of evidence. The farm consisted of about 60 morgs of land of which about 40 morgs was productive. Since the farm was not mechanized, the work of running it was peasant work, that is, done with scythe or axe in hand, and therefore labor-intensive. There being few servants or hired hands, the whole family pitched in: Tadeusz, Artur, their sister Zosia, another sister, the oldest of the children, now Mrs. Tokarzowa, widow of a

---

[1] About 12 km from the station at Stróże.

[2] Steinhaus's mother-in-law.

[3] Wincenty Witos (1874–1945), a prominent member of the Polish People's Party from 1895, member of parliament in the Galician Sejm 1908–1914. Thrice prime minister of Poland: in 1920–1921, 1923, and 1926.

worker in the railroad warehouse, and Artur's wife, a woman of ordinary peasant stock who achieved twice as much as anyone else. Even Tadeusz' wife Stefa, a cashier in the railroad station at Stróże, devoted every spare minute of her time to tending the garden. The cottage in which we were to live had once served as the living quarters of hired farm laborers.

We were very warmly received. The relations between my mother-in-law and brother-in-law on the one hand and the Cieluch family in the "manor house" on the other had been settled to everyone's satisfaction a year earlier. It was Stefan Dobrzański, brother to Jan Dobrzański, who had found this place of refuge for them. Our connection to the latter was through Krysia Szmoszowa, the wife of my brother-in-law and sister to Jan Dobrzański's wife Maniuta. She had died in 1935. When serving in the Polish army pre-war, Stefan had been on friendly terms with a subordinate hailing from Stróże by the name of Schoenemann, and it was he who originally suggested the farm as a possible place of refuge for my mother-in-law and brother-in-law. Of Witold Schoenemann there remained no trace: knowing he was under investigation by the German police, he had fled, and when they came looking for him to the house of his wife's family, the Igielskis in Stróże, they set to interrogating his wife Stasia in their usual boorish manner, addressing her with the familiar "thou", and boasting that they could arrest her. Alas, she lost her head and said "Then go ahead and arrest me!", so they did arrest her, and she ended up in the Ravensbrück women's concentration camp.[4]

The Igielski household was run by Mrs. Dela Ochalikowa, the second oldest of the Igielski girls, whose husband, a former Polish officer, was in a prisoner-of-war camp. A lady by the name of Mrs. Tosia Michałowska was living with them; her husband was also in a prisoner-of-war camp, but a different one from that where Dela's husband languished. In addition to the elder Igielskis, their daughter Dela, and Tosia, we also met the young Mr. Igielski, Janek, who had served in the Polish War Ministry before the war, but of necessity had had to move back to his parents' place with his wife, a young lady from Warsaw, and now worked as a clerk in a local sawmill called "Kindler and Red." There were also Witold Schoenemann's two daughters, the grown-up Irka, and Kocia, just beginning school, and 12-year-old Janusz, Dela's son. If I spend too much time mentioning all the members of this family, it is because they were all so extremely kind to us, helping us orient ourselves, showing us how and where we could obtain the necessities of life, receiving us hospitably, and, above all, advising us on the crucially important matter of appropriate documents.

The closest house to the one we were staying in was occupied by a Mr. Stanisław Woźnica with his wife and two children. He had been removed from his farm

---

[4]This camp, planned by SS leader Heinrich Himmler himself, was located near the village of Ravensbrück, about 90 km north of Berlin, and was operational from 1939 to 1945. It is estimated that over 130,000 female prisoners, from almost all European nations (Polish women accounting for around 25 %), passed through the camp, of whom only between 15,000 and 30,000 survived. Medical experiments were conducted on the inmates. (See Stasia's report below.)

at Koło, near Kalisz, and given the job of buying eggs and vegetables for the *Ernährungsamt*.[5] The house he lived in, at a crossroads, had formerly belonged to a Jew, who, along with all other Jews of the surrounding villages, had been ghettoized in Bobowa, the chief town of the parish. Every now and then a contingent of German policemen would go there for their sport, namely to shoot a dozen people in the street, after which they would enjoy a feast—to which end tables and chairs, tablecloths and napkins, and food and drink were to be supplied by the sequestered Jews—before returning to Gorlice. Most of those Polish locals who actually witnessed these goings-on were horrified, but those who did not see such events with their own eyes were by and large indifferent. Some there were who even expressed satisfaction, sensing the approach of the long-anticipated time when businesses in Poland would at last be run by aboriginal Poles—such as Mr. Woźnica.

Summer being in full force, we went several times on outings to the river with Dela and Janek Igielski and his wife. On the river bank there was a dense growth of willows, between which grew long-stemmed flowers, competing for space with vines and fragrant growths of thyme and mint. The far bank was thickly wooded with a variety of trees. The clear water flowed rapidly over a rocky bed, and a dip seemed to remove a film from the eyes just as a swim in the Wisłoka used to do. And if one lay on one's back staring at the sky, and forgot for the moment about the Germans, one could indeed imagine one was back in Jasło lazing by the Wisłoka.

Mr. Ząbek, a friend of the Igielskis, came to visit, with news of Jasło. We already knew that the Germans had paved over the marketplace and asphalted the main streets, because my brother-in-law had been in the area in the summer of 1941. Ząbek relayed the circumstances of the arrest of Janek Adamski, that loyal friend, boon companion, and favorite son, tortured to death in Oświęcim. It seems he heard that his uncle's servants had reported to the Germans that he, the uncle, had weapons—probably hunting rifles—buried on his farm at Wolica. He had gone to the railroad station to warn his uncle of the danger of arrest and himself been stopped and arrested.

* * *

I observed from a safe distance—second hand—the official activities of the Germans in the general area. At that time—mid-1942—they had not yet given up on their ultimate goals. German officials in civvies were constantly travelling about in cars or buggies photographing ragged village children—by way of documenting Polish beggary—and registering those peasants with names of German origin. In fact they consider as German not only obviously non-Polish names like "Augustyn" or "Kielor", but also those like "Kowal",[6] presumably taking this to be a Polonization of "Schmidt". It was feared that everyone would be issued with identity cards identifying some as acceptable, and others as inferior or even expendable. These busy census-takers also visited parish registries and archives looking for births registered in Germanic names and records proving that certain families or

---

[5]"Nutrition Office"

[6]"Smith"

villages were of German origin. The larger farm holdings were examined—possibly with a view to offering them as rewards to officers who had served on the front lines. Such inferences were easy to make if one read articles in the *Krakauer Zeitung*[7] about the German hunger for *Lebensraum*[8] east of the Vistula. However, this bureaucratic folderol was the more innocent side of their activity, the arrests and murders continuing apace in Gorlice, Grybów, Nowy Sącz, and Jasło constituting the other half of their coin. The district head in Nowy Sącz was the infamous Hamann,[9] an appalling torturer, who assembled as henchmen a set of unspeakable thugs and killers.

In the Bobowa parish where we were living these horrors were less in evidence because villages and towns were relatively much sparser—in addition to which we were blessed with an official who employed all the ingenuity and courage he could muster to frustrate German aims, saving hundreds of people from prison, torture, and death. This was Mr. Laska, district police commissioner for Bobowa. He took it upon himself to warn a quarry of the impending arrival of the dog-catchers, so that he might absent himself—who knew where?—when they called. Old Mr. Cieluch and his son Artur were among those he periodically warned, following which Artur would do a vanishing trick and his father go off to a hiding place a short distance away. However, he found it hard to endure the exile for very long, and, despite repeated warnings, returned home too soon. Thus it was that on one October morning a Polish police officer called Kubala took a shot at a 75-year-old man who had climbed out of a window and was fleeing towards the forest, and brought him down, though not fatally. A few years younger and he'd have saved himself in the dense woods. His son Jan was also arrested in Kraków. These events depressed us greatly. We were told that the Gestapo had been searching for old Mr. Cieluch's son-in-law Mr. Garlicki since 1939, but he had always managed to evade them. His wife remained in Zawoja with their son Maryś, who at war's end had risen to the rank of lieutenant,[10] while their daughter Lidka and little son Oleś stayed with their grandparents in Berdechów.

Ever more terrible crimes were being perpetrated by the German occupiers. In Rzepiennik, a village not far from Gorlice, the Germans discovered an unregistered doctor hiding in a barn, and by way of punishment immediately had every member of the family hiding him shot. That the Nazis indeed planned a "final solution of the Jewish question" had by now become abundantly clear. The district administrator for

---

[7]"The Kraków Times", a German-language newspaper published intermittently in Kraków from 1799 to 1945. Over the period 1939–1945 it was controlled by the occupying Nazis.

[8]"living space"

[9]Heinrich Hamann (1908–1993), Border Police Commissioner in Nowy Sącz from late 1939 to mid-1943, whence he was transferred to Jasło to work with the Security Police. Sentenced to life imprisonment in 1966 by the District Court of Bochum for the murder of hundreds of Jews and others, and for the deportation of people living in the ghetto in Nowy Sącz to Bełżec extermination camp in August 1942.

[10]Possibly in the AK.

the region around Rzepiennik witnessed one more such execution. It was organized with typical German thoroughness, practicality, and perverted love of order: the inhabitants were divided up into groups, and while one group was being executed, the next were digging their graves, while a third undressed, and a fourth was being marched to the place of execution. The elderly, young girls, and children all went to their deaths naked, because, having taken all their possessions, the Germans must strip them even of their clothes.

Another massacre was carried out in the village Biała, near Tarnów, on the river Biała. There the Germans ordered their intended victims to take money and some possessions in a suitcase with them, lulling them into thinking they would be pursuing some kind of life in the camps. Then on arrival at their grave, they were forced to hand over the money and their few possessions and subjected to the sickeningly crude irony of such statements as "You won't need money where you're going since Jehovah doesn't allow beer." Many groups of Jews were trucked to Gorlice, and from there to various secret destinations, many to Bełżec, where a facility was being built for the wholesale destruction of people. Since apart from the driver there was at most a single soldier riding in the cabin of each truck, it was naturally asked why the victims did not disarm their solitary guard and make a run for it. However, these people had suffered so much over the previous few years, continually humiliated and threatened from all sides, robbed of their possessions, constantly exposed to rumors of what lay in store for them, that their will had been completely undermined.

Apart from those who witnessed from close at hand how these people were led to an undeserved death—experiencing first-hand the horror of this unimaginable hecatomb[11]—most of Polish society was indifferent to their fate. And in the annals of evil done by mankind a large chapter will be devoted to the behavior of the Polish police during the German occupation, which was such as to make one think they were trying to outdo their masters in perfidy. They were zealous in beating prisoners confined in narrow spaces, hunting down and yanking refugees from the garrets, cellars, and even toilets where they were hiding, and delivering them up to the Germans. The Polish police officer Kubala, whom I mentioned earlier, in addition to distinguishing himself by his cruelty, used his position to extort money from his victims, or the gold and bits of jewellery that they had been hoarding against the ultimate emergency. A lesser goon, Krzeszowski, a policeman in Polna,[12] also accumulated a fortune by means of extortion and bribery. Once when a gentile woman brought butter into the Jewish ghetto in Polna, she was caught by Krzeszowski's men, shut up for the night in a synagogue with an old Jew, and finally, in the morning, paraded through the market place with two ragged Jews to the accompaniment of the laughter and catcalling of the rabble. When her husband

---

[11] In Ancient Greece, a sacrifice of a hundred cattle to the gods. It now means any large-scale slaughter or massacre.

[12] A village about 7 km from Berdechów.

offered a few thousand złotys to buy her out of captivity, he was refused. She soon lost her mind.

That summer of 1942 we were visited by my niece Basia Zglińska, the daughter of my sister Irena. This was the first time I had actually gotten together with a member of my family since we left Lwów. Of course, Basia had much to tell us about what had befallen others of our family. She told us that my Mother had moved from Krzeszowice to Zabierzów, that she—Basia—and her mother Irena visited her there quite often, and that my Mother and eldest sister Felicja[13] were doing quite well. My daughter Lidka regularly sent them money from Warsaw. We often had letters from Warsaw, so we knew what was going on there. Lidka had had to change apartments several times. Janek's friends—especially Ryszard Matuszewski—helped them a great deal, although of course life remained difficult and dangerous.

I gathered that my Mother did not have a very clear conception of the dire predicament we all found ourselves in. She apparently constantly hoped to be able soon to return to her home in Jasło, thinking the war must end in just a few more months at most. Yet, like her, we too tried every day to peer through the darkness of our present situation in hopes of glimpsing the dawning of a brighter future. My brother-in-law and I discussed endlessly—occasionally also with various other members of the Cieluch and Igielski families and others so inclined—the question as to when the war might end. However, since we were ignorant of the productive capacities of the warring sides, and of the extent of the damage being inflicted by allied bombing raids or by German U-boats, we could hardly get beyond mere speculation tinged with a little morale-boosting wishful thinking.

But there was one thing about the war I could definitely say was the case, namely that everyone was merely biding his or her time, waiting it out, living in anticipation of the time when they would emerge from the accursed well of despair they found themselves trapped in. All were of the opinion—or nourished the hope—that the people of distant lands—Chinese and Americans, Africans and Australians—must surely rise up *en masse* to put an end to this ridiculous nightmare, and that these millions of people backed by mountains of funds would arrange our return to our multi-roomed apartments with washrooms, and to cream cakes and comfortable retirements. At this time we hadn't yet heard of the AK. When I mentioned that beyond the creek someone had run into three armed Polish soldiers in the dead of night, I was told that it was make-believe.

In early November, old Mr. Cieluch returned unexpectedly from prison. There was great joy and we all gathered at the farmhouse and drank home-made currant wine to celebrate the return of the "old patriarch." And our spirits were bolstered further next morning by the news of the Anglo-American landings in Algeria and Morocco.[14] This together with the news of the German imbroglio in the Caucasus— as predicted by Fuliński—were clear signs of the beginning of the end for the Germans.

---

[13] Felicja's married name was Müller.

[14] "Operation Torch", begun on November 8, 1942.

However, people were still preoccupied with present difficulties, such as the raids on train travellers conducted by squads of policemen from Jasło or Gorlice or Nowy Sącz. These would suddenly descend on a train when stopped at a station and take the passengers' butter, eggs, meat, vodka, other commodities such as thread, cloth, leather, and often also any baggage they had with them—and if they caught a few more Jews that was also not to be sneezed at.

There was one man in Stróże who lived differently from everybody else. This was Dywan, owner of a hut on a few morgs of land. During the interwar years he had been considered a petty thief, suspected of stealing the occasional rooster from his neighbors, and coal from the railroad. However, now things were different: the police stole the cocks and everyone stole coal. The police periodically came to his hut to arrest him as a known thief, but he was always too clever for them and hid out when he got wind of their approach. He was extremely resourceful: if a component needed replacing in a radio, say, he would know how to get it by stealth from German equipment, and moreover knew how to carry out the repair. He normally worked early in the morning as a butcher's assistant. On one such predawn morning a pair of *Bahnschutz* guards were alerted by the light visible in the butcher shop, and, going up to the door, saw Dywan there. Dywan shouted to the butcher "Leon, don't be afraid!" and, picking him up bodily, hurled him at one of the Germans, then leaped over the two bodies on the ground, and before the second *Bahnschutz* guard knew what had happened, was already well away. A shot wounded his hand, but he managed to reach the woods across the river before the two railroad guardsmen had their wits properly about them.

Stefan Dobrzański often came from Kraków to visit. In Kraków his wife, Zosia, used to make cookies which he hawked to pastry shops around the town. He told us how one evening he arrived home in a terrible mood. He had caught the eye of a certain character on a streetcar, and when he alighted, this fellow had also got off, and, coming right up to him, said: "Don't try running away. I have armed agents observing us. You are a Jew and you will come with us. Don't be surprised; I must avenge my father's murder by the Bolshevik Jews." So he was taken by this man, who called himself Heinrich, and his henchmen to the police station and then to the Kraków ghetto, where he was examined by Jewish "experts" who were able to clear him of the accusation of Jewishness. He thus had the opportunity of viewing the ghetto from within. There he saw elegantly uniformed Jewish militia men reporting to Polish overseers, and young ladies laughing as they related to their parents—just the way all teenage girls do—some adventures that had befallen them—all as if they were oblivious of the death sentence hanging over their heads.

Thus finally he was released, but told that he must submit certain documents. "Although you are not a Jew," said the man Heinrich, "you have unworthy deeds on your conscience." Stefan therefore travelled to Lwów and then on to Warsaw for the requisite documents, spending much of his slender financial resources in the process, but in the end was unable to find out where he should take them, there being no record of any appointment for him to hand them over. Thus the demand for further documentation was an instance of the mere habitual politeness of Heinrich the dog-catcher. As a result of this ordeal he moved to Jankowa, some six kilometers from us.

One day, when we were sitting on a bench in front of our hut with Stefan, who happened to be visiting us, there suddenly appeared before us an infamous *Bahnschutz* guard whom Dywan called *Bomba* (The Bomb), wielding a thick stave and accompanied by a Polish policeman. They demanded Stefan's passport, and said that there were rumors to the effect that the Cieluchs were employing a Jewish farm laborer by the name of Józek. It is possible that someone in Stróże had denounced us out of ill will towards the Cieluchs, or perhaps some garrulous drunk in the railroad station restaurant in Stróże had said too much while fraternizing with Germans there. It was fortunate that Józek, said to be a Ukrainian, was not around at that time. For some time the local people brought food to the forest for him, and then he vanished without trace. (He survived.)

From February 1, 1942, as in Osiczyna I continued jotting down scientific notes as ideas occurred to me, but was unable to expand any of them into complete articles because the lack of inner calm prevented me from concentrating consistently on any single theme. Here in Berdechów I was kept busy in other ways. My brother-in-law had had a student named Zbigniew Pająk, the son of a railroad man, whom he had taught in the senior class of a mathematical and natural science *lycée*. Since Zbigniew was living in the neighborhood, I took over his instruction. I also took on the job of teaching Januszek, Dela's son, with the aim of having him master the new Gymnasium curriculum. I began with the humanities, and Januszek turned out to be a capable student, especially in French. At first the instruction proceeded without much of a plan since we had no textbooks, but then Stefan brought us some from Kraków where he had managed to unearth a few schoolbooks with French texts.

My teaching job brought with it various benefits. Our neighbors the Igielskis repaid me in pasteurized milk which they sometimes enriched by adding fresh milk straight from their own cow. Mr. Igielski also loaned us money, and supplied firewood at less than the official price, among other favors. Zbyszek Pająk's father furnished us with kerosene and calcium carbide for our carbide lamps, and also brought us the things we had left behind in Osiczyna, saving us a difficult trip. On the way he had been held up by some Ukrainians in charge of a stretch of railroad, and ordered to go to Lwów to pay some fee or other amounting to a few hundred złotys.

On the other hand, my occupation as teacher carried with it added danger since private schools were forbidden. The fact of our having *Kennkarten* with false names was a continual worry, lest our change of identity be discovered somehow as a result of the additional contact with people that giving lessons entailed. We produced our own application forms and photographs ourselves, without recourse to official documents, for, as Igielski correctly maintained, we had best keep our activities as much as possible away from official purview. However, we did meet an official by the name of Szafraniec, Laska's sidekick, who, in concert with the latter, looked after all those in need of a little of a certain kind of care from officialdom. It was he who corrected "misleading" —not to say incriminating—entries in the records of registrations. Thus if, for example, someone from Galicia wished—as we had— to move in 1942 to the Kraków district, then it was he who made the necessary emendations.

Our help was needed to bring in the harvest. At first I found tying up the small sheaves difficult, but finally learned to do it with tolerable efficiency, and then to throw them so as to form layers. I did not attempt to scythe since this was a skill one acquired only as a youngster. The most straightforward work, raking hay, gave me a great deal of satisfaction. Around the hut there was also physical work to do, in particular, sawing wood and then splitting it with an axe. Although not difficult work, in winter it required more than an hour's labor each day.

Our son-in-law continued to make moderate amounts of money, most of which he smuggled to Warsaw, whence our daughter sent us and also my Mother regular monthly remittances sufficient only for the most modest of lifestyles. Taking into account the changes in prices of such necessities as flour, butter, eggs, and milk, etc., in prewar terms this monthly allowance would be equivalent to the wages of a janitor, and considerably less than those of a streetcar driver. Of a morning we washed black bread down with coffee made from rye, and in the cold weather drank a sour vegetable soup, while for midday dinner we had pea or potato soup followed by dumplings, or, more rarely, potatoes with cabbage or eggs, and, very rarely, meat. Our evening meal consisted of a potato gruel or *purée*, and sour milk in summer and ersatz tea in winter. Obtaining those food items not produced on the farm was often difficult; fatty foods were especially hard to come by. The slaughter of swine was punishable by death, and the constant confiscatory raids along the roads and railroads meant that the peasants hid their supplies of pig fat, or sold it only to the most trustworthy of buyers. We were lucky in that we did have such suppliers, and furthermore the Cieluchs never slaughtered a pig without sending us some beautifully prepared liverwurst. The Igielskis also kept us in mind, and occasionally Mr. Szafraniec brought us a few extra rations of flour, sugar, and preserves from Bobowa. Since we had no ration cards, my wife Stefa had to bake bread at least twice a week. She learned to do it very well indeed, despite the lack of a bread oven.

Thus life would have been tolerable if not for the constant dread hanging over us—in particular, dread of whatever new order the Germans might issue, and even if, for a week or two, there were no such orders, then rumors would surface to keep us sleepless with worry. One such rumor had it that every farm worker would soon have to obtain an *Arbeitskarte*,[15] which we would certainly be unable to obtain—at least legally. And then a certain village elder hinted that the Germans were constantly demanding that he supply them with a list of all members of the intelligentsia living in the area. In fact, we held nothing against this particular official—quite the reverse!—since he always managed somehow to foil the Germans' intentions. Every few weeks the Germans needed more slaves, and to this end imposed on each village a quota of souls to be transported to Germany as forced labor. Such rulings met with subtle but powerful resistance from the general populace and through the ingenuity of officials such as Laska, Szafraniec, and the above-mentioned village elder in circumventing them.

---

[15] "work card"

One of the worst aspects of the situation consisted in the help given to the Germans by locals acting as spies. For example, in Bobowa there was a certain Polish *Volksdeutscher* by the name of Müller who was assiduous in aiding the German police in every conceivable way. He would travel with the police to villages around Bobowa to help them round up Jews, and beat the victims as they were being led away to their death—such at least was his boast. He was given the job of carrying out an inventory of the farm laborers in the area around Bobowa, and at many of the farms he visited he demanded a sumptuous feast be prepared for him—of chicken with all the trimmings and vodka—and even forced the village elders to bring pretty girls to his quarters in Bobowa. Another example: in Stróże itself there was a certain Szczypta, a former postal worker, who combined the functions of snitch, backstreet clerk, and *Volksdeutscher*. In the interwar period he had had a reputation as a con-artist and extortionist, but now he seized on the chance to play the more significant role of spy. In addition to his denunciations, he took substantial payments in wheat and potatoes from the families of arrested peasants, and from peasants earmarked for transportation, maintaining that he was in a position to intercede with the Germans—but did absolutely nothing of what he promised. And at the railroad station in Stróże, there were, in addition to the German *Bahnschutz* police, their Polish counterparts. For instance, a certain local by the name of Marian Proszek, who prior to the war had been considered mildly retarded, now wore a German *Bahnschutz* uniform and wholeheartedly played the role of an official of the Third Reich, surpassing in zeal his fellow police officers in exacting fines from travellers for trifling misdemeanors, identifying passengers transporting prohibited goods, and arresting those without identity cards. He was aided in this by the greatest numbskulls among the railroad workers, such as the porter Serafin. He denounced to the police a certain doctor and his wife who had escaped from the Kraków ghetto, and they were soon apprehended on the road to Gorlice and shot. A certain Emil, formerly a butcher from Grybów, distinguished himself in similar ways, as did another butcher by the name of Durlak. These pet poodles of the Nazis hardly got a bone for all the help they showed them, since victims meeting their end in the basement of the railroad station were first stripped bare by their German killers, and even their clothing, soaked in sweat as it was, was sent to Germany. Thus the embroidered blouses of murdered peasant and Jewish women reappeared in the clothing stores of German towns and were bought by unsuspecting German men as lovers' gifts.

* * *

We decided to carry out a useful service on the farm, namely, to produce for our benefactor Mr. Cieluch a map of the arable areas of his farm. We did this without instruments, using only wooden pegs and a steel measuring tape, dissecting the land into trapezoids and then triangles and calculating the areas of the triangles using

Heron's rule.[16] Then in the spring of 1943 my brother-in-law used a special drawing ink to make a beautiful plan drawing of the whole farm, with the triangles of the complex labelled with their areas. I estimated the error in our computations at not greater than one percent. I then conceived the project of making a sundial for the farm. I first had to make precise drawings of the sundial's components: first of the horizontal disc-shaped base for the dial and as gnomon a right-angled triangular shape whose hypotenuse was to be parallel to the earth's axis of rotation. I found a way of calculating the position of the edge of the shadow that the gnomon would throw on the dial from the table of values of the tangent function in a book of school mathematics written by A. Łomnicki, and carried out detailed calculations of corresponding times using the known values of the latitude and longitude of Stróże. However, my project remained in my notebook, that is, at the planning stage, because I couldn't find suitable rain-proof material to fashion it out of. I would have liked to see on my identity card the characterization "retired sundial designer". Somewhat later I did build my sundial—with considerable help from Zbigniew Pająk.

But there were other indicators of the passing of time. Stalingrad[17] loomed over the Germans like an omen of their inevitable doom. We heard more and more of the activities of the AK—in particular, of their assassinations of Polish collaborators and spies. This in turn lashed the Germans into an ultimate fury, so that they raised to fever pitch their hostage-taking, reprisals, murders, burnings, ferreting-out of Jews, and deportations. The rate of arrests became frenetic; in Grybów, Gorlice, and Nowy Sącz there seemed no end to them, and people visibly quailed. Adam Igielski, a relative of our Igielski neighbors, was among the recent arrestees. He had been a locksmith working for the railroad, a model father and husband, a Pole of character and integrity, and there would appear to have been no grounds for arresting him. With my own eyes I saw the Polish policeman Kubala driving a horse and cart with Dr. Luster, who had been hiding incognito in a village near us, sitting bound in the cart, *en route* to his death. However, Kubala survived this particular victim only by a few months. The court of the underground government of the Republic of Poland passed sentence on him, and he was assassinated in broad daylight on the street in Nowy Sącz as he left the station. Of course, his assassin vanished at once. Once while on my way to see the shoemaker Derechowski, whose workshop lay about a kilometer and a half from Berdechów, I heard five gunshots ring out. Sometime later I heard that these were the sound effects, so to speak, of the assassination of the policeman Krzeszowski, on whom a guilty verdict had also been handed down

---

[16]Heron's formula for the area of a triangle with sides of lengths a, b, c is $\sqrt{s(s-a)(s-b)(s-c)}$ where $s = (a+b+c)/2$, the semi-perimeter. Heron of Alexandria lived around 60 AD. However, the rule attributed to him was actually discovered by Archimedes (287–212 BC).

[17]The Battle of Stalingrad fought by Germany and its allies against the USSR for control of Stalingrad (formerly Tsaritsyn, and subsequently Volgograd) lasted from August 1942 till February 1943. The German defeat was a turning point in World War II, rendering a German victory in the East impossible. Considered among the bloodiest battles in the history of warfare, with an estimated two million combined casualties.

by the Polish underground government. Thus he died in Polna, on his own property, near his own house. The spies were terrified. The policeman Proszek was under sentence of death, as well as another who found out beforehand of his death sentence and wanted to flee, but was not permitted to do so by his German bosses. For the moment only Szczypta was operating actively, mostly around Tarnów.

Ordinary folk talked about the war a lot, but only as perceptively as they knew, which is to say in an extremely uninformed way. Since they had only the roughest idea of geography, they wondered why the British didn't make a landing in Gdańsk, and having no understanding of international relations they took such rumors as that the British had an army in Turkey seriously. Not knowing what the word "ultimatum" meant, they nevertheless used it freely, for example, maintaining that America delivered an "ultimatum" to the Germans every month. They read things in newspapers that weren't there—for instance, that Japan had declared war against the USSR. Incurable optimists, they believed that "in the modern world everything is possible." They talked about strategy in war as if the movements of armies were like excursions organized by the tourist agency "Orbis". I heard someone ask in a surprised tone: "Since there is a straight road from Rome to Vienna, why don't the British and Americans, already in Rome,[18] march right along it." Educated people did turn up from time to time, but I avoided them as best I could since I viewed them as potentially dangerous. Almost the only person with whom I had serious intellectual discussions during this time was Dr. Knaster in Lwów; we exchanged long letters, but not too frequently lest they attract attention. Most of our correspondence was taken up with our running "argument" about whether or not there is a scientific solution to the problems of life, but we begged to differ also on other questions. For instance, Knaster held the opinion that waging war is a matter of skill, and that the Germans had this skill while the Allies didn't. I countered this with Karl Kraus' parable: In a certain knockabout act by a clown at a vaudeville show, the clown, the hero-villain, shows off by smashing crockery, turning over tables and wardrobes, and knocking down walls, and stands triumphantly amid the ruins. But then a feather rises gently from the floor, so he steps on it. But the moment he moves it flies up again. So he dumps a wardrobe on it. But when, out of curiosity, he lifts the wardrobe a little to look under it, the feather flies up again.... We also sent each other mathematical questions. One such that we discussed was the problem of practical subdivision: a layer cake, or apple pie, is to be divided up among n people to the satisfaction of all n of them. The case $n = 2$ has the well known "Solomonic" solution: "One cuts, the other chooses", rendering the division independent of argument since each can ensure he gets at least half independently

---

[18]Rome fell to allied troops, mainly American and British, but including those of several other Allied nations, on June 4, 1944, after an extremely arduous nine-month push up the peninsula from Sicily.

of the other. Knaster told me that he and Banach had managed to solve the general case of n people dividing up a cake,[19] something I had thought impossible for $n > 2$.

The letters we received from our daughter Lidka in Warsaw were a constant solace. She and Janek were managing quite well, and although I didn't fully understand where they got their income, they had enough to be able to send money regularly to us and to my Mother. Janek had moreover to provide for his mother and sister in Kraków. In the spring of 1943, Janek's father was arrested over some minor matter, and then we heard that he was to be released. However, instead of being freed he was killed. We also heard the horrible news that Stasia Blumenfeld[20] had been arrested. It seems that initially she had been arrested for smuggling alcohol, and then was to be freed—and at her home they had prepared dinner, expecting her at any moment—but at the last minute she was sent to a camp on the grounds that she was of Jewish descent, and there she was murdered. Her son was also arrested but released when he contracted tuberculosis. Knaster told me sometime later that all she had earned from her trading enterprise went to the AK, the Polish underground army. To add to this there was the fact that we had been unable to extricate the Grünhuts from Truskawiec, all of Lidka's efforts to persuade them to leave having been of no avail. Despite my efforts to make contact, a silence descended on that quarter, never to be broken.

In the spring of 1943 I met a Mr. Rząca, owner of a farm of a few morgs, with five sons. Three of the five wanted to study, and at first I taught all three, following the first-year Gymnasium syllabus. However, after a while I had to give instruction to the youngest separately since he wasn't able to keep up with his brothers. I had to take him through the subject-matter of the elementary seventh grade, which had been eliminated under the system imposed by the *Generalgouvernement*. I came to an agreement with Mr. Rząca that he would supply us milk at a reduced price by way of payment. He fulfilled this obligation to the letter, and in addition every few weeks sent us gifts in the form of chicken, duck, and sometimes sides of pork or ham, as well as fruit and vegetables when in season. His boys were capable learners, especially the elder two, and the lessons went well.

Through my work as teacher I became familiar with the schoolbooks in use in the Gymnasia. I saw that the history textbooks were actually much better than the ones in use when I went to school. They contained beautiful reproductions of works of art and relevant documents, and pictures of objects, buildings, documents, and coins from the epochs under discussion, as well as significant passages quoted verbatim from the appropriate sources. The geography textbooks were also not bad. However, I cannot praise Pawłowski's textbook, where, for example, the passages dealing

---

[19]The solution is essentially as follows: Someone slowly moves a knife over the surface of the cake until one of the contestants thinks the portion thus far demarcated represents at least $1/n$ of the whole; at this moment he calls a halt, takes this part as his cut, and the division continues with the remaining $n-1$ contestants.

[20]See Chapter 10. Steinhaus and his wife had stayed with the Blumenfelds in Lwów for a short time after they fled their own apartment.

with coal-burning and hydroelectric power stations were simply nonsensical. As far as Klemensiewicz's[21] textbook of Polish phonetics and grammar is concerned, I can only say, first, that it is very badly written, second, that it is without any value for learning correct Polish speech, and, third, that the point of view it adheres to, taken from Nietzsche, that Polish arose as a phenomenon of natural history, is highly questionable. The Latin textbook of Ewa Appel and Żmigryder-Konopka[22] for the third-year Gymnasium class of the reformed system provided such an overwhelming mass of information about every standard Latin text, and so much historical, philological, and grammatical commentary, that neither teacher nor student had anything left to do. Although I had to admire the book, I had doubts then as to whether the wealth of material did not somehow obscure its pedagogical objective.

* * *

Once Stefan Dobrzański had organized his new abode—not far from us, appropriately—he brought over his wife and child from Kraków. It was very agreeable having him living close enough all through 1943 for him to be able to drop in on us or us on him when the mood so took us. Now I had someone with whom I could talk sensibly about the war and matters generally. Although he was basically of good cheer, he was far from repeating the standard optimistic rubbish in the air. Of course, he was not fooled by German propaganda any more than I; we did indeed read the *Krakauer Zeitung*—since it was almost our only source of information about the outside world—but not just for any surface information we might glean since we could also read between the lines to discover what was behind the outbursts of propaganda. In the spring of 1943, Stalingrad was lost, and the Germans were now in the process of descending from the peak of their success, and were unable to hide their disappointment and fury. Goebbels was absolutely beside himself, apoplectic, it would seem, with frustration. For three years he had been loudly maintaining that a team that didn't score goals in the first hour and a half was not going to make up for it in the last ten minutes, yet now he was proclaiming the opposite—that the final battle of the war would be decisive. It was becoming clear that Hitler had made his people complicit in the most heinous of crimes partly with the aim of excluding any possibility of surrender. The average German believed in the instinct for revenge and knew very well that the enemy nations had reason enough to wish to exact vengeance! Naturally the German media singled out [illegible name] and Lippmann[23] as illustrating the bloodthirstiness

---

[21]Zenon Klemensiewicz (1891–1969), Polish linguist, specializing in the Polish language. Professor at the Jagiellonian University of Kraków. Fought in both world wars, and participated in underground education in Poland during the German occupation.

[22]Zdzisław Żmigryder-Konopka (1897–1939), professor of ancient history at Warsaw University, and senator in the Polish parliament in the interwar period.

[23]The reference might be to the influential American journalist Walter Lippmann (1889–1974), but nothing is known about his being a special target for Nazi hatred. More probably Steinhaus

of the Allies. The time had arrived in Germany when the devastation from the bombardment had perforce to be openly acknowledged, and the newspapers began publishing long lists of churches, hospitals, schools, museums, etc. destroyed by British bombs. Artistic types decried the losses to the great German cultural achievement. Press reports were constantly being published to the effect that the Allies were aiming their bombs intentionally at the temples of religion, art, and science. Those who had turned a blind eye when their own soldiers smashed babies' heads against walls now complained bitterly when an American bomber pilot 3000 meters up failed to spot children in a playground. Sheeplike followers of a *Führer* proposing the murder or subjugation of all of "non-Aryan" humankind now bemoaned the destruction of their suburbs. Having justified the area bombing of London as politically appropriate, Hitler suddenly remembered that on September 4, 1939 he had proposed an agreement between the warring parties that the air war be restricted.[24] But to make the argument stick, he had to fudge the beginning of the war, shifting it to a later date by some days, no doubt hoping that people would forget how during the first days of September the *Luftwaffe* had bombed passenger trains, strafed refugee columns—including women and children—, farmers in their fields, and even cattle—not to mention the extensive bombing of Warsaw of September 1939.[25] He also seemed to have conveniently forgotten how his propaganda machine in its usual heavy-handed way had mocked the British for dropping leaflets over Germany in September 1939, warning that "now the paper bombing is over, the real bombing will commence," and adding "They are dropping these idiotic leaflets because they have no bombs." And then when the bombs came he could only remember his proposals for an aerial truce! Despite the entrenched racism of Hitler and his henchmen, they now somehow saw their way to complimenting the Arabs and forming an alliance with Japan. Earlier propaganda shots showing Russian prisoners-of-war with the caption "Asiatic hordes", gave way in 1943 to photographs of detachments of Georgians and Kurds in their national dress under the command of German officers, marching briskly off to war against the USSR. That their choice lay between dying of hunger or joining the Germans could be learned only directly from the lips of these "freedom fighters." Of course, such recruitment represented a violation of the Geneva Convention,[26] but, Hitler

---

meant to refer to Frederick Lindemann, First Count Cherwell (1886–1957), English physicist. As scientific advisor to the British government during World War II, he strongly advocated "area bombing" of German cities. In this case the other person cited might well be Sir Arthur Harris, First Baronet ("Bomber Harris") (1892–1984), Air Officer Commanding-in-Chief of RAF Bomber Command during the latter half of World War II. It was he who was given the task of implementing Churchill's policy of "area bombing" of German cities.

[24] From September 1939 Hitler had repeatedly proposed that aerial bombing be restricted to zones of military operations, in accordance with president Roosevelt's proposal that civilians and unfortified cities be excluded as targets of aerial bombardment.

[25] And the bombing of Guernica in 1937 during the Spanish Civil War.

[26] The "Geneva Convention relative to the treatment of prisoners-of-war", effective as of 1931.

being who he was, this war was carried out by the Germans with a total cynical disregard for any rules or conventions relating to the warfare of past epochs.

After Stalingrad German propaganda changed tack. While the German war machine was advancing, and the promise of unlimited oil and wheat seemed more than merely plausible, there was next to no questioning of the decision to wage war on the USSR. But the situation post-Stalingrad called for different forms of justification. Now, for example, the need to curb the amassing of armaments by the Soviet Union was offered as sufficient reason for invasion. The tortuous argument ran thus: such stock-piling of weaponry and soldiery by the Soviets was greater than anyone knew back in 1941, so the waging of war to put an end to the concomitant danger of such a large and well-armed foe represented yet another indicator of the *Führer's* military and political genius. Racists who had hanged Poles for having intimate relations with German women, now, their hubris blunted, turned a blind eye to marriages between German soldiers and Polish women, and, instead of commending men of the Polish resistance for shaving the heads of Polish women who consorted with Germans, hunted them down as fair game for scapegoating. The moral bankruptcy of Hitler's regime, its political and ideational vacuity, became more and more evident. The German leadership began to look like what it in fact was: a band of well-armed thieves maintaining professional propagandists to befuddle fools at home and abroad. Even on a practical military level, the emptiness at the top, its lack of intelligence and professionalism, came more and more to the fore, with military strategy reduced to the ingenious World War I lance-corporal's single and oft-repeated command: under pain of death do not retreat under any circumstances!—the brilliantly single-minded key to victory. He held the childish conviction that to retreat—even for purely tactical reasons—was to become the loser.

* * *

Lidka's letters continued to bring us news of our friends and acquaintances. We heard, for instance, that Dr. Ziemilski had been killed, but his wife and son had found refuge in Warsaw. Ludwik Oberländer had also perished, but his wife was lying low in Kraków, and his son had found a place for himself in Warsaw working at his profession, that of weaver. The Skulskis, Horowitzes, and Lila Holzer had also all gone to Warsaw and were coping one way or another there, but the latter's husband had succumbed to the fate of millions. Naturally my daughter's circle of acquaintance consisted chiefly of Janek's friends and associates, who were for the most part poets or other literary types, so there were many of our old friends whose fate remained unknown to us till much later. Thus Dr. Blumenfeld, the last to abandon the factory "Laokoon", had been diagnosed with cancer and taken to Warsaw for treatment. My cousin Dyk and one of his daughters had been arrested, never to be heard from again.

It was at that time, that is, in mid-1943, that I began to keep a diary, partly with the idea of keeping my mind occupied. Earlier, from November of the previous year, my wife had found a hobby that appealed to her, namely learning English.

My brother-in-law had a few English books, and Stefa[27] worked through them with my help. These included some of Edgar Allan Poe's stories, an excellent novel by Walpole,[28] and Bernard Shaw's play *Saint Joan*, which he borrowed from Mrs. Kulinowska, wife of the Bobowa veterinarian. She—Mrs. Kulinowska—also felt a yearning to learn English and asked my brother-in-law and me if we would give her lessons. These turned out very profitable because Mr. Kulinowski gave us a kilogram of meat each week by way of payment. It needs hardly to be said that it was very difficult to obtain meat at that time, since German administrative measures were still very much in place in 1943, and all livestock was earmarked for military consumption. As I think I have already mentioned, for a certain time the illegal slaughter of swine was punishable by death. But assaults had already begun on the German measures: dairies were raided, and parish offices were broken into and registers burned, peasants herding cattle to delivery points were attacked and the cattle dispersed, and village administrators serving as flunkies of the Germans were beaten up.

Once done with the books available locally, we obtained more English reading matter from Kraków. We were helped in this by Stefan Dobrzański, who periodically travelled thither, and by Mrs. Michalakówna, a former administrator in the steelworks "Zgoda", who periodically came to see my brother-in-law. We craved news of Kraków, and she was able to satisfy us to some extent in that regard. Once we had exhausted local and Kraków sources of English books, I thought of having our daughter send us new ones from Warsaw. She knew the address of a Mr. Miętus in Warsaw who could arrange delivery of such books, and at our end Mr. Szafraniec helped us take receipt of the shipments without calling attention to ourselves.

From time to time we got hold of some of the smaller newspapers. The officially permitted ones among these were, though well edited, not very forthcoming, but once we obtained a copy of an underground newspaper published in Lwów, written in unusually sophisticated language. From it we learned that on *Piaskowa Góra* in Lwów the Germans had killed hundreds of thousands of Jews, and later, in the Fall of 1943, hundreds of Vlachs[29] and tens of thousands of Ukrainians who had followed the retreating German army from beyond the Don to Galicia hoping to escape from the USSR.

That spring Maniuta Dobrzańska, wife of Jan Dobrzański, moved to Jankowa[30] with her children and sister Wanda, and from her we learned that Lwów had become a hellish jungle where Ukrainians murdered Poles, Germans murdered Jews, and

---

[27] The author's wife, *née* Stefa Szmoszówna.

[28] Sir Hugh Seymour Walpole (1884–1941), prolific English writer. A best-selling author in the 1920s and 1930s, his works have since fallen into neglect.

[29] The Vlachs, or Wallachians, were a people stemming from the latinized population of the Balkan Peninsula. In medieval times, groups of Vlach shepherds settled in various regions of the Carpathian Mountains, e.g., in Moravia, Slovakia, Poland, and Ukraine. The modern Romanians are of Wallachian origin.

[30] Where Stefan Dobrzański was living at the time.

Polish underground organizations murdered Germans and their collaborators. To leave one's quarters after nightfall was to take one's life in one's hands.

This was the era of the great German flight westwards, beginning at the end of January 1943. By the autumn of that year the Bolsheviks were at the Polish border, but there their march westwards halted for a time, and nothing very decisive occurred during the winter of 1943–1944. That winter we were often at a loss as to how to keep ourselves busy, so we played chess for hours on end. The three chief players were my brother-in-law, our student Zbyszek Pająk, and I. Chesswise, Zbyszek was in a class by himself. I have never been able to determine what the mental capabilities of a talented chess player are away from the chess board, or how lack of talent at chess relates to mental abilities in other areas. Mr. Woźnica, for instance, quickly picked up the game and rose to a fairly high level of competence, whereas I made no progress over 30 years—although it's true that from the time I was 20 years old I played very infrequently.

Since the German occupiers were on the verge of retreating, their administrative supervision became more lax: the supply of the prescribed quota of corn and of fresh milk to their separators, and their cattle drives, as well as other levies, were now fulfilled tardily, with an eye to the imminent end of German rule. This also explains the joyful mood of the harvest festival held on the farm at Berdechów in August 1943. The whole Cieluch family, together with the servants, the hired reapers, us their guests, and the Cieluchs' friends the Biernats from Śnietnica, amounted to some 30 people taking part. There was as much *bigos*,[31] chicken, beer, and vodka as we could consume, followed by *tortes*, ordinary cakes, and tea. The Cieluchs' homemade sausages were better than any I have eaten before or since. The boys sang, Mrs. Biernatowa competing with them, and Mr. Biernat and Tadeusz Cieluch drank toasts of *Eiercognac*,[32] and Wilk, a reaper, argued passionately with Koszyk about who was the master reaper. Next to old Cieluch sat our elderly neighbor Mrs. Krokowa, full of dignity, and growing redder and redder from the beer she drank. The children gobbled up quantities of cake at a separate table, and the young Cieluch ladies worked in the kitchen, brought out plates of food and removed empty ones. On this occasion we were left in peace, as the frequency of visits from German police and quartermasters—though not of the Polish police—had abated somewhat. Szczypta represented our only real threat, since he could be relied upon to resort to blackmail in the case of a minor infraction, and was continually hoping to discover a major one so as to advance himself in his self-appointed role as political agent for the Germans. But for the moment he seemed to be dormant.

Yes, the Polish police still showed up regularly. Once a whole squad of them came to the door while the Cieluchs and some guests were all sitting out in front of the farmhouse enjoying the warm evening weather. Chief Laska began chatting to

---

[31] A traditional, hearty, long-simmered meat-and-sauerkraut stew, considered the national dish of Poland. Variation is possible: fresh cabbage instead of sauerkraut, and the addition of mushrooms, for example.

[32] A drink made with eggs, sugar, vanilla, schnapps, cognac, and sometimes cream or milk.

Mr. Cieluch, while his subordinates sat around in the yard. Music could be heard wafting in from across the road, where Mr. L., owner of a farm adjoining the Cieluchs', was hosting his harvest festival. Dywan was one of his guests. Although he was aware there were policemen a hundred paces off, this only heightened his devil-may-care attitude: "It's not them that are on my tracks, it's me that's on theirs." After a while two of the younger policemen, tired of waiting, decided to see what was going on at Mr. L's, all unawares of the presence of Dywan since he was indeed on the lookout for them much more than the other way around. The police launched their sweep towards morning, but of those brought in almost all were released, even though several among them were known to be professional thieves. Commandant Laska's opinion, disclosed in a discussion with a colleague that evening, was that even in the present extreme circumstances a thief had by law the right to explain his actions and otherwise defend himself in an open court, whereas handing him over to the Germans would not be to deliver him up to due process and punishment, but to undue torture and death.

As I may already have mentioned, in Berdechów I met a certain Mr. Ząbek, a friend of the Igielskis, who had been living in Jasło. He told me that the Germans had paved Jasło's streets beautifully, that they had built a passage connecting the *starosta*'s office with the adjoining house, and that they had torn down the more run-down dwellings. For paving stones they had used, he said, granite and marble headstones from the Jewish cemetery. Thus pedestrians now trod on the inscriptions of the bereaved, or might stop to read the names of departed friends and acquaintances. Through the Igielskis I also became acquainted with two teachers, Miss Brach and Miss Stiller. These ladies had been teachers for many years, and were dedicated and enthusiastic about their profession. In addition to the officially authorized curriculum, reduced by the Germans to the reading of *Ster*,[33] an idiotic little utilitarian publication designed to fit in with their agronomic program, they secretly taught the subjects that had been dropped from the school syllabus.

Our student Janusz Ochalik made excellent progress, especially in Polish, French, and geography, while Zbyszek Pająk turned out to have a mathematical gift. His sister, an attractive brunette, was to be married, and came with her *fiancé* and mother to invite us to the wedding. We were not keen to go, but accepted so as not to offend the young couple. But the wedding was a great success, and we ended up enjoying ourselves. There was dancing till dawn to music made by a fiddler and a pianist. The young couple turned out to be superb dancers, and my wife and I derived much pleasure from recalling the art, neglected by us since 1939.

A few months later Miss Pająkówna as was, now Mrs. Dziedziakowa, came to us on some business that I don't now recall—perhaps with one of the parcels of

---

[33]"Helm", the official German newspaper published in Polish in the *Generalgouvernement*. Himmler had ordained that the non-German population in the East should not be educated above the fourth grade of elementary school. Above that level only education in trades or factory work was on offer. All Polish institutions of higher learning were closed, and their facilities either exported to Germany or converted to military barracks or offices. Teaching in Polish was punishable by death.

meat which we so welcomed. It was morning and I was still in my dressing-gown over a night-shirt and pajama pants tucked into high boots, preparing to go thus accoutered for my morning walk. When our visitor arrived, I set off, as usual, in the direction of a nearby wood. Walking on the edge of the road, I ran into her again, and we continued on together a little way. Suddenly a police car appeared from behind us. We hastily said goodbye, and I turned off the road, following my habitual route back home through the woods, when the car stopped. It contained six bruisers in green uniforms, with yellowed whites of the eyes and faces bloated and red from overindulgence in drinking. They were armed with rifles and pistols. One of them called out to me to come over to the car and identify myself. I went over and began to explain that I lived in the house visible from where we were. "Where is your identity card?" I indicated that I had gone out in my nightshirt, and naturally had left the card at home. Thankfully, they were satisfied with this answer, and continued on their way. I later found out that they were returning from a village about two kilometers away where they had set fire to the cottage of a certain Mrs. Kostrzewina, killing her three daughters and a girl from a neighboring hut. We also learned that the murders were in reprisal for an attack on the German unit installed on an estate in Jeżów, in which Mrs. Kostrzewina's son had allegedly participated, when in fact he had already been sent to Germany in a forced labor contingent. Thus this horrible crime lacked all justification, even of the most perverted kind. Much later we heard that the ringleader, a certain Filiber, perished in a bombing raid on Berlin. German deeds like this always brought to mind the following words of the German *Generalgouverneur* of Poland: *Es wurde in diesem Raum niemals so klar und so sauber regiert wie jetzt.*[34]

All through 1943 quotas of deportees were imposed on every district. German newspapers termed these impressments "recruitments of volunteers", but in practice the "recruitments" took the form of police arrests. In our district, fortunately, Mr. Laska always strove to arrange matters in such a way—by selecting only those well known as socially or mentally defective—that the majority of "recruits", so-called, were sent back from Kraków as unfit. There were also those who managed to escape before crossing the border, and others who made it back all the way from Germany. However, the German raids, or round-ups, on railroads, roads, and, especially in the larger towns, on homes and places of work, were slowly but surely emptying Poland of its people. The aims of such deportations were fivefold: to supply cheap labor for the German armaments industry and agriculture; to swell the depleted ranks of German infantry with Polish recruits; to remove from Poland that element of the population capable of rebellion; to incidentally destroy that same element by forcing its members to work in areas susceptible to allied bombardment; and, finally, to acquire Polish hostages in case, at the war's close, the Poles should seek to revenge themselves on Germans left behind in the *Generalgouvernement*. Not being aware of this multiplicity of objectives, our peasants were initially slow to realize that they should resist deportation with all available strength and cunning.

---

[34]"These lands were never governed in so clear and clean a manner as now."

It was only through the increase of attacks by the AK on village officials that the servile eagerness with which the Polish administration helped to organize the supply of "slaves" for deportation to Germany was dampened. Bribery of such officials also increased in 1943, and Polish officials working in the *Arbeitsamt* were more inclined to issue direct exemptions from deportation, or work cards validating such exemption, in return for money, or, even better, a flitch of bacon.

Of course, one of the best ways of ensuring that one not be deported was to keep one's name off official lists, and out of registries of whatever sort. Thus we did not apply for ration cards, and, as a general principle, strove to hold ourselves apart from any activity or occasion that might result in our names being taken down. In late 1943 we received a letter from Janek lacking the usual postscript by our daughter. This was alarming in itself, but the contents of the letter were much more so. In it Janek related how once, while on her way out of Warsaw, Lidka had been stopped and sent to the police station, but eventually freed thanks to her indomitable *sang froid*, provided she furnish certain documents next day. So next day Janek went to the police with the requisite documents, only to meet with incomprehension, since the accusatory party was not to be found. We had a terrible few moments before we realized that Lidka was no longer under any threat. It was only about a year later that we learned the details of what had transpired and why.

We were invited on New Year's Day 1944 to visit with Stefan Dobrzański and his wife in Jankowa, within reach by foot. The time was passed agreeably and we met some new people there, among them Mr. L. and his wife. Next day, January 2, I went to Miss Stiller's apartment; she had told me earlier that a certain unnamed person wanted to see me, and, after consulting with my wife, I had decided to meet the person in question. This turned out to be a Mr. Zabierowski, who had been deputed[35] to assess the underground teaching going on in the vicinity of Stróże, in particular to ascertain who was giving lessons at the Gymnasium level. He instructed me as to the precise Gymnasium program I should adhere to, and supplied materials to help in this. The extent to which the Polish underground educational enterprise was organized came as a surprise to me. From that day on I was to be paid 300 złotys a month, although for some reason the payments ceased in May 1944. The second instalment, that is, for February, was paid by a Mr. Jan Zaremba, owner of a property in Stróże Niżne,[36] a person of considerable intelligence, a polonist and former teacher at the Cieszyn[37] Gymnasium. His position with the underground educational organization included the monitoring of the teaching being carried on clandestinely, a task he applied himself to with the utmost diligence and selflessness. At last there

---

[35] Presumably by the "Secret Teaching Organization" (*Tajna Organizacja Nauczycielska*) responsible for most of the underground elementary and secondary education conducted during the German occupation. It is estimated that by 1942 about a million and a half children were being educated by such organizations, as well as certain adult individuals, at the elementary level, a hundred thousand at the secondary level, and ten thousand at university level.

[36] "Lower Stróże"

[37] Town in the Silesian Voivodship, on the border with the present-day Czech Republic.

was someone with whom one could discuss the situation without having to explain that Turkey did not have a common border with Hungary, or that Sikorski[38] had not been murdered by the English.

By this time—early 1944—it had become abundantly clear that the Germans were on the defensive. Their propagandists were tying themselves in knots producing new fodder for the gullible—often to the extent of reversing the order of cause and effect. Although, through complicity, every member of German society had the blood of innocent people on his or her hands by the bucketful, their rulers now had the temerity to decry Allied barbarity, publishing documents showing that the Allies had indeed taken the decision to destroy Germany through bombardment—as if that decision was not at least in part prompted by the years of terrible butchery by the Germans, on such a scale that it could hardly be kept secret. Dr. Goebbels and his staff made much of this Allied strategy in the media, concluding from it that Germany had no choice but to defend itself to the very last. Death on the field of battle was preferable to death meted out as punishment by one's enemies. Continued resistance held out hope of a change in fortunes, or some chance, however slim, whereas capitulation would lead to the extinction of the whole nation, women and children included, at the hands of the bloodthirsty barbarian. To this main torrent of vituperation, lesser tributaries were united—such as the contention that the bombing of German cities provided proof of the collaboration of Bolshevik atheism with Jewish capital in a conspiracy to destroy Christian churches and cathedrals.

I couldn't deny myself the pleasure of cutting out some of the more exquisite samples of German propaganda from their newspapers. For example, in the *Krakauer Zeitung* of January 27, 1944, there appeared an article under the caption *Warum führt USA Krieg?*[39] The introductory sentence asserted that the United States had no material reason to fight Germany, that the conflict arose from an ideological opposition between the materialistic American spirit and the idealistic spirit of Germany—possessed also by Japan, naturally. According to this article, those in power in America were afraid that German ideas would capture the minds of the American people, enabling them to see who their true foes were, namely, the sixty or so leading capitalist families on Park Avenue, for whose sake millions of American soldiers were being sent to their deaths. Thus Roosevelt was not at all concerned with the welfare of his people, but was instead intent on spending billions to kill women and children and destroy irreplaceable monuments to the greatness of European culture in order that the capital investments of the big trusts and concerns secure a good return.

In early 1944 one could not help but be struck by the fact that in the German press the topic of the supremacy of the Aryan race or the ethnic primacy of

---

[38] Stalin broke off relations with Sikorski's government in 1943 when Sikorski asked the Red Cross to investigate the Katyń Forest massacre. In July 1943, a plane carrying Sikorski plunged into the sea after takeoff from Gibraltar, giving rise to a number of conspiracy theories relating to the circumstances of his death.

[39] "Why does the USA make war?"

Germans was no longer being written about. The clique surrounding the *Führer* was beginning to appreciate that the racialist refrain might sound discordant to the ears of their Romanian, Italian, Hungarian, and Slavic allies, so they changed their tune of domination of the master race to a lullaby promising protection against the Bolshevist bogeyman threatening the "New Europe". But a few months later when even the most unobservant dolt could see that the touted "new order" was rather a "new disorder", the "New" was dropped from "New Europe". They also stopped calling Churchill and Eden[40] idiots, perfumed sybarites, dilettantes, ignoramuses, etc., perhaps appreciating that such dismissive insults did not gibe well with the respect for British arms engendered among the Germans by the weight of their bombs.

Thus gradually the war for more space for the master race was being transformed into a defensive war of attrition. As the area occupied by the Germans shrank, so, apparently, did the resources of their archpropagandist. Then, to cap it all, they began to insist that the war had broken out only because Poland would not permit a connecting highway to be built from Germany to Gdańsk.[41]

We had kept in touch with Ms. Kamila Speidlówna, our friend from Zimna Woda, while the postal service was still operating, even managing to get parcels to her. From her we heard of the terrible things that had been going on there, for instance, that Witold Otto had perished in a camp or in prison, having been arrested, together with his wife and daughter, probably for listening to foreign news broadcasts. Occasionally we wrote also to Halina Szpondrowska, from whom we heard that her son Staś had been staying at the house of one of her friends, a Mrs. Olszewska, in a village in the neighborhood of Dębieniec. However this lady had also fallen into German hands and perished, so that Staś returned to his mother and his younger brother. They were pursuing their studies, and doing well, while their mother, as always, worked hard to make ends meet.

* * *

In order not to be considered heretics of some sort, we sometimes went to church. My brother-in-law and I attended old Mr. Igielski's funeral service, held in the church in Wilczyska. This was wooden throughout, decorated with beautiful paintings all of a dark red and bronze tone, very much in harmony with the color of the wood and the subdued light from the stained glass windows. The church in Stróże was another matter altogether: an example of twentieth century Gothic illustrative of the Latin proverb *Duo cum faciunt idem, non est idem.*[42] It was in

---

[40]Robert Anthony Eden, 1st Earl of Avon (1897–1977), British conservative politician. British Foreign Secretary during World War II, and Prime Minister 1955–1957.

[41]In the aftermath of World War I the victorious powers had established what was to be called a "Polish corridor" between Germany proper and East Prussia, allowing Poland access to the Baltic Sea. The city of Danzig (now Gdańsk) was designated a free city, under the protection of the League of Nations and independent of both Poland and Germany.

[42]"Even if two do the same, it isn't the same."

what passed for the Gothic style, but built of brick and concrete, and situated next to the station amongst huts and piles of debris, so that it looked as if it might have some function relating to the station. But its interior was even more surpassingly ugly: all baroque gilding and bluish stars, factory-made of some tinselly material, little molded figures like gingerbread men or those seen on wedding cakes, small pictures and fake columns stuck to the walls without regard to the overall effect or the design of the building. One might very well be surprised at how unerringly the original founders, architects, and patrons collaborated in maximising the overall repellently gimcrack effect.

From hearing the sermons and pondering the Latin text of the mass when I attended church, I arrived at the conclusion that the Roman Catholic faith is at odds with the Polish psyche. The foreign God of the Catholics, born out of the sands of the Arabian Peninsula, and transmitted via Hebrew, Greek, and then Latin, moves the faithful only through a pathos of great distance, and of an obscure miracle play. The natural spirituality of our peasants, old people, and simple working class folk searches in vain in the Catholic Church for something akin to its own, essentially pagan, God, which has very little if anything in common with the divinity of Catholic dogma. Of course, here I'm not thinking of the young ladies who view attendance at church as, even though a duty, a nonetheless pleasant occasion for showing off their best dresses and newest hairdos: these are mere accessories to the complex ritual of Roman Catholic worship. The Polish people can never become Catholic in the rooted sense in which the French or Spanish nations are Catholic, and with us Roman Catholicism will of necessity rigidify and become caustic, reduced to ceremonies understood only by those who don't believe in them.

Like most of the priests we listened to, our local priest Father Boratyński adhered closely to theological texts in his sermons. This meant that the sermons were largely agitational, since they involved exhortations, quotations from scripture or dogma, comparisons, and, above all, were delivered in a voice of the gravest solemnity inimical to any doubt parishioners may have had as to the truth of the God-thesis dreamed up by the shamans of a desert tribe of ancient Semites. And the parishioners, worn down by poverty, anxious over the fate of their families, consumed with fear of the Germans, drew consolation from the very incomprehensibility of words which evidently promised much. The teaching of practical morality, of how to solve the problems inevitably imposed by wartime realities—in particular, the Christian attitude to the murders, thefts, lies, and terror generally—was evidently out of the question. Of course, the Germans caught on quickly to the usefulness of the Church in keeping people occupied with processions and other rituals—keeping them quiet, as it were. They also found them useful in a more immediately practical way, namely as storehouses of munitions, gasoline, and food, since they considered them safe from intentional bombardment, at least.

Almost at the same time as Stalin began making concessions to religion,[43] pious articles began to appear nearly every day in the official German-controlled Polish press—such as episodes from the lives of the Saints, discussions of the architecture of Gothic cathedrals, and news of church and parochial life. However, in the German-language newspapers such as the *Krakauer Zeitung* the attitude to religion remained as before: religion is *für den polnischen Bauern*,[44] *das Herrenvolk* having finished with this non-Germanic notion a long time ago. This cynically ambiguous official attitude to religion depended for its effectiveness on the mistaken assumption that the Polish intelligentsia could not read German very well.

* * *

On March 19, 1944, we observed large military trucks on the main arterial road nearby. We found out that they were headed for Hungary to forestall capitulation by the puppet government there. A few days later long trains of closed freight cars began passing through the Stróże station. At the gratings over the windows one could see hands and faces of people taking a breath of fresh air on their way to martyrdom. These were Hungarian Jews travelling to their deaths in the gas chambers at Auschwitz.[45] The trains, each of about 30 or 40 cars, each car carrying from 70 to 100 people, continued for over two months. Initially just two trains passed through each day, but later on the daily number increased to as many as ten, with similar numbers returning empty back to Hungary. Some of the more arithmetically inclined railroad workers calculated that from late March to late June over half a million people passed through. Although they were at the time unaware of the destination, they could hazard a guess and when asked would sometimes run their finger over their throats. Teenagers amused themselves by throwing stones at the wagons as they passed. We later heard from the few survivors that most of those being transported had had some belongings with them: a suitcase, and often money or wrist-watches or pieces of gold or silver jewellery. They had no source of water, and when the weather turned warm, people began dying of thirst in the closeness of the railcars. Occasionally, when the train was stopped, a guard with a canteen of water would go up to a car, helmet in hand, and return with a helmetful of money, rings, watches, etc., the price of a few mouthfuls of water. There was a story of a

---

[43]During the German invasion of the USSR Stalin, realizing that, despite the persecution of the Russian Orthodox Church by the Soviet authorities from immediately after the revolution, that institution still had a strong hold on much of the populace, began to make concessions to religion. From that time the Russian Orthodox Church was administered as an official arm of the Soviet state, although severely restricted in its operations. During the period of de-Stalinization following Stalin's death, many of the concessions were withdrawn, and anti-religious propaganda re-intensified.

[44]"[Religion is] for the Polish peasant."

[45]Starting in late March 1944, over 400,000 Jews were transported to Auschwitz from Hungary over a period of 2 months. Most of the deportees were gassed immediately upon arrival, following selection by doctors such as Mengele and Thilo. See: *Holocaust: A History* by Debórah Dwork and Robert Jan Pelt, W. W. Norton & Co., 2002.

railroad man at Stróże station who went from car to car collecting, for a price, metal containers, which he promised to fill—but never did. Could a Christian churchgoer find it in himself to cheat out of a cup of water desperately thirsty people faced with death? Alas, it seems so. Cases of transportees fleeing while the train was stopped or attempting something like a mutiny were extremely rare. Thus 20 soldiers per train, comfortably ensconced and drinking heartily, escorted two or three thousand people a few hundred kilometers to their deaths without meeting significant resistance.

It was around this time that the Germans adopted a hostage system—like all their methods of dealing with the populace of conquered lands, a system of extreme cruelty, made more macabre by the admixture of lying and hypocrisy. It worked as follows: they began by arresting people at random, mostly in the cities and larger towns, and had lists of their names with alleged transgressions inscribed on placards posted up according to administrative district for all to see. The transgressions included such things as "membership in a secret organization", "robbery", and "sabotage of the rebuilding of German-occupied Poland" —this last having been invented for those arrestees on whom nothing else could be made to stick. The arrestees were then subjected to a mock trial where they were inevitably found guilty and sentenced to death, but then conditionally reprieved, the condition in question being that there should be no attempts on the lives of Germans or their Polish spies, nor sabotage of military of administrative infrastructure in the administrative district of the guilty party. This subterfuge—namely the insistence that the death penalty was being carried out only on the guilty—was intended to circumvent criticism to the effect that they were using the barbaric system condemned by all civilized peoples of reprisals against innocent hostages. But of course in reality the court proceedings and passing of sentences were a farce, so there was little if any difference between this system and out-and-out hostage-taking. In any case murder in plain daylight was still resorted to. We heard of raids of German police and Ukrainian auxiliaries on small towns near Jasło—Kołaczyce, Dąbrowiec, and Frysztak among them—where scores of members of the intelligentsia were killed on the spot, and others arrested to supply the hostage system.

In view of the intensified hostage-taking, the presence in Stróże of Szczypta, the informer mentioned earlier, was threatening to us and others. He had recruited a staff of teenagers and even children to inform him as to the goings-on in the villages around about. One day he got word that four strangers were living in a hut near the river belonging to a certain Mogilski, so he informed the *Bahnschutz* police, and a squad of these with Emil Druśko at their head quickly surrounded the hut in question and began tossing grenades through the windows. Four men burst out firing, but Druśko was ready with his machine gun. Two escaped and two were killed: Dywan, at last, and an Englishman who had presumably escaped from a prisoner-of-war camp. The house burned down. This episode called attention to Szczypta, and soon led to his demise. A few months later two young men in alpine gear came knocking at his home. His wife answered, and while one of the men waited outside the other went in, giving out that he had heard that Szczypta supplied official work applications. The scent of money in her nostrils, Szczypcina told him that her husband had just returned from business in Tarnów and was resting on a sofa in another room. The visitor then went straight in and finished off Szczypta with a few shots, after which he warned Szczypcina not to move for 20 minutes on pain of death and left. The two were seen walking calmly in an easterly direction by farm laborers working in the fields. After half an hour had passed Szczypcina ran to the oil refinery a hundred paces away where a unit of soldiers was stationed, but was unable to rouse them to action. It took the head *Bahnschutz* guard another hour to organize a chase, and the perpetrators escaped. However, this successful assassination was paid for with the deaths of ten hostages. On the other hand, had he not been killed, Szczypta would have destroyed twice as many, in addition to which this killing put the wind up other collaborators, such as Emil Druśko and Marian Proszek.

Nevertheless, Proszek soon provided new evidence of his loyalty to the Third Reich. A certain Mr. Wiatr, the son of a local peasant, and formerly an important figure in an organization devoted to Polish folklore, a serious man of merit, was wanted by the Germans. In view of this he moved about using a false identity card. Proszek came across him one day, and asked to see his identity card, immediately recognizing it as false since he and Wiatr were old acquaintances. Said Wiatr: "Marian, you know me, your old compatriot!" Replied Proszek: "Yes, I do know you, and your identity card is false." And he handed Wiatr over to the Germans, who sent him to Jasło where he was tortured to death. It appeared the Germans failed to get any other names from him since no arrests of his contacts occurred subsequently. Proszek had been sentenced to death by the AK, and when he got wind of his sentence abruptly ceased his espionage for the Germans. Emil Druśko, on the other hand, kept at it, and was responsible for several more murders. An example: someone reported that in a hovel in Stróże that had formerly belonged to Jews, a Jew by the name of Sroka was hiding in the attic. And indeed a well-insulated hiding-place had been organized there for him by a servant, who continued bringing him food. When Druśko and his henchmen arrived the servant made a run for it, but they caught Sroka and shut him up in the cellar. Druśko was later heard bragging about how he had set his dog on him and tortured him before finally killing him: "I toyed with him the way a cat torments a mouse," he boasted. In addition to Proszek and Druśko there was a third *Bahnschutz* auxiliary called Bruno, similarly sanguinary, who, to take just one instance, shot a mountain-dweller dead on the station platform in Bobowa for being without an identity card.

I sometimes conveyed information to the AK. Thus for example when Mrs. Michalakówna, who was considered above suspicion, came to visit for a few days, I would give her a message written in "klinography", the code used by the AK, for her to pass on to "the son of the organ grinder." I also asked her to write to Olga Pamm, my wife's cousin, in Switzerland, asking her to cease writing to Mrs. Paulina Sz., since recipients of letters from abroad were sometimes summoned by the political arm of the police to explain their contents, and Paulina was understandably afraid of being called to account. The fact that people safe in Switzerland wrote to those in German-occupied Poland showed an ignorance on their part of the conditions in the *Generalgouvernement*. The existence of a regime that was at the same time so callous of human life and expert at brazen dissimulation simply defied belief if one were not subject to it oneself.

In the spring of 1944 Jan or Stefan Dobrzański occasionally visited us with their respective broods, and we held a feast in our apartment on the occasion of the christening of baby Agnieszka, daughter of Stefan and Zosia.

* * *

The Russians were then still gathering strength, it seems, for the big push that came on June 22, 1944. One could tell something was afoot from the panicky flight of the Ukrainians and *Volksdeutsche* from the East: train upon train passed through crammed with people, furniture, grain, animals—in short, everything but the kitchen sink. Poles were fleeing before the Ukrainians who in their savage fury set fire to whole villages and slaughtered the inhabitants by way of a parting salutation, and these in turn were fleeing before the Soviets from whom they could expect no mercy. The further west the trains travelled the less fearful the Poles felt and the less secure the Ukrainians sitting opposite them or cheek by jowl in the railcars. Among the disparate passengers there were policemen and men in the uniforms of officers of minor municipal militias, displaced persons from beyond the Zbruch[46] already in their second year of slow trekking along the roads, Volga Germans,[47] and Polish farmers who had settled lands acquired via the Polish-Soviet war. Many of these refugees—especially those accompanied by children—periodically lost their way since which railcar was attached to which locomotive often depended on accident or bribery.

When I went walking along the river bank where the railroad line ran parallel to it for a while, I would often pass idled railcars packed with fleeing Ukrainians and *Volksdeutsche*—"rabble", as they were called by Poles—and a smattering of Poles, many of whom were presumably collaborators and spies. In fact, most of the passengers were Polish speaking, but nonetheless anxious to avoid the reckoning awaiting them should they stay. There were also ordinary Polish refugees, with stories of confusion to tell: how they had been saved from Ukrainian bands armed by the Germans by what seemed to be Hungarian detachments, how the German military police had intervened on the side of the Ukrainians—although some Ukrainian bands were now attacking Germans—how the ranks of rampaging Soviet partisans were being swelled by escaped Soviet prisoners-of-war—called "Kałmuks"[48] by the locals—and how these sought to avoid the SS and German military police, who were likewise at this stage of the war not inclined to carry out sweeps against them. The picture of a sort of Balkanization of the region into

---

[46] In Polish, Zbrucz, a left tributary of the Dniester forming part of the border between Poland and the USSR established by the Treaty of Riga following the Soviet-Polish war of 1919–1921.

[47] Ethnic Germans originally invited as immigrants to Russia in the eighteenth century, living along the Volga around Saratov and further south. In tsarist Russia they had been allowed to maintain their German culture.

[48] The Kalmyks are a western Mongolic people, now forming a majority in the autonomous Republic of Kalmykia on the western shore of the Caspian Sea. In Polish common parlance, *Kałmuk* refers to Asiatic people generally.

territories temporarily controlled by various warring bands did not augur well for the future of postwar Europe.

Thus the Soviets renewed their offensive on June 22, 1944, were in Lwów within a month, and then quickly took Przemyśl and Jarosław. In July Leonard Zgliński, the husband of my sister Irena, came to us from Kraków. He told us that their daughter Basia and Irena herself had been blackmailed, Irena by a Polish policeman, and Basia by a brute egged on by a gang outfitted in SS uniforms, but possibly not authentic SS men. It had been decided that Basia should go to work for a family by the name of Świeżawski near Przemyśl, in the belief that this would be a safe haven, but the approach of the Red Army now made this doubtful, and Leonard was regretting the move. We later learned that her employers had gone to a property called Myłan they owned near Żmigród, as a place more likely to be safe in this time of turmoil. Leonard supposed he would soon be called on to return to his position as manager of the electric power station in Kraków. This did indeed seem an imminent possibility what with the Soviets crossing the Vistula without meeting significant resistance, and the attempt—albeit a failed one—on Hitler's life at the *Wolfsschanze*[49] near Kętrzyn, his East Prussian field headquarters, on July 20. It seemed then that the war must end in a matter of weeks. The threat of a break in communications with Kraków determined Leonard on returning thither on July 22 after a stay of two weeks with us, during which he had gained a much needed rest, although his nerves were obviously still frayed.

We could now hear the war, and sometimes even see direct evidence of battle. We felt the rumble of column after column of army vehicles returning from the front lines and train after train laden with troops, guns, and tanks passing through from Jasło. The retreat was in full swing. The last letter we received that summer from Lidka was dated July 11, shortly after which the post office ceased functioning. Then on August 1, the Warsaw uprising[50] erupted, and we knew we would not be hearing from her for some time. This situation caused us much anxiety—and there were other sources of worry.

On August 4, as I was returning home from my customary evening stroll, I stopped on the meadow next to the park to talk to a neighbor by the name of Dudzik on his way to the station to start his shift. As we parted a German motorcyclist passed by us going in the direction of Polna, and a few seconds later I heard gunshots. Clearly someone was shooting at the motorcyclist at the other end of the park. I quickly returned home and, since I was wearing only a light jacket, took a

---

[49] "Wolf's Lair"

[50] An attempt by the Polish Home Army (*Armia Krajowa*) to liberate Warsaw as the Red Army approached the eastern suburbs of the city. The Soviet advance halted on the eastern side of the Vistula until, after 63 days of fighting, during which about 16,000 members of the Polish resistance army and some 180,000 civilians were killed, the uprising was put down. Before retreating west, the Germans systematically razed over a third of what was left of the city, block by block. The Red Army entered the city only in January 1945.

warm coat, and with my brother-in-law set off in the direction of Wyskitna,[51] leaving the ladies behind. By this time the menfolk at the farmhouse had already herded the cattle over to Koszyk's property, the nearest to our farm, and left, abandoning, as it were, their womenfolk. We had gone scarcely two hundred paces or so when a man with a gun caught up with us and called out "Hands up!" Then a few others came up, all armed. They were dressed in a variety of clothes, mostly civilian. The leader demanded my *Kennkarte*—incidentally, this was the first time anyone had ever looked at it in an official capacity, so to speak—and, concluding we were not the enemy, began conversing with me, regretting that the "cursed German got away." He explained that the main purpose of the action had been to seize the German's weapon, but they had failed. He proposed that we sell them weapons, but of course we had none. Then he and his companions walked slowly towards the wood, and disappeared into it.

We two decided to spend the night at Stefan Dobrzański's, so we turned our steps thither, and, walking through the fields and keeping clear of the roads, we got there around nightfall. Stefan was not at home but his wife Zosia made up beds for us on the kitchen floor. In the morning Stefa came from the farm with bread and extra clothes, and told us that no one had come looking for us. So we waited till afternoon, and returned to the farm. We were lucky: the attack on the motorcyclist had not led to any reprisals against local males.

A few days later, while we were sunning ourselves on the bank of the Biała, my wife immersed as so often those days in an English novel, and I myself just idling the time away, we heard shots but were unable to determine their direction. We put on our jackets, and set off home. On the way we ran into Mrs. Rząca[52] near her house. She told us she had seen on the road, near the cross-roads and then again near the park, two horses at full gallop drawing a cart, and then a little later a squadron of soldiers running from Stróże towards Berdechów, but giving up the chase at the edge of town. At home we learned the details: soldiers from the AK had attacked the cart as it was travelling to Stróże, bearing the schoolteacher Miss Brach and three German soldiers, who had first taken her furniture off to Polna, and were now in the process of abducting her. The AK men had startled the horses, which reared up, causing one of the soldiers to fall under the wheels and have his legs broken. A second soldier had let himself be disarmed, but the third had run off in the direction of Stróże. Although no one had been killed, this time the Germans reacted. In the late afternoon of that same day, while I was gathering blackberries in a nearby wood and my brother-in-law and the other men were helping with the harvest, a detachment of German police showed up. They fanned out in front of the wood and swept across the field, finally surrounding our dwelling. One policeman stood by an open window and ordered it closed, while two others went inside and asked where the menfolk were. When they heard that the men were all working in the fields, they looked into the wardrobe, told the ladies—Stefa and her mother—not to be

---

[51] Part of Stróże.

[52] Wife of a neighboring farmer, whose sons Steinhaus was teaching. See above.

upset, and went over to the farmhouse, where they settled down to consume a large quantity of bread and butter and kefir, delivered themselves of the principle that "one should always look to one's proper nourishment and not allow other considerations to bother one overmuch," and, after instructing the ladies to tell the owner of the farm that he should report to the police station to inform them about local "gangs", departed. Thus once again we were lucky. The moment when the policeman had told them to close the window had been an especially bad one for the ladies since on more than one occasion the Germans had burned down huts with their inhabitants shut inside.

* * *

The war seemed to be at a standstill again. Although every day we could hear the sound of artillery, now in the direction of Jasło, now over towards Tarnów, there were no further visible signs of a German retreat. Soviet planes passed overhead during the day but at such a great height that they could scarcely be made out even against the clearest of skies. During the Fall of 1944, at night we heard the droning of great fleets of planes, which we presumed to be flying to Slovakia to drop men and armaments in forest clearings. These overflights then increased in frequency, but were never accompanied by dogfights or bombing—which was somewhat surprising since there were nearly always military trains standing at the Stróże railroad station on their way to or from Tarnów or Jasło, although nobody knew for certain exactly to what purpose. It subsequently became clear that the trains were transporting supplies and troops to Jasło.

On August 20, a whole detachment of Ukrainian police, consisting of around 100 men and 50 horses, arrived in our district, and were billeted in various huts and houses around about, including those on the Cieluchs' property. They were under the command of a unit of German police, including an *Unteroffizier* and an officer. The German quarters were strictly separate from the Ukrainian ones— presumably because the Germans did not relish hobnobbing with the Ukrainians. This was evidently not a formation intended for fighting at the front, since they had only horse-drawn peasant carts for transporting themselves and their belongings, which they kept in boxes and trunks. Nevertheless, they styled themselves a "supply column." They—the Ukrainians—were forever checking their belongings to make sure their comrades-in-arms hadn't filched them. They had money, gold, and jewellery stolen from Jews, and cloth and other booty taken from others they had murdered. They always kept their revolvers about them, even when walking around in their bathing trunks. The Germans had mustered locals to dig trenches, and sometimes the Ukrainian policemen idled away the time observing this work. They were used by the Germans in peacekeeping, searching for thieves, and on raids against unfortunates who had attracted their attention in some untoward way or other. They had an air of resignation about them, perhaps having to do with the realization that they had joined their fortunes to that of the losing side in the war. Some thought of running off into the woods, but feared they might be recognized by locals as Ukrainian collaborators. Their families had been sent off to Germany, or Austria, or Slovakia, but the news they had of them was not good: they were either relegated to camps or sent back to where they came from. They did a lot of damage while at the farm: they fed their horses with precious grain from the barn, let them run free in the fields through the crops, left their carts all over the place without regard for the farmwork needing to be done, and were constantly coming into kitchens demanding that something be cooked for them, even though they had their own field kitchen. And in the rest of the neighborhood they were similarly destructive and inconsiderate, feeding their horses wherever they pleased, taking bales of hay and clover at will from farmers and peasants, and threatening them with their guns if they objected. I found out that, although some of them had been actual policemen, or at least worked as farriers or grooms, there were also Gymnasium students among them who had had their education rudely interrupted, and even teachers of such, and a few who insisted they had been owners of estates or belonged to the wealthy classes. Polish refugees from the East brought stories of the Ukrainians as organizers and perpetrators of massacres of Poles found on Ukrainian territory. Of course, one cannot always believe such tales; however, the coarseness evident in the faces of this uncouth, ruffianly lot would incline one to believe them.

Everyone was relieved when, on August 31, 1944, the whole detachment departed, leaving behind the unmistakable signs of their stay: mountains of horse manure, pastures trodden down to the bare earth, and a few bloated carcasses. No longer did the scuffle of horses tied up at the corner of the house, or loud cries of "Jałny, Jałny!" —Jałny being the detachment's factotum, head groom, all-round intendant, and chief broker—wake us at night. However, this positive development was more than balanced by the terrible news of the fate of Warsaw, and the complete absence of news of Lidka and Janek.

Our neighbor Mr. Woźnica had set up a sort of booth on the edge of the road, where information might sometimes be obtained. One day I noticed a typewritten card pinned up there informing the populace that the Germans were permitting the establishment of a citizen's committee, which, conjointly with the Polish police, was to be responsible for safety and order in the district. Incidentally, Woźnica's booth had become a notorious drinking den, and he himself was now active in just those pursuits he had formerly blamed the Jews for: selling diluted beer, bartering vodka for army blankets, trading in gold, and so on.

By this time the Allies were making good progress against the Germans in France,[53] but in our part of the world the front seemed to have become permanently bogged down near Jasło. In early September a unit of ten German soldiers appeared at the farm, and began digging trenches for machine gun emplacements near the road, but the very next day, as if suddenly scared off by something, vanished. Then on September 13 a column of trucks—bearing ammunition, I think—drew up. Soldiers set to work preparing a rough road cutting through the park, but again by late the next day they were gone. It was on September 15 that I saw flares sent downwards from passing airplanes during the night. Over the period September 17–21, we were again visited by a column of trucks; this was a military engineering

---

[53]Since "D-Day", June 6, 1944, when a vast Allied armada landed American, British, Canadian, and Free French troops on the beaches at Normandy.

unit responsible for repairs. One of their company buried a box of ammunition in the barn, which was later dug up and handed over to the AK via a certain Mr. Kossowski.

* * *

It was time for the school examinations. All three Rząca boys and Dela's son Janusz had completed the Gymnasium level first grade back in the spring, and since then the two older Rząca boys had completed the second grade. My brother-in-law and I had taken it in turns to teach the youngest Rząca boy, Tadeusz, and another boy, Dudzik, who had also completed the first grade that Fall, while I taught the older Rząca boys unaided. I also helped the young Irena Betuchówna, a fourth grade student, with her French, and taught English to the Zaremba children. As the official district examiner in mathematics and German, at specified times I had also to examine students of other teachers.

From July 23, supplies from Gorlice were cut off. People were now afraid to travel, especially in the direction of Jasło, in case the railroad bridges were mined. By that Fall, the Germans had even cut openings for explosives at the bases of telegraph poles near the railroad lines. Their attitude towards the Polish populace went through a one-hundred-and-eighty-degree turn; now, instead of berating us, they sought our help, so they said, in combating Bolshevism. In practice this meant helping them dig trenches, and from July on, they had people of all sorts digging trenches on the far side of the Biała River. The trenches eventually stretched as far as the eye could see towards Grybów, and along the bank of the river all the way to Tarnów. The willows and other trees on both sides of the river were cut down, so that the copse that had hidden us from intruders while we sunned ourselves or sat reading on the bank, raising our heads from time to time to gaze at the soothing view of the steep bank opposite with its dense growth of trees, was no more. We were bereft also of this remaining idyll.

We still had no news of our daughter. In mid-September we heard that Janek was in Milanówek,[54] but that he also had heard nothing of Lidka for over six weeks. We began to think that perhaps she had after all been sent as a forced laborer to Germany. Then at last a postcard with no return address arrived[55] informing us that she had been reunited with Janek in Milanówek, and instructing us to write to them through Irena, in Kraków, because that was where they were going.

In the meantime a full-scale evacuation was in progress in Jasło. The inhabitants were instructed to quit their dwellings by a certain date, and anyone failing to comply found himself thrown out bodily with—if he was lucky—a suitcase of belongings. Once the town was emptied of people, the German military, and especially the SS, went through it stealing every moveable object of any value. Everything remaining was broken and soaked in naphtha to await the flames. The

---

[54] A town very close to Warsaw.

[55] Since at this time there was no postal service, presumably such communications were effected via contacts travelling to and fro.

larger buildings were mined. The final conflagration was, however, postponed till November, perhaps to coincide more or less with the end of the Warsaw Uprising.

While on the one hand the Germans seemed to be bent on destroying as much as they could of Poland before retreating, on the other hand official propaganda took a new, friendly, turn. The *Führer* was heard to praise his Polish allies, who spontaneously and gladly took up shovels and set to work digging lines of trenches to serve as last-ditch, so to speak, lines of defense of the Fatherland against the advancing Bolshevik hordes. Furthermore, the German *Generalgouverneur* of Poland announced that in every Polish town the position of *starosta* would be supplemented by a town council made up of Poles authorized to offer advice. The finale of this crescendo of concessions consisted in an offer to recruit Polish volunteers for "auxiliary service" in resisting the Red onslaught, with the note struck repeatedly *fortissimo* of Britain's sacrifice of Poland to Stalin, accompanied by chords of sympathy for poor Mikołajczyk.[56]

* * *

It was no longer remotely possible for the Germans to talk of a victory to be secured by the genius of the great leader, with any hope of being believed. A new aim had to be invented to replace the old one. Since the Slavic *Untermenschen* had for some time been seen to be in a winning position, the thesis of German racial superiority had had to be abandoned. Once the Allies had landed in Normandy, the V-1 and V-2 rockets, Hitler's secret weapons,[57] hardly held up the progress of the war. The retreat had now to be sounded. Goebbels hammered away at Germany's purported historical mission of defending Europe from Bolshevism and Americanism: "Once Germany fails in this mission, history will have lost all meaning," he wrote. But since this sublime metaphysical argument could hardly be repeated by your rank and file *Parteigenosse*[58] without an accompanying wink, attached as he was mostly to beer, sausage, and other such concrete realities, it was necessary to reduce it to something more down-to-earth. Thus since Germans had always been fighting as much for their rightful existence as a nation as for victory over others, so it was claimed, the war was now called a *Verteidigungskrieg*. Like the *dénouement* of an amazing magic trick, the war was revealed for what it had always been: a *defensive* one. The phrase "defensive war" was very cleverly chosen, since each of the two words comprising it lends the other justification.

---

[56] Stanisław Mikołajczyk (1901–1966), Polish politician, Prime Minister of the Polish Government-in-Exile 1943–1944, and Deputy Prime Minister 1945–1947 in postwar Poland before the USSR took full political control of Poland.

[57] V-1 rockets were used to bomb southeast England from June 1944, and, towards the end of the war, sites in Belgium. They were a pulse-jet-powered archetype of what would later be called "cruise missiles." The V-2 rocket, a ballistic missile, and the prototype of all modern rockets, was first used against London in September 1944. Over 3000 V-2s were launched by the *Wehrmacht* at Allied targets, mostly London and later Antwerp, causing the deaths of an estimated 7250 people. It is estimated that some 12,000 forced laborers died working at the production sites.

[58] "party member"

As I have already noted, this official change in tone was accompanied by a change in attitude towards Poles. There was no more face-slapping on station platforms or kicking of passers-by in the street: these and the other demonstrations of the contemptuous German style of the "fat" years ceased. On the other hand, the *Generalgouvernement* Police Chief Koppe's[59] program,[60] and Sauckel's[61] promise to leave no stone on stone for the Soviets to benefit by, remained in force. This is typical: the Germans did not recognize any "this or that" exclusive alternatives; confronted with such a choice they always chose "this *and* that." Another example: Following on the mass executions of hostages on the streets of Warsaw and Kraków and in the Pawiak[62] and in Niepołomice,[63] there came the uprising, followed by the utter destruction of Warsaw. The inhabitants of whole suburbs of Warsaw were murdered during this last phase. Even hospitals were not spared: the hospitalized wounded, doctors, sisters of charity working in the hospitals—all were killed. Many thousands were transported to Germany, or to Auschwitz to be gassed. And meanwhile the *Generalgouverneur* made a general request for alms in the form of foodstuffs, medicines, and money for the benefit of displaced persons—victims, he claims, of the "insane activities of the degenerate Warsaw intelligentsia."

Lidka and Janek arrived in Kraków on October 8, 1944, and, leaving her husband there at his mother's, she came on to us in Berdechów on October 22 with Mrs. Garlicka. They came by train, and since there were no passenger trains, rode in the brakeman's caboose, whither they were escorted by kindly German soldiers past train guards who would normally have made difficulties for them, civilians being forbidden access to military railcars. Not long after her arrival, we heard that Janek had been admitted to hospital in Kraków with a disease which the doctors were unable to diagnose, but suspected might be leukemia. We were, naturally, overjoyed to see our daughter after three years of separation. We listened intently to her tale of those years in Warsaw, learning of the fate of many friends, relatives, and acquaintances. I will limit myself here to just those events of greatest significance for us.

---

[59] Wilhelm Koppe (1896–1975), German Nazi Commander responsible for numerous atrocities against Poles and Jews in *Reichsgau Wartheland*—the region of west-central Poland annexed by the Germans in 1939, with Poznań as its chief city—and the *Generalgouvernement* during the German occupation of Poland in World War II. In October 1943 he was made *Höherer SS- und Polizeiführer* in the *Generalgouvernement*, with headquarters in Kraków.

[60] Koppe ordered all prisoners executed so as to avoid their being freed by the advancing Red Army.

[61] Fritz Sauckel (1894–1946), Nazi politician. Hitler's chief recruiter of slave labor during World War II. Sentenced to death by the International Military Tribunal at Nuremberg.

[62] A prison built in 1835, used by the occupying Germans as their largest political prison and as part of the Warsaw concentration camp system. Some 37,000 prisoners were executed there and another 60,000 sent to extermination and concentration camps. Shortly before the Warsaw Uprising 2000 men and the remaining 400 women prisoners were transported to Gross-Rosen and Ravensbrück. The prison was destroyed by the Germans in 1944.

[63] A town on the Vistula, 25 km east of Kraków.

Finding a flat in Warsaw hadn't been easy—in fact would have been well nigh impossible without the help of Janek's friends and the general attitude of absolute intransigence of most members of the Warsaw intelligentsia *vis-à-vis* the Germans. These made a point of aiding those sought by the German police, sharing what they possessed with those in need, and opposing the German occupiers— their laws, language, edicts, absolutely everything issuing from them—in every possible way. In particular, they boycotted concert houses, theaters, and cinemas. Their policy of isolating themselves from everything German frustrated the German plan of fostering among the indigenous inhabitants a sense of the superiority of the colonizers' culture—on the English pattern, as it were. Exactly the opposite occurred: the Polish educated class managed to convey their sense of German inferiority to any German attempting to associate with them.

To make ends meet Janek and Lidka ran various small businesses. One such involved the sale of colored scarves made in the Warsaw ghetto, which Janek hired people to distribute, or distributed himself. When this business dryed up he turned to selling gold items—mainly gold watches—supplied by Mr. Rudolf, the husband of my wife's sister-in-law Irka, and a Mr. Juliusz P. These two had arranged transportation to Warsaw from Lwów of a whole suitcase of such items— watches, gold chains, gold cigarette cases, etc.—on behalf of the owners, a Mr. and Mrs. Zipper. In payment for this service they were temporarily loaned a few of the Zippers' articles, to be returned after the war. Janek was commissioned to sell these precious keepsakes to jewellers in Warsaw, since for obvious reasons Mr. P. did not relish the idea of traipsing about Warsaw hawking their wares.

While travelling in connection with a related such business in May 1942, he was robbed, and we only now heard the details of what had happened. He had coins and little gold articles with him in a small bag which he was transporting from Lwów to Warsaw, and had dropped in on us in Osiczyna *en route* on May 16. The car he was travelling in, hired for the purpose, was being shared by a Mr. Zaborski. When they saw a police roadblock up ahead, Zaborski advised Janek to hide the bag between the seat cushions, but after the police had finished with them and waved them on, he couldn't find it. When, in Warsaw, he told Lidka of the loss, she immediately saw Zaborski's hand in the matter. Janek went to the police, promising them 30% of the worth of the valuables, and the police did indeed find half of them in Zaborski's flat. The inscriptions on the little chains and crosses matched Janek's description so accurately that Zaborski had no choice but to admit he had taken the bag, but tried to pass off the action as just a "joke". By coincidence, Zaborski turned up in Stróże shortly after Lidka arrived on her visit to us. By then she had told me about his "joke", so-called, and also that by way of revenge he had sought to denounce Janek to the Germans for money-changing. Fortunately, he didn't know their new address in Kraków. It seems he'd come to Stróże for a few days to see a Mr. Piotrowski, a Lwów stationer and local administrator, to obtain a document certifying that an attempt had been made on his life by Ukrainians, but he was turned down because he was known in Lwów as having spied for the Germans, and Piotrowski suspected that the assassination attempt had most probably been made by the AK.

The engineer Kossowski, Piotrowski's son-in-law, knew about Zaborski's shady character, but continued—out of gullibility, perhaps, or because he secretly fed on them—to believe in some of the latter's fantastic stories—for instance, his claim to have been a victim of the NKVD and the only person to have been rescued alive from one of their dungeons in Lwów a few days before the arrival of the Germans, the rest—so he said—having been machine-gunned through the narrow basement windows by Jewish women wearing Red Cross armbands. He further claimed to have then denounced these Jewish women to the Germans.... I tell this mad story as an example of what supposedly intelligent people such as Kossowski were prepared to believe about Jews. That same Kossowski maintained that the Bolsheviks deported only two Jews from Lwów during the whole year and a half of their occupation, even though it was well known that the figure was in the many thousands. He also repeated the racist hearsay that the student of the Lwów Polytechnic who had been killed in 1938 had in fact been murdered by Jews because of his support for the laws condoning separate seating for Jews in lecture halls. One easily guesses the tendency of Kossowski's fantasies.

From Lidka we learned the fates of many of our old friends and other people we knew. She told us about Dr. Blumenfeld: his assistant Dr. Ostern had suicided by taking poison, and Mrs. Ostern had fled to Warsaw with their little daughter, so he was left alone in Lwów. A Ukrainian militiaman had come for him and taken him to the Gestapo headquarters on Pełczyńska Street. But on seeing his bandaged face—cancer had eaten away part of his nose so he kept his face bandaged to hide the damage—and looking into his eyes, one of the Gestapo men said *Was machen Sie hier? Gehen Sie sofort nach Hause.*[64] The Ukrainian, fearful of forfeiting his bounty, protested, but the Germans merely kicked him and repeated the injunction that Ignacy go home. He went to stay in a nearby village, but as the cancer was worsening, eventually decided to go to Warsaw to have it treated. What happened to him there, in particular during the uprising, remained unknown. His elder son was serving in the AK. His younger son was imprisoned, caught pneumonia, was released, and when last heard from was in Czortków, in all likelihood waiting for the Red Army. Much later we heard that he had perished.

Dr. Ziemilski[65] left Lwów for Warsaw with his wife and mother-in-law in 1942. He travelled in a separate compartment from the two ladies as a precaution, was recognized and possibly denounced by another passenger, and was killed. However, his wife and mother-in-law made it to Warsaw, and survived the Warsaw Uprising. It was generally assumed that Ziemilski's son Jędruś was serving in the AK somewhere.

Our friend Mrs. Loewenstein and her son, a very talented young man of around 20, continued living in Lwów during the German occupation, making no effort to escape. Then one day someone called to summon her to the police station, and, failing to find her home, left a note. Can she have failed to understand the dreadful

---

[64] "What are you doing here? Go home at once."

[65] A friend from Lwów; see Chapter 10.

import of such a summons for herself and her son, or was it just that comprehension was by now beyond her? She dutifully went along with her son as ordered, never to return. The mind feels ineradicably blighted by the recurring thought of how this beautiful, cultivated lady and her bright son were led to frightful, squalid deaths.

Following the murder of her husband, our friend Mrs. Achender had found herself alone in Warsaw with her child. She kept inside as much as possible, but once when she ventured forth children began chanting after her: "A Jewish woman, a Jewish woman!" A Polish policeman all in blue[66] stopped her and demanded 500 złotys to let her go. She went back into the apartment house, but unfortunately her landlady had no ready money, and while people were hurrying off to get the money from acquaintances, the policeman, too impatient to wait more than the 15 minutes he had set arbitrarily as a time limit, took her off and handed her over to the Germans. Her child survived, however.

Fredek Borowicz[67] had a very lucky—and odd—escape. He was living in Warsaw with all of his family, working in a factory as electrician. He told the story of how a coworker from Poznań began telling him of his experiences in Lwów under the Soviets. "You had to be on your guard all the time," he said, "because there were Jews in disguise everywhere." "In disguise?" asked Fredek. "Yes, they were disguised as streetcar conductors, Ukrainian soldiers, postmen, you name it." But to get to his narrow escape: he was walking along the street minding his own business when he was stopped by a Polish auxiliary, a spy, who called to others nearby, including a German officer, and they grabbed him and shoved him into a droshky.[68] Fredek kept his nerve and told them in a level voice to leave him be because they'd be wasting their time with him. The German then ordered the droshky to stop, and the spy took Fredek over to the wall at the edge of the road, frisked him, and then, clutching his head dramatically, said quietly "You have the hide to walk around Warsaw in broad daylight! What can one do with such people?" "My life is in your hands," said Fredek. "OK," was all he answered, after which he went over to the waiting pack, and shouted *Alles in Ordnung*.[69] Nevertheless, they took Fredek to the *Kripo* headquarters in Aleje Ujazdowskie,[70] where he spent the night in a cell in the basement. In the morning the same Polish auxiliary arrived, had him freed, and they left the building together. By sheerest chance they ran into Lila Holzer on the street. Fredek pretended not to know her, but, suspecting nothing, she called out "Don't you recognize me?" The Polish policeman asked Fredek who she was, and then said unexpectedly "Should you need my help, call me," and gave Fredek a name and telephone number. Somewhat later, when a cousin of Fredek's was arrested, he tried

---

[66] Polish policemen of the *Generalgouvernement* wore a navy blue uniform, whence their nickname the "Blue Police" (*Granatowa Policja*).

[67] Possibly Alfred Borowicz (originally Horowitz) (1909–2006), brother to Olga Pamm and Irena Beczkowska (*née* Horowitz).

[68] A two-wheeled or four-wheeled public carriage.

[69] "All is in order."

[70] Ujazdowskie Avenue, a street parallel to the Vistula, belonging to the "Royal Route" in Warsaw.

the number, and was answered by a woman who said she was the man's mother, and that he had been sent to Auschwitz and died there. Fredek could only conclude that the man who had saved him had been working in the *Kripo* as an undercover agent of the AK.

I shuddered as I listened to Lidka's elaboration of her "brush with death" of a year before, which I mentioned earlier, but about which we had then only the general outline. In the autumn of 1943, more than a year ago (I'm writing this on December 8, 1944), she had taken a bus destined for a village outside Warsaw, where she was to work as a governess. However, at the edge of the city the bus was stopped by Polish policemen, who ordered all the passengers off in order to determine if any of them were Jews. Although the passengers jeered at them, the policemen were zealous, and ended up taking two people with Lwów identity cards—one of whom was our Lidka—to the Gestapo. She could have bought the police off with a payment of 500 złotys, but didn't, and she and the other passenger were taken to the Gestapo headquarters. On the way one of the policemen explained to her that there they had a "doctor", an "expert" in racial characteristics. "Of course," said Lidka, "Aryan blood being altogether of a type by itself, he simply takes a blood sample." The gendarme looked somewhat abashed at this, and later on, in front of the Germans, he lied, telling them that he heard her say *Oy vai.*[71] They were taken into a room where two handsome, blue-eyed, blonde young men in uniform sat at a table facing a cluster of Jews, standing, evidently quite badly beaten, and dressed in rags. She gathered they had recently been rounded up, many from their hiding places in basements, and had been given away partly by their inability to speak Polish well. The two smooth young Germans ran insolent appraising eyes over Lidka. One of them claimed, in a voice showing utter indifference to everyone present but his companion, that a Jewish woman could not have such a slim build or such shapely legs, and the other enunciated some nonsense about the "Euroasiatic type". When Lidka came before the collaborationist "doctor", he questioned her about the shape of the wafer used in the Roman Catholic communion, about the "Ave Maria" and "The Apostle's Creed".[72] He also asked why her destination, the village Kotowa, was not on her *Kennkarte* as the place she was staying at, to which Lidka responded that it was an ancestral village formerly belonging to her husband's family, and that she herself was descended from the Wóycicki[73] family, of Tatar origin. Finally, the so-called doctor declared that no Jewish woman could possibly speak Polish as well as Lidka, and they took note of her address, ordered her to return in two days with her husband and certain documents, and let her go. Janek did go there with the requisite papers, but was unable to locate the "commission" in question. In any case they had somehow gotten hold of the wrong address, because Lidka heard that they had come looking for her at an old address she had quit long before. Thus it seemed

---

[71] Yiddish for "Woe is me".

[72] Traditional Catholic prayers.

[73] Possibly Kazimierz Władysław Wóycicki (1807–1879), Polish ethnologist, folklorist, writer, and publisher. Historian of Warsaw.

that she and Janek had extricated themselves completely from what might have been a bad situation.

Lidka told us that Tadzio Hollender had been killed in Warsaw, having been denounced as a communist even though he was one of the very few who had had enough integrity not to join the Soviet Writers' Union in the days of the Soviet occupation of Lwów, not wanting to be considered a Bolshevik sympathizer. Stanisław Saks was also killed, having come under suspicion as a former Polish officer in mufti chiefly because of his military-style mustache. And Ginczanka had been sent to Auschwitz.

Through Janek, Lidka had met many literary figures during those years in Warsaw. She actually stayed at Jerzy Zagórski's place and then Jerzy Andrzejewski's[74] during the last period of her Warsaw sojourn. In between bouts of trying to make enough money from trading in gold to support himself and Lidka, Janek wrote essays, for example, one about Conrad,[75] and others about Stendhal,[76] and Tacitus.[77] As I have already mentioned, the Warsaw intelligentsia held themselves strictly apart from the Germans: this psychological isolation and the concomitant spiritual independence was perhaps the greatest wartime achievement of this class. Their attitude had nothing to do with snobbishness, however: they maintained relations with the Warsaw populace, frequently intervening to save Jews and others, in addition to engaging in other activities such as boycotting German cinemas and theaters. There was complete solidarity among the members of this class as far as their anti-German attitude was concerned: thus Ferdynand Goetel[78] was just as much an enemy of the Germans as Kotarbiński,[79] and that Kossak-Szczucka[80] hid Jewish children was nothing exceptional among members of the intelligentsia.

---

[74] Jerzy Zagórski (1907–1984), Polish poet, essayist, and translator. Published in the underground magazine *Kultura Jutra* during World War II. He and his wife Maryna, also a translator, fought in the Warsaw Uprising in different units. Jerzy Andrzejewski (1909–1983), prolific Polish writer. Joined the communist party in 1950, but left in 1956 at the time of the uprising in Hungary. From the 1960s a strong supporter of Poland's anti-communist movement.

[75] Joseph Conrad (originally Józef Teodor Konrad Korzeniowski) (1857–1924), Polish-born English novelist. Considered one of the greatest novelists writing in English.

[76] Marie-Henri Beyle (pen-name Stendhal) (1783–1842), French realist writer, known for his acute analysis of his characters' psychology.

[77] Publius Cornelius Tacitus (AD 56–AD 117), Roman senator and historian of the Roman Empire. His *Annals* and *Histories* span the period from AD 14 to AD 96, that is, from the death of Augustus to the death of Domitian.

[78] Polish novelist, playwright, journalist, and political activist. Lived from 1890 to 1960.

[79] Tadeusz Kotarbiński (1886–1981), Polish philosopher and logician. Studied under Kazimierz Twardowski. A representative figure of the Lwów-Warsaw school. Created a philosophical theory called "reism".

[80] Zofia Kossak-Szczucka (1889–1968), Polish writer—mainly for the Catholic press—and World War II resistance fighter. Co-founded the wartime Polish organization "Żegota" set up to assist Poland's Jews escape the Holocaust. She was arrested and sent to Auschwitz in 1943, but survived the war.

For their subversive and disdainful attitude they earned the especial hatred of the Germans, and in the speeches of the *Generalgouverneur*, he often sounded a particularly hysterical note of vituperation when this "degenerate" class was touched on. For this reason members of that class were constantly being singled out for arrest. Thus Rettinger[81] was pointed out in the street by a Uniate[82] priest and sent to Dachau,[83] where he died, while Hanka Ż. was sent to Auschwitz, and Mrs. Dora Blum, mistaken for someone else, was also sent to Auschwitz. In this way were we bereft of these and many others of our circle of acquaintance as it had been prior to late 1939—such as Saks, Janek Adamski, Herzberg, Hollender, and Lulu Oberländer.

Lidka also told me that Krahelska[84] and Handelsman[85] had probably been denounced by the NSZ.[86] If true, then this was an unpardonable crime on the part of the NSZ, one tending to aid rather than hinder the enemy, since Krahelska and Handelsman were shining pillars of Polish resistance. Kotarbiński took their arrests as a timely warning to himself, and quit Warsaw.

During the occupation of Warsaw, some scholarly seminars were run. Extensive teaching was also carried on underground at both the Gymnasium and university levels, and volumes of illicit poetry were published. There were several hundred underground newspapers and magazines, including an illustrated one devoted mainly to humor. Private economic life—the underground market—accounted for more than 95 % of all economic transactions, so that the official system imposed by the Germans was reduced to a mere surface activity. Even the Germans traded on the black market, since one could obtain anything one's heart desired, although at a price determined by the gold standard. As a result, the złoty current during the

---

[81] Mieczysław Karol Rettinger (1890–1944?), Polish literary critic and political activist. Frequenter of the "Scottish Café" in Lwów.

[82] Term applying to "eastern Catholic churches", in Poland meaning mainly the Ukrainian-Greek Catholic Church or the Slavic Catholic Church.

[83] The Dachau concentration camp, the first of the Nazis' concentration camps, located near the south German town of Dachau, began operating in 1933. Originally intended for political prisoners and Jews, it came to be used for prisoners from all German-occupied territories, but mostly from Poland. It is estimated that from 1933 to 1945 between 200,000 and 250,000 prisoners passed through the camp, with the number of those killed there anywhere from 33,000 to 148,000 (if those perishing in sub-camps are included).

[84] Halina Krahelska (1892–1945), Polish writer and political activist. A member of the early Polish Socialist Party, she took part in the 1917 Russian Revolution. She returned to interwar Poland where she was active in supporting measures for social improvement. During World War II she was in the Polish resistance. She perished in Ravensbrück concentration camp.

[85] Marceli Handelsman (1882–1945), Polish historian at Warsaw University. In hiding during World War II because of his Jewish background, he took an active part in underground education, serving as a professor of the underground Warsaw University. Also served in the AK. Arrested by the Gestapo in 1944, he was murdered at the Mittelbau-Dora concentration camp on March 20, 1945.

[86] *Narodowe Siły Zbrojne* (National Armed Forces), an anti-Soviet, anti-German paramilitary organization forming the right wing of the Polish resistance during World War II, and taking part in the Warsaw Uprising.

war was by 1944 worth only a hundredth or less of the pre-war złoty. The new, wartime, banknotes, bearing the signature of Młynarski,[87] represented a currency *sui generis*, lacking any emblem of the state, not guaranteed, or backed, by any holdings of any kind, printed in Polish as if to persuade Poles that it was in some way independent of the machinations of the *Generalgouvernement* and the Third Reich. These banknotes represented in concrete form just one more proof of the stupid cunning, the perverse obtuseness, of the German leadership: they thought that once Poles had obtained these worthless bits of paper in exchange for their cattle and corn, milk and wool, time and work, they would somehow feel they were worth something merely because their compatriot, poor innocent Młynarski, was issuing them. However, soon even the *Generalgouvernement* was seen to make tacit admission of the primacy of the black market when it ordered its vodka monopoly to sell vodka at the black market price of 140 złotys a liter rather than the official price. At the same time they began to print vast quantities of 100 złoty and 500 złoty banknotes to cope with the inflation induced by the low estimation in which their money was held by the public. And the more the official currency was distrusted, the more inflation rose, pushed up by the underground economy—the official above-board economy having negligible impact.

Every day, conversation with Lidka elicited news of more deaths in the long litany of losses of friends and acquaintances. The Krzemickis were arrested in Warsaw essentially through their obstinate refusal to leave their apartment, and in the end perished. The lawyer Szumański was arrested once but released a year later. But then he was re-arrested, and this time he was done for. Janusz Korczak,[88] a cousin of my sister Irena's husband Zgliński, had been running an orphanage, and decided to move it into the ghetto. When it had become clear what fate the Germans had in store for himself and the children under his tutelage, he decided to stay with them when they were ordered out of the ghetto, ignoring friends imploring him to make a break for it. So he went with the little children, tidily and elegantly dressed as if going for an ordinary stroll, and helped them board the train that took him with them to their deaths.

---

[87]Feliks Młynarski, a Polish economist, was appointed head of the *Emissionsbank in Polen* ("Issuing Bank in Poland") by Hans Frank, the *Generalgouverneur* of German-occupied Poland, in 1940, answerable to the representative of the *Reichsbank* Fritz Paersch.

[88]Janusz Korczak (pen name Henryk Goldszmit) (1878 or 1879–August 1942), Polish-Jewish pedagogue, children's writer, publicist, pediatrician, and social activist. During the *Grossaktion Warschau* (July 22 to September 21, 1942) he refused to forsake his orphans, staying with them when they were sent from the Warsaw Ghetto to Treblinka extermination camp. Over the indicated period approximately 300,000 Jews were removed from the Warsaw Ghetto, most to Treblinka. When the Nazis entered the ghetto in April 1943, the Jewish remnants staged a revolt, lasting a little under a month. The ghetto was then liquidated, with a further 49,000 being transported to camps, and around 7000 shot on capture.

It seems Winawer died of a heart attack, and Szenwald died mysteriously in Russia while fighting in the Polish division commanded by Berling.[89]

I made an attempt to estimate the losses to Polish mathematics occasioned by the war. If we include descriptive geometry and logic, then, as of the time of writing, December 13, 1944, the losses due to death, deportation, emigration, and murder, supplemented by the number merely missing, amount to 15 professors, 12 docents, and many other mathematicians not involved directly in tertiary institutions. Inferring from this estimate that the loss in mathematical productivity was in the vicinity of 70 % of its pre-war value, I calculated that we would need at least 25 years of strenuous recruitment and training to make up the loss—assuming, of course, peace and favorable economic conditions in postwar Poland. Although we had managed to bring out just one more issue of *Studia Mathematica* in Lwów in 1940/41, all other Polish scientific journals ceased publication from 1939, and of course we received no further issues of foreign scientific journals.[90]

The contents of Polish libraries were in part carried off by the Germans, and in part stolen by others. The wonderful library of the Warsaw University Mathematical Seminarium shared this fate. I don't know what happened to the library of the Jagiellonian University Seminarium in Kraków, but I do know that Orlicz managed to save about half of the books and journals of the Lwów mathematical library by having them transported to some private storehouse in 1942. The losses in Polish books generally were tremendous, from private as well as public libraries. For example, the several hundred books Lidka and Janek acquired when they first went to Warsaw had had to be abandoned when the war forced them to quit their apartment.

The last place they lived in before leaving Warsaw was at 81 Szreger Street in Bielany,[91] and Lidka told us how she had been able to witness some of the uprising from their window, and from the roof of the house. She saw great explosions which seemed to toss whole buildings into the air, fires everywhere, and at night the play of searchlights and flares, and could even distinguish the different caliber artillery by their roar. She and Janek had become separated since Janek was out somewhere when the uprising erupted, and no longer able to get back. Following the collapse of the uprising, the entire civilian population of Warsaw was expelled from the city, so Lidka had to quit the apartment, and, taking a small travelling trunk, found

---

[89]Zygmunt Henryk Berling (1896–1980), Polish general and politician. Fought for Polish independence in the early part of the twentieth century. During World War II, after deserting the Polish Army of General Władysław Anders, he later became commander of the 1st Polish Army in the USSR. Played a secondary role in the postwar Polish government.

[90]The losses to Polish mathematics—and to Polish science generally—have never been fully evaluated. Such an evaluation was rendered especially problematical by the annexation of a large part of eastern Poland by the USSR following the war, and subsequent Soviet politics inimical to Poland. However, a few times attempts were made to estimate losses of mathematicians, in particular in *Fundamenta Mathematicae* 33 (1945). See also below. *Note added by Roman Duda, editor of the second Polish edition*

[91]A northern suburb of Warsaw, on the left bank of the Vistula.

herself in the street near a group of women being ushered off by soldiers. One of the women, who had been selling butter and milk on the street, came up to her, and taking advantage of the inattention of the soldiers, wrapped a shawl about her shoulders, and had her mingle with the other women. She walked off with them, abandoning her travelling case on the sidewalk. Fortunately, someone later retrieved it for her. Finally, she left Warsaw on foot with the Andrzejewskis, with just the travelling case of possessions. They spent nights sleeping rough or in abandoned shacks in villages, hoping that the frontline would pass over their heads without sweeping them up with it. The frontline wavered capriciously, and people caught near it had to run from village to village to evade the combat zone. At last they reached Milanówek, where they met up with Mrs. Sawicka and Mrs. Mroczkiewicz, *née* Skulska, who had been hiding there already for a considerable time.

As for Janek, when the uprising broke out he found himself about to be herded onto a transport train headed for Pruszków.[92] Pretending to be drunk, he went up to a railroad worker who looked sympathetic, and gave him a gold ring in exchange for his cap and railroad jacket. He then circled around Warsaw via the outskirts, and was for a short time involved with Fredek in fighting with a partisan AK detachment commanded by Captain Lanca. Eventually, he was able to get to Milanówek, where he met up with Lidka.

A few days later they moved on to Kraków, to the apartment where Janek's mother and sister were staying. Lidka left to come to us, and, as I mentioned earlier, a little later Janek fell ill, and was hospitalized. He felt better, checked himself out, but then relapsed and had to return. The doctors were unable to diagnose his illness. They suspected leukemia, then paratyphoid fever,[93] then pneumonia. He was lucky to have Mrs. Wyszomirska and his sister, who was working as nurse in the hospital, to look out for him. My sister Irena Zglińska also visited him to see that he was being treated appropriately.

* * *

From mid-1944 the chain of authority in the surrounding villages was reorganized, the former *starostas*, tax collectors, administrators of the so-called "trading centers", and so on, being replaced by police of various kinds: German, Polish or Ukrainian, ordinary or specialized. The change seemed to have to do with a new and urgent task, namely that of digging defensive trenches, to which end the new officialdom was to round up willing or unwilling trench-diggers. The recruitment system varied from village to village. Thus in Gródek, all males from 16 to 35 were compelled to register for the work, the first batch being taken to a camp in Biała, while later ones who turned up with gifts of vodka and sausage were exempted. But in the villages neighboring on ours a different system was used. The laborers

---

[92] A town on the western boundary of the Warsaw urban area. During World War II, a large transit camp was organized in the Train Repair Depot there to house evacuees expelled from Warsaw, many of whom were subsequently sent to concentration camps.

[93] An enteric illness related to typhoid fever, but caused by a strain of salmonella.

were to be paid ten złotys a day, or, occasionally, with bread or other items such as shirts or jackets. No one was compelled to work except those who had been singled out earlier by the local administration, and many—especially boys and even some girls—volunteered. Each village had two policemen billeted on it in a supervisory capacity whose upkeep had to be met by the villagers. The whole project came under the umbrella of the *Organisation Todt*,[94] which sent more supervisory personnel. However, these tended to be elderly men, in some cases so old as to be semi-invalids, and the demands they made of the workers were merely token, so that the diggers spent much of the time smoking, chatting, and walking between the campfires kept lit at intervals along the trenches. But once in a while an officer with overall responsibility for the defenses would come, and the ditch-diggers would affect zeal. Now and then the police organized raids on railroad stations, trains, churches, and village dwellings with the aim of swelling the numbers working on the defenses. Those thus abducted were sent to *Arbeitslager*,[95] but a great many escaped as soon as they had consumed whatever food they had with them: it was not difficult to run off since naturally the work sites were spread over a large area and could not all be guarded effectively. There was a catch, however, since the workers' identity cards were kept at a central location in each camp, but then again at the cost of a little vodka or bacon it was possible, using the keepers of local drinking dens as intermediaries, to retrieve one's identity card.

All this activity meant that the locals were bothered more often than before by policemen coming to the door for some reason or other: to demand something to eat, for corn, to suss out a suitable billet, or—ostensibly, at least—to check that the hut had the required list of inhabitants pinned to the door jamb. They would sit themselves down at the table and wait till vodka and sausage appeared, and if this were breakfast-time, they might extend the visit till lunchtime, or even till afternoon tea. There was, however, one good thing about these visits, namely, that one could take advantage of the opportunity offered by having policemen made mellow by hospitality to arrange troubling official matters more smoothly than might otherwise have been possible. Not only policemen but sometimes higher ranking German officials of one kind or another also invited themselves to dinner in this way. These episodes began to take on an air of mendicancy. Those same Germans who had had Polish peasants or POWs working as forced laborers put to death for fraternizing with German women were now in the position of indigents inviting themselves not only to Polish manorial tables, but to those of peasants. Once they had used up their slender allowance, the rank and file German soldiers were now reduced to bartering their boots or stolen truck parts for eggs, milk, warm clothing, tobacco,

---

[94] A Third Reich civil and military engineering organization founded by Fritz Todt, an engineer and senior Nazi figure. It was responsible for a wide range of engineering projects both in pre-World War II Germany and in the occupied territories during the war, when it exploited forced labor. When Todt was killed in a plane crash in early 1942, the organization was absorbed into the renamed and expanded Ministry for Armaments and War Production (*Reichsministerium für Rüstung und Kriegsproduktion*) with Albert Speer as minister.

[95] "work camps"

etc., and there were frequent reports of offers of ammunition, revolvers, and even rifles. Those with nothing to barter became beggars, descending to such pitiful levels of humility that some of them even took to saluting or bowing to the locals. The more arrogant ones resorted to blackmail, pretending they were authorized to check lists of inhabitants of dwellings, or search for partisans.

At that time there were indeed independent bands of partisans prowling about, as well as other groups with their own particular axes to grind: pseudo-communists, nationalists splintered from the AK, German deserters, fugitives from concentration and other camps, and of course ordinary bandits. Clashes between partisans and Germans were becoming ever rarer as both groups looked more and more to the basics: food, vodka, and thievery. Demoralization among the Germans was becoming ever more obvious. They had long ceased believing in victory, and expressed their feelings on this point loudly. The ordinary German soldier and even many German officers seemed to have no idea of the terrible debt of blood contracted by their fellows in Poland. Having read in the newspapers that the Polish population was loyally cooperating in "the defense of Europe," on Christmas Eve they went asking for gifts for the soldiery to those manor houses where the gentry had somehow maintained possession. Some of those Germans sure of Germany's defeat were so credulous—or plain ignorant—as to believe they might remain in Poland after the war. This lent credibility to the newspaper item that I read in one of the little underground Polish publications, to the effect that after the Americans had taken Aachen,[96] the captured German military personnel asked the American command if they would be getting Christmas presents. There were other such instances of total failure to grasp the enormity of what Germany had wrought. For instance, the paymaster of *Organisation Todt*, busy acquiring gold, diamonds, silver coins, etc. from local Poles in return for favors, boldly asserted that when it comes to the point that Adolf Hitler is caught hiding under the table, he, the paymaster, will gladly lay his riches on the table.[97] He frequently plied the soldiers with vodka, and when all were in their cups ordered them to sing songs in praise of the Allies. Thus he was very frank in his contempt for the German endeavor—now that it was foundering—in the presence of the rank and file soldiers under him and the Poles, but if an unfamiliar German officer or, especially, an SS man, hove into view, he clammed up and went out of his way to avoid them; he had worked out exactly which officers from nearby commands he could talk freely to. His attitude demonstrated once again ignorance of the dimensions of Germany's burden of guilt. Perhaps he was taken in by the final note of propaganda sounded in 1944, which to the victims of Germany's perfidy rang so false as to register as merely another gradation of contempt for a conquered nation. One could plunder them, murder them, raze their cities, decimate their intelligentsia, devastate their forests, ruin their industry and

---

[96] Aachen, or Aix-la-Chapelle, is a town in North Rhine-Westphalia, along the German border with Belgium and the Netherlands. It was the favorite place of residence of Charlemagne.

[97] Meaning that he would give them up, or openly use them for himself?

agriculture—and then win them over by uttering a few condescending words during the harvest festival!

In December news reached us that Jasło had been completely levelled with the ground. Anything left standing after the mines had gone off was knocked over or burned, and then peasants from neighboring villages were made to clean up the rubble blocking the main streets so that the army could move freely through them. Thus in the space of a few days, the building that had been my family's home for generations was reduced to dust, and my dream of spending my old age there was set at nought. Our house, built 200 years before, was no longer—with its thick eighteenth century walls and strongly built vaulted ceilings it would have served as a home and family mainstay for another 200 years at least.

By late Fall the *Generalgouvernement* had shrunk to a region the size of Slovakia, the Red Army having penetrated to beyond Jasło, Szczecin, and Warsaw. Thus the border of the Reich had retracted to just a little east of Kraków and Częstochowa. German politics with respect to this little region took another strange turn: they now announced that the German authorities would henceforth be serving the local authorities in merely an advisory capacity. However, there was no time for such arrangements to be put into effect, and in any case the Germans continued with their scorched-earth policy, razing towns and cities and emptying villages of their inhabitants. The mad propaganda continued, with two Polish writers—Skiwski and Burdecki[98]—doing their utmost in that respect. Their Polish-language papers *Gazeta Częstochowska* and *Przełom* were even dropped from planes by the Germans. In issue No. 13 of the latter, of November 1944, they continued to harp on the theme of Britain's betrayal of Poland. The howl of protest at the loss to the Soviets of the eastern borderlands of Poland was given its fullest expression. Yet, although the ink, paper, and the writers themselves were financed by the "Reich Ministry of Public Enlightenment and Propaganda",[99] nowhere could one discover what the boundaries of Poland would have been if Germany had been victorious. Although these writers and their National Socialist *confrères* wrote freely about their country—Poland—as having a great future, with a significant political role to play in the world, they left entirely unaddressed, first, the question as to why they alone of all Poland's writers were privileged to have their posturings printed and distributed, and, second, the much more important and practical question as to how the reader of their blather might avoid being sent to a concentration camp. They wrote that in pre-war Poland, our towns and villages "swarmed" both on the Sabbath and on other days with the long coats worn by orthodox Jewry, and that the interwar Polish Government had connived at Jewish exploitation of the misery of the Polish people.

---

[98] Jan Emil Skiwski (1894–1956), Polish writer, journalist, and literary critic. Collaborated with Nazi Germany, publishing pro-Nazi newspapers in occupied Poland during World War II. Escaped to South America after the war. Feliks Burdecki (1904–1991), Polish writer of science fiction and popularizer of science. Collaborated with the Nazis during their occupation of Poland during World War II. Fled to South Africa after the war.

[99] *Reichsministerium für Volksaufklärung und Propaganda*.

But on the other hand they were deafeningly silent on the recent liquidation of some four million Polish Jews—men, women, and children, young and old alike— and on the methods used by the Germans to force these people to work prior to being exterminated. They hinted at a British conspiracy in the death of Sikorski in a feeble way, without having the chutzpah to come right out with it. And, finally, they pointed to Lwów, Wilno, Red Ruthenia, and Volhynia[100] as Polish lands seized and despoiled by the Soviets, conveniently forgetting the Molotov–Ribbentrop pact of 1939 whereby the Germans acknowledged these lands as lying in the Soviet sphere. Not only did they wilfully ignore the recent past, they somehow manage to forget the present!—by which I mean the agreement between Ribbentrop and Vlasov,[101] according to which Russia, including the lost Ukrainian lands, would be returned to the "White Russians". Thus if we were to take them literally, Skiwski and Burdecki's bosses, having moved our western borders eastwards, had now formed an agreement with Vlasov that our eastern borders should be moved westwards— the same arrangement as that made with Stalin some years before—resulting in us Poles' being confined to a region the size of Slovakia. It was as if the Germans wanted to donate eastern Poland to the Soviets or to Vlasov, or whomever you like, just not to the Poles. Were they blind, deaf, and dumb, these two Polish writers, to continue, at this late juncture of the war, to commend the visibly expiring, though still malevolent and vindictive, Nazi regime?

The indeterminate character of Poland's situation was, of course, favorable to propaganda of every sort. What was clear was that the USSR was not about to yield on the Polish-Ukrainian border established in September 1939, but much else was in a state of fluidity. Mikołajczyk was replaced as Prime Minister by Arciszewski,[102] who was absolutely opposed to territorial concessions. Britain was encountering factional difficulties in Greece, the Americans had become bogged down in the Ardennes,[103] and the Allies had to tread carefully where Stalin was concerned since his troops stood, in overwhelming numbers, near Budapest, in eastern Slovakia, near the Austrian border with Serbia, and had finally moved into Warsaw and beyond.

---

[100]Polish cities and regions occupied by the USSR from September 1939 to June 1941, and incorporated into the Ukrainian or Lithuanian Soviet Socialist Republics at war's end.

[101]Andreï Andreevich Vlasov (1900–1946), Russian Red Army general. After distinguishing himself in several critical actions following the German invasion of the USSR, he was put in command of the 2nd Shock Army of the Volkhov Front and ordered to lead the attempt to lift the German siege of Leningrad. However, his army became isolated and was destroyed by the Germans in June 1942, Vlasov refusing to abandon his troops. He was captured by the Germans and, eventually, professing himself an anti-Bolshevist and anti-Stalinist, headed a German-sponsored project to create a Russian Liberation Army in Germany—a project not strongly supported by Hitler since he had other plans for Russia. At war's end Vlasov was ultimately captured by the Soviets, and after a year's imprisonment in the Lubyanka, tried for treason and hanged.

[102]Tomasz Arciszewski (1877–1955), Polish socialist politician. Prime Minister of the Polish Government-in-Exile in London from 1944 to 1947, the period when the Western powers ceased recognizing this government.

[103]The last major German offensive, called by the Americans the "Battle of the Bulge", lasting from December 16, 1944 to January 25, 1945.

12 Stróże

In the villages of our district, the police now did nothing but drink themselves into a stupor from morn till night; any vestigial official functions remaining to them they performed in the bars and pubs. Thus anyone wishing to petition to quit digging trenches, should bring a side of bacon, say, to the officer-in-charge in the drinking-hut where he was imbibing, a camp-follower on his knee, a hunk of sausage in one hand, and a tumbler of vodka in the other. The services of the official Polish-German interpreter, a certain Szczerbski, seemed no longer to be needed, and he slept all day after drinking all night. The *Bahnschutz* auxiliary Emil Druśko, in love with his work, apparently, killed a certain Wojtaczek and a certain Leśniowski, whom he accused of giving protection to Mogilski, accused in turn of having sheltered Dywan's band of four. After their arrest Wojtaczek and Leśniowski had been locked up, and later released by another *Bahnschutz* guard. But then Druśko, drunk, ran into them near the railroad station, and, incensed, shot them dead. For this he was transferred to Bochnia[104] where he ran afoul of a band of blackmarketeers, who, taking advantage of him when drunk, knifed him to death.

---

[104] A town on the river Raba, about halfway between Tarnów to the east and Kraków to the west.

Chapter 13
Diary Entries

DECEMBER 20, 1944. Yesterday the post brought news of Janek. Just two days out of hospital, he again came down with paratyphoid fever, and checked himself back in. But now they think it's diphtheria. He says the inflammation has subsided, and, since his heart and lungs have been affected by the disease, he is thinking of leaving the hospital and going to Krościenko[1] for the cure. Lidka would then have to go there to look after him, and how would she get there? The weather is below freezing, and the conditions in the unheated train compartments are extremely cramped, with passengers treading on each other's feet and squabbling over space. And, much worse, any young woman with a Warsaw identity card risks being deported to Germany, or at the very least to a local work camp. Furthermore, Lidka's shoes are almost worn out and she coughs continually. Then there is the worry over money. They left their savings in Warsaw with the Matuszewskis, but these fled to the countryside when the uprising erupted, and the money would most likely have been stolen or destroyed in the ensuing *mêlée*. Since the beginning of the month Stefa and Lidka have been making batches of ersatz chocolate balls out of roasted flour, marmalade, butter, and sugar with a view to selling them to grocery stores in the neighborhood. They are very tasty, only too small, I would say. I would have to help carry them to Grybów, which is about six kilometers away, and what with the frost and the likelihood that the store owners would not be interested in buying the confections, I felt some reluctance. Luckily, it seems Janek was able to get hold of the money needed to pay for his convalescence.

DECEMBER 21, 1944. It's twelve degrees[2] below zero. Yesterday, a beautiful, clear day, I went to the Zarembas' in the afternoon to give a lesson, and found them

---

[1] Krościenko nad Dunajcem, a village in southern Poland close to the mountains, about 78 km south-east of Kraków.

[2] Celsius.

Fig. 13.1 Poland after World War II (Map courtesy of Carolyn King, Department of Geography, York University, Toronto.)

greatly dispirited because a band of a dozen or so armed men had raided the farm during the night and stolen an ox and a cow. They are now faced with a dilemma: if, on the one hand, they report the theft to the German authorities, then not only will the thieves return and take off everything they can, but they will also beat up the farmer, that is, Zaremba himself, while on the other hand if they don't report it, then the Germans will find out anyway from others, and suspect the Zarembas of collaborating with the AK. It is true that there is circumstantial evidence indicating that the robbers are not affiliated with any military or paramilitary grouping, but just your basic bandits. On the way back I saw a splendid glowing Venus by the slim crescent of the new moon, and in the west a sheen the color of willow-green. Last night we heard Soviet planes overhead, and distant detonations sounded from midday till late at night. We heard that German military columns continue to march northwards.

Fig. 13.2 Hugo Steinhaus and wife Stefania in the 1940s (Courtesy of the HSC Archive, Wrocław University of Technology.)

DECEMBER 22, 1944. Mrs. Garlicka returned today via Tarnów from her visit to the Kielce[3] district, and told us of some of the goings-on there. She said she saw three Red Army soldiers enter a farmhouse near where she was staying and help themselves to vodka in the owner's absence. It seems the owner had collaborated with the AK and was in hiding because communist bands wanted him dead. She also talked to two soldiers who were spending a night in the house of a woman who sold butter. They wore the remnants of German uniforms, yet one of them was French and the other spoke Polish—evidently they were recruits who had deserted. In Kraków, where she had been staying before coming to us, she had once seen a great mass of such unwilling recruits—"volunteers", so-called—from the Pruszków transit camp, being shepherded by a sizable escort of German soldiers taking care to prevent them communicating with any of the townspeople watching. She had also seen people with the letter "P" sewn onto their jacket lapels—apparently former forced labor workers who had been sent to work in factories in Stuttgart for four months, where they had witnessed an Allied bombing sortie of some two thousand planes. She had also heard how, in reprisal for the assassination of a high German official in Bochnia and the kidnapping of two others, the perpetrators having escaped, forty people from the region around Bochnia had been chosen at random and executed. While all about him there reigns an atmosphere of insanity, violence, and dire need, the *Generalgouverneur* Hans Frank issues the prognosis of a significant rise in the standard of living for Poles.

The productivity of the land has gone down by some 30 % over the last year, our farmer acquaintances tell us, yet the quotas levied by the occupiers have

---

[3] A city in south-central Poland.

risen by some 80%. The Germans have stopped giving out ration cards for flour, and now issue them only for rice bran and rye coffee. All goes on as before: compulsory conscription is still being enforced, and impressment of forced laborers for deportation to Germany continues, as does the work of digging trenches—not to mention the transportation and murder of innocents. And the propagandists try ineffectually to cover this up with twisted logic and idiotic forecasts of good times ahead. Poles are, however, used to the mismatch. What is of greatest concern to them now is rather what is happening in the regions of Poland already occupied—or re-occupied—by the Soviets. Those who manage to get through the front from those areas tend to be interested only in the prices of flour, sugar, meat, etc., but from what they say it seems that in the region of eastern Poland that Stalin is demanding be annexed to Ukraine, Soviet rule is as we experienced it from September 1939 to June 1941. There are rumors that the Poles from that region are to be resettled in the area around Lublin—which I personally think not such a bad idea.

DECEMBER 23, 1944. All day yesterday we could hear intensive anti-aircraft artillery fire over towards Gorlice aimed at the Soviet planes continually passing overhead heading west. In the evening Mr. Woźnica dropped in and entertained us with stories of the shenanigans in his "saloon"—for example, how a drunk calling himself "Moskal",[4] a collaborator, threw fistfuls of banknotes about to show how well off he was, at the same time calling Woźnica, the publican, so to speak, all the bad names he could lay his tongue to. Woźnica just coolly gathered up the money and sent for the police.

The German propaganda machine tells everyone, including their own soldiers, that the Soviets first reduced Jasło by bombardment and artillery fire, and then, when they captured the town, spent a few days levelling it completely. The truth of the matter is that only three or four bombs fell on the town, and Soviet tanks lingered there for only a few hours, since the place had already been reduced to ashes by the Germans. There are at least ten thousand people who could bear witness to the fact that the Germans mined the larger buildings and set fire to the rest, sometimes timing these actions with the passing of Soviet planes overhead. The naivety of these criminals is extraordinary: Sauckel raves about "no stone remaining on stone", and after exerting colossal—and, incidentally, wasteful—efforts to achieve this goal—a monstrous venting of spite—they try to shift the blame onto the Soviets.

Late yesterday afternoon, two German artillerists came to the door asking for linden wood because they wanted to carve children's toys.

Miss Lidka Garlicka heard from neighbors that a certain Poczobut had just been shot dead by Germans in Grybów for consorting with bandits.

DECEMBER 24, 1944. Janek arrived unexpectedly yesterday. He was discharged from the hospital in Kraków two days ago. At the end of the Christmas and New Year

---

[4]That is, a Russian, or, by extension, a Soviet citizen.

holidays he is to go to Poronin[5] to convalesce, having received, as a known writer, a grant-in-aid[6] for this purpose.

He told us how the Germans approached several hundred young people in Kraków, took away their *Kennkarten*, and told them to appear at the police station at a certain prescribed time. When they were all gathered there they were photographed together. The photographs were published in the newspapers as evidence of the "mass reporting of Polish volunteers" anxious to help the Germans make a stand. Apparently they didn't even bother to return the *Kennkarten*. He also related how people were recruited by *Organisation Todt*[7] to work on a project in Sweden, but that as soon as they were fitted out with uniforms and boots in Berlin, they returned to Poland and sold them, thus making around 5000 złotys each. Other newsworthy items: in Warsaw German units were now evicting women and children from their homes and using them as a human shield ahead of German contingents to ward off attacks by the AK; some German tank units had gone into battle with Polish civilians lashed to the tanks with barbed wire; more Varsovians were executed by the Germans than died during military operations; and, finally, most of the destruction of Warsaw was carried out after the uprising had been put down—and after buildings had been looted of all food supplies, clothing, and useful materials of any kind, to help the Germans get through the winter.

DECEMBER 26, 1944. Yesterday Dolek,[8] Lidka, Janek, and I enjoyed a Christmas breakfast at the Cieluchs'. Janek told us a great many things about occupied Warsaw, and I urged him to write it all down. We heard loud cannonades to the north-east.

DECEMBER 27, 1944. More interesting news items communicated by Janek: the post-uprising destruction of the Radium Institute in Warsaw together with its valuable instrumentation, which might have been put to good use in Germany— one more example of the lack of basic common sense pertaining to many of the Germans' actions; the systematic abortion of pregnancies of female members of the forced labor brigades transported to Germany; and the widespread ruse for avoiding such recruitment: have a 500 złoty banknote folded into your identity card. (So common was this as a means of purchasing one's exemption from forced labor, that it had spawned a market: on the Kercelak[9] one could buy counterfeit 500-złoty banknotes for 50 złotys—and these were first-class fakes, easily good enough to fool the police.) Janek also told us that in the Ravensbrück concentration camp for women, the German doctors have been using the inmates as experimental subjects, injecting them with the germs of a variety of infectious diseases in order to try out hypothetical cures. A lady known to the Matuszewskis smuggled out a letter to them

---

[5] A village about 80 km south of Kraków.

[6] From the Polish underground.

[7] The Third Reich civil and engineering group founded by Fritz Todt in 1933. See earlier footnote.

[8] Adolf Szmosz, the brother of Steinhaus's wife Stefa. He was a talented engineer educated at the Lwów Polytechnic, and before World War II worked at the Zgoda steel plant in Lwów.

[9] A square in Warsaw functioning as a marketplace from 1867 to 1944, the largest marketplace of pre-war Warsaw.

in which she lamented that her prospects for surviving the camp were negligible in view of the fact that she had contracted an incurable disease in this way.

JANUARY 1, 1945. Yesterday I went to the Zarembas to give lessons. Mr. Zaremba told me that in Błażkowa, in the foothills behind Jasło, the Gestapo coordinated a raid on the property of a certain Mr. Czerkawski with a gang of forest bandits. They sent the bandits ahead to do the dirty work, then stripped them of their loot and shot them. When the bandits' *confrères* got wind of this, they took Czerkawski's life in revenge, believing him to be in cahoots with the Gestapo.

A Flash of Memory

Mr. Piotrowski, a Lwów stationer, told me the following story. In the interwar period there was a clerk working in the paper factory run by Aleksandrowicz, whom, as a known communist, the police routinely locked up for a week prior to every May Day. When, in September 1939, the Bolsheviks moved in, he was elevated to the post of the firm's delegate to meetings with the Soviets, and generally well treated by the usurpers. But then all of a sudden he was arrested. When his friends enquired as to the reason, they were told that a search of his home had turned up 24 sheets of paper presumed to have been taken from the firm. When it was explained to the Soviets that when Aleksandrowicz had been boss—he had since fled Lwów—he had allowed his officials as much paper as they wanted, the clerk was released. However, a few weeks later he was again arrested, once again for reasons with every appearance of being trumped up. Finally, the real reason came to light. A Soviet official assigned to the factory told Piotrowski informally that since the clerk had been dissatisfied before the Bolshevik takeover, when things were so good in Lwów, they expected him to be even more dissatisfied now they were in charge.

Diary Entries (Continued)

JANUARY 4, 1945. Yesterday Mr. Woźnica told us that German soldiers hailing from around Poznań,[10] who had recently served on the frontlines near Jasło, had come to his home seeking hospitality. They had been in the habit of fraternizing with Poles, speaking to each other in Polish—naturally, since Polish was their first language by far over German—, and since they had gone so far as to organize, among other things, the singing of Polish Christmas carols and Christmas eve festivities, their German superior officers now felt a need to nip the collaboration in the bud by

---

[10]Since this region had for some time (most of the period 1793–1918) been under Prussian and then German rule, it had a large ethnic German population.

disarming them and sending them off to join a cadre in Wadowice.[11] They were predicting that this year they'd all get back home. What is interesting here is how the behavior of German officers in the field—such as those who disarmed these conscripts—is at variance with the messages on the placards posted up all over the place calling for recruitment of auxiliaries to swell the depleted German ranks. It seems that the succession of sudden political *volte-faces* at the top was out of phase with developments at the base of the organizational machine—and that the usual fiendish cunning was no longer of any avail.

A uniformed German soldier turned up in our neighborhood, apparently on the run. He said he had been a prisoner-of-war of the Germans, and had been set to work on fortifications, that is, digging ditches, often knee-deep in water. After more than a year of this, he had volunteered to join the army, and was given appropriate papers, though with new, German, first and last names.

Mr. Ząbek[12] told us that the oil refinery in Niegłowice had been stripped of its oil and gas pipelines and cables—in short, everything that could be carted away—and then levelled to the ground.

JANUARY 7, 1945. Janek left for Kraków yesterday. Soviet planes fly overhead every day, strafing and bombing, and the German anti-aircraft artillery in Gorlice seems to have little effect. Yesterday three German railroad workers in Wola[13] were wounded. The whine of katyushas,[14] the thud of howitzers and mortars, and the rumble of explosions continued uninterrupted to the south—apparently there is fighting in Slovakia.

Yesterday a certain Mrs. Zarańska, accompanied by her husband Mr. Szczęśniewicz, came to us from Polna to ask if I would take on the instruction of their three daughters. This brings to thirteen the number of pupils I have, occupying me in all around 24 hours a week, but the pay is still inadequate to provide for the three of us.

The so-called Polish Committee of National Liberation (PKWN)[15] in Lublin, also known as the "Lublin Committee", is calling itself a government, and as its first action has demoted Mikołajczyk, and snubbed Raczkiewicz[16] and Arciszewski.

---

[11] A town in southern Poland, 50 km from Kraków.

[12] A friend of the Igielskis. See earlier.

[13] A village near Gorlice.

[14] Soviet rocket launchers.

[15] *Polski Komitet Wyzwolenia Narodowego*, a self-proclaimed interim executive authority in Poland, established in Moscow in July 1944 under Stalin's control, with its headquarters in Lublin, and the socialist activist Edward Osóbka-Morawski (1909–1997) as its chairman. On December 31, 1944, the PKWN was renamed the "Provisional Government of the Republic of Poland" (*Rząd Tymczasowy Rzeczypospolitej Polskiej*) with Osóbka-Morawski as Prime Minister, lasting from January to June 1945. He was also prime minister in the next incarnation of the Polish government, namely the "Provisional Government of National Unity", from June 1945 to 1947.

[16] Władysław Raczkiewicz (1885–1947), president of the Polish Government-in-Exile from 1939 to 1947. At Stalin's urging, the Allies withdrew their recognition of this as Poland's official government in February 1945, at the Yalta Conference.

This snub was alleged to be in response to the killings of communists carried out by the AK in the neighborhood of Lublin. Doubtless this heralds the beginning of a civil war.

I have been trying to compile a list of Polish mathematicians who, for some time before the war—say, from around 1935—and perhaps for some time during the war, had been professors or docents, but were lost to us through emigration, deportation, or death. Here is my list:

Kazimierz Bartel, professor of descriptive geometry at the Lwów Polytechnic. Murdered by the Nazis in July 1941.

Antoni Łomnicki, mathematics professor at the Lwów Polytechnic. Murdered by the Nazis in July 1941.

Włodzimierz Stożek, mathematics professor at the Lwów Polytechnic. Murdered together with his sons in July 1941.

Stanisław Ruziewicz, professor at the Jan Kazimierz University, and Rector of the Lwów Business Academy. Murdered by the Nazis in July 1941.

Leon Chwistek, professor of logic at Lwów University. Fled Lwów to evade German captivity in June 1941, but died in Russia in 1944.

Samuel Dickstein, mathematics professor at Warsaw University. Died at the time of the German invasion of Warsaw in 1939.

Stanisław Zaremba, mathematics professor at the Jagiellonian University of Kraków. Died during the war.

Antoni Hoborski,[17] mathematics professor at the Academy of Mining and Metallurgy in Kraków. Died as a result of frostbitten legs in the Sachsenhausen concentration camp in 1940.

Witold Wilkosz,[18] mathematics professor at the Jagiellonian University. Died during the war.

Kazimierz Abramowicz,[19] mathematics professor at the University of Poznań. Died before the war.

Alfred Rosenblatt,[20] mathematics professor at the Jagiellonian University. Emigrated to Peru before the war.

Jerzy Spława-Neyman emigrated to the US in 1937, and in 1938 moved to Berkeley, where he held a chair in mathematics.

Stanisław Ulam went to Harvard University in Cambridge, Massachusetts, two years before the war, where he remained.

Alfred Tarski, docent in logic at Warsaw University. Emigrated to the US in 1939.

---

[17] Antoni Hoborski (1879–1940), organizer and first Rector of the Academy of Mining and Metallurgy in Kraków.

[18] Witold Wilkosz was arrested together with other professors in Kraków by the Gestapo in 1939, and later released because of ill health, dying in 1941.

[19] Kazimierz Abramowicz lived from 1889 to 1936.

[20] Alfred Rosenblatt (1880–1947) emigrated to Lima, Peru, in 1936.

Antoni Zygmund, mathematics professor at the Stefan Batory University of Wilno. When the Soviets invaded Wilno in 1940, he escaped to Sweden, later moving to the US.

Stefan Kempisty,[21] mathematics professor at the Stefan Batory University of Wilno. Deported to Russia in 1940.

Józef Marcinkiewicz, mathematics docent at the Stefan Batory University in Wilno. Taken captive by the Soviets in 1939 and sent to Kozielsk, where he was murdered.

Aleksander Rajchman, professor at the Free Polish University in Warsaw. Perished in the concentration camp in Dachau[22] in 1940.

Stanisław Saks, docent at the University of Warsaw, professor at Lwów University. Arrested and executed by the Gestapo in Warsaw in 1942.

Juliusz Schauder, professor at Lwów University 1940–1941. Murdered by the Gestapo in 1943.

Herman Auerbach, mathematics professor at Lwów University 1940–1941. Murdered by the Nazis in 1942.

M. Jacob, professor at Lwów University 1940–1941. Left for Finland[23] in 1943.

Władysław Hetper, docent of mathematical logic at the University of Lwów. Deported to the USSR in 1939, he vanished without trace.

Stanisław Krystyn Zaremba (son of Stanisław Zaremba Sr.), mathematics docent at the University of Wilno. Went to Tashkent in 1940, and eventually to England.

Arnold Walfisz,[24] mathematics docent at Warsaw University. Took up a position in Tbilisi, Georgia, before World War II.

Dr. Witold Hurewicz,[25] docent in Amsterdam. Accepted a position in Beijing but didn't leave for China because of the Second Sino-Japanese War (1937–1945). Eventually obtained a position at the Massachusetts Institute of Technology.

Szolem Mandelbrojt, worked at the Collège de France from 1938.

Mieczysław Biernacki,[26] mathematics professor at the University of Poznań. His fate is unknown.

---

[21] Stefan Kempisty (1892–1940) died in a Soviet prison in Wilno.

[22] Aleksander Rajchman actually perished in the Sachsenhausen concentration camp in 1940.

[23] Marian Mojżesz Jacob did not in fact escape to Finland, but perished in Warsaw in 1944.

[24] Arnold Walfisz (1892–1962) remained in Tbilisi for the rest of his life.

[25] Witold Hurewicz (1904–1956), eminent Polish topologist. Took his doctorate in Vienna under Hans Hahn and Karl Menger. Assistant to Brouwer in Amsterdam 1928–1936, after which he went to the US. Considered the discoverer of the higher homotopy groups.

[26] Mieczysław Biernacki (1891–1959), Polish mathematician and chemist. From 1929 associate professor at the University of Poznań, and from 1937 professor. He spent the war in Lublin, where he had been born, working at the Maria Curie-Skłodowska University there.

Dr. Adolf Lindenbaum,[27] docent at the University of Warsaw, and subsequently professor at the university established by the Soviets in Białystok. Together with his wife he was murdered by the Nazis.

Stefan Bergman,[28] professor in Tbilisi. Went to Paris before the war, then to the Massachusetts Institute of Technology.

Antoni Przeborski, mathematics professor at Warsaw University. Deceased.

Stefan Kaczmarz, mathematics docent at the Jan Kazimierz University in Lwów.

The fate of those of a younger generation gone from Poland was likewise varied:

Dr. Jan Herzberg, assistant in mathematical logic at Lwów University. Went missing in June 1941.

Dr. Ludwik Sternbach, mathematics assistant at Lwów University 1940–1941. Killed by the Nazis.

Dr. Zygmunt Birnbaum[29] went to the US before the war, where he took up a professorship in applied mathematics in Seattle.

Dr. Marek Kac left for the US in 1938, eventually obtaining a professorship at Cornell University in Ithaca, NY.

Kazimierz Borek, doctoral student at Lwów University. Investigated applications of statistics to pharmacology. Arrested by the NKVD in 1941, was never heard from again.

Dr. Samuel Eilenberg,[30] topologist, left for the US before the war, becoming a professor at Columbia University.

S. Mosler, mathematics docent at Lwów University 1940–1941. Went missing.

Furthermore, it is known that Aleksander Wundheiler[31] is in Washington, and that Sokół-Sokołowski[32] perished in Katyń.

---

[27] Adolf Lindenbaum (1904–1941), Polish logician and set-theorist. He was murdered by the Nazis at Ponary, near Vilnius (Wilno), in 1941. His wife, Janina Hosiasson-Lindenbaum, a mathematician and philosopher, perished in Wilno in 1942.

[28] Stefan Bergman (1895–1977), Polish-born American mathematician. Born in Częstochowa, he taught in Tbilisi from 1934, and from there went, via Paris, to the US, where he secured a position at the Massachusetts Institute of Technology. His last position was at Stanford University. Worked in the fields of several complex variables, partial differential equations, and applications.

[29] Zygmunt Birnbaum (1903–2000) obtained his doctorate in mathematics from Lwów University in 1929, and studied in Göttingen 1929–1930. Worked as a life insurance actuary for the Phoenix Life Insurance Company in Vienna and Lwów. Left for the US in 1937, where he eventually obtained a position at the University of Washington, Seattle. Worked mainly in functional and complex analysis and statistics.

[30] Samuel Eilenberg (1913–1998) obtained his Ph.D. from the University of Warsaw under Borsuk in 1936. Emigrated to the US before the war. Co-founder of category theory with Saunders Mac Lane.

[31] Aleksander Wundheiler (1902–1957), member of the Lwów-Warsaw school of mathematics. Escaped to the US at the beginning of World War II.

[32] Konstanty Sokół-Sokołowski (1906–1940) was a mathematician in Vilnius. Was in a POW camp in Starobielsk, then perished in Kharkiv.

Others who perished, probably by the hand of the Nazis: Dr. Zygmunt Zalcwasser,[33] D. Weinlösówna, Dr. S. Braunówna,[34] Mr. Czubiński, Dr. Salomon Lubelski,[35] mathematics professor at Białystok, D. Pepis,[36] a docent at Lwów University 1940–1941, and Dr. A. Stechel.

A complete list would contain the names of around fifty mathematicians, representing about 70 % of the total of creative mathematicians from between 1935 and 1945 who were either Polish or of Polish origin, but who were lost to Poland.

JANUARY 8, 1945. Yesterday a woman, Mrs. Z., told us that four Soviet soldiers had come late at night to the door of Mr. Rylski's farmhouse and asked for the *khozain*.[37] His wife told them he was asleep and that she was afraid to wake him and frighten him. They did not insist, and she set out a meal of vodka, bread, and sausage for them. After eating their full, they made to leave, but one became adamant about wanting to stay. After a few vain attempts to persuade him otherwise, the others shot their comrade and quickly left. The shot woke Rylski, and he summoned the German police, who merely had the body interred and forbade them in harshest terms to spread word of the incident. Strangely, a few days later the Russian soldiers returned and apologized to Rylski for their impropriety.

Another violent incident, this one indicative, perhaps, of the internecine war we have to look forward to: the head of Soblik's[38] band, presumably a member of the AK, shot the commandant of a group of Peasants' Battalions[39] in Wilczyska, for "threatening a Polish officer with an axe."

The Soviet *desanty*[40] are clearly intended to prepare the ground for the infantry following them, to assess relations amongst the locals, and facilitate the later liquidation of potential centers of political resistance. Thus I don't think the *desanty* are intended to engage in any military actions, but rather to take an interest in everything, to suss out the lie of the land.

---

[33] Zygmunt Zalcwasser (1898–1943), professor at the Free Polish University in Warsaw. Perished in Treblinka.

[34] Stefania Braun (1908–1943), a Warsaw mathematician. Circumstances of death unknown.

[35] Salomon Lubelski (1902–1941?), mathematics professor at Białystok 1939–1941. Co-founder, with Arnold Walfisz, of *Acta Arithmetica*. Died in the Majdanek concentration camp.

[36] Józef Pepis (1919–1941), outstanding logician. Died just after the Germans entered Lwów.

[37] Russian for head of the house.

[38] Possibly a sobriquet.

[39] "Peasants' Battalions" (*Bataliony Chłopskie*, BCh) were formed in mid-1940 by the Polish agrarian political party *Stronnictwo Ludowe* (The People's Party), and by 1944 had been partially integrated with the AK. They were formally disbanded in September 1945.

[40] Airborne troops, paratroopers.

JANUARY 10, 1945. Today there was a dogfight high overhead between German and Soviet fighters. Stefa had been preparing to go to Grybów, but there was a round-up going on in Stróże, so she didn't go. I borrowed a copy of a translation of Erasmus's[41] *In Praise of Folly* published in 1875. Here I found a variety of surprisingly radical opinions about the origins of Christianity, about whether one really should feel obliged to honor debts incurred at the card table, about the state of things in Europe in the nineteenth century, and so on—and not in the form of footnotes added by the translator, but as part of what was presumably Erasmus' original text!

JANUARY 12, 1945. Yesterday Stróże was subjected to a heavy bombardment by Soviet planes for the first time. A little before noon, a few planes flew overhead, apparently scouting the area, and were fired on ineffectually by German anti-aircraft artillery. Then at about 1:30 pm I saw six planes—but there must have been more—flying over us towards Stróże, again subject to feeble anti-aircraft fire from the German guns at Gorlice. When they were almost directly overhead, we could see them break formation, and swoop in low in strafing and dive-bombing runs. Seven people were killed, and more than a dozen wounded, including a few German soldiers and policemen. A few houses had their windows blown in, one lost its shingles, and some sheds were burnt down. Although most of the bombs were relatively light, a couple must have been in the hundred kilogram range. Some of them failed to explode, and later in the park we found fragments of the bomb fins and casings, as well as ordinary machine gun shells. The sound of the guns and bombs and answering artillery was so loud it seemed that the battle was taking place directly overhead. The Soviet bombers were armed with machine guns, and, so we were told, protected with armor-plating coated with rubber. Luckily Stefa got back from Grybów at noon, well before the raid commenced.

JANUARY 13, 1945. The fiasco of standardized prices revealed in all its farcicality: now the post offices sell postcards only to persons working in some sort of official capacity. But the post office charges them 50 groszy, whereas the set price is 12 groszy. But officials[42] sell cards to others for only a złoty a card!

JANUARY 16, 1945. On Sunday, January 14, we learned that the Soviet winter offensive was about to begin. Yesterday, Monday, from early morning on we heard the unceasing thunder of battle over to the east and north, and sometimes even to the south. Soviet planes flew continually overhead in squadrons of nine, but there was no more bombing of Stróże for the time being. Sometime during the day we heard that Wiślica and Kielce had been taken by the Red Army. The signs of retreat are ever clearer. The SS and the police have abandoned Stróże, leaving behind their Ukrainian collaborators and a few overseers from *Organisation Todt*, but in Grybów and elsewhere the Germans were apparently still forcibly recruiting men into the army as they retreated. Today we heard a huge explosion over towards Gorlice and

---

[41]Desiderius Erasmus of Rotterdam (1466–1536), Dutch renaissance humanist, Catholic priest, and theologian. Called "Prince of Humanists" for his enlightened views.

[42]Dishonest officials with access to the cards.

smoke billowed into the sky; it seems the Germans had blown up the factory in Glinik Mariampolski[43] (about 7:45 am).

Long trains of military wagons continue passing through Stróże from the direction of Jasło. The weather is beautiful, with fresh unpolluted snow blindingly bright in the sunshine. One hears periodic thumps just over the horizon—probably from the heavy artillery in Gorlice attempting to stave off the retreat a little longer. It is said that the Red Army has already captured Tunel, near Miechów.[44] A train labelled "Flak"[45] left the railroad station last night.

JANUARY 19, 1945. In the afternoon of January 16 the bombing of Stróże resumed in earnest, lasting for 35 minutes at its maximum intensity. A train standing by for the final evacuation was not touched since it was on a siding behind the station, but the tracks were destroyed, and that put an end to railroad traffic.

Before the bombing started we had seen a hundred and fifty or more German soldiers cross the road, evidently heading cross country. When the bombing commenced, they took shelter in the little wood near our dwelling. One of them came to talk to us. He spoke of the war in a tentative and neutral way. Two others with whom the Cieluchs spoke at the farmhouse said that their officers had abandoned them, that many German soldiers had been taken prisoner in the neighborhood of Jasło and Gorlice, and, to put it briefly, *Deutschland ist kaputt.*[46]

A little after sundown I went to the field beyond the barn. It was dry and still, with just a touch of frost, and Venus shone like a suspended lantern. The moon was very close to full, and one could make out a thin crescent of the dark side. Beyond Stróże I could see a fiery glow on the horizon. The little bridge on the road between us and Stróże had been blown up by the Germans, so we were cut off. Peace and quiet reigned. Later in the evening while sitting at home, we heard the put-put of a motorcycle, and surmised that its German rider had not known that the bridge was no more.

At about 9:00 pm we heard tanks heading in our direction along the road from Polna, and soon heard voices speaking Russian, female as well as male. The tanks appeared to take up a position before launching a barrage, with answering German fire. The din was frightful, and our windows blew in. It was only when morning broke that we realized that the German shells had been bursting quite close to our hut, which was so flimsy that a single shell would have been enough to destroy it completely. We dressed and went outside, to be greeted by the sight of Soviet soldiers walking about between the farm buildings, emerging from who knows where. They were calm and wore white sheepskin greatcoats. One of them had been wounded and came to us to have his wound dressed, and his comrade, aided

---

[43] The former name of a rural municipality, now absorbed into the municipality of Gorlice. An oil refinery and a factory manufacturing drilling machines and other tools was established there in the late nineteenth century by the Canadian William Henry MacGarvey.

[44] A town about 40 km north of Kraków.

[45] Short for *Flugzeugabwehrkanone*, that is, anti-aircraft defense artillery.

[46] "Germany is kaput."

by my mother-in-law and Mrs. Woźnica, did an excellent job. Another went to the farmhouse to ask for hot milk, and soon there was a crowd of them keeping the Cieluch ladies busy making coffee with milk. We noticed that under their sheepskin greatcoats they had woolen garments. Some wore waterproofs over their greatcoats.

Thus was Berdechów occupied by the Red Army on January 17, 1945.

By the middle of that day they had set up a field telephone, and were clearly not about to move on. Then a captain told us that the farm would have to be vacated because it was about to become the site of a battle. Although some of the Soviet soldiers tried to tell us it was the wrong direction, we—Stefa, Lidka, and I—decided to go to Polna. Thus we set off in a blizzard, each carrying a suitcase, struggling through snowdrifts the whole way. When we arrived, following the example of Mr. and Mrs. Szczęśniewicz, who had also been forced out of their house, we took refuge in the rectory, where by turns we sat or lay all day and night in the basement listening to the sounds of the battle raging all about. Only during those periods when the struggle seemed to subside did we venture upstairs. By a miracle the rectory remained unscathed. Upstairs it was thronging with Soviet soldiery. They had changed since Lwów times. While they indeed drank heavily and accosted the womenfolk, they at least indulged in only a little empty political agitation. "Did you know that you already have your own government in Lublin?" they said, or, equally vacuous: "After we have gone on to defeat Germany, we will return home and leave you to yourselves." They took whatever they wanted. Mrs. Szczęśniewicz told us that they had left their house in ruins, taking the crockery, cutlery, flour, and poultry, and anything else they thought might be of use, drunkenly filthying the bedding, breaking the beds, and wrecking the piano. It was much the same at the rectory, although one could discern that they were making an effort to hold themselves in check to some small extent out of respect for Syzdek, the priest. There was a Russian general there who knew some Polish; he claimed to have read Mickiewicz in the original. He talked to us with alacrity and told us how things lay—among much else how some German prisoners of war held in the Ukraine had received letters from family members containing plans of Soviet cities with certain areas marked and the inscription: "I would like to have a house here when the war's over." Early on Thursday morning—yesterday, that is—a Polish-speaking soldier came to the basement. He said that he was from Warsaw, but had become a Soviet citizen during the war. He took down our names and subjected us to a political lecture: about the legitimacy of the government in Lublin appointed by Stalin, the new borders of Poland, the dire shortage of food, the strength of German resistance, and the need for us to help the Red Army overcome it. Although he spoke with only a slight peasant accent, that he was indeed from a peasant family was clear from the way he overused the word "village". He told us that at the Majdanek Lubelski concentration camp the Germans had indulged in a spree of killing: "Poles, Belgians, Frenchmen,...." It was noticeable that he left out Jews. Great numbers of infantry and columns of military vehicles, including tanks, moved through all day and into the night, carrying various kinds of artillery, including katyusha rocket launchers. It was noticeable that they did not douse their lights—presumably since a German air attack was no longer to be feared.

Late on Thursday we decided to return to Berdechów. There we found our home standing but turned inside out and filthied almost beyond description. The floor was strewn with the books and papers from our desk, among which I found this diary intact. Our clothing, coats, linen, watches—in a word everything we had left behind—was gone, so we were reduced to a state of destitution. We discovered that my mother-in-law had remained in the house. We had thought she had taken refuge in a neighbor's basement, but she had stayed there only an hour and a half before returning home, where she had run into the plundering bands of Soviet soldiers. She had taken some things to her room, but had naturally been unable to prevent the looting, so that not only were our belongings gone but also many of my brother-in-law's things. We were left in peace the rest of that night, but today, Friday, soldiers came and went from noon to nightfall to use the cooking and dining facilities, or catnap on the floor, with the accompanying constant loud banging of the door. By evening we were left blessedly alone, although in a house with broken windows, and doors which wouldn't close against the cold, and in the midst of dirt and disorder. Needless to say there was no question of shaving as all razors and razor blades had been taken.

I myself spent the day trying to analyse the differences between the specimens of Soviet manhood that we were now dealing with and those of three and a half years ago in Lwów. Two nineteen-year-old *tankisty*[47], Nikolaĭ and Volodya, who were manning a tank repair station set up on the farm, and who dropped in to chat, made a large impression in this regard. The stony-faced NKVD automatons with their piles of documents that we had had to deal with in Lwów could not have contrasted more with these two *russkie soldaty*[48], as they called themselves—big and strong, hard as nails, inured to all adversity, taking frost, hunger, and even bullets in their stride. Like all the Soviet soldiers we saw they were admirably clothed for the purpose they served—in padded undergarments, and by way of outer wear long sheepskin coats or woolen greatcoats covered with waterproofs. They were fed with American canned food,[49] and armed with machine guns and everything else they might have need of on the road ahead—except, it seems, for sufficient quantities of vodka and tobacco. Supplies were in such abundance that they could afford to be wasteful with bullets, grenades, shells, and gasoline. They explained that when advancing, the forward line of troops had perforce to "capture" food supplies from peasant huts or farmhouses, since the convoy of military vehicles following in their wake, including the army kitchens, could not keep pace with them. They needed neither pep talks nor political harangues, and a minimum of poring over maps and tables: they wanted to fight and needed no egging on by *politruki* or anybody else. They drank, took what they needed from the locals, and, when drunk, accosted and raped local women. A major said to us: "We can't shoot them all for looting!" These two

---

[47]"tank crewmen"

[48]"Russian soldiers"

[49]Under the auspices of the program called "Lend-Lease", the US supplied the UK, the USSR, China, and other Allied nations with war *matériel* from March 1941 to September 1945.

and the others we met had altogether stopped reciting Marxist and Leninist homilies, or bragging of the achievements of the Soviet people, or condemning capitalists. Their first business was no longer the production of propaganda but the implacable waging of a brutal war. They themselves had erected crosses on the graves of a few among the vast multitudes of their fallen comrades. When I asked about the Jews, they said that although it was well known that these had been shirking the hard frontline business of war, safe in commissariats, general staff headquarters, workshops, etc., they would be brought to account after the war. They predicted that in postwar USSR there would be the same prosperity as they had had glimpses of in Poland, and heard about in Britain and America. They had now been beyond the Soviet borders, and had had their eyes opened.

I was most amused to hear that Nikolaï and Volodya had befriended the family of Stanisław Klobassa's[50] son, chief administrator in Zręcin,[51] and had celebrated New Year's Eve in his home. Of Klobassa Sr. they were told that he had "his horses shod with gold." They would say little about Stalin except to insist that he had a straight nose and had nothing in common with caricatures of him. They expressed an intense hatred of the Germans for, they said, starving prisoners of war to death, drowning children in village wells, slapping the faces of peasants and workers, and for razing towns and villages in Belarus, Ukraine, and Russia—among much else. They passionately desired to get to Germany and exact vengeance the Russian way, sure of being approved in this by their *nachal'stvo*.[52] In my modest opinion it would suffice to quarter them in Germany for a year for the people of that country to either flee abroad or hang themselves.

JANUARY 26, 1945. The new regime has been in place for ten days, but we have as yet very little idea of what's in store for us. The front is now so far to the west of us—somewhere near the Oder—that we are no longer within earshot of the cannons' roar. The convoys of military vehicles have petered out.

Laska[53] and Karol Zaremba[54] have been arrested, but already hundreds of locals have signed a petition to free them. As before, the arresting is done by the NKVD; we haven't as yet seen any Polish authorities in action.

Soviet soldiers continue dropping in sporadically. One such attached himself to us for several hours and behaved rather arrogantly, perhaps because he was more than a little drunk. He waxed on inconsequentially about how he didn't believe in icons since God was invisible, although he does manifest Himself to us in the rising and setting sun. He also heatedly—and quite correctly—maintained that thunder

---

[50]In Chapter 1 it was related how when Steinhaus was still a boy, Klobassa Sr. discovered oil near Jasło and became exceedingly rich.

[51]A village in Krosno County, Subcarpathian Voivodeship, south-eastern Poland.

[52]Superiors, ultimately Stalin.

[53]District police commissioner for Bobowa. Helped locals escape Nazi persecution. See Chapter 12.

[54]A brother of Jan Zaremba, a landowner working in the Polish underground educational enterprise during the war. See above.

was not caused by friction between clouds, but denied that the relative positions of moon and sun as these appear to us have anything to do with the earth's rotation, on the grounds that the motions of sun and moon relative to a person standing on the earth's surface are such as to cancel the effect of the rotational motion of the earth—or some such unlikely farrago. Conversation with him left me with the impression that the stresses of the war on top of Marxist-Leninist ideology have resulted in Soviet heads being stuffed with a peculiarly hybrid species of nonsense.

This soldier also asserted that the Soviets will not occupy Germany but will instead set up a new government there. One could detect in him—and others we spoke to—a tacit envy mixed with hatred of Poles, for the comfortable homes we had lived in and the conveniences we had at our disposal before the war, and with a touch also of admiration that we had so effortlessly, it seemed, generated the wherewithal of this well-being. When they get to Germany, this envious hatred, combined with vengefulness, will certainly find an outlet in wanton despoliation—in particular of any delicate figurines, porcelain or human. They were beside themselves with ire over the fact—or perhaps it was merely rumor—that food packages were being smuggled to Germany by Soviet citizens of German background.

JANUARY 29, 1945. Below zero temperatures and snow. The railroads and post offices no longer function. The impressment of all able-bodied males born between 1914 and 1925 into the Red Army, and the expropriation of all farms of more than fifty hectares—these are the only new developments, though rather radical ones, to be sure. Although I have recovered my diary and notes, I remain inconvenienced by the lack of a hairbrush or comb and razor blades. A ring engraved with my Father's initials, CBS, which I valued, is also gone. We heard that the front line is now a mere 140 km from Berlin, but all such facts about the outside world come from Red Army soldiers, so may or may not be true. Laska and Karol Zaremba remain locked up, while Marian Proszek walks about freely.

My health is not so good. Soon we will have no income as my teaching job may shortly be coming to an end: in two weeks, so they say, a Gymnasium will begin functioning in Grybów, first with Forms I and II, and then two weeks later Form III will open. A few Jews have come out of hiding and returned to Bobowa, Grybów, and Łużna. In most cases they had been hidden by sympathetic peasants. Having lived through a frightful time, they have now returned to a world emptied of families, friends and acquaintances, and with their former homes occupied by others or in ruins. And instead of sympathy, by and large they meet with sneers from their former Polish neighbors, as if expressing disappointment that they had not succumbed. A strange people, the Poles.

FEBRUARY 2, 1945. Proszek and Szczypcina have been arrested and taken to Gorlice. It now seems that Laska and Karol Zaremba will be judged by Polish courts. Jan Zaremba visited us yesterday with the news that he has obtained a position as teacher of Polish in a Gymnasium in Gorlice run by Mr. Zabierowski, who also wished to offer me a position.

The newly appointed officials are very young: thus the *starosta* of Gorlice, apparently a former barber's apprentice, is a mere 26 years old, and the police chief in Bobowa just 22, and moreover in a constant state of intoxication. They say he won't listen to the two sides in a squabble unless each supplies him with half a liter of vodka. We are told that postal service is to resume shortly, though for the time being without stamps.

We have heard that Młynarski's banknotes[55] have now become worthless. Some Jews who had been hidden by a peasant in Moszczenica went to Gorlice to see what they could retrieve of their belongings and perhaps return to their former apartments, but were rounded up and executed by the Soviets. A similar massacre occurred in Kowel: the Soviets came across around twenty Jews and summarily executed them on the grounds that they must have been Hitler's spies since "how else could they have survived?"

There is fresh news about the front: the Red Army has crossed the Oder and occupied Gdańsk, Poznań, and Katowice. The German civilian population of the adjacent regions have fled west in great numbers.

Janek has returned from three weeks in Szczawnica.[56] He brought his belongings on a sledge, having somehow managed to get a ride for most of the way in a military vehicle. He talked about the deportation of Varsovians[57] from Kraków, which was being carried out by the Wrocław Gestapo when he was in Kraków, by direct order of Himmler. He had been told that the Germans posted up placards stating that "the Warsaw contingent [in Kraków] is rife with criminal elements causing the rise in prices of necessities and the propagation of flea-borne spotted fever (typhoid) and other diseases." This text is interesting in that it is word for word the same as that on the placards concerning Jews that I remember reading in Lwów in the Fall of 1941. Thus Nazi propaganda was also mass-produced, it seems: one expression of revilement fits all.

There was a rumor that the Załuskis[58] had been arrested in Iwonicz, but that all had been released except for one of the brothers.

FEBRUARY 9, 1945. Today Janek set off for Kraków intending to get there by what one may perhaps called "combinatorial" means: first by horse and cart, then hitchhiking, or, failing that, on foot. Although our political discussions have not led to accord, at least we have agreed amicably to differ. However, yesterday, when in the January 19 issue of *Głos Ludu*[59] he came across articles in the best Lwów style of the period 1940–1941 of the Soviet occupation, especially one concerning the Polish-Soviet cultural alliance—with tributes from Professor Józef Wasowski[60]

---

[55]The official currency under the Germans. See Chapter 12.

[56]A resort town in Lesser Poland Voivodeship, in southern Poland. Because of the presence of thermal springs and favorable climatic conditions, frequented since the mid-nineteenth century for treatment of respiratory and digestive illnesses.

[57]When the Warsaw Uprising failed, about 22,000 Varsovians fled from Warsaw to Kraków.

[58]The dominant, and aristocratic, family in the spa town Iwonicz during Steinhaus's childhood. See Chapter 1.

[59]"Voice of the People"

[60]First president of the Polish-Soviet Friendship Society, founded in 1944, a vehicle for organized propaganda events, such as celebrations of the October Revolution, as well as exchange programs, promotion of Soviet culture, etc. Lived from 1885 to 1947.

to the high artistic level of painting in the Soviet Union—and another about the printing of five million copies of parts of Wasilewska's novel about dear Joseph[61] as the tried and true friend of the Poles, he felt constrained to express a wish to go as far away as possible—to Mexico, say—or if that were not feasible, then to volunteer for the army in the hope of having no leisure to take in nonsense. As far as I am concerned the stuff in *Głos Ludu* indicates all too clearly what the political situation will be six months hence. The very phraseology gives them away: praise for the Moscow Metro,[62] Wasilewska and her adherence to the party line, insults hurled at the Polish Government-in-Exile in London, in conjunction with the overall pretence that their ideas are supported by a large majority of people, and the fear of censure by the overseeing *politruk* evident in every line. Six months hence pathetic requests will be addressed to the great Marshal urging him to respect the modest wish of "the Polish peasant, worker, and working member of the intelligentsia" to unite with their Soviet brothers and sisters. There were some new words in use, such as "Slav" and "pandemocratic", and some old words were notable for their absence: "communism", "bolshevism", and "godlessness", for example, but apart from that it was the same twaddle all too familiar to us from the Soviet occupation of 1939–1941.

FEBRUARY 11, 1945. Yesterday a Serb dropped by at the Cieluchs'. He had been freed—by partisans, he said—from the camp at Oświęcim, and had been trekking homewards through Slovakia before reaching us. He estimated that during the 21 months he had been incarcerated about three and a half million people had died there. He said that he had spent a great deal of the time digging with a shovel as part of labor gangs, on a bowl of soup a day, and often in fifteen degree frosts without a coat. Any apparent malingering was rewarded with a blow to the head with a rifle butt, and if the culprit was then unable to get to his feet, he was shot. There was a two-man committee consisting of a Gestapo officer and a doctor who officiated at the periodic "selections". New arrivals and sometimes camp veterans would have to run naked between these two, and when the doctor pointed at one of them, the Gestapo man would yank him from among the runners by means of an iron hook around the arm. A clerk would note the number tattooed on the skin of the prisoner thus singled out, and in this way a list of immediate victims compiled. Any inmate who was sick for over a month was likewise entered into the list of those destined for immediate liquidation. From among the prisoners the Germans chose privileged *Sonderkommando*[63] members, mostly German criminals, but including other prisoners,[64] who were given supervisory powers over the cells

---

[61] Stalin.

[62] Construction of this splendid—some would say extravagant—subway system was begun in 1931 as a showpiece of the Soviet system. By 1935 thirteen stations had been completed and linked up. Construction of extensions continues to this day.

[63] "special unit"

[64] Many sources assert rather that the *Sonderkommando* units were comprised mainly of Jews, forced to aid with the disposal of the bodies of those killed in the gas chambers.

and barracks. Some trains packed with victims—mostly Jews—were directed onto a special siding, where the already half-dead human freight was off-loaded and herded naked in groups of a few hundred into large chambers equipped with shower heads to allay suspicions, after which the doors were sealed and poison gas piped into the chamber. The mass murder achieved, the *Sonderkommando* members removed gold fillings and crowns from the teeth of the dead, loaded the corpses onto wagons and took them to the crematoria. By way of privilege they were allowed to take for themselves certain objects belonging to the victims, and they ate better than most prisoners. Each such *Sonderkommando* group functioned for about ninety days after which time as a rule they themselves were done away with. There were people from all over Europe in Auschwitz, our visitor said: Poles, Serbs, Jews aplenty, Frenchmen, Czechs, and so on—all destined for liquidation, some earlier, some later. He told us that prostitutes were brought in from Germany, relatively well fed and well dressed, access to whom was permitted the German, Polish, Serbian, and certain other prisoners, but not Jews or Russians. These also were murdered in batches every few months and a fresh lot shipped in. Torture was routine. Our Serb had heard all of this from one of the torturers who had fallen ill in the course of his work, been admitted to hospital, and then returned to the camp as an inmate. He had told him also of an experimental medical center set up in Oświęcim where they infected prisoners with strains of certain virulent diseases and then injected them with drugs in search of a cure, tested the efficacy of certain poisons on prisoners, induced and then aborted pregnancies, maimed one twin child, keeping the other as control, and so on. None of the human guinea pigs survived beyond a few months, needless to say.

Our Serb visitor told us that when defeat became clear, although the Germans hardly relaxed their work of extermination, they nonetheless made an attempt at destroying the evidence of their hellish endeavour. For instance they commenced covering the crematoria over with earth and planting grass and bushes. However, such was the celerity with which the Russian juggernaut advanced on them that they were taken by surprise, and unable to effect a satisfactory cover-up in time.

FEBRUARY 14, 1945. Yesterday a sort of army staff headquarters was established in the Cieluchs' farmhouse. Officers arrived in a cortege of cars, set up a field telephone, and installed a sergeant in our hut. This man, a modest Ukrainian from Uman'[65] of some 30 years, had been an ostler in a *kolkhoz*[66] before the war, but on account of the war he had learned to drive. He had attended school to age 11. He

---

[65] Uman' (in Polish Humań) is a city in central Ukraine.
[66] The Russian portmanteau word for collective farm.

had no desire to get to Berlin. He informed us that we Poles would have everything we need "as soon as Żymierski[67] liberates Berlin."

From him and other Red Army soldiers one formed a general impression of that army as one of very great, but on the whole elementary, practicality. Its members are not particularly daunted by such obstacles as the steppe, deserts, hunger, cold, etc.— they simply confront and then deal with such difficulties in primitive but effective ways. They form very strong bonds with one another and the network of these is very extensive, so that when faced with an impediment of any kind they know who has a map, or where the expertise is to be found for the erection of temporary bridges or building of roads in their own, often unexpectedly simple, style. Their great weakness is for the bottle; they'll do absolutely anything for a hundred grams of vodka.

Yesterday the place was alive with the rumor that after all Lwów was to remain in Poland, and that the Polish Government-in-Exile in London is to be welcomed in Poland. But this must be merely wishful thinking, since the conference of the "Big Three" is still in progress.[68] What is certain is that the Red Army has crossed the Oder along a 160 km front. A Polish officer who had witnessed the event told us that a prisoner-of-war camp near Berlin whose inmates were chiefly Polish officers who had earlier been prisoners-of-war in the *Oflag IIC Woldenberg*, had changed hands several times during the fighting, with the result that the unfortunate prisoners had been subject to fire from both sides. This news caused Mrs. Michałowska[69] to become rather agitated since her husband had been incarcerated somewhere there.

FEBRUARY 17, 1945. Tosia Michałowska has heard that her husband had somehow reached Kutno,[70] was in reasonable health, and that many of those arrested with him had also survived. Jan Dobrzański went back to Kraków some time back, and now he writes that he has managed to get a ride to Katowice, where he is to take over the management of an iron works. One has to be careful: three young people from Lublin on their way to Śląsk to take over the management of state forests came close to being shot by a Red Army patrol, and of six British airmen who crashlanded their bomber near Żmigród,[71] two were killed by marauding Ukrainians, and of the

---

[67] Michał Rola-Żymierski (1890–1989), Polish communist military commander and Marshal of Poland from 1945 till his death. Returning to Poland from France in 1943, he became deputy commander of the Soviet-backed *Gwardia Ludowa* (People's Guard), and then commander of the *Armia Ludowa*. He subsequently became commander-in-chief of the Polish army fighting alongside the Red Army. From 1946 he was head of the "Commission for State Security", and was responsible for repressions carried out against former non-communist members of the resistance.

[68] The Yalta Conference of Churchill, Roosevelt, and Stalin, at which the re-establishment of the nations of war-torn Europe, especially Germany, was discussed, in fact went from February 4 to February 11, 1945.

[69] Tosia Michałowska was living with the Igielskis, neighbors of the Cieluchs. Her husband had been arrested. See Chapter 12.

[70] A town in central Poland. Two main railroad lines cross there, Łódź–Toruń and Warsaw–Poznań.

[71] A town in the present Lower Silesian Voivodeship in south-western Poland. Prior to World War II it had been part of Germany, and known under the name of Trachenberg.

remaining four two were shot by their presumed allies[72] before they could identify themselves, the surviving two being allowed to return to their base after payment by Poles of a ransom in liqueur.

Money as a medium of exchange has ceased functioning. There has been talk of the banknotes being officially stamped to distinguish the guaranteed ones from the unrecognized ones. Pensions have not been paid for months.

Yesterday we read of the outcome of the Yalta Conference. The two main points are, first, that the Provisional Government of the Republic of Poland will be reorganized, with representatives of Polish emigrants having seats, and, second, that the eastern boundary of Poland will henceforth be the Curzon Line.[73] There were also conditions concerning the occupation of Germany by the Allies, and the political status of Greece and Serbia, among other things. This we read in a newspaper that is now being distributed, but which, unfortunately, is in Russian. We also read there that Budapest has been captured and that Berlin is being heavily bombed.

FEBRUARY 18, 1945. Yesterday we were visited by some Russian officers who had come from Bielsk.[74] They turned out to be intelligent people. One of them (a lieutenant) criticized Sholokhov, and his opinions were neither boastful nor naive. The peasant sergeant from Uman' billeted on us keeps changing his story: now he says he had 11 years of schooling and worked in an auto-repair shop as a mechanic. He says his mother was Polish. I told him how forty years ago one could do such things as travel without a passport to Berlin, telegraph money to Paris, and order shoes on account from London by postcard, but now at every turn one needs bits of paper to do anything: to get through passport controls and others of the proliferating official barriers. He said he shared my concern. Then I tried the remark on him that the USSR has everything nature can supply in overabundance: vast fertile lands, coal and oil, precious minerals, and so on, but that it lacks the one crucial ingredient, namely someone who tells the truth. To this he rejoined: "I knew that already in grade seven." He had an excellent earthy peasant sense of humor. When I told him of the return of several Poles who had emigrated to Canada before the war, he said that if a Russian walks at a lively pace, and gets overheated, he will say to himself "Well walked!", while a Jew will say instead "Better to travel badly by means of

---

[72] That is, Soviet soldiers.

[73] A demarcation line between the Second Polish Republic and Bolshevik Russia formally published in 1919 in an Allied Supreme Council declaration in the wake of World War I, but disputed by Poland (George Curzon (1859–1925) was British Foreign Secretary at the time). In the Treaty of Riga ending the Polish-Soviet war of 1919–1921, Poland was conceded about 135,000 km^2 of land east of the line. At the Teheran Conference it was agreed, and confirmed at Yalta, that post-World War II Poland's eastern border should, with minor variations, follow the Curzon Line. Stalin successfully resisted Churchill's proposal that certain parts of East Galicia, including Lwów, be retained as Polish territory. Poland's territorial losses were compensated for by the incorporation of former German areas to the west. Thus the net effect of the postwar territorial settlement was that Poland as a whole was translated west by some hundreds of kilometers.

[74] Possibly the present-day Bielsko-Biała, a village in southern Poland.

any form of transportation than go on foot, even if well."—by which he meant, I think, that it's better to lead a difficult life in Canada than an easy one in Poland. His knowledge extended to his having a vague idea that European culture owed a great deal to the ancient Greeks. When I told him of the pre-war high standard of living in France, he said: "I don't look to people like you and me, whose only care is the immediate one that they and their families should do well in the here and now, but to those who strive to do something worthwhile for which they will be known a hundred or two hundred years in the future. Does France have such people?" I doubt that there are any people left in our country capable of thinking as swiftly or as generally as this man. Stalin will have his work cut out for him curbing such minds.

FEBRUARY 19, 1945. Yesterday Dolek returned from Jankowa, where he had been awaiting the return of Jan Dobrzański from Kraków, bearing a letter for Lidka from Joanna Guzówna.[75] Joanna wrote that she had had extraordinary adventures in Russia, that she had most recently been a commandant of a military unit stationed in Puławy,[76] but had now been appointed director of a department charged with the caretaking of cultural monuments in the Kraków Voivodeship. The letter contained the offer of a job for Lidka, and expressed a desire to come and visit us, the governor of the Kraków Voivodeship, Adam Ostrowski,[77] having offered the services of a chauffeured car to that end. My son-in-law Janek has been flitting about between Łódź, Warsaw, and Lublin; it seems that like just about everyone else these days he has become some sort of official dignitary.

The Kraków newspaper of February 14 has reprinted the text of the Yalta agreement. That the Yalta Conference was at the highest level—attended as it was by the "Big Three"—is evident in the great importance and definitiveness of the decisions taken, as itemized in the agreement. British resistance to Stalin's proposals for Poland was easily beaten down. Thus although in the document the Lublin government is characterized as temporary and not fully representative, the Government-in-Exile in London is mentioned only periphrastically as run by "certain Polish emigrants." And elections to the *Sejm*, the freedom to advance candidates, and the institution of universal voting by secret ballot are mentioned in an equivocal manner. For the moment the outcome of the Crimean lucubrations has been to reduce the authority of the Lublin government to promulgate independently conceived measures. What has not, however, been reduced in any way as yet is the prevailing impoverishment and hunger all about us.

---

[75] Joanna Guze, Polish writer and translator and friend of Lidka. She had gone off with the departing Soviets as the Germans approached Lwów in June 1941. See Chapter 8.

[76] A town in Lublin Province in eastern Poland, situated at the confluence of the Vistula and Kurówka rivers.

[77] Polish lawyer and diplomat. Educated in Lwów. Arrested by the NKVD in Lwów in 1944 and then released. Appointed Governor of Kraków Province in February 1945, and later Polish ambassador to Sweden and then Italy. From 1954 to 1967 director of the Polish State Publishing Institute (*Państwowy Instytut Wydawniczy*). Lived from 1911 to 1977.

The officers billeted at the Cieluchs' farmhouse have finally departed, as has also our sergeant, with fond farewells. He gave Granny[78] his tobacco allowance as a parting gift.

In the war bulletins one hears already that Cottbus[79] and Żagań[80] have been overrun, and that Dresden has served as the stage of a gruesome spectacle. The flights of thousands of Allied bombers bringing death to many tens of thousands of civilians in Dresden[81] represent perhaps the penultimate act of the insane tragedy perpetrated by the Hitlerites.

FEBRUARY 23, 1945. Two days ago Dolek went to "Matusiczka"[82] to get some butter and ran into a group of Jews returning home to Slovakia from Oświęcim. I caught up with them to give them a dozen eggs. They related several things, including an episode where an SS man took pity on a boy of fifteen and held him back from entering the gas chamber at the last moment. They had escaped from a first contingent of eighty thousand prisoners being transferred by the Germans from Oświęcim towards Gliwice,[83] and while they were hiding in a wood the Soviet offensive had caught up with them. Seeing some of his tormentors—members of the Gestapo—being loaded onto the back of a truck—presumably for transportation to Siberia—a Jew ran up and slapped the nearest one as hard as he could. Said the wretch: "You hit me when I'm unarmed?", to which the Jew retorted "And when you beat me was I armed?" The scene they thus described to me is priceless in that it is so revealing of the Hitlerite's psyche, which has become so inured to the meting out of punishment and death that his truncheon or pistol has become like a grotesque organic appendage. Those who lack this extra limb, as it were, are the objects on which he exercises it. I imagined the surprise of the Gestapo man—perhaps as astonished as a marksman at a rifle range would be if the object dummy with a target where his heart should be were to suddenly rush up and hit him in the face.

We sometimes read articles from the newly emerged press. There is the Kraków *Dziennik Polski*,[84] obviously a vehicle for Lublin propaganda. A typical article there is like the wolf disguised as the grandmother in "Little Red Riding Hood". It contains a lot of blather about the "ancient Piast lands" and democracy. An issue

---

[78] Steinhaus's mother-in-law.

[79] A city in Brandenburg, situated on the river Spree about 125 km southeast of Berlin.

[80] A town on the Bóbr river in western Poland.

[81] Between February 13 and February 15, 1945, a total of 3600 RAF and USAAF planes flew four raids over Dresden, dropping around 4000 tons of high-explosive bombs and incendiary devices on the city. The resulting firestorm razed around 39 km^2 of the city center. Recent estimates put the resulting number of deaths at 25,000.

[82] Presumably a woman of the neighborhood.

[83] A city in Upper Silesia, in southern Poland, near Katowice.

[84] "The Polish Daily"

of the weekly *Nowa Epoka*[85] contains an article about Poles who were officials of the Austrian Empire before World War I, but then joined the Polish civil service on the outbreak of war in 1918.[86] It lumps Stanisław Łoś,[87] Aleksander Skrzyński,[88] and a certain Habsburg noble from Żywiec[89] together with Leon Kozłowski,[90] thus placing some of the best Poles in the company of the traitor Kozłowski. I note that the Habsburg archduke to whom I refer here never registered himself as a *Reichsdeutscher* or *Volksdeutscher*. Apart from this misleading article, this self-styled democratic clarion contains not a word against communism, nothing on the dubious history of the triumphantly godless Bolshevik regime, nary a whisper about the deportations and mass executions carried out by the NKVD—in fact nothing whatsoever about the NKVD. But wait, yes, I do see something written there about the Katyń massacre, however entirely in the style of German propaganda, which naturally detracts from its plausibility. This is obviously intentional—the wolf's muffled voice ill-disguised as Grandma's, the truth deadened by the Kremlin. The *Nowa Epoka* is but a small step away from toeing the Kremlin line.

FEBRUARY 25, 1945. Two days ago Michałowski arrived from the officers' camp in Woldenberg.[91] He was terribly emaciated, so weak that he had barely managed to get to us. On the way he had come across a German storehouse that must have

---

[85] "The New Epoch", the main organ of the Democratic Party at the time, publishing articles on politics and culture.

[86] This was an unattributed article printed in the very first issue of *Nowa Epoka* of January 21, 1945, p. 6, generally critical of the interwar *Sanacja* government of Poland. *Note added by Aleksandra Zgorzelska, editor of the first Polish edition*

[87] Jan Stanisław Łoś (1890–1974), landowner and Polish diplomat of the interwar period, professor at the Catholic University of Lublin, and head of the Humanities Faculty of this university 1957–1959.

[88] Aleksander Józef Skrzyński (1882–1931), Polish politician. Prime minister of Poland from 1925 to 1926.

[89] This was Archduke Karl Albrecht of Austria-Teschen (1888–1951), from 1919 to 1949 Karol Olbracht Habsburg-Lotaryński, owner of an estate at Żywiec (a town in southern Poland) who, in spite of his Austrian connections felt himself to be Polish, and enlisted in the Polish army in 1918. On the outbreak of World War II in September 1939, at the age of 51, he once more enlisted in the Polish army. For refusing to sign the German *Volksliste* he was imprisoned till near the end of the war. After liberation he moved first to Kraków then to Sweden.

[90] Leon Tadeusz Kozłowski (1892–1944), Polish archaeologist and politician. Prime minister of Poland 1934–1935. Served with distinction in the Polish-Soviet war of 1919–1921. Held various posts in the interwar *Sanacja* government. Arrested by the NKVD in Lwów at the beginning of World War II, he was released two years later and eventually found his way to Germany and was taken to Berlin, where he may or may not have collaborated with the Nazis. In any case in 1943 he was taken to the site of the Katyń massacre as an expert on behalf of the Germans, and is known to have assisted them in their propaganda war with the Soviet Union. He died from wounds received during an Allied air raid on Berlin.

[91] A German town till 1945, it is now in northwestern Poland, and has been re-named Dobiegniew. The POW camp was first established by the Germans in September 1939 for ordinary soldiers and non-commissioned officers, but in May 1940 was re-designated as *Oflag II-C*, a POW camp for officers. It ceased functioning in January 1945.

been provisioned extraordinarily abundantly since it had not been entirely depleted even by the Soviet soldiers. He told us that he had also seen a farm with as many as 800 dairy cows. The camp had been visited by a representative of the International Red Cross, whom the camp commandant had refused to admit to the prisoners' barracks, despite the fact that he had a letter of permission from the Ministry of War in Berlin. He also refused to allow him to take anything the prisoners had written, quoting standard camp regulations. A true Hitlerite in not truckling to any non-Nazi authority.

Perhaps Karol Zaremba and Laska have been sent to Sambor,[92] and so beyond the new border of Poland. If this is so then they may be lost to us.[93]

Yesterday the commandant of the railroad station held a celebration honoring the Red Army. A large portrait of Stalin was displayed prominently, flanked with two small pictures of the great leader, in accordance with the finest NKVD aesthetics. There were many impromptu speeches—all at the commandant's orders!

FEBRUARY 26, 1945. We have heard that in Nowy Sącz the Soviets jailed a group of former inmates of Auschwitz as vagrants, and also that the Soviets are already busy dismantling electrical installations, sawmills, and what have you, and shipping the industrial equipment off to the USSR.

MARCH 2, 1945. There is not a single person in the Lublin government that one would not characterize as anything but a nincompoop. They printed money, but the banknotes stuck together when they tried to cut them to size. They produced postage stamps depicting the pedestal of the monument to Jagiełło,[94] most probably the first postage stamps ever printed showing the plinth with its inscription, but without the actual heroic statue—as if they wanted to celebrate the victory at Grunwald without Jagiełło!

One feels there is no way out of the mess: there is hardly any currency in use,[95] no communications network, no materials, no flour, no lard. By Spring there will be widespread starvation. The mayor of Gorlice is one Florek, a former house painter, and allegedly a bad one at that, while the *starosta* is the former owner of a restaurant which catered to the Germans. No one knows who the governor of the province is, or who the Lublin government's minister for internal affairs.

A Soviet military driver stole the watch of a railroad man K., who took the registration number of the vehicle and rushed to Grybów to complain to the relevant

---

[92] A town not far from Lwów, called Sambir in Ukrainian.

[93] This hypothesis turned out to be false; at the time of writing they were safe in Sanok and Rzeszów respectively. *Author's Note*

[94] Władysław Jagiełło (Ladislaus II Jagiello) (1362?–1434), King of Poland and Grand Duke of Lithuania. Founded a dynastic union of Poland and Lithuania by defeating the knights of the Teutonic Order at Grunwald in 1410. An impressive bronze equestrian monument to him is located in Central Park, New York City. It was sculpted by Stanisław K. Ostrowski (1879–1947) for the Polish pavilion of the 1939 New York World's Fair, and is a replica of a monument in Warsaw which the Germans melted down into bullets after taking Warsaw at the beginning of World War II.

[95] New złoty banknotes were first introduced in late 1944 by the *Narodowy Bank Polski* (The Polish People's Bank), in circulation till 1945.

Soviet authority. The thief was apprehended and shot, and an NKVD officer returned the watch to its rightful owner K., apologizing on behalf of the USSR, and adding: "This driver must be a Jew, since only Jews and Tatars steal!" From this I gather that the role of Jews as the source of all mischief—*Cherchez le Juif!*—is appreciated also in the USSR.[96]

MARCH 3, 1945. A week ago a new student from Lwów by the name of Misiakiewicz arrived. He hopes to enter a lyceum.[97] And now another student has turned up called Tarasek, a latecomer to the first grade of the Gymnasium. Thus altogether I now have nine students. My teaching and the yeast that Stefa produces and barters for eggs and butter are our only means of eking out a living. No one seems to know yet if the foreign language of choice at the Grybów Gymnasium will be German or English. So-called "free" tuition (in the Gymnasium) is to be paid for at the rate of 200 złotys per month.

Lately there has been little news from the front apart from the capture of Wrocław and Szczecin. People freed from the German camps pass through heading eastwards—not only Poles and Jews, but Serbs, and even some Frenchmen, Englishmen, and Americans. We have heard that many people have settled on the outskirts of Jasło in makeshift dwellings. There's not a murmur about the reconstitution of the government promised in the Yalta accord.[98]

MARCH 4, 1945. Yesterday I informed my lively seventeen-year-old students from the Witalis family that Turkey had declared war on Germany.[99] They expressed surprise that "Poland had not done likewise." It turned out that they were all totally ignorant about Polish contributions to the war effort: about the Government-in-

---

[96] The former Russian Empire contained the largest population of Jews of the diaspora (around five million in the nineteenth century). Anti-Semitic policies and periods of persecution—pogroms—were frequent. Through extensive emigration, especially at the turn of the nineteenth century (to the US) and in the 1980s (to Israel and the US), it has been reduced to something of the order of 300,000.

[97] The level of highschool after the Gymnasium.

[98] At Yalta Stalin agreed to the holding of free elections in the Soviet-occupied states of Eastern Europe, and also that some members of the Polish Government-in-Exile in London would be permitted to join the Lublin government.

[99] After maintaining steadfast neutrality throughout the war, the Grand National Assembly of the Turkish Republic—which in 1923 had succeeded to the Ottoman Empire, Germany's ally in World War I—voted unanimously to declare war on Germany and Japan on February 23, 1945.

Exile in London, about Sikorski, Anders and Monte Cassino,[100] the Polish Navy,[101] the heroism of the Polish troops at Narvik, and even about the army of Berling and Żymierski. I was reminded of a simile from a poem in the book *Shi-King*[102]: "Military glory is like a red flower on the steppe whose petals have been stripped off by the wind."

The issue dated February 25, 1945, of the new literary periodical *Odrodzenie* (Revival) came into my hands. It's the thirteenth issue, but it's evidently the magazine which is having a spell of bad luck, not me. Apart from the promising first chapter of Nałkowska's[103] *Living Ties*, what is written about in this magazine is of less interest to me than what is not written about. For instance, the London Government-in-Exile of Poland is barely mentioned, but the little that is written about it has a decidedly ironical tone, hinting, in particular, that its members are not democrats. They avoid charging that government with the passing of the "bench ghetto" law[104] passed by Ozon,[105] but doubtless only because it was now considered not quite *bon ton* to talk about Jews. On the other hand, the magazine lauds

---

[100] Władysław Anders (1892–1970), Polish general. In 1939 the Polish cavalry brigade he commanded was forced to withdraw before the overwhelming German onslaught. Captured by the Soviets, he was imprisoned in the infamous Moscow Lubyanka, where he was tortured. When Germany invaded the USSR he was freed and appointed commanding general of an army comprised of Poles who had been deported to the USSR in 1939–1941. Only a few hundred thousand of the original million and a half deportees emerged from Siberia alive. Anders managed to transform this half-starved, ragged mob into one of the most feared Allied fighting forces of World War II, undefeated in battle. Anders' army, together with a sizable contingent of Polish civilians, managed an exodus from the USSR via the Persian Corridor into Iran, Iraq, and Palestine, and then further west, where they fought alongside the Western Allies. One of their most notable battles took place at Monte Cassino, a crucially strategic hill in the mountains south of Rome (crowned with a medieval Benedictine monastery), almost impregnably fortified by the Germans. After attacks on the position by various US, French, British, New Zealand, and Indian divisions all failed to dislodge the Germans, Anders and his troops took the hilltop after seven days of bitter fighting, on May 18, 1944. However, he and his army were betrayed by the Western Allies at Yalta. After the war Anders fled to England where he lived out the rest of his life in exile.

[101] Most of which escaped to Britain.

[102] *The Book of Odes*, one of the "Five Classics" of ancient China.

[103] Zofia Nałkowska (1884–1954), Polish novelist. Her early novels showed the influence of the Young Poland movement and focused on exploring the feminine psyche, while her later ones were more concerned with social problems.

[104] Compelling Jewish tertiary students to sit in specially designated areas of lecture halls. See Chapter 7.

[105] The National Unity Camp (in Polish *Obóz Zjednoczenia Narodowego* or OZN), a Polish political party founded in 1937 by leaders of the Sanation movement. Its stated aims were to improve national defense and safeguard the 1935 Constitution. It sought to portray Rydz-Śmigły as Piłsudski's political heir and the "second person in the country" after President Mościcki. Perhaps the point here is that the Government-in-Exile had some continuity with the prewar Ozon government, so might easily have been linked to the passing of the "bench ghetto" law.

Miłosz,[106] Sandauer,[107] and Putrament[108] for supporting agricultural reform.[109] The German corporal who frankly and feelingly told Lila Holzer in Lwów in 1941 of the deaths of so many young Jews in the Lwów Citadel is to my way of thinking far superior to the likes of such mealy-mouthed political fellow travellers as Putrament and company.

MARCH 11, 1945. Yesterday all morning one could hear the rumbling of big guns in the distance. They say that there is a battle going on near Żywiec.

Winter is back—oodles of snow and below-freezing temperatures—making travel more problematic since only possible on foot. Hence we here in the countryside are for the time being out of reach both of administrators and of business people.

Ridiculous events occur or are rumored to do so: we have it on good authority that a stranger showed up at the office of the deputy *starosta* in Gorlice, declared himself the deputy *starosta*, and was given the post.

The Soviet conscription process is such that a few thousand show up to enlist at each site, but only a few hundred daily are accepted. Famished men line up for days in the cold, willing to become cannon fodder in order to get fed.

The break-up of large holdings is having negative effects: for example, in Stróżna[110] a proprietor quit his farm, a former manorial holding, to pre-empt eviction, but was then ordered to return "because the farm must be kept running." And it's clear that the uniform application of the parcelization law to places like, say, Lower Silesia or Pomorze,[111] where the wealth of the country is concentrated in large industrialized holdings, will be as fully effective as mining and bombing in laying waste the economy.

In Silesia and Kraków they say that soon people will be dying of hunger. In Bielsk roaming squads of German soldiers continue killing Poles.

---

[106] Czesław Miłosz (1911–2004), Polish writer, poet, and translator, of Lithuanian background. In Warsaw during World War II, postwar he was appointed cultural *attaché* to the Paris embassy of the communist People's Republic of Poland. In 1951 he defected, ultimately to the US. In 1953 he received the *Prix Littéraire Européen*. In 1961 appointed professor of Polish Literature at the University of California at Berkeley. Between 1951 and 1980 his works were banned in Poland by the communist government. Awarded the Nobel Prize for literature in 1980. At last, in 1993, returned to Poland.

[107] Artur Sandauer (1913–1989), Polish essayist, literary critic, and translator. Worked on the staff of the weekly *Odrodzenie* in Warsaw 1948–1949. Attempted to oppose the doctrine of "social realism" in art. Although never a member of the Communist Party, was a member of the National Council of Culture (*Narodowa Rada Kultury*), which supported General Jaruzelski in the 1980s.

[108] Jerzy Putrament, Polish writer, poet, publicist, and politician. A communist supporter, after World War II he was by turns diplomat, deputy to the Polish Parliament (1952–1961), and member of the Central Committee of the Polish United Workers' Party (1964–1981).

[109] In the sense of a more equitable distribution of land—see below.

[110] A village southeast of Kraków.

[111] Pomerania in Latin. Historical region along the south shore of the Baltic Sea.

On February 21, Bierut[112] returned from Moscow, and suddenly he went from Head of the National Council to President of Poland. But still nothing much is happening: the establishment of a government by free elections agreed to by Stalin at Yalta is apparently not exactly in the offing; Zaremba and Laska remain in prison; we see no evidence of the reorganization of the judicial system; and the Gymnasium in Grybów cannot start lessons because the building itself has no glass in its windows, and no benches or desks for the students. Moreover, because of the war there are many latecomers.

On the other hand, the Western Allies have crossed the Rhine, and it seems to me the Soviets are less insolent than before.

MARCH 18, 1945. Janek was here for three days, and put us in the picture. He says that the government in Lublin consists of a dozen or so people for whom a Soviet's word is their command: all merely semi-intelligent, they are a mix of ideologues and opportunists. Thus the so-called government's role seems to consist largely in relaying commands from Moscow to subsidiary centers like Kraków and Łódź. Janek says that around four out of every five positions in the Presidium of the Council of Ministers are occupied by Jews, even among the higher-level ones. He agrees with me that the agricultural reform will not only serve to strip the big landowner of his property, but will also have the unintended effect of impoverishing those who replace them, since no one can make a living by farming half a hectare of land. Thus in the end the reform will lead merely to an increase in social disaffection. Since parcelization was never the policy followed in the Soviet Union,[113] one wonders why it is being pursued here.

At the present time everyone is starving or on the brink of starvation except for provincial governors, directors of government enterprises, and writers. The latter, together with actors and journalists, are given the best apartments, sumptuously furnished, of those formerly occupied by top-ranking Germans, are paid "extremely well", says Janek, and have other privileges—and all thanks to their slavish service to the propaganda machine of the Lublin government, which is wearing itself out covering up the fact that at most one in ten of the populace supports it.

It's interesting that on returning from a sojourn in the Soviet Union most of our so-called communists complain about the distorted societal norms resulting from the tremendous spiritual pressure of the ideological straight-jacketing of people.

Dziutka Schoenborn[114] arrived just after Janek. She had been to Jasło. She said that there are now only about 3000 people living there even if one includes the adjoining villages of Ułaszowice and Sobniów. There are many living in basements in the town proper. Dziutka's father Dyk became insane, apparently because—but

---

[112]Bolesław Bierut (1892–1956), Polish communist leader, NKVD agent, and hard-line Stalinist. President of Poland 1947–1952, Chairman of the Central Committee of the Polish Communist Party 1948–1956, and Prime Minister of Poland 1952–1954.

[113]Where Stalin pushed continually for full collectivization, with disastrous results for Soviet agriculture—not to mention the millions who starved to death in its initial stages.

[114]Daughter of Steinhaus's cousin Józef (Dyk) Schoenborn. See e.g. Chapter 2.

this may have been just the last straw—he had considerable difficulty explaining his nickname "Dyk" of forty years to his captors the first time he was arrested, and then when he was arrested a second time they asked him about it again, and he was somehow unable to make them understand. He died later in a prison hospital. Dziutka related how she had been transported along with other workers and much of the machinery of several Lwów automobile workshops to Opava[115] in the *Sudetenland*[116]. She managed to escape and return to Lwów using false papers obtained with the help of some German soldiers. She visited Jasło a few times when it was occupied by the Germans, and managed to transfer some of her belongings from there to Lwów. Dyk's wife Wanda, Dziutka's mother, had fled Lwów for Drohobycz,[117] but had been forced to work in some capacity for the German occupiers there until she was killed. Presumably someone informed against her. Once Zosia[118] was arrested by the Ukrainian militia in Lwów, but when they telephoned the *starosta* Barnaś in Jasło, he assured them that the Schoenborns are rock solid Polish Catholics, and she was freed. She is now living with Dziutka in Kraków.

Thanks to her friend Joanna, Lidka has obtained a position in the Kraków district office in charge of preserving historical relics. Józef Nacht has surfaced; it seems he manages a theater for soldiers in Łódź or Lublin, I forget which. We're told that some of the actresses are captains, that is to say they are captains offstage but women onstage. We weren't enlightened as to their actual military duties. One doesn't need to resort to the theater for fantasy these days, since the amount of offstage dressing up and play-acting exceeds the bounds of the wildest imaginings.

Dziutka also told us that my old friend Staszek Adamski[119] had turned up; she had happened upon a notice he had pinned up in an office of the Savings Association requesting information about surviving members of his family. She also had news of my niece Muta Müller: she is working in a Polish children's hospital in Stalinabad,[120] and badly wants to come home.

Today a car came from Nowy Sącz to take Lidka to Kraków. Janek went with her, intending to continue on to Łódź, where he plans to begin publication of his literary magazine *Kuźnica*[121] with Borejsza. The chauffeur told me he is paid 5000 złotys a month. The car also bore a character in a military uniform who said he was in charge of printing shops in the province. He talked down to us in an arrogant way, querying us as if we were local peasants.

---

[115] A town in northern Moravia (in the Czech Republic, as it is now).

[116] The former German name for the regions of Czechoslovakia inhabited mostly by ethnic Germans until the end of World War II.

[117] Drogobich in Ukrainian. A city in the present L'viv Province of western Ukraine.

[118] Dziutka's sister.

[119] Steinhaus's friend Stanisław Adamski from Jasło days.

[120] In Tajik Dushanbe, renamed Stalinabad from 1929 to 1961. Capital of Tajikistan.

[121] "The Forge", a Marxist literary magazine founded in 1945, promoting Social Realism. Jan Kott was its chief editor for a period after the war.

Dziutka is also leaving; she plans to go to Jasło and see if she can lease the land on which her parents' house once stood. Such optimism!

Janek tells me that Leon Chwistek was in 1944 received by Stalin, who, observing what he took to be symptoms of dementia, promised to send him a doctor. And indeed on the morrow two doctors came to see the patient, who expired soon after.

It seems that Olga, together with Alina and Alina's husband Stanisław Dawidowicz[122] and their daughter Agnieszka are travelling to Jasło—returning home—unaware of the fact that Jasło is no more.

MARCH 19, 1945. Yesterday afternoon a little boy delivered a card to us. It was from Dziutka, informing us that she had been arrested by the NKVD, who had handed her over to the local militia to be taken to Grybów. Apparently she was arrested because her papers identified her as residing in Gorlice, whereas she was picked up in Stróże. At any rate she was back with us by evening. She said that the NKVD lieutenant in Grybów, who was quick on the uptake and very courteous towards her, called the man who'd arrested her an idiot. She could have made a break for it on the way to Grybów, but hadn't wanted to expose the militiamen to possible repercussions.

MARCH 21, 1945. A few days ago the Woźnicas, husband and wife, left, reducing the population of our hut by half, and today a local refugee family by the name of Bul left to return to their home town in Zamojszczyzna,[123] whence they had fled a year ago out of fear of the Ukrainians.

Two of the more adventurous of my students from the Witalis family found an unexploded shell, experimented with it, and have been badly mutilated as a result. One lost eight fingers and the other has a badly lacerated face and may lose the sight of one eye.

The bulk of the former manorial lands in Biała belonging to the Pieńkoś family had been leased out for cultivation to tenant farmers—peasants, essentially—for fifty years. The fact that these leaseholders, long residents of the district, are now to be thrown off those lands in the name of fair redistribution will not tend to enamor them of the government that passed that law.

I have noticed that the Soviets have deleted the first two names from their scroll of the four evangelists Marx, Engels, Lenin, and Stalin—because they were Germans? The press reports issuing from Lublin never mention Marx. Instead, there is a lot of nonsense about Pan-Slavism and the historical importance of Łużyce[124] in the style of the most simple-minded of patriotic fables.

---

[122]Stanisław Dawidowicz, a professor of engineering in the Częstochowa University of Technology in the 1950s.

[123]Region in southeastern Poland, near Zamość.

[124]In Latin, Lusatia, a border region between Poland and Saxony bounded by the Kwisa and Łaba (Elbe) rivers, originally inhabited by Celts and, following their expulsion, long contended over by Slavs and Germans.

The Jagiellonian University of Kraków has reopened for business, announcing the beginning of a new academic year. A Kraków Polytechnic is to be founded.

I received a letter from Stefan Dobrzański telling me, among other things, that his former office manager, now become his boss, has summoned him to Poznań.

MARCH 22, 1945. Spring is in the air.

For the past three days the blood-curdling howl of the katyushas could be heard continually in the south-west, but today it is quiet.

Poland has not been invited to participate in the San Francisco Conference.[125] At a meeting of his Conservative Party, Churchill reminded his listeners of the twelve months when Britain was all alone in the war against Hitler.

Our newspapers are filled with the fabrications of fools: Bolshevik propaganda, but minus materialism and communism, now that Marx is banished. The "democracy" they tout is of the Asiatic sort, where the state holds monopolies on salt, matches, political views, history, and—well, everything, and it is all served up in standard IKC Polish,[126] and with much use of quotation marks in case—God forbid!—someone should miss their feeble little witticisms.

The government is shameless in the way it praises itself—in particular, by means of posters pasted up everywhere claiming that "The government gives peasants land!", as if the land were really theirs to give. Other posters proclaim recruitment into the Polish army, yet the majority of volunteers are sent away since in reality there are neither enough uniforms nor bread for Polish recruits. The Red Army, on the other hand, occasionally—as the need arises, perhaps—pressgangs people into enlisting.

Yesterday a Russian soldier came around asking for something to eat, even a little cabbage, he said, like a beggar. He is the Soviet stand-in for the train dispatcher at the railroad station in Stróże. Soldiers stole a pig from Dudzik, the father of one of my students. Such is everybody's mental state that such events interest them more than such weighty ones as the Allied crossing of the Rhine, or massive British and American bombing raids.

MARCH 25, 1945. For a few days distant explosions have been audible.

I received a letter dated March 14 from Mother and Fela.[127] In a letter to Alina Dawidowicz,[128] Muta[129] expressed a desperate homesickness. It seems that in order to be allowed to return home she must be summoned by her family. We have learned that the position in the hospital in Stalinabad was organized for her by young Zakrzewski,[130] who had joined the Polish army and was now in Persia.

---

[125]The United Nations Conference on International Organization (April 25–June 26, 1945), an international meeting at which the United Nations Organization was established. It was attended by delegations from 46 countries.

[126]That is, Polish as used by the *Ilustrowany Kurier Codzienny* (The Illustrated Polish Daily).

[127]Felicja, the eldest of the author's sisters.

[128]*née* Chwistek.

[129]Muta Müller, Felicja's daughter.

[130]Possibly from the Zakrzewski family of Rozprza—see Chapter 6.

Another beautiful spring day! Dolek had a letter from Jan Dobrzański[131] in which he writes, among other news, that although the choicest jobs are going to young engineers, the former managerial people are being looked after appropriately in that regard.

MARCH 26, 1945. Along with thousands of others Mr. Laska has been sent east, probably to the Donbass.[132] Instead of duly arraigning him and other Polish citizens before a Polish court,[133] they are shipped across the border like so much meat. The fact that this goes on with impunity under the nose of the Lublin government says more about that government than anything that might be written or said against them.

Dudzik found clues as to who stole his pig, provided by Polish militiamen, the chief suspects. I myself saw a couple of them lugging what looked very much like several dozen kilograms of sausages into the hut serving as their headquarters. There's really nothing to be done as they're all in cahoots.

MARCH 27, 1945. The sound of explosions is now coming from further off. The Red Army detachment stationed in Grybów has moved on to Sucha.[134]

Yesterday a solemn funeral was held for Wiatr[135] of blessed memory, betrayed to the Germans by Proszek, and ten peasants from Wilczyska also murdered by the Germans. Wiatr's body was exhumed in Gorlice. Of course, delegates from the Polish Workers' Party[136] gave speeches blaming the Polish Government-in-Exile and the AK, with which Wiatr and the peasants were connected.

Yesterday bandits wielding automatics robbed a former restaurateur of 10,000 złotys and then went to the tailor Janik to steal pork fat. They terrorized those present at the tailor's, took the lard, and went on their way. Meanwhile the militia whiles away the time shooting at crows.

We are out of coal and I have no shoes.

APRIL 3, 1945. Yesterday Dolek saw Jan Dobrzański, who had come to Jankowa for Easter. Apparently he has been appointed governmental overseer of the metals-processing industry in Katowice. He reports that the Soviets are stripping all industrial installations in the Opole[137] and Katowice basins of their machinery—in fact, of everything they can remove—regarding everything the Germans installed,

---

[131] Obtained a position as manager in an iron works in Katowice—see earlier.

[132] Short for *Donets basseĭn* in Russian. The Donets Basin is a heavily industrialized region of eastern Ukraine. Rich coal deposits were discovered there in the late nineteenth century. The Donets is a river flowing through the region, the largest tributary of the Don.

[133] Even Poles who had declared themselves *Volksdeutsche* were vouchsafed this right, since for the charge that they betrayed their country—that is, Poland—to stick, they had to be tried in Poland. *Author's Note*

[134] Now called Sucha Beskidzka. A town in the far south of Poland.

[135] This seems to be a brother of the more famous Narcyz Wiatr referred to below.

[136] In Polish, *Polska Partia Robotnicza* or PPR.

[137] A city in south-western Poland, in the region of Silesia, noted for its rich mineral deposits from the Cretaceous period.

in particular, as rightful spoils of war. These were considerable since the Germans had built up a large military-industrial complex in the area as well as investing in the necessary peripheral industries. However, the denuding of the factories of their plant surely shows that the Soviet Union is not really concerned that Poland regain any lost industrial strength—or perhaps they want to be sure that Poland is unable to re-arm itself.

There is nothing to eat in the country. Every day 15,000 starving people arrive in Poznań looking for food, and it's the same in Katowice. People stream westwards in great herds.

Woźnica came home from a visit to Izbica and hired ten people as servants and farmhands on his family's farm, former manorial lands. Today the Zarembas were given less than 48 hours notice of eviction, yet I believe the expropriation of a holding of this size—around 37 morgs—runs contrary to the Lublin decree. The Kossowskis are also to be evicted, together with the Piotrowski family, which has been living in a hut on their land. It seems the expropriators intend to knock down the old barns and other farm buildings, ostensibly to make way for the application of improved methods. This is very dubious. All this taking of property from the propertied is clearly aimed at currying favor with the landless, in the hope of inspiring some sort of spontaneous "people's movement", which simply refuses to come into being, however. Meanwhile, said "landless" take the land offered them by the government while bearing in mind that said government is likely to be short-lived.

The new local administration is comprised of nobodies from the militia, ex-saloon keepers, and former ne'er-do-wells, and if one applies for a certificate or permit of some sort, one must needs dictate the text for them, they being semi-literate at best. Train travel by day is impossible because of lack of space, and at night is dangerous since Red Army soldiers use the cover of darkness to rob passengers, considering the latter's property as merely their due.

The distant bangs have stopped; apparently many of them were caused by boys meddling with unexploded bombs and shells.

The Allied crossing of the Rhine began on March 22, and the next day there were massive crossings by British and US troops in the Rees-Wesel-Dinslaken area, and then a few days later the US Seventh Army crossed near Worms. Thus great numbers of soldiers, supported by quantities of light tanks and artillery, now occupy the Palatinate,[138] as well as parts of Hessen, and even of Westphalia and Bavaria. What exactly is happening is somewhat hush-hush, but by now the western Allies must already be deep inside Germany. We have heard of American landings in Iwo Jima,[139] which is part of Japan proper.

---

[138] In German *Rheinpfalz*, a region in south-western Germany occupying more than a quarter of the territory of the German federal state of *Rheinland-Pfalz*.

[139] The island of Iwo Jima was part of the Empire of Japan. Thus the Battle of Iwo Jima (February 19–March 26, 1945) was the first American attack on Japanese home territory.

No news from Kraków. I have begun to extract what I can from my notes of these past years, first gathering together everything I have written pertaining to random variables.

I have been contacted by a Father Klawek, who lived out the occupation in a Dominican monastery in Biała. I don't know how he found out about me, but in any case I received a message from Father Sępowicz in Polna asking if he might talk with me. Yesterday we dined together. He was ignorant of what had occurred in Lwów after December 1939. He has been offered a position as Professor of Biblical Studies at the Jagiellonian University in Kraków. He said there was a rumor that the former faculty contingent at the Jagiellonian University would be transferred holus-bolus to Wrocław. He talked interestingly about the literary and philosophical values inherent in the books of the prophets, especially Isaiah.

APRIL 5, 1945. We continue to be mostly in the dark as to what's going on in the world. There are rumors about the Americans entering Czechoslovakia,[140] and that now they are rushing forward, now holding back.[141] In Silesia Oskar Maresch, a cousin of the Dobrzańskis, has been killed. He had worked as a representative of the Polish Government-in-Exile, in a supervisory capacity in connection with steelworks and other industrial plants during the German occupation, and presumably he has been liquidated by order of the PPR.

The newspapers—or perhaps one should say non-newspapers—have taken to reiterating like lunatic optimists the motto "All's well!" as if reassuring children, and adopted a special style to match. There are articles praising "wholehearted" work, assuring readers that reconstruction is "on the boil", and so on. They contain idiotic caricatures of Hitler, print a diatribe, lacking both wit and art, launched by the attorney Dobromęski at the *Volksdeutscher* Słowik, etc., yet articles addressing the reality of Soviet soldiers' robbing stores at night, or how the Soviets, in a slighting gesture of disregard of his position as Poland's Prime Minister—and no doubt indicative of Poland's actual puppet status *vis-à-vis* the Soviet Union—requisitioned the car in which Osóbka-Morawski, Prime Minister of the Lublin Government, was returning from Silesia, so that he had to finish the trip by bicycle.

We hear that the region of Poznań and beyond is an agricultural and industrial *tabula rasa*: the little the Germans left behind has been removed eastwards by the Soviets. Thousands of cattle died in barns for lack of feed and water.

In Jasło the population is now about 7000.

APRIL 7, 1945. From the newspapers I see that our new style of journalists resemble young girls of the Victorian Age in having to find out for themselves things considered not quite proper for them to know. The list of such things is

---

[140] American troops were welcomed by Czech civilians in Pilsen on April 7, 1945. Prague was liberated from German occupation by the Red Army in May in the last great European battle of World War II.

[141] General Eisenhower, the Allied Supreme Commander, decided to halt the Allied advance—against the urgings of Montgomery, Patton, and Churchill—at the Elbe since postwar zones of occupation of Germany and Berlin had already been agreed to at Yalta, and he wished to spare the troops under him from unnecessary further fighting.

long, containing in particular: Marxism, *A Short Course...*, Polish anti-Semitism, the financing of Jews by the Polish government in London, the upper part of the Grunwald Monument, Osóbka-Morawski's bicycle, atheism, the destruction of the telephone lines, the cost-of-living index, deportation to Siberia, the Jan Kazimierz University in Lwów, the Ossolineum, the Soviet love of watches, Katyń, American aid, censorship, and so on and so on. On the other hand, it is proper for them to know that "people are now hard at work in every place where over recent years German bands and the AK were engaged in destroying democracy."

APRIL 10, 1945. The day before yesterday Dolek left for Katowice with Woźnica. They spent the previous two days looking for a willing peasant with a horse and cart. Woźnica whiled away the time drinking with militiamen until at last two horses, belonging to different owners, were found. They left at 10 am and were in Tarnów by 8 pm, where Woźnica's contacts were expecting them. They didn't, however, manage to board the train since there were four times as many passengers as seats.

We have had no news of Lidka since her departure on March 18.

APRIL 14, 1945. The lack of any communication from Lidka has us worried.

We are bombarded by unsettling rumors: that mass arrests are impending, that Fredek Krzysztoń, head of the local committee, is threatening to expropriate the Cieluchs, who, according to my estimate, have only around 23 hectares of arable land and perhaps another ten hectares of woodland, and that they are now calling up former Polish officers—for God alone knows what war, since there's nothing much left to do against the Germans.[142]

Positive political propaganda doesn't cheer people up since what they really want is to be told that tomorrow all present difficulties will vanish like an ephemeral nightmare. Neither do they want to hear that the wholesale destruction of central Europe, the most thorough in 300 years, bodes ill for the foreseeable future. They've no feeling for or interest in such things as the wider scope of the war: they dismiss the war with Japan as too distant to be relevant to them, the arming of China[143] as the exclusive concern of American financiers, the talk of bestowing India with the status of a "Dominion" of the British Commonwealth as a British diplomatic subterfuge, and Roosevelt's death[144] as an assassination carried out by the Soviets. Let them fight elsewhere to their heart's content, just so long as peace prevails where we are! It all has the aspect of a game with highly convoluted rules which they don't begin to want to master, since all they wish for is a sizable slab of lard, shoes, clothing, half-way decent living quarters, and a paying job—although who was to furnish all this was still a matter of wild conjecture.

Last night our local militia shot and killed a Russian soldier who, with a boon companion, had been terrorizing the residents of Stróże in a desperate search

---

[142]Except for the battles for Prague and Berlin.

[143]China under Chiang Kai-shek was considered an Allied Power during World War II, despite the continuing internecine struggle with the Chinese Communists. Chiang's regime, the "Kuomintang", continued to receive "Lend-lease" armaments from the US till their suspension in 1946.

[144]On April 12, 1945.

for vodka. He had been the first to fire, blazing away at the militiamen with his automatic rifle. The NKVD was not overly concerned, merely arresting his companion. In Stary Sącz[145] a Russian soldier drowned. When Polish bystanders rushed to save him, his comrades stopped them, saying "Why get wet? There are plenty of us to go round!"

APRIL 19, 1945. Yesterday Mr. Pająk[146] came to us from Kraków with his son Zbyszek. Although he brought us no letter from Lidka, Zbyszek said he'd seen her and she looked fine. They tell us that Berlin is being shelled from the outskirts, and that Red Army troops are about to enter the city. Thus one may suppose that the Western Allied forces and the Red Army have between them essentially occupied Germany, and that the war is just about over.

Molotov has issued an explanation of the Soviets' renunciation[147] of their non-aggression pact with Japan of 1941,[148] to the effect that Japan's conflict with the Western Allies helps Germany by drawing Allied forces away from the European arena. This reeks of disingenuousness since Japan had signed the Three-Power Pact[149] well before the signing of the non-aggression pact with the USSR, and when Japan started the war with America in December 1941, the Soviets made no protest to the Japanese.

In Łódź Polish factory workers offered resistance when the Soviets attempted to remove machinery, and soldiers from the Polish army entered the conflict on the side of the workers, arguing that if the equipment represented spoils of war, then it was theirs more than the Soviets'. An impending recruitment in five age-groups was announced in the Nowy Sącz *Powiat* but not in the Gorlice *Powiat*, where presumably the news was suppressed.

Fathers Klawek and Sępowicz visited us yesterday. Klawek informed me that Professor Kulczyński[150] has been put in charge of reorganizing what's left of the Lwów faculty. Some of these have been formally registered as faculty members, but as yet there is no teaching going on. They told me that Professor Krzemieniewski is ill with cancer, and that Bulanda was asked to sign a declaration of Soviet citizenship[151] but refused, and was removed from Lwów, probably at the same time

---

[145]Town in the Lesser Poland Voivodeship, in the far south of Poland. One of the oldest of Polish towns.

[146]Father of Zbigniew, one of Steinhaus's students.

[147]On April 5, 1945.

[148]Later, on August 8, 1945, the USSR declared war on Japan, thereby honoring a promise made to the Allies at Yalta.

[149]Between Germany, Italy, and Japan, signed in Berlin on September 27, 1940, explicitly aimed at the Communist International.

[150]A former Lwów University colleague of Steinhaus.

[151]Presumably because Lwów had now become part of the USSR.

as they removed Burzyński[152] and Fryze.[153] It would seem that the Soviets are not favorably disposed towards Lwów professors.

The explosions we hear are those of mines and unexploded bombs, which the militia is trying to disarm. The commandant of the Grybów militia lost his life doing this work.

APRIL 29, 1945. On April 26, Mrs. Wanda Żukotyńska[154] brought us a letter from Lidka dated April 10. She told me that Bronisław Knaster is to move to Kraków from Lwów, and that Rogala, Krzemieniewski, Ingarden,[155] and Kulczyński are already there. Alas, Kulczyński's wife has been arrested in Lwów.

Lidka wrote that she is working, and that she has registered me as a Lwów professor and collected 1700 złotys on my behalf, representing the February and March instalments of my pension. It appears that professors at the former Jan Kazimierz University in Lwów are now to be settled in Wrocław.

The Red Army is in Berlin, and the Allies are in the vicinity of Munich, and have crossed both the Elbe and the Danube and entered Czechoslovakia. There are reports of the most terrible reprisals being carried out by Soviet soldiers in Germany: murder and rapine are the order of the day. Many Poles deported as forced labor to Germany fall victims to the indiscriminate violence.

Nearer at hand, the Soviets continue with their removal of anything moveable, for example, the three hundred wagonloads of paper from Aleksandrowicz's firm in Kraków, which was originally transferred there from Lwów by the Germans. As a result Piotrowski, the paper merchant, is returning to Stróże.

In Kraków the lawyer Narcyz Wiatr,[156] brother of Proszek's victim, also, like his brother, active in organizing underground Peasants' Battalions during the German occupation, was murdered after a sentence of death allegedly handed down by the

---

[152] Włodzimierz Stanisław Burzyński (1900–1970), Polish professor of engineering. Professor at the Lwów Polytechnic from 1934. In January 1945 he and other professors were arrested by the NKVD. After seven months' imprisonment, he returned to teaching and research in Lwów, but in July 1946 moved to the Silesian Institute of Technology in Gliwice.

[153] Stanisław Fryze (1885–1964), Polish electrical engineer. Professor at Lwów Polytechnic. Arrested by the NKVD in January 1945, he was forced to work as a coal miner in the Donbass for several months. After his release he returned to Lwów, but left in June 1946 to take up the position of head of the Department of Electrical Engineering at the Silesian Institute of Technology in Gliwice.

[154] Sister-in-law of Steinhaus's wife's brother. See Chapter 10.

[155] Roman Witold Ingarden (1893–1970), Polish philosopher, working in phenomenology, ontology, and aesthetics. Professor at Lwów University. From 1945 held positions briefly at the Nicolaus Copernicus University in Toruń and the University of Kraków, from both of which he was banned for his idealism. In 1957 the ban was lifted and he was reappointed to the University of Kraków.

[156] Narcyz Wiatr (*noms-de-guerre* Zawojna and Władysław Brzoza) (1907–1945), Polish lawyer and populist political activist, member of the agrarian Polish People's Party. During World War II he led Peasants' Battalions in the underground resistance movement. After the formation of the communist Lublin government, advised members of his organization against compliance, and himself remained in hiding. However, he was shot by the Myślenice communist secret police in Planty Park in Kraków on April 24, 1945.

AK was read to him. A lady friend of his was present when, a quarter-hour after the murder, during a search of Wiatr's apartment by the NKVD, the designated hitman changed his militiaman's uniform for civvies so as to divert attention away from the NKVD. People are saying that the PPR won't get away with this.[157]

The Lublin government has signed a twenty-year agreement with the USSR to act as allies against Germany. Molotov visited Washington twice,[158] and was twice closely questioned on the issue of Poland, once by Truman and again by Stettinius[159] and Eden.

MAY 1, 1945. The American and Soviet armies met along the Elbe River near Torgau on April 25. Yesterday I learned that the *Gauleiter*[160] of Bavaria capitulated to the Western Allies, preferring them to the Soviets, and likewise in parts of northern Germany. Hitler's forces must surely be swamped by the Red Army in Berlin by now. Assuming the Germans surrendered on April 30, the war will have lasted exactly 68 months!

At the San Francisco conference there was considerable pressure on Molotov to explain why there was no Polish delegation. It is even rumored that Truman declared aggressively that the US would represent the interests of Poland at the conference, and forty of the 47 delegations demanded that Poland retain her pre-war eastern border. Since the Yugoslav and Czechoslovak delegations, together with de Gaulle,[161] the new friend of the USSR, supported the Soviet position, and the US and Britain had their hands tied by the Yalta accord, it follows that there were really just two countries whose delegations aligned themselves with the USSR on this question. Perhaps Molotov was also reproached concerning the Polish-Soviet pact as violating Soviet undertakings *vis-à-vis* the western Allies. In any case the conference was interrupted by Molotov's having to return to Moscow for further instructions.[162]

---

[157] In 1996 the Kraków office of the Institute of National Remembrance charged a certain Stanisław P., a former member of the Kraków communist secret police, with participating in Wiatr's murder. The trial was postponed because of the ill-health of the defendant.

[158] In May 1942, when he held talks with Roosevelt, and in late April 1945, when he spoke with Truman before going on to the conference in San Francisco where Truman bluntly repudiated the Soviet "Eastern Europe-United Nations nexus."

[159] Edward Reilly Stettinius (1900–1949), US Secretary of State from 1944 to 1945 under Presidents Roosevelt and Truman.

[160] Nazi term for leader of a region of the Reich.

[161] Charles de Gaulle (1890–1970), leader of the Free French Forces June 1940–July 1944. President of the Provisional Government of the French Republic August 20, 1944–January 20, 1946.

[162] According to another source, the number of delegations was 46, and Stalin sent Andreï Gromyko back in Molotov's place. The main issue leading to Molotov's replacement was the dispute over whether Argentina should be admitted to membership, strongly supported by other South American delegations but opposed by Molotov, and whether Ukraine and Belarus should be admitted, given that Poland was not represented.

I have been reading issue No. 20 of the magazine *Odrodzenie*. It is like a textbook for writers wishing to acquire the latest puffed-up journalistic jargon and style: thus they use "depict" when "show" would do, "participate" instead of "take part", "expresses himself" for "talks", and suchlike, as well as grating German constructions such as "told in a way that he viewed as proper his narration." One never found such barbarisms in the *Wiadomości Literackie*.[163] Since those writing for *Odrodzenie* can't write, one infers that their expertise lies elsewhere!

MAY 4, 1945. Last night we heard that Germany has unofficially capitulated. This means the war lasted 2069 days.

I have been writing a book on independent functions[164]. It will be necessary to recast the statements and proofs of many of the theorems on this theme in the literature, in particular because many of them are stated imprecisely.

The day before yesterday Mr. Pająk brought us a letter from Lidka dated May 1. She informs us that Olga and her family are now in Kraków living with Irena and her family. She also writes that Banach is to have a position in the university in Kraków, and that Bronisław Knaster has one already. As for myself, nothing attracts me to that city, and, in any case, as I wrote in my reply to Lidka, people like me are no longer needed.

My wife's enterprise making and selling yeast is doing quite well.

MAY 9, 1945. Yesterday we received the news that at 2 pm on May 7 a cease-fire on all European fronts was called, and at that very moment at Rheims, France, in a schoolhouse serving as Eisenhower's headquarters, the final instrument of surrender was signed by the new Reich President, Grand Admiral Karl Dönitz.[165] It is said that Hitler and his closest cronies, including Goebbels, have committed suicide in Berlin, and that Himmler has been arrested,[166] etc.

Meanwhile, the weather is wonderful. On May 7, Jan Igielski, Schoenemann and his son, Jan Dobrzański, and the Witmans arrived by car from Silesia. According to Dobrzański, who is now an important figure in the central organization in charge of

---

[163]"Literary News", an extremely influential high quality Polish literary magazine, published in Warsaw between 1924 and 1939.

[164]This was never completed, although Steinhaus returns to the idea several times. *Note added by Roman Duda, editor of the second Polish edition*

[165]In fact, the surrender was signed at 01:41 on May 7 by *Generaloberst* Alfred Jodl on behalf of the *Wehrmacht* and President Karl Dönitz. The surrender was to take effect at 23:01 on May 8. Considering, in particular, that the USSR was under-represented at Rheims, and not sufficiently forewarned, Stalin demanded that another ceremony be held in Berlin next day—May 9 Moscow time—with Field-Marshal Wilhelm Keitel officially capitulating to Marshal Georgiĭ Zhukov in the presence of representatives of the western Allies. Victory Day has since then always been celebrated in Russia on May 9.

[166]Heinrich Himmler, head of the SS, was captured on May 21, 1945, and suicided two days later by biting down on a cyanide capsule in a gap between his teeth.

the Polish metals industry, the Soviets continue to strip our industrial plant in Śląsk Opolski[167] of machinery, rolling stock, even desks and telephones.

I asked Stanisław Rząca how many of his fellow students like the Russians, to which he replied "Not a single one!" There are many former communist sympathizers whose love for the Soviets—never requited, by the way—has evaporated.

MAY 11, 1945. Poles are coming back home from the defunct Reich. Along the roads trek people returning from Saxony, Potsdam, Vienna, etc. Some of them have acquired a horse and cart. Along the way many have been stripped of their possessions and the womenfolk raped *en masse* by Red Army soldiers, their Polish identity cards notwithstanding. Any objecting too strenuously are done away with.

The car in which Schoenemann and his son came has broken down. Schoenemann bribed the station commandant with 500 złotys to arrange for it to be shipped by train. However, the commandant promptly demanded to see the car's registration papers, saying that he would have to submit them for scrutiny to someone in Nowy Sącz to determine whether the car might be classified officially as spoils of war, in which case it would be confiscated. Fortunately for Schoenemann it was returned to him, but not the 500 złotys. Yesterday they departed, the car being towed by a truck hired from the plant at Glinik Mariampolski.

Train travel remains hazardous since Soviet soldiers frequently rob the passengers, and occasionally throw anyone objecting out of the speeding train; this was done to a poor woman one night as her train was crossing a bridge over the Wisłoka.[168]

At last we have warm weather.

The Zarembas' land was parcelled up, yet only half of it has been sown.

I encountered serious difficulties when I tried to sort through the proof in Marek Kac's paper in *Studia Mathematica* of the so-called Gauss–Laplace theorem on probabilities.[169]

MAY 12, 1945. Today we obtained a newspaper dated May 10 reporting the details of the German capitulation. The surrender was signed by Keitel and took effect at 23:01 Central European time on May 8.[170]

---

[167] A region of Upper Silesia centered around Opole, transferred from Germany to Poland after World War II.

[168] A tributary of the Vistula in south-eastern Poland.

[169] Possibly the Central Limit Theorem, which states conditions under which the mean of a sufficiently large number of independent random variables will be approximately normally distributed. The theorem is especially important in that it provides justification for using the normal distribution to approximate large sample statistics obtained from controlled experiments.

[170] This was the official surrender organized in Berlin by the Soviets. There was another held earlier in Rheims, France, with the western Allies predominating. See above and below.

MAY 14, 1945. It turns out that our Soviet-controlled newspapers have purposely distorted the history of Germany's capitulation. The instrument of surrender signed ten minutes before midnight on May 8 at the Red Army headquarters in Berlin-Karlshorst was in fact merely a ratification[171] of the official capitulation by General Jodl at Eisenhower's headquarters at around 2 pm on May 7, to take effect at 23:01 Central European time on May 8, that is, at 00:01 on May 9 Moscow time. This view of the matter is as expressed in Churchill's speech of May 8 preceding the Berlin ceremony. The newspapers present a highly slanted version of the two rituals, stressing the role of the Soviets over that of the western Allies in winning the war, and praising Zhukov, the marshal who brought Berlin low, and to whom Keitel capitulated, far above all others.

The sputteringly ineffectual expressions of hatred for Germany that one sees in the newspapers are revealing of the journalists' lack of imagination: the "German hydra",[172] "German thugs", "fascist reptiles": these represent the limits of their invective, while at the same time illustrating perfectly the content of everything the columnists and upstart *feuilletonistes*[173] concoct. For example in one article, its author brags about how, revolver in hand, he compelled a German to fetch wine and sardines from the cellar. Such fatuously smallminded scribblings have the odd—and doubtless unfortunate—effect of awakening an emphatically undeserved sympathy for the Germans.

We are informed that Truman and Stalin are to meet.

MAY 17, 1945. My student Staszek Rząca has been called to a military school for officers, and three days ago he came to say goodbye. The authorities at first asked for volunteers, and then called up everyone considered military material, volunteering them on their behalf, so to speak. Thus reserve officers and those who had seen active war service were enlisted in special officer corps, the aim being to isolate any dissident members of the old guard from the new recruits. However, as Staszek said, they needn't have worried since all are likeminded, though not in the way the Lublin government would have it: they detest the word "democracy" precisely because of the way it is used—or rather abused—by the fraudulent political mouthpieces of that government.

Rumors are flying that the Americans are coming to Poland, but these are in all likelihood false, being prompted by the American aid in the form of foodstuffs transported to Łódź and Silesia, escorted, we are told, by Polish-American soldiers of the US army. Certain Soviets tried to steal the goods at the border with American occupied territory, but were bought off with 30 kilograms of lard.

---

[171]This may not be quite accurate. See the relevant footnote above.

[172]The Lernaean Hydra of Greek mythology was a serpent-like chthonic water-beast with many heads, with the interesting peculiarity that for each head cut off it grew two more. It was the second labor of Heracles to kill this beast.

[173]Here in the sense of writers of improbable stories.

It is also rumored that the Czech Army is pushing into Żywiec, Sucha, and the Nowotarski district,[174] but surely if this were happening the Soviets would be up in arms.

MAY 24, 1945. We have no way of confirming news fed to us about the Czech Army. As to the matter of the Polish sixteen,[175] it seems that Stalin has finally admitted to their arrest. These were considered representatives of the Polish Government-in-Exile in London, and believed their immunity to be fully backed by Britain and the US,[176] and were fooled, furthermore, by the Soviet offer to place a plane at their disposal to bring them to Moscow. So they went, and from the time they left in March till now there has been no word as to what's become of them. The Government-in-Exile asked in vain for information, and then Eden[177] submitted a note of protest on their behalf through the British ambassador in Moscow, and, finally, the great Panjandrum himself came out with it: the sixteen were saboteurs and will undergo a court trial. The matter was broached at the San Francisco Conference, with the British delegation stoutly maintaining that these were the very people qualified to make the provisional government workable. Of course, Stalin simply ignored this and other protests, and the substance of the Yalta accord, a key point of which was the holding of free elections in liberated states, has gone by the board, while lip-service continues to be paid to it. Thus, however the provisional government might be constituted, it is now accepted in the international forum that it represents Poland. What can anyone do against a monster who holds all the cards?

In an issue of the literary magazine *Odrodzenie*[178] my son-in-law Jan Kott had an article taking issue with the title of a Parisian production of *Hamlet* advertised as *Hamlet ou le distrait, pièce de M. Shakespeare*, which had been translated as "Hamlet, or the Absentminded One, a Play by Mr. Shakespeare". I doubt whether the translation of *distrait* as "absentminded" conveys what the French composer of the title actually had in mind.

Recently the authorities issued everybody with blank forms printed with a list of the kinds of depredation for which damages might be sought. People quickly filled

---

[174]Two towns and a district in southern Poland close to the then Czechoslovak border.

[175]In March 1945 sixteen members of the Polish Underground State, operative throughout the German occupation, were lured to Moscow, ostensibly to discuss their eventual inclusion in the Soviet-controlled Provisional Government of Poland. Instead they were imprisoned in the Lubyanka where they were submitted to brutal interrogation and torture. Only after somewhat feeble protests from Washington and London did Stalin admit to the arrests, reassuring his former Allies that the prisoners would receive light sentences. The show trial took place in June 1945, all but one of the victims admitting to alleged "crimes". Only two of the sixteen survived beyond the next six years.

[176]Which, however, withdrew their recognition of the Government-in-Exile on July 6, 1945.

[177]Anthony Eden, then Foreign Secretary in Churchill's government.

[178]No. 22, of April 22, 1945. *Author's Note*

out the forms, banded together, and appointed a commissar to represent them—and suddenly the forms were withdrawn from circulation. I imagine that so many of the submitted forms included complaints—allowable according to the form—about assault and robbery by Red Army soldiers, that the heap of forms soon rose to such a height as to represent an all too obvious unpleasant reminder to the Polish powers-that-be of the actions of the subordinates of their mongoloid superiors.

For several days railcars carrying mostly former Russian and Ukrainian forced laborers returning eastwards from Germany have been standing along the tracks outside the railroad station. Since most lines are clogged with wagons piled high with coal and machinery in transit to Russia from Silesia and Germany, and this cargo has all to be transferred from standard to wide-gauge tracks, there are strings of railcars that have been standing idle at the Biała and near the former manorial property near Tarnów for more than three weeks. The passengers barter articles of clothing—shirts, socks, etc.—, as well as shoes, thread, rings, and other such items for eggs, butter, milk, lard, vodka, and, most of all, bread. People from round about bring whatever food items they can spare since there is a desperate shortage of clothing and footwear. As Stefa aptly observed, the people from the railcars resemble Gypsies in their lifestyle. They are not bothered by promiscuity, they are used to camping out in the fields, and are apparently not greatly inconvenienced by rain or a chill wind. During their oft-delayed slow migration eastwards they conclude wild marriages, sing and dance to the harmonica, and show contempt for steady work and the bourgeois lifestyle, together with a certain charm and haughtiness quite different from ours. There is a great deal of drinking, and sometimes they are driven to pilfering. I think some of them may have acquired weapons along the way. Perhaps a quarter of all Europeans are now reduced to living from hand to mouth as they do.

MAY 26, 1945. Marysia Igielska gave birth to a boy this week. Recently we had news from Stasia Schoenemann that she has been reunited with her husband in Katowice, and that they plan to visit us together. There is still no news of Mr. Ochalik.[179] It may be that he is in a camp in the British or American zone of occupation of Germany. Between the western occupation zone and the Soviet one there is apparently a kilometer-wide no man's land. According to our underground press, thousands of Poles are applying to the British and Americans hoping to get out of the Soviet zone, and, presumably, never return to Poland.

---

[179] A former Polish officer sent to a POW camp by the Germans. Husband to Dela, eldest of the Igielski girls. See Chapter 12.

The international political situation is strained over Trieste,[180] Poland, and the island Bornholm.[181] There are also difficulties over Vienna,[182] where the Soviets are employing their usual high-handed tactics. Each such conflict brings out ever so clearly the character of the Provisional Government as a mere puppet of the Soviets. One anticipates them signing over to their masters in the Kremlin all of the moral and material capital that might be said to constitute "Poland".

Railcars keep arriving at the station, are shunted onto a siding, and stand there for weeks. The Vlasov people[183] left to be replaced by the Lemkos.[184] A band of Red Army soldiers carried out night raids on peasant huts, and the NKVD settled the matter by confiscating the stolen goods, leaving no one satisfied. In this way Poland is being balkanized into disjoint sets of discontents. Moryś[185] escaped.

MAY 28, 1945. To get an idea of what lies in store, it is instructive to observe how the Soviets treat the Lemkos, that is, the Rusyns from around Nowy Sącz. They first take away their livestock, mainly cows, then transport them from there to the region around Sanok,[186] sometimes holding them in the railcars for weeks. Once arrived they separate off the able-bodied men and dispatch them to Slovakia, most probably to the army, and the rest—women, children, and the elderly—eventually wind up at best in some *kolkhoz*, at worst behind barbed wire in a so-called labor camp.[187]

---

[180] By May 1, 1945, most of Trieste had been captured from the Germans by the Yugoslav Partisans' 8th Dalmatian Corps, which held it till June 12, when Tito agreed to hand administration of the city over to the British and Americans—but not before many real or imagined opponents of Yugoslav communism in Trieste had been imprisoned or killed. The status of Trieste as a "free territory" was established as of September 1947 by the Paris Peace Treaty, signed in February 1947.

[181] Bornholm island, in the Baltic between Sweden and Poland, had belonged to Denmark since the seventeenth century. During World War II it was taken over first by the Germans and then by the Soviets, who eventually agreed to hand the island back to Denmark in return for suitable concessions.

[182] As in the case of Berlin, Stalin was forced to honor his previous agreements according to which Vienna, after being overrun by the Red Army, was subdivided into five zones, with the four peripheral zones administered separately by Britain, the US, France, and the USSR, and the central zone by a joint Allied Commission. This was the situation till 1955, when the Soviet Union agreed to relinquish their zones of occupation in eastern Austria in return for guarantees of Austria's permanent neutrality.

[183] The term "Vlasov people" here refers colloquially not only to the soldiers of the former Russian Liberation Army (ROA), or "Vlasov army", but includes former members of certain military or police units made up of Russians, mainly prisoners-of-war and emigrants, and formed under German auspices.

[184] In Polish, *Łemkowie*. See below.

[185] Maurycy (Morcio, Moryś) Bloch (1917–1995), a fellow student of Steinhaus's daughter Lidka. Went to the US in 1938, remaining there.

[186] Town in south-eastern Poland, in the Subcarpathian Voivodeship, formerly, from 1340 to 1772, the Ruthenian Voivodeship.

[187] In the USSR, many such deportations of entire peoples, mostly Muslims, were carried out at Stalin's behest during and after World War II.

There seems to be a dearth of competent officers, since, in addition to our young educated men, the authorities are calling up former Polish reserve officers for active service. But the procedure for verifying their competence is bizarre: after a certain lapse of time the militia,[188] with whom the officers have to closely liaise, are asked for their estimate of the officers. A negative response means a term in a labor camp, while if positive they are compelled to swear an oath of loyalty to the Lublin government, under pain, again, of being sent to a labor camp. Some of our young men have been commissioned to officer duty as far away as Kharkov and Odessa!

In Koźle, a community in the Opole Voivodeship,[189] some folk wanted to found a Polish Gymnasium to accommodate a few dozen aspiring students. However, the students couldn't attend since they were afraid of being abducted in the street on the way to or from school. The Soviets round up the lads and put them to work, and the lasses simply vanish. In Gorlice several students have been arrested.

The deputy *starosta* in Gorlice has been arrested—perhaps for the tenth time—and taken to Kraków. So he escaped again, but was recaptured and re-arrested.

With a Moskal I bartered two kilograms of bread, half a liter of butter, and ten eggs for a pair of shoes with rubber soles. They had been worn previously, but the soles were in good condition.

We now receive flour rationed by means of coupons. Presumably it comes from the Americans.

MAY 31, 1945. Yesterday Stasia Schoenemann arrived here after four years' absence, mostly in the Ravensbrück concentration camp for women. She told us that over that period about five thousand prisoners had been gassed there. They were subject to periodic medical checks, and any showing signs of debilitation—such as swollen legs or poor complexion—were selected for gassing. At the end, when they were being transported west, she managed to slip away with 67 others without being noticed by the escort. She also managed to get hold of a horse and cart, so that even the weakest survived the trip home.

The Rząca family has had bad news. The so-called officers' school to which Staszek was summoned, operated for only a week, after which all the officer cadets were locked up in quarters in Montelupich prison in Kraków, without pallets to sleep on, let alone uniforms, drilling, or instruction. Their treatment is a good illustration of the incompetence of the Lublin government: these young would-be cadets were just the innocent sons of peasants, most without secondary education, but with an honest desire to prove worthy of officer rank, and one can imagine the disillusionment fostered among them by such dereliction.

For vodka Soviets barter pilfered two-stroke motors with an aggregate output of one, two, or four horsepower. There are a great many of them, spoils of war from Germany, brand new, complete with spare parts and full instructions. No one knows what their intended use was. Some of our railroad workers stole some of these

---

[188] In the USSR (and modern Russia) the word "militia" means the police force. Perhaps this is the appropriate way to interpret the word from this point on, with Poland within the Soviet ambit.

[189] A province in Silesia.

motors off the wagons when no one was looking. There are also typewriters, engine accumulators, and other technological items in abundance, all or any of which can be had in return for vodka.

I watched a transport pass through carrying Polish soldiers from Germany to Rzeszów. As they passed they yelled out that they were AK troops, and that they hated the Bolsheviks. Chaos reigns.

Subcarpathian Rus'[190] has been annexed to the Ukraine, and therefore to the USSR, without asking anyone's by your leave, let alone organizing a conference to discuss the issue.

A new parliament is to be convoked in London in July. It is rumored that Sosnkowski[191] is again to be a member of the Government-in-Exile.

JUNE 2, 1945. I have had another letter from Jan Dobrzański, in which he describes a trip he recently made from Gliwice to Wrocław. He says that all the way on either side of the road one sees just smouldering ruins amid general desolation, and that Wrocław itself is a uniform landscape of rubble, with starving people wandering listlessly and aimlessly about.

Train after train passes through the Stróże station carrying some of the dozen or so millions of displaced people back to their homelands from German camps, factories, farms, etc. There are about seventy railcars that have been standing idle on sidings for some time now, packed with Ukrainians from the region of the Dnieper. A few days ago a few of their children sickened, and yesterday some of them died. The Ukrainians are now blaming their death on the milk they obtained locally, saying it was poisoned, yet no one else has fallen ill after drinking it. Belief in ill will, and acts of poisoning motivated by it, is deeply ingrained in the Eastern mind, in particular in the Soviet version.

Elections of councillors to local administrations have been held. There was a notable lack of fanfare, but the voter turnout was large indeed. Both the new and old candidates seemed to me to be of acceptable caliber. But then the *starostas* declared the election invalid! So that's what our so-called "democracy" looks like!

The education system is to be reformed. Starting in the next school year, there will be eight-grade elementary or primary schools, and three-year secondary schools or "lyceums". Latin will not be taught in the elementary grades, and in universities there are to be two streams, one for ordinary students and the other for so-called "scholars". This two-tier system is reminiscent of the parcelization of land. And if

---

[190] A small region of eastern Europe, now mostly in Ukraine, with smaller subregions in Slovakia and Poland. In Ukraine it is more usually called "Trans-Carpathia" since it is on the opposite side of the Carpathian Mountains from most of Ukraine. A more neutral term is "Carpathian Ruthenia."

[191] Kazimierz Sosnkowski (1885–1969), Polish independence fighter, politician, and Polish Army general. During World War II he escaped to the West where he held various posts in the Polish Government-in-Exile in London. On the death of Sikorski he became Commander-in-Chief of Polish Armed Forces in the West. He was eventually dismissed because of his criticism of the western Allies for their apparent abandonment of Poland to the Soviets. Lived in Canada from the end of the war.

each parish is to boast an elementary school, where are all the teachers to come from?

Polish soldiers travelling eastwards ask the locals for the names of those who might safely be enlisted as partisans and armed. Although they are being forced to go further east, they say that under no circumstances will they consent to going beyond Lwów, and will desert and hide in the woods if it should come to that. They are of one mind on this, their experience with their Soviet comrades-in-arms and in Russia having opened their eyes once and for all. As I think I mentioned earlier, the Polish officer cadets were fed very well during the first week following recruitment, but when they refused to sign declarations of party adherence were shut up in Montelupich prison in Kraków.

The moral bankruptcy of the Provisional Government is becoming more evident by the day: they are simple drained of all initiative. The impending rupture with the former western Allies represents an especially trying predicament for them, since, despite their commitment to communism, the idea of an armed conflict with Britain, for example, cannot be very palatable even to Osóbka-Morawski. His position *vis-à-vis* Stalin is comparable to that of Krzeptowski's[192] *vis-à-vis* Hitler.

Everyone's in the dark as to what's going on in the world. Trainloads of wheat from Germany pass through, and the railroad workers help themselves or use vodka as a medium of exchange.

I hope to have completed my logarithms by mid-June, which is the time agreed to.[193] Then I would like to write a short note, in Polish, on the quadratic tariff, that is, the schedule of charges for electricity,[194] complete with tables and graphs.

We've run out of both money and supplies.

JUNE 6, 1945. Late on Saturday, June 2, Stasia Schoenemann came to visit again, and told us in detail about her ordeal in Ravensbrück. There were women of 28 different nationalities incarcerated there, totalling 40,000. She herself had been put in charge of a block of 2000 women. Roll call was at four and work began at six. Periodically groups of women from other camps, including Auschwitz, arrived, and were immediately assigned to workbenches in the camp workshops or factories outside the camp. For the most part the inmates worked as seamstresses, shoemakers, and in some cases did artistic, decorative work such as embroidery.

---

[192] Wacław Krzeptowski (1897–1945), political leader of a highland people—called *Goralenvolk* by the Germans—from Podhale, in the Tatra Mountains near the border with Slovakia. Hans Frank, Nazi governor of Poland, allowed Krzeptowski to establish an independent state for this ethnic group in southern Poland, on the grounds that they were racially closer to Aryan, and Krzeptowski reciprocated by attempting unsuccessfully to recruit a "legion" to fight with Germany. After the war he was sentenced to death by the Home Army, and executed.

[193] This is the first mention of this enterprise. It would seem that Steinhaus had at some earlier time contracted to produce a table of logarithms—perhaps in connection with his work as a teacher in the underground school system.

[194] The French version of the paper on the quadratic tariff is: H. Steinhaus, "Sur le problème du tarif électrique", *Sprawozdania WTN Seria B*, No. 1 (1946), pp. 21–22. The Polish version appeared in the same journal in 1947 (No. 2, pp. 80–81).

In addition to male administrators, they were closely supervised by a contingent of a few hundred female SS guards. Mornings and evenings they were given unsweetened coffee and dry bread, and at noon boiled rutabaga[195] or unpeeled potatoes, with water to drink. Occasionally food packages would arrive, and although the ones containing better quality items were usually intercepted by the female overseers or the SS guard, sometimes something got through.

There was a POW camp not far off, and women assigned work outside the camp would sometimes find an opportunity to communicate with the prisoners of war. Since these had an undetected transmitter, the women were able to get information about what was going on at Ravensbrück to the outside world.

Mild violations incurred slaps to the face or a few kicks, and more serious ones a thorough beating. Anyone leaving their barrack at night was penalized with a few months in isolation in the dark, and subsequently a lengthened workday and a decreased food allowance. A certain Dr. Gebhardt[196] had an experimental surgery in Ravensbrück. He operated on certain chosen inmates, cutting open their skull or abdomen—without the use of anaesthesia—in order to observe how long they survived before succumbing to shock or sepsis. He drew off the pus from the patients' suppurating organs and wounds for analysis. There was no lying-in hospital: any such "rabbits", as they were called by the experimenters, who happened to survive more than a day were returned to their barracks without dressings on their wounds or any care whatsoever. In 1945, thanks to the Swiss Red Cross, an exchange of prisoners was arranged, and, by switching documents, some of the more enterprising and courageous of the inmates managed to smuggle one such "rabbit" out with the lucky ones chosen to leave, and in this way the grisly goings-on inside the camp became known to the wider world. Gebhardt's subjects were intended to die in order to prevent information about his ghoulish experiments being leaked to the outside, but when the inmates notified the camp commandant that a "rabbit" had escaped to Switzerland, the rationale for ending their lives lapsed, and a few capable of surviving were reprieved.

Dr. Z. Mączka[197] gave a lecture about Ravensbrück on "Polish radio," broadcast from London. She mentioned, in particular, that once a week a committee of doctors would carry out a "selection" of the inmates, separating off any with symptoms of illness—for example, swollen ankles—for gassing. It is estimated that a total of around 5000 Polish women were exterminated. How many perished altogether over

---

[195] Or yellow turnip, a cross between the cabbage and the turnip.

[196] Karl Gebhardt (1897–1948), German medical doctor. Personal physician to Himmler, and one of the chief coordinators and perpetrators of surgical experiments on inmates of the concentration camps at Ravensbrück and Auschwitz. Found guilty of war crimes and crimes against humanity by a US military tribunal, Gebhardt was sentenced to death. He was hanged at Landsberg Prison, Bavaria, on June 2, 1948.

[197] Dr. Zofia Mączka (1905–1992) was a prisoner in Ravensbrück camp assigned to work as X-ray technician. At the end of the war she testified to and wrote about the horrifically cruel experiments performed in the camp, in particular about experiments where bones were fractured, pieces of bone removed or transplanted, and the periosteum removed.

the period of the war is unknown, but Stasia's guess is that it must be in the several tens of thousands.

Among the inmates there were *Sonderhäftlinge*[198]: a cousin of Roosevelt's, Mme Spitz, former owner of a diamond firm in Paris, two Countesses from the Potocki family,[199] and Countess Karolina Lanckorońska.[200] This last refused all privileges at the camp, demanding to be quartered in the most crowded barracks. She did, however, get out of the camp earlier than some, being exchanged for a German general.

Stasia managed to save perhaps as many as a few hundred women in her block. She showed us a notebook in which many of her fellow inmates recorded expressions of gratitude for her dedication to defending them and shielding them from the worst. For instance, she saved a Croatian woman from death by telling the pertinent female German SS overseer *Was wir hier bekommen, wird euch doppelt zurückgegeben,*[201] with a hint that not only she but also her family would be in for it.

One of the worst things was the lot of mothers, whose newborn babies died after a week or so because of a lack of milk.

JUNE 9, 1945. Two days ago Wanda Żukotyńska arrived from Kraków with a letter from Lidka and 1500 złotys for me. There was sad news in the letter: Maria Knastrowa, wife to Bronisław Knaster, and one of those few instrumental to our survival, has died of heart failure. Lidka writes that she tried to convince Bronisław to come and see us, but he was unreceptive to the idea.

It seems that *Fundamenta Mathematicae*[202] will remunerate authors of accepted papers. I have at last completed my table of logarithms. In order to get six-figure accuracy I had to devise certain arithmetical shortcuts. Now I will devote myself to my note on quadratic tariffs. In this connection the following idea occurred to me: that a quadratic tariff is unique in the sense that in that case the change

---

[198]"special prisoners"

[199]One of them was Countess Zofia Potocka, *née* Tarnowska (1901–1963), wife of Andrzej Potocki (1900–1939), who fought against the Germans in September 1939, and was murdered by Ukrainian peasants during the retreat of the Polish forces.

[200]Polish historian and art historian. Daughter of Count Karol Lanckoroński, a Polish nobleman from Galicia. Countess Karolina fought in the Polish resistance in World War II. Lived from 1898 to 2002.

[201]"What happens to us here will be repaid to you double."

[202]The story of Volume 33 of *Fundamenta Mathematicae* is dramatic. Ready for printing before the Germans invaded, it was ordered destroyed by the German director of the publishing house, but a printer, risking his life, hid the plates. In the summer of 1945, Wacław Sierpiński retrieved them, added a few notes composed during the war—including one by Steinhaus—and, at the beginning of the volume, a list of names of those lost to Polish mathematics, and with the great help of Bronisław Knaster, arranged for its publication. The appearance of this new volume of *Fundamenta* in December 1945 served as a powerful boost to the low morale of surviving Polish mathematicians, and a demonstration to the world of their still unbroken will to continue. *Note added by Roman Duda, editor of the second Polish edition*

in consumption—of electricity, say—per unit increase in price is proportional to consumption, so provides a direct measure of the economic pressure brought to bear by the imposition of such a price schedule.

I cannot think that working conditions in a town environment could be better than they are here. To live in the summer in what are in effect splendid natural parklands, and not to have truck with authorities of any kind, and not be bothered by committee meetings, conferences, etc., and the booming of loudspeakers—could there be a more ideal working environment?

Yesterday I came across young Stanisław Rząca at the river. He and his brother Adam have run away from the officers' school. They arrived home yesterday evening, but, fearing that the military police would come looking for them, they spent the night in the woods, where dinner and breakfast was brought to them. Adam went to enlist in the AK and Staszek took refuge with a nearby family. They said they had run away because they could not endure Soviet propaganda. Later I found out from Adam that the AK have adopted a policy of treating PPR[203] people like *Volksdeutsche*. As a result there is some fighting going on; we can hear the booming of artillery over towards Rzeszów.

Wanda Żukotyńska was at Irena's and saw my whole family there safe and sound. They were surprised that I hadn't yet been offered a professorial position at the Jagiellonian University of Kraków.

Although the newspapers contain little factual news, one can infer from their tone that there is a significant amount of political tension between East and West. They complain about the large British presence in the Near East—600,000 British, they say—and the disturbances caused by British agents there. They also mention de Gaulle's complainings about the Arabs in Damascus,[204] and his observation that Egypt, Syria, and Iraq are not genuine states.[205] This is interesting, since, if true, it follows that Britain might have united the whole of the Arab Middle East in a single powerful Islamic bloc.

I have discovered that a bird I often come across in the woods, a little bigger than a starling and with a spotted breast, is called a "mistle thrush." Our neighbor Mrs. Krokowa says that this is just the old name for the fieldfare.

JUNE 11, 1945. Mr. Motyka,[206] former assistant lecturer in botany, dropped in yesterday. He is now teaching in the Gymnasium in Grybów. He had not heard

---

[203]The Polish Workers' Party, the Polish communist party active from 1942 to 1948, when it merged with the Polish Socialist Party. See earlier.

[204]From 1939 there was a struggle for independence from the French in Syria, which succeeded in 1945, but not before the French had bombed Damascus.

[205]Iraq's modern borders were determined by the League of Nations in 1920, after the fall of the Ottoman Empire. It was ruled by Britain as the British Mandate of Mesopotamia until 1932, when it gained independence as the Kingdom of Iraq. In 1958 the monarchy was overthrown and the Republic of Iraq proclaimed, ruled by the Arab Socialist Ba'ath Party.

[206]Józef Motyka (1900–1984), Polish botanist and lichenologist. During the war he worked in the Lwów Botanical Gardens, returning in 1944 to the region where he had been born and raised to teach in the Gymnasium at Grybów. In 1945 he was made Director of the Department of Plant

of the deaths of Łomnicki, Stożek, and Ruziewicz, even though he had remained in Lwów till the Spring of 1944. He said that generally speaking the Ukrainian allies of the Germans were more determined on making life difficult or impossible than even their Nazi overlords. During the war he taught *Fachkurse*[207] under a German head, a professor from Leipzig. This man was embezzling funds allotted to him for field workers, so that Motyka had a hold over him which he used to have about a hundred people employed in a make-work project involving digging in Cetnerówka,[208] thereby saving them from deportation or worse.

He told me that botany had suffered irreparable losses as a result of bombings, which destroyed irreplaceable collections of "originals", that is, specimens defining classes or other subdivisions of the scientific taxonomy. In particular, the famous African herbarium in Berlin was utterly destroyed, as well as other formerly world-renowned herbaria in Munich, London, Vienna, Warsaw, and, as now seemed likely, Budapest.

I showed him my graphical representation of Haeckel's[209] biogenetic law. He told me that he had never seen such a representation, and that Haeckel's law had largely been borne out by his own observations.

I walked with him most of the way to Grybów, and we conversed about the issue of duplication among collections of biological specimens. He also told me that although Zabierowski, Father Klawek, and Father Dziedzioch all admitted to knowing me, only the first was willing to say exactly where I could be found. As we parted I suggested to him that I might be able to contribute something to the application of statistics to plants.

JUNE 13, 1945. Yesterday Mr. Szczęśniewicz, a teacher from Polna, came by. He has just returned from an extensive tour of the Poznań area looking for a teaching position. He said that, while the farmers are managing, there is absolutely nothing for anyone with the slightest intellectual bent to do there. There was a splendid sugar refinery in the region to be annexed to Poland, and the Ministry of Industry sent a team of experts there to restore it to working order. However, two weeks after it begins operating along comes the Red Army and strips it of all its machinery and other installations.

In Śląsk Opolski and Brandenburg,[210] although the *Wehrmacht* has left, the Soviets seem to be in no hurry to do anything about the German people left behind. They are even allowing them to form professional associations such as workers'

Geography and Systematics at the newly established University of Maria Curie-Skłodowska in Lublin.

[207] Courses providing technical training.

[208] A former estate close to Lwów.

[209] Haeckel formulated the theory that "Ontogeny recapitulates phylogeny", that is, that the embryological development in any species reflects the species' evolutionary history. The theory is now largely discredited, though acknowledged as partly accurate.

[210] After World War II, Neumark, the part of the German federal state of Brandenburg east of the Oder-Neisse Line, was transferred to Poland, and its German population expelled.

councils, and since the hiring of labor is regulated by these, Poles looking for work are left in an untenable situation.[211]

Szczęśniewicz also said he had met prisoners freed from Mauthausen,[212] to whom the official *Dziennik Polski* recently devoted some space, describing the horrendous obstacles the British had placed before them and which they had so valiantly overcome in order to get back to Poland. The poor fools had believed the propaganda and crossed the boundary between the British-occupied zone and the Russian zone. Of course, the Soviet border control immediately robbed them of everything of any use in their possession. They were now cursing their fate, since they had heard how the Americans treated ex-inmates like themselves, providing them with the best quarters in former SS rest homes, sending them to convalesce at seaside resorts, and providing excellent food and first-rate conveniences to facilitate their recovery. They will thus henceforth form a bulwark against Lublin propaganda.

JUNE 16, 1945. Yesterday I went to Bobowa to get food coupons. Upon my return I found Wanda Żukotyńska in our hut; she had come to say goodbye as she has had a telegram from Dolek[213] informing her that he has found a position for her. Tadeusz Cieluch received a similar telegram, so he's also leaving. And Mrs. Garlicka left a week ago to re-unite with her husband, so the Berdechów band is shrinking.

An armored train transporting technical equipment of various kinds had been standing at a siding for several days, but today it finally departed—however, not before Dudzik[214] had sneaked on board in the wee hours and thrown off a receiver, a transmitter and a transformer.

Last Friday I was given a haircut by a different barber, my usual barber, the saloon master, being absent. The substitute told me some of his experiences as a non-commissioned officer imprisoned in various Nazi camps. In the last one the prisoners had been forced to carry boxes of stones, one work party in one direction and another back. Tardiness was rewarded with jabs with rifle butts. This is how the Germans respected the Geneva Convention.[215]

A few days ago a conference in Moscow was announced for June 17 to determine the composition of the Polish government. The delegates from Lublin will be Osóbka-Morawski, Bierut, Gomułka, and Władysław Kowalski; from London,

---

[211] The expulsions of ethnic Germans from various regions of eastern Europe after World War II took place in three stages: a spontaneous flight before the advancing Red Army (1944–1945), a disorganized mass expulsion by the new rulers of the territories in question immediately following the war, and a putatively more ordered transfer of people following the tripartite Potsdam Agreement of July–August 1945. By 1950 there remained only about 2.5 million ethnic Germans in the regions in question, around 12 % of the pre-war total.

[212] The Mauthausen-Gusen concentration camp built in the vicinity of the villages of Mauthausen and Gusen in Upper Austria, 20 km east of Linz. Liberated by soldiers of the US Third Army in May 1945, the last camp to be liberated by the western Allies.

[213] Dolek, Stefa's brother, had gone to Katowice. See above.

[214] The father of one of Steinhaus's pupils. See above.

[215] The Third Geneva Convention, adopted in 1929, stipulated conditions of humane treatment of prisoners of war.

Stanisław Mikołajczyk, Jan Stańczyk, and, as a replacement for Józef Żakowski, Antoni Kołodziej; and, as representatives of the State National Council (*Krajowa Rada Narodowa*, KRN), Adam Krzyżanowski, Stanisław Kutrzeba, Zygmunt Żuławski, Henryk Kołodziejski, and Wincenty Witos.[216] Meanwhile fifteen of the sixteen delegates arrested earlier have been freed[217]—presumably through the intercession of Harry Hopkins.[218]

I have now completely finished my table of logarithms, and am beginning my note on the quadratic tariff.

JUNE 18, 1945. I was wrong: it turns out the delegates from the underground government were not freed. Their trial will take place at the same time as the deliberations as to our new provisional government. One member of the "aboveground" delegation, Henryk Kołodziejski,[219] I know as a fellow student from my days in Göttingen.

Mrs. Garlicka came for a visit from Opole, where she has been assigned a beautiful "post-German" apartment, furnished and with kitchenware. She has also hired two German girls to do the housework. Egged on by hunger, they beg to be given the work, even if only for food. However, Mrs. Garlicka is a little unhappy about living in an apartment confiscated from its previous owners.

The Congress of the Academy of Sciences of the USSR, convened to mark the 220th anniversary of that Academy (formerly the Russian Academy), was attended by Jacques Hadamard. So he's still alive!

The *Dziennik Polski* gives the names of two more scientists murdered in Lwów in July 1941: Docent Jerzy Grzędzielski and Docent Stanisław Mączewski. So these names should be added to the list I compiled earlier.[220]

JUNE 19, 1945. Something else Stasia Schoenemann told me about Ravensbrück: at one period there were among the prisoners there 500 female Soviet soldiers, who, to a woman, behaved perfectly as a unit. They obeyed only orders issuing from their captain. If she regarded a command given by an overseer as inappropriate, the whole five hundred would obstinately refuse to comply. Once they went on a hunger strike, and even when Polish inmates brought them something they refused to accept it. It was customary to chase all inmates outside in the frost after a hot bath,[221] and as a result even some of the Soviet women came down with pneumonia.

---

[216] Replaced by Władysław Kiernik, after Witos fell ill.

[217] Steinhaus was misinformed here. The show trial of fifteen of the leaders of the wartime Polish underground government lured to Moscow on false pretences began on June 18. See earlier footnote.

[218] Harry Lloyd Hopkins (1890–1946), a close advisor to Roosevelt. One of the architects of the New Deal, and of the Lend Lease aid program in support of the Allies.

[219] Polish statesman, historian, economist, and journalist. From 1945 member of the State National Council and then the *Sejm* Legislature.

[220] See Chapter 9.

[221] This is, in fact, considered part of the ritual of the Russian bathhouse.

I am trying to persuade Stasia to write a book about her experiences in Ravensbrück concentration camp. The New York publishing house G. E. Stechert might be interested in an English version!

There are rumors concerning the make-up of the "Government of National Unity" which is in the process of being formed. The names of Krzyżanowski, Kutrzeba, and others are bandied about as those of possible ministers.

Yesterday in Grybów two Polish sappers were killed disarming mines, and today another one. Why don't they use Germans to do this work?

JUNE 22, 1945. Yesterday another Polish soldier was killed by a mine, reportedly after having successfully disarmed a few thousand of them. This was told to me by Mr. Motyka, who spent the night here. We talked a great deal about taxonomical groupings of plants.

Kossowski dropped in yesterday. He told us of an exchange of shots between the militia and an unidentified armed band in Tarnów. Before the militia arrived the bandits had been in the process of attacking a house which had recently become a place of refuge for some Jews who had fled Rzeszów when threatened by a pogrom. It seems that in Rzeszów, along with a certain section of the populace, Polish soldiers had been persecuting—in particular, beating up—Jews. The ostensible cause was the unexplained disappearance of some girls aged ten and less: the Jews were accused, in accordance with the age-old myth, of ritual murder. We see that the large-scale ritual murder of Jews of the last ten years has still not put an end to the evil witch-hunting of yore. Perhaps the myth will outlive the Jews themselves.

There is to be a three-power conference in Berlin[222] at a date yet to be determined. They must have decided not to wait for the end of the Japanese war.

I have had a card from Wacław Sierpiński, now in Kraków, accepting my paper for Volume XXXIII of *Fundamenta Mathematicae*, of which he is chief editor. I will be paid 1500 złotys for the paper.

Stefa says she wants to go to Wrocław to acquire a house.

JUNE 29, 1945. The past week has been very busy. On Tuesday morning Dolek arrived from Silesia by truck. He says that many of the factories there are up and running, although the productivity per worker is down to 0.5 or 0.7 of its pre-war value. Some of the engineers spend all their time looting former German apartments. All German males aged between seventeen and fifty are being deported beyond the new border to the contracted territory of Germany, but the women are allowed to stay for the time being. They do all the tidying up, sweeping, and similar low-grade domestic work, and the shrewd ones wear Red Cross armbands to deflect enmity.

---

[222]The conference was held at Cecilienhof, in Potsdam, home of Crown Prince Wilhelm Hohenzollern, scion of the Prussian royal line, from July 16 to August 2, 1945. The chief participants were Harry Truman, Joseph Stalin, and Churchill, replaced by Clement Attlee when it was learned that Churchill had lost the British general elections. Some of the issues discussed: appropriate punishment of the Nazis; reconstruction of Europe; ensuring future peace; war reparations to the USSR, and indirectly to Poland; expulsion of Germans from territories now established as no longer German; and the recognition of a Provisional Government of National Unity in Poland (and the withdrawal of recognition of the London Government-in-Exile).

13 Diary Entries   441

There is a shortage of labor. Dolek praises the policies of the Minister of Industry Minc,[223] to whom he attributes, in particular, the obedience of the German female workers—*tout comme chez eux*.[224] Having taken what they wanted of Polish and German plant, the Soviets are now proffering used cars as gifts to the management of Silesian factories.

The vacillation over Zaolzie boils down simply to the Soviets wanting to teach Beneš[225] a lesson, since after the annexation of former Czechoslovak territory in the Subcarpathian Rus' to the Ukraine he sought to orient his government more towards Britain and the US.

The Red Army has already been withdrawn from Czechoslovakia, and it is possible that the Soviets will agree also to relinquish control of their zone of occupation of Austria. Several German *Länder*, including Saxony and Thuringia,[226] will remain under Soviet hegemony. The Red Army is supposed to quit Poland by July 10.

Since yesterday three divisions of Cossack cavalry have been riding through eastwards. Four soldiers came in a car looking for a billet for the night. In view of our past experience, we hid all the moveable objects in our room, and gave them Dolek's room. Today, by late afternoon they had packed up their telephone and driven off. Later I went over to Dudzik's and found them there. Dudzik told me that they had said of their stay with us that "A fat old man turned himself inside out to accommodate us." Among the cavalry are regimental bands, and the Cossacks' mounts are decorated with little flags. There are lines of artillery and wagons drawn by splendid Mecklenburgers[227] and Demiperchesons,[228] and the riders are mounted on beautiful saddle horses, many apparent thoroughbreds, some with elaborate caparisons of what look like Persian rugs. There are also soldiers on foot, some of whom drive herds of cows, while others ride in horse-drawn carts, some of which carry pigs, while others are covered in canvas, doubtless laden with German goods

---

[223] Hilary Minc (1905–1974), Polish communist politician, and pro-Soviet Marxist economist. From 1944 to 1956 a member of the Politburo of the PPR (the Polish Workers' Party, or, from 1948, when it merged with the Polish Socialist Party, the Polish United Workers' Party).

[224] "Just as back home [in Germany]"

[225] Edvard Beneš (1884–1948), leader of the Czechoslovak independence movement and second President of Czechoslovakia. At the end of World War II he returned to the newly reconstituted Czechoslovakia from exile in England, and resumed his post as president. He presided over a coalition government, from 1947 headed by communist leader Klement Gottwald, which, in particular, oversaw the expulsion of ethnic Germans from Czechoslovak territory. Beneš resigned when the communists organized a *coup d'état* in February 1948, a move ratified by the people at the elections of May 1948, thereby aligning Czechoslovakia with the communist bloc.

[226] And also Sachsen-Anhalt, Mecklenburg-Vorpommern, and Brandenburg, which contains Berlin, from 1949 subdivided into East Berlin and West Berlin.

[227] A breed of horse from the Mecklenburg-Vorpommern area of north-eastern Germany, a tough "warmblood" breed with good physical and mental stamina.

[228] The Percheron is a breed of heavy draft horse, deriving originally from a region of France called "Le Perche." A breed with great muscular development, and of a determined and willing character.

of one kind or another. Of course, most of the horses, and all of the cattle and pigs have been acquired in Germany. The whole procession is accompanied by packs of dogs. The Cossacks and other soldiers look well fed, well clad, and content with their lot. There is no raping of the locals or stealing—on the contrary, they now keep their distance and don't talk of their matchless leader. Our locals look upon them with indifference, unmoved by the charm of a Cossack's songs or the dashing figure he cuts with his cap askew, seated high on a magnificent stallion. But the might of the army and its first-class organization are clear to all.

It seems that Mikołajczyk, Stańczyk, Żuławski, and possibly Kołodziejski are to be ministers in the new united government[229] and that Witos and Grabski will be members of the presidium of the State National Council.

It appears that wartime censorship is to cease and one may now own a radio receiver. It is also rumored that an agreement has been reached in Moscow whereby Zaolzie is ceded to Poland.[230] *Ad maiorem Russiae gloriam!*[231]

Before Dolek departed—in a truck carrying at least 30 people!—Staszek Rząca dropped in. He wants to continue with his studies, and will do so as soon as he finds a safe place to stay, since he is still on the run from the officers' school.

Fredek Borowicz became a member of the PPR out of solidarity with members of the party who, when he was arrested by the UB,[232] secured his release. Pająk saw Lidka in Kraków, but since he was there for only two hours there was no time to get a letter for us.

I've been warned that Wrocław is in a dreadful state: mines everywhere, mounds of rotting corpses, the countryside laid waste for miles around, people living on canned preserves. I'd rather go to Prague.

JULY 3, 1945. We have finally learned from Jan Zaremba that in the Provisional Government of National Unity, whose membership and structure received Stalin's *imprimatur* in Moscow, Władysław Kiernik will serve as Minister of Public Administration, Mikołajczyk as Minister of Agriculture, Mieczysław Thugutt as Minister of Transport and Telegraph, and Czesław Wycech[233] as Minister of Education, while Stanisław Grabski and Wincenty Witos will be deputy heads of the presidium of

---

[229]The Provisional Government of National Unity (*Tymczasowy Rząd Jedności Narodowej*, TRJN) was officially inaugurated on June 28, 1945 by a decree of the State National Council. It was made up of Polish communists and former members of the Government-in-Exile. The Prime Minister was Osóbka-Morawski, Deputy Prime Ministers Gomułka and Mikołajczyk.

[230]This turned out to be false. Zaolzie was returned to Czechoslovakia.

[231]"To the greater glory of Russia" (Lat.). Compare *Ad maiorem Dei gloriam!*

[232]*Urząd Bezpieczeństwa Publicznego* (briefly, UB) (Department of the Security Service), branch of the Ministry of Public Security of Poland (MBP), the Polish communist secret police, intelligence, and counterespionage service operating from 1945 to 1954. Its main goal was the eradication of the anti-communist groupings and the socio-economic base of the wartime Polish Secret State and of the Polish Home Army.

[233]Czesław Wycech (1899–1977), Polish politician and historian. As a member of the Polish Underground Government during World War II, he was responsible for organizing underground education. Minister of Education in the Provisional Government of National Unity 1945–1947.

the State National Council.[234] The remnants of the Polish Government-in-Exile in London are refusing to recognize this cabinet, and are making seven demands, including one for three-epithet elections.[235] When Mikołajczyk came to Kraków he was greeted ecstatically. The underground battalion *Las* (Forest) is still operational, but since the AK was dissolved from London, the NSZ[236] grows automatically in its place, like underbrush. It is conceivable that they are the ones responsible for the pogrom in Rzeszów.

There are rumors to the effect that just east of the San there are Ukrainian rebel bands led by former SS men. The border in that region has been moved 46 km to the east, a small concession.

Columns of Soviet cavalry passed through for three days in a row. All told there must have been several tens of thousands of horses, of all breeds and hues, many sick and lame, but also many splendid ones in good condition: huculs and other mountain breeds, draft horses of several breeds, Hungarian strains, and even some of Arab and English stock. Herds of livestock came through with them, including white Hungarian oxen for fattening, and cattle taken from Polish farms by the Germans during the war. Rather than leave the horses on their last legs for us to tend, they push them till they fall. They ride well. The soldiers are neater looking than those we encountered when the Red Army was advancing through Poland, given more to washing, eating out of decent vessels, and expecting bedclothes at their billets. Some that I spoke to predict that once Japan is defeated, Britain and America will help with the reconstruction of their country.[237] The locals try to have as little to do with them as possible, although the only thing they now steal is hay. Since contact with them can lead to misunderstandings, this may be a good thing. They themselves have evidently been told to hold themselves apart from the local people, in particular to avoid involvement in political discussions.

---

[234] Mieczysław Thugutt, who had run the Polish Ministry of Internal Affairs in the Government-in-Exile in Stockholm, rejected the offer to head a ministry, and remained abroad. In his place Tadeusz Kapeliński became Minister of Transport and Telegraph.

[235] Free, fair, and democratic.

[236] *Narodowe Siły Zbrojne* (or NSZ) (National Armed Forces), a Polish anti-Soviet and anti-German paramilitary organization, part of the Polish resistance movement during World War II, later working underground against the postwar Soviet puppet government of Poland. It was associated pre-war with the political camp of National Democracy.

[237] In April 1948 the US launched the European Recovery Program (also called the "Marshall Plan" after American Secretary of State George Marshall), a program of financial support to help rebuild European economies and combat the spread of Soviet-style communism. Aid was offered to the USSR and the countries of the Soviet bloc, but was refused. (An approximately equal amount of aid ($13 billion) was given between the end of the war and the implementation of the plan.) By 1952, when funding ended, the economy of every participating state had surpassed pre-war levels. The plan is thought to have been largely responsible for the total political and economic recovery of postwar western Europe.

Jan Dobrzański's wife Maniuta came to see us. She told us of a conversation she had had with a former inmate of Mauthausen concentration camp. This man had related how when they were liberated they were advised by agents of the western Allies to go to Belgium, where they would find work, even if only in the mines, while other, Soviet, agents told them they should return to Poland. There were heated debates as to which alternative was better, and when a certain Polish engineer protested loudly—to the point of hysteria—about being coerced to go to Belgium, the Americans had wanted to place him under arrest, but had not done so because the pitiful crowd of inmates objected. So most of them came back here, and are now regretting their decision.

We are indeed permitted to own a radio.

The trial of Okulicki[238] and his fifteen companions is over. Okulicki "confessed", that is, gave a speech full of contrition, apologizing for all sorts of trumped-up political crimes. The foreign press was not admitted to the courtroom.[239]

Osóbka-Morawski continues to make scandalous agitprop speeches in which the word "democratic", with which he is obsessed, is subject to the most thorough-going perversion of meaning. He accuses those who would oppose him as "undemocratic", as if none of his lot have observed that in the US there are two opposing parties, Republican and Democratic, yet no one in the US would think of claiming that voting for the Republicans is undemocratic. Is he really unaware that Britain also is a democracy even though it has several parties, including a Conservative one? Kutrzeba, who is now President of the Polish Academy of Arts and Sciences[240] (*Polska Akademia Umiejętności*, PAU), writes articles for Kraków Catholic newspapers which would do President Mościcki,[241] author of the most foolish of lucubrations, proud. In the cultural weekly *Odrodzenie* Jan Kott has an article demanding that newspapers be allowed to publish caricatures of political figures, including ministers. But he needn't have bothered because it's not possible to caricature caricatures.

---

[238] General Leopold Okulicki (1898–1946), Polish army general, and the last commander of the anti-German underground Home Army during World War II. He served in the Polish Legions in World War I, and fought in the Polish-Bolshevik War of 1919–1921. In Warsaw during the German siege of that city, he evaded capture and joined an early underground resistance organization. Arrested by the NKVD in 1941, he was imprisoned and tortured in various Soviet prisons. Released after the signing of the Sikorski–Mayski agreement, he eventually became the commander of the 2nd echelon of the underground Polish Home Army. On the Soviet takeover of Poland, he became one of the "Polish sixteen" lured to Moscow under false pretences. Although sentenced to ten years in the Gulag at the Trial of the Sixteen (June 18–21, 1945), he was murdered in the infamous Butyrka Prison in Moscow on December 24, 1946.

[239] One source claims that members of the foreign press and observers from Britain and the US were present. The dates of the trial were chosen carefully to overlap with a conference on the creation of a Soviet-backed Polish puppet government.

[240] Also translated as the "Polish Academy of Learning".

[241] Mościcki had been forced to resign his office in September 1939, and after a period of internment he was allowed to go to Switzerland, where he remained throughout the war, dying there in 1946.

JULY 5, 1945. Two days ago we had a bumper crop of letters. Kama Speidlówna[242] wrote that she has survived by a miracle, and Halina Szpondrowska[243] wrote from Częstochowa that she and her sons are well. She says that Tosia and Elza Otto[244] have also turned up in Częstochowa, but although Tosia spent three years in a camp she is as impossible as ever.

We also had a letter from Lidka, via Stasia Schoenemann, telling us that she was paid my salary for June and another 1500 złotys for my paper for *Fundamenta Mathematicae*. Janek, who at the moment is with her in Kraków, added an optimistic postscript, contradicting Lidka's expressed feeling that nothing has changed, that all will be as it was during the earlier Soviet occupation 1939–1941—although she thinks that travel abroad may be possible. I get a sense that she and Janek will go overseas before us. Included in her letter was one from Knaster. He remains disconsolate over the death of his wife, and is looking for an apartment for himself, and also for us. He says it is tropically hot and humid in Kraków and that everything is "colonially" expensive. He mentions again that Auerbach, the Sternbachs, and the former Rector Beck swallowed cyanide pills at the Jewish hospital in Lwów to avoid being transported to their deaths.

There was a letter from Sierpiński. He says that his son escaped to the West through Ankara. He also tells me that I am soon to be made a member of PAU—this time unconditionally—and also of the Warsaw Scientific Society (*Towarzystwo Naukowe Warszawskie*), but in the meantime, as a member of the Lwów Scientific Society, I should vote for the elevation of the Warsaw Society to the status of an academy.

I also had a letter from Jan Dobrzański, who is as pessimistic as Lidka about what we can expect. He mentions that Mikołajczyk gave a speech in Kraków which was greeted with wild enthusiasm. Apparently there were calls of "Long live London!" and "Down with the occupation!"

The cavalry has already passed, but we are told to expect infantry columns through shortly. Trainloads of machinery from the west continue rattling past. The "Big Three" are about to sit down together once more.

Kulczyński has returned from Wrocław. According to him the re-establishment of the university there touted by government and press is a pipe-dream. There are still a quarter of a million Germans there. I'm afraid that in Poland's future there do not so much figure "many" universities, as *mnogo*.[245]

The deputy *starosta* of Jasło went on an official tour of the district in a car decorated with little red and white flags, but the Soviets decided they needed the car and took it off him.

De Gaulle has suddenly jumped up and appointed an ambassador to Warsaw.

---

[242] A lady Steinhaus met in Osiczyna.

[243] Another friend made in Osiczyna.

[244] Tosia was the wife of Witold Otto, in whose house Steinhaus and his wife stayed in Osiczyna. She tended to make their lives difficult there. Elza was her daughter.

[245] Russian for "many".

It's been raining steadily for four days.

I'm unable to determine if Henryk Kołodziejski is or is not a member of the Provisional Government of National Unity.[246]

JULY 10, 1945. Tomorrow will mark three years since we came to Berdechów. I have a feeling we won't be here much longer.

It rains day after day, and this will delay the harvest.

On July 7 what's left of the Government-in-Exile[247] made a last plaintive broadcast, claiming that Poland has been returned to the situation of September 1939. The new "united" government has now been recognized by the US and China. It is said—but this may be just the grinding of the rumor-mill—that, in addition to Mikołajczyk, three other members of the Government-in-Exile will join it. Jan Zaremba claims that most of the officer class in exile will merge with the British armed forces, most of the vast numbers of Polish emigrants and refugees will not return,[248] and our politicians abroad will either stay in London or emigrate further to Ireland or Canada.

Now Soviet officers are saying that Red Army garrisons will after all remain in Czechoslovakia and Poland. The Battalion *Las* will then be forced to cease its operations. I have a sense of darkness spread over the land.

There has been an attempt on Mikołajczyk's life.

People move westwards *en masse*.

JULY 13, 1945. At last we got some newspapers, brought by Wanda Żukotyńska, but they contained mostly stale news. There is a report of an interview with Kulczyński.[249] Once one has made adjustments for undue optimism, it would seem that about one-third of the traditional residential area for students and professors of Wrocław University is more or less intact, and a few of the actual university buildings have also been left standing, while others are only partly destroyed. I see also that Henryk Kołodziejski is indeed a member of the State National Council. Parnas attended the Moscow conference on the terms of reference of our new Provisional Government of National Unity, and spoke to Kutrzeba and Spławiński[250] there. I also learned that Mazurkiewicz has died.

---

[246] Apparently he was.

[247] This final broadcast was actually on July 5. The Polish Government-in-Exile, despite no longer being recognized by Britain or the US as Poland's legitimate government, maintained a formal existence in London till 1990, when it returned the symbols of the Polish Republic to Lech Wałęsa, the first post-communist president of Poland.

[248] For example, by the end of World War II about six million Poles were living in the US.

[249] Stanisław Kulczyński took part in underground education in Poland during World War II. Postwar, he was active in the university and polytechnic in Wrocław.

[250] This may be Tadeusz Lehr-Spławiński (1891–1965), Polish linguist specializing in Slavic, Baltic, and Celtic languages. Taught in Lwów 1922–1929 and in Kraków 1929–1962. During World War II he was incarcerated for several months in Sachsenhausen concentration camp. Rector of the Jagiellonian University in Kraków 1945–1946.

I've had another letter from Knaster. He is still pining over the demise of his wife—a broken man! Of those to whom we owe our lives, Mrs. Knaster, Witold Otto, and Benedykt Fuliński are no longer alive!

I'm writing to Dolek to get him to ask Dobromęski[251] to intervene on behalf of Laska, and to Lidka to ask Kołodziejski's wife to do the same.

On July 7 troops of the US army marched through the Brandenburg Gate.[252] It is rumored that the Soviet garrisons remaining in Germany are to be reinforced by Polish and Yugoslav troops. The British and Americans have only half of Germany, but they have the better half. They say I. G. Farben[253] is already hard at work on their behalf. They have all of the radium produced by Germany—all 25 grams of it. The coal mines are already in production in the Ruhr. They found Ribbentrop's hoard in Salzburg: seven million British pounds' worth.[254] They will do more for themselves and for Germany because they are not taking the narrow view.

The engineer Kossowski has finished painting my sun dial. When it's done, along its rim I'll have him make the inscription: "G. Krochmalny,[255] sun dial maker," and have it installed in a sunny spot in the garden.

I have just about finished my paper on the quadratic tariff as applied to electricity rates. I conclude it with remarks to the effect that my results foreshadow the future of the scientific use of price scheduling to regulate consumption, which, as far as I know, is something entirely new. Up till now this has only been done using rough intuition.

Today is the first day of real July heat.

Columns of Soviet infantry are trudging through Grybów, and a few stragglers pass through here as well. Yesterday in Stróże one of these was shot dead by the militia. He had terrorized some neighbors of the Rzącas, demanding vodka and threatening to shoot them if they weren't forthcoming. One of his three comrades had a machine gun, and when the militia arrived began shooting at them. The end result of the ensuing fracas was that the leader of the four lay dead.

---

[251] Possibly Mieczysław Dobromęski, public prosecutor in a criminal court in Katowice after the war.

[252] A former city gate and now one of the chief historical symbols of Berlin and Germany. It is the monumental entry to *Unter den Linden*, the boulevard lined with lime trees which formerly led to the city palace of the Prussian monarchs. The *Reichstag* is one block to the north.

[253] A German chemical industry conglomerate, formed in 1925 when a number of companies were merged. The company collaborated closely with the Nazis during World War II. It held the patent for the pesticide Zyklon B used in the concentration camp gas chambers, and owned over 40% of the company that manufactured it. Thirteen of the 24 directors were sentenced to prison terms by a US military tribunal (the 1947–1948 "IG Farben Trial"), but some of these were eventually reinstated as leaders of the postwar companies that split off from IG Farben.

[254] Two iron boxes containing gold coins valued at $10,000 were found near the villa "Schloss Fuschl", owned by Ribbentrop. It is conjectured that much more Nazi treasure still lies hidden in Austrian lakes and mountain fastnesses.

[255] Grzegorz Krochmalny was Steinhaus's assumed name during most of the war.

People wonder constantly over the fate of Lwów, not able to believe that it's no longer part of Poland. However, it seems that Szczecin[256] is to be given to Poland.

The evil perpetrator of the Katyń massacre[257] made a speech in which he asserted that the Polish delegates to the conference on the composition of the provisional government to succeed the Lublin one, had reached, without any pressure being exerted on them, unanimous accord. No, of course they weren't being pressured! The show trial of Okulicki and others of the "Polish sixteen" had been arranged to overlap with the conference merely to "focus their minds" so that they might quickly reach an appropriate decision!

When Beneš became angry about the annexation of Subcarpathian Rus' to Ukraine, Fierlinger[258] trotted off to Moscow to see what could be done. In the meantime Rola-Żymierski went to Cieszyn to give a rousing speech *à la* Rydz-Śmigły about the essential Polishness of Zaolzie. The upshot of all this was that Fierlinger gave up on Subcarpathian Rus' "in the common interests of both Slav nations,"[259] and was promptly rewarded with Zaolzie, while Rola-Żymierski was left out in the cold. Then the lever in Moscow regulating these territorial squabbles was turned to "Stop", the poor Marshal[260] returned home empty-handed, and the horrid pseudo-patriotic journalese mercifully ceased.

We hear that the Soviet soldiers returning home to the USSR are now more and more being subjected to careful searches when they arrive at the region between the San and the Zbrucz. Any spoils of war they picked up in Germany or later are now confiscated. This failure of the authorities to adhere to a tacit agreement infuriates the soldiers, whence the periodic shooting matches.

One can sense the urge to "go west" strengthening. Many of the landless or those with little fixed property are moving to the former German lands of Lower Silesia or former western Prussia. Some are considering leasing one farm in our locale and another in the new Polish territories, in this way circumventing the agricultural reform law—but this is hardly feasible since there is no one in those territories with whom to negotiate a lease. There are farms abandoned by Lemkos, who left grenades hidden in the farmhouse fireplace or the chimney to explode when a fire is lit.

---

[256] The capital of the West Pomeranian Voivodeship, on the Baltic Sea, in north-western Poland. Poland's largest seaport. Dates from the eighth century. Belonging to Prussia since the early eighteenth century—in German it was called Stettin—it was transferred to Poland after World War II, and the German population expelled.

[257] Stalin.

[258] Zdeněk Fierlinger (1891–1976), Czech politician. Prime Minister of Czechoslovakia from 1944 to 1946, at first in the London-based exiled government, and then in liberated Czechoslovakia. Between 1946 and 1948 he was leader of the Czech Social Democratic Party, and chief proponent of unification of this party with the Communists. In 1948 became a member of the Central Committee of the unified Czechoslovak Communist Party.

[259] That is, Czechoslovakia and Ukraine.

[260] Rola-Żymierski.

JULY 16, 1945. Finally we have hot weather. However, one can't get to the river because oil leaked from the idle refinery covers the grass and stones on its banks, and the water is likewise polluted.

Late last night I aligned my sun dial with the pole star in prescribed fashion, but today the time by engineer Kossowski's wrist watch, synchronized with the time announced by the Tarnów radio station, was twenty minutes ahead of that indicated on the sun dial.

Today I wrote a draft of a letter to be sent to both Kiernik and Witos concerning the cases of Laska and Karol Zaremba. The letter will be copied by Jan Cieluch, as it should come from an established resident of the area.

Yesterday a telegram dated July 9 came from Lidka: "Ignore yesterday's telegram. There is no apartment." Then today we received a card dated July 7, to the effect that Bronek[261] had found a five-room apartment for us, and that Stefa should go to Kraków to secure it. The first telegram, presumably dated July 8, never did arrive, so we may thank the Post Office for saving Stefa an unnecessary journey.

The British predict that the rather indeterminate, chaotic situation in Poland will last three years, and, in particular, that next spring will see extreme hunger in the land. The Soviet soldiers here react by threatening war against Britain.

The planned Potsdam meeting will deal—among other issues—with the return to Poland of Polish soldiers who happened to be abroad at the cessation of hostilities.

Although they have declared war on Japan, the Soviets do not seem to be helping the Americans much in their struggle in the Pacific.

JULY 17, 1945. A couple of days ago I went to see Marian Szafraniec.[262] He had come by train from Katowice for two days. Since the railcars were jam-packed with Soviet soldiers he had had to make the trip standing on the buffers between two cars. He said that in Gliwice one risks one's life if one goes about at night unarmed, and that at an inn in Katowice, Soviet soldiers put their clumsy big booted feet up on the small tables and broke them, to their huge amusement.

Letters are censored—evidently by semi-analphabetic girls.

The aunt of the Witalis students came today to ask if they might resume. Thus I now have seven students. It would have been eight except that the Witalis girl is away for two weeks.

Yesterday a Red Army soldier who had dropped in at the Cieluchs' expressed enmity towards the British, saying they would have to be taught a lesson. This is a new note: the Germans have given way to the British as the chief enemy. If forgiveness of the Germans continues they may perhaps start handing out arms to German prisoners of war! But this is ridiculous, since there is a preponderance of Germans in the part of Germany held by the western Allies.

Jan Cieluch took it on himself to omit my sentences about how Laska saved the lives of several Jews, on the grounds that "it is better not to write such things."

---

[261] Bronisław Knaster.

[262] Assistant to Laska.

There is probably no more undemocratic nation than the Poles, yet the word "democracy"—if not the practice—continues to be drummed into them. The democratic Post Office was issuing democratic postage stamps with representations of such beacons of liberty as Kościuszko and Traugutt[263] on them, but they were soon withdrawn. Some democratic stamp enthusiast in the democratic ministry in charge of the post made some money out of this.

Engineer Kossowski promised me some decent paper to write on, since I will soon have to resort to palimpsests. I'm running out of the backs of used scrap paper.

JULY 19, 1945. We have our own "Wild West" in Lower Silesia. Thousands of families are moving either there or to the lands formerly belonging to western Prussia. They squat in abandoned apartments, take whatever's left, and either settle down or move on to greener pastures—or give up and return home. In most cases those who choose to stay are evicted from the apartments by some PPR authority on the say-so of some government higher-up.

The Germans have discovered the Soviets' weakness for "vodochka", and some manage to obtain permission to stay by lubricating them appropriately. In similar vein, German girls go with Red Army soldiers.

There are constant rumors that we won't after all get Wrocław, and that the Oder rather than the Nysa[264] will be the new border. The "Big Three" got together in Potsdam the day before yesterday. It is said that Truman came with a list of 30 agenda items concerning postwar European borders. They have agreed to a Peace Conference to take place in Paris next year.[265]

Just today have I started listening to the radio in Dudzik's hut. It seems they have stopped censoring the mail. The Kossowskis leave Stróże tomorrow, and the Zarembas are straining at the bit.

JULY 20, 1945. Mrs. Szczęśniewicz (*née* Zarańska) was here relaying horror stories of Mauthausen concentration camp as told to her by her brother, Mr. Gomółka, liberated from there just a few months ago. She said that they would put several sick people together in a single bed, and if one died the others kept quiet about it for as long as possible in order to get his portion of soup. Her brother told her that some of the Tatar prisoners ate the livers of the dead. It was Himmler himself who invented the variety of labor-torture in use there: carrying heavy rocks up an incline.[266] Whoever fell was beaten. The wife of the camp commandant liked

---

[263]Romuald Traugutt (1826–1864), Polish general and war hero, best known for his leadership of the "January Uprising" against Russia 1863–1864. He was sentenced to death and hanged on August 5, 1864, by the then Russian rulers of north-eastern Poland.

[264]The river Nysa Łużycka (Lausitzer Neisse in German, Lusatian Neisse in English) is a tributary of the Oder, lying mostly well to the west of the Oder. The postwar Polish border with Germany lies along the Oder–Nysa Łużycka system. Wrocław (Breslau in German) is situated on the Oder, but upstream from its junction with the Nysa Łużycka.

[265]The Paris Peace Conference, July 29–October 15, 1946, aimed at finalising peace settlements with Italy, Bulgaria, Rumania, Hungary, and Finland.

[266]Many of the inmates worked in a nearby rock-quarry, where they were forced to carry rough-hewn blocks of stone of up to 50 kg weight up 186 "steps of death."

to view the agony of the dying through a small glassed-in window in the wall of the gas chamber. Inmates who fell sick received medical attention but nothing to eat. Convalescents were then gassed. The camp commandant, *SS-Sturmbannführer* Ziereis, was caught by the Americans in the mountains of Upper Austria. He testified that in other camps the skin of victims was used to make handbags and shoes for the wives of SS men.[267]

The rumor is going the rounds that the Warsaw government has agreed with the USSR that five Soviet divisions should remain in Poland for five years "to render assistance." We hear commentary over the radio oddly complimentary of the Germans—for example, for preventing former prisoners from the Mauthausen camp from robbing local peasants. They had been made to work under German supervision on just one scanty meal a day.

Some would-be settlers in Lower Silesia are returning home because of the establishment—by the Soviets, one can only presume—of German patrol squads.

Today Lidka Garlicka passed the *matura* examination.[268]

JULY 21, 1945. Yesterday the Cieluchs held a festive tea party to celebrate Lidka Garlicka's graduation. Later, towards evening, Stasia Schoenemann dropped in to tell us that, thanks to Dolek's kind intercession, Zygmunt Michałowski[269] had obtained the post of chief administrator in a foundry in Wałbrzych,[270] and only awaits a telegram of confirmation. But he may have to wait as much as two weeks, since that's about how long it took for Lidka's telegram of July 8 to get to us. (It did arrive at last, two days ago.)

Stasia also told us of a row at the Pająks' place.[271] Apparently Pająk lunged at his son-in-law with a knife, so the latter decided to go by train to his parents' place in Grodek. While waiting at the station he put it about that his father-in-law writes anonymous tale-bearing letters to the authorities about people he doesn't like. In fact, there had been a case of an engineer from Nowy Sącz notifying a local councillor by the name of Podgórski that an anonymous letter had been sent from our district to the Polish version of the NKVD informing them that said engineer had been a member of the AK. The letter was ignored, and the identity of its author remains unknown.

JULY 22, 1945. No one knows if Szczecin is to be in Poland or just Szczecinek.[272]

---

[267]Franz Ziereis fled with his wife in May 1945, but was discovered in the mountains of Upper Austria and wounded while trying to escape. He was returned to Mauthausen, where he died.

[268]The high-school leaving examination, qualifying the student to enter university.

[269]Zygmunt had spent the war in an unspecified prison. His wife Tosia had found refuge with the Igielskis, neighbors of the Cieluchs.

[270]In German, Waldenburg, a city in Lower Silesia.

[271]Zbigniew (Zbyszek) Pająk was one of Steinhaus's students.

[272]Szczecinek is a town in Middle Pomerania, northwestern Poland. The castle and town were built in 1310 on the model of Szczecin, which is about 150 km to the west, on the Baltic Sea, at the mouth of the Oder.

It seems that priests know one big thing: Poland has not suffered enough. The prelate from Biała was at the Cieluchs' and explained that the fact that some Polish women became pregnant while their husbands were away fighting against the Germans shows that we deserve even greater calamity. And the catechist in Grybów told his flock that Warsaw, as the focal point of the general havoc, merited its destruction, and, furthermore, all the world's cities, centers of corruption that they are, should likewise be destroyed—like Sodom and Gomorrah.

Yesterday Zbyszek Pająk brought us three letters: one from Joanna Guzówna dated July 11, one from Lidka dated July 10, and one from Bronisław Knaster dated July 12. Lidka wrote that Nacht and his family are all alive, and that Nacht himself is directing a cabaret somewhere on the Oder. Knaster advised me to move to Toruń,[273] complained about having been used by people, and enclosed a copy of a summary of Herman Auerbach's 1933 paper "On the Probability of Error in a Sum of Decimals,"[274] from the *Zeitschrift für angewandte Mathematik und Mechanik*. But mostly he wrote about his wife, whose loss has affected him very deeply. Joanna wrote on behalf of Kulczyński and Loria[275] to ask me if I would be prepared to "found" a mathematical and natural sciences department in Wrocław. They had hinted—at least so Joanna said—that the permanent incorporation of Wrocław into Poland depends on visible successes in re-establishing and re-organizing institutions in the area around that city. Today I composed a letter of acceptance.

Zbyszek Pająk tells me that Nikodým[276] is giving a course of lectures on quantifiers at the Polytechnic.

JULY 26, 1945. Yesterday Stefa read in the newspaper that I have been nominated "corresponding member" of the Polish Academy of Arts and Sciences. This occurred at the meeting of the Academy of July 21, when there took place a debate on the effect of the war on Polish science. The author of the article in the *Dziennik Polski* reporting on the meeting, writes "Polish science accuses the German nation!" And I accuse Polish science for having, apparently, completely forgotten about the twenty-two Lwów professors murdered by the Nazis in Lwów in July 1941, and about the many others murdered subsequently. At that same meeting Jan Hirschler was officially expelled from the Academy. One might think of this as a final settling of an issue going as far back as my *Habilitation*[277] in 1917.

Recently there was an afternoon meeting of those officials concerned with the welfare of the peasantry, called, I think, by Mikołajczyk and Kiernik. At the beginning of the meeting security personnel came into the room and checked everyone's papers, including those of Minister Kiernik, then sat down at the back.

---

[273] Ancient Polish city in northern Poland, on the Vistula.

[274] "Über die Fehlerwahrscheinlichkeit einer Summe von Dezimalzahlen".

[275] Stanisław Loria, Polish physicist from Lwów. Founded the Department of Physics at the University of Wrocław in 1945, and was vice-Rector till 1946. Moved to Poznań in 1951.

[276] After the war, Otto Nikodým taught at the Polytechnic in Kraków. He emigrated in 1946, moving to the US in 1948.

[277] Described in Chapter 8.

Since they showed no inclination to leave, and cast such a pall over the proceedings, the meeting was adjourned at 2 pm without much, if any, debate!

The rumor that Japan has capitulated is as yet unconfirmed.[278] Bazhan,[279] the Deputy Chairman of Ukraine, has promised to repay Poland for the "cultural riches" lost to her through having the eastern part of Poland, including Lwów, torn away.

My first duty in connection with my new position at the University of Wrocław was to note down the names of Gołdówna (medicine) and Wiatr (political economy) as candidates for bursaries.

Motyka is supposed to come and see me today.

JULY 29, 1945. Two days ago Mrs. Tokarzowa[280] and Mrs. Garlicka, together with her children Lidka, Maryś, and Olek,[281] and their dog Bej, left Berdechów. A "repatriation" railcar had been arranged for them, and they were thus able to spread themselves out and travel comfortably.

We have a new regime here in Berdechów now that old Mr. Cieluch's wife has returned after several years' separation.

Kossowski told me that Poland has an agreement with the USSR whereby all those who were Polish residents as of 1939, of either Polish or Jewish nationality, have the right to return to the new Poland. One detects here a subterfuge, a return to the tsarist policy of driving Jews into Poland. Of course, on the State National Council there is much dissent over this, counterbalanced by speeches by members of the Bund[282] and the Left Poale Zion.[283] But everyone is polite—in fact, their voices are smooth as honey—and much of the dissent is *sub rosa*. There must be a lot of clenching of fists in pockets going on.

---

[278] Emperor Hirohito announced the surrender on August 15, 1945, so this was indeed just a rumor.

[279] Mykola Bazhan (1904–1983), Ukrainian poet, writer, and translator, and Soviet Ukrainian political figure. Deputy chairman of the Council of Ministers of the Ukrainian SSR 1943–1948, deputy of the Supreme Soviet of the USSR, and head of the Writers' Union of Ukraine 1953–1959.

[280] A widowed daughter of the Cieluchs.

[281] Short for Aleksander.

[282] A Jewish socialist organization founded in Vilnius in 1897 aimed in particular at overthrowing regimes encouraging anti-Semitism, and promoting the creation of a Jewish proletariat. During World War II the Bund went underground, taking part in ghetto revolts and in partisan activity. The Bundist representative in the Government-in-Exile in London committed suicide in protest at Allied inaction in helping the Jews. The Bund continued to function in Poland after the war till 1949, when it was repressed by the Stalinist Polish government.

[283] Poale Zion (Workers of Zion), a socialist Jewish workers' organization, was founded in 1901, when the Bund rejected Zionism, with centers in various cities of the Russian Empire. It later split into leftist and rightist halves, which were permitted to function as political parties in postwar Poland until they were disbanded in February 1950.

There is a Catholic newspaper called *Tygodnik Powszechny*[284] (General Weekly) put out at the behest of the Polish archbishop Sapieha. In a recent issue I saw a list of Polish cultural figures lost to Poland between 1939 and 1945, compiled by Bolesław Olszewicz, a docent in geography at the Jagiellonian University of Kraków. The list includes Beck and Auerbach, but the date of Auerbach's death is given incorrectly as July 1941. I will send a supplementary list to Olszewicz of more names that should be on the list. In Issue 18 of July 22 there was an article by Arkadiusz Piekara[285] entitled "Indeterminism Rules the World of Atoms." The Church considers this, Heisenberg's Uncertainty Principle,[286] as—at last!—a scientific discovery in its favor. This is reminiscent of the Church's faulty logic in its fight against Darwin's theory of evolution. Whenever an experimental result requires the modification of a theory T the Church would like to deny, they triumphantly announce that T has fallen in its entirety, as if not(not T)≡ T for theories. I should write something up about this: one should explain to ordinary folk that the only subtlety about the Church's stance is to be found in its lies. It reminds me of how Professor Hadała used to say to us in the Gymnasium that "$a - b = c$, therefore God exists"—a plagiarized and deformed version of something attributed to d'Alembert or Euler.[287] Piekara seems to believe that from the assertion that the existence of a set of observations implying that a coordinate x of an electron's position at time t satisfies $a < x < b$, entails the non-existence of a set of observations implying $c < mv < d$, where v is the component of the velocity of the electron in the x-direction at time t, and $(b - a)(d - c) < h$, Planck's constant, it follows that human activities are not determined. But I see in this claim determinism having its way with a physicist: it's just what he *would* say. However—I admit it—that may be beside the point. More to the point is my sense that just because I cannot determine the position and velocity of an electron at a given time does not at all mean it doesn't have a definite

---

[284] The weekly *Tygodnik Powszechny* was—and still is—a Roman Catholic magazine, established in 1945 under the auspices of Cardinal Adam Stefan Sapieha. In communist Poland it was the one magazine allowed to some extent to express views critical of the ruling regime.

[285] Arkadiusz Henryk Piekara (1904–1989), Polish physicist. Imprisoned in Sachsenhausen and Dachau, he was freed in 1940, and spent the rest of the war working in a nitrogen plant in Mościce. After the war he held academic positions in Kraków, Gdańsk, Poznań, and Warsaw. The Polish title of the article in question was "Indeterminizm panuje w świecie atomów".

[286] Stating that certain pairs of physical properties of a system—such as momentum and position of an electron—cannot both be simultaneously determined to arbitrarily high precision. First published by Werner Heisenberg in 1927, it is not a statement about experimental limitations, but a fundamental physical law postulating an intrinsic "fuzziness" at the quantum level, where probabilistic estimates are the best one can expect.

[287] The anecdote, doubtless apocryphal, is as follows: Denis Diderot, editor of the great *Encyclopédie*, was visiting St. Petersburg at the invitation of Catherine the Great, empress of Russia. Alarmed by the Frenchman's arguments for atheism, the Empress appealed to the devout Euler, then doing his second stint at the St. Petersburg Academy of Sciences. Euler advanced towards Diderot and said in a tone of complete seriousness: "Sir, $\frac{a+b^n}{n} = x$; hence God exists. Reply!"— striking Diderot dumb, so the story goes.

position and velocity at that time.[288] Yes, I must write about these matters. To know the truth is no sin.

AUGUST 3, 1945. Mr. Ząbek, whom I met through the Igielskis, visited. He advised me to submit a request for restitution of the family property in Jasło, including the land where my Father had his brickworks. Of course, the title deeds and mortgage papers have all been burned. He told me that someone has taken up residence in our warehouse on the edge of the market place. He also informed us that Dr. Żelazny and his wife, Dr. Warchałowski, Balicki, Professor Pyrek, and Karpiński all survived the war.

I had a letter from Sierpiński with an enclosed pamphlet in which Banachiewicz protests that the "topologist" (my term, not Banachiewicz's) has immediately assumed the role of an active member of the Academy.[289]

Knaster advises me not to look for difficulties in Wrocław, to be positive, if possible. Motyka was here again today; he also is to go to Wrocław.

Laska and Karol Zaremba have returned—presumably from the Gulag—along with about 150 others whom the NKVD decided were after all not subversives.

In Sowina, a village near Jasło, a Jewish family who wanted to take their land back from peasants who had settled on it, were killed.

Dolek wrote to say that he will come for his mother at the end of August. Lidka writes that there is now definitely an apartment for us in Kraków, and suggests Stefa go there and begin getting it habitable. Stefa is all packed up and just waiting for the word that she should set off.

Stanisław Rząca passed the fourth grade final Gymnasium examination; in fact, all of my students have passed their year and will be promoted.

The newspapers are full of the outpourings of would-be writers. There are articles by Krzyżanowski and Kutrzeba[290] that are noticeably and refreshingly frank and free of special pleading. Krzyżanowski explains that the concessions made in Moscow were necessary in order for the vestiges, at least, of the conditions laid down by the "Big Three" to remain in place. Polish intransigence would simply have led to a renewal of hostilities, though now between West and East. Thus, as the saying goes, "Better that Pociej be indebted to Maciej than the other way around."[291]

---

[288] This would seem to go against the general view of the meaning of Heisenberg's Uncertainty Principle.

[289] Probably a sarcastic reference to Steinhaus's fresh appointment to the Academy.

[290] Members of the Provisional Government of National Unity. See earlier.

[291] Quotation from Mickiewicz' *Pan Tadeusz*, become a proverb. Since Pociej is a count and Maciej (Matthew) a commoner, presumably the sense is that one should avoid being beholden to the great and powerful.

At a congress of Poles in Canada, Infeld[292] and Julek Stawiński[293] spoke.

AUGUST 10, 1945. Over the past week quite a lot has happened. First, I found out why my sun dial was out: Motyka reminded me that the pole star is inclined two degrees to true north. In the evening the pole star lies east of the pole, so that my alignment of the gnomon with it resulted in an error of a quarter of an hour.

Then I went to Wyskitna to have my photograph taken for my new legitimate identity card, which I shall need if I am going to be living in Kraków. While there I met an acquaintance Poznański, and elaborated to him my idea of a clock with its (vertical) face divided in two by a horizontal line, and showing summer time on the upper half and winter time on the lower half.

I also had to go to Bobowa to register my departure. To this end I borrowed a bicycle from Kossowski, and, despite not having ridden a bike for 43 years, am proud to relate that I completed the round trip without significant mishap. While there I visited Laska's family. Although the "commandant", as they call him affectionately, has not yet shown up, the policeman Hebda has returned, so presumably all those innocents wrongfully arrested by the NKVD will surface eventually.

The resolutions of the "Big Three" at Potsdam have been announced. Thus the Oder and the Nysa Łużycka will henceforth form the border between Poland and Germany, so that the cities Szczecin and Zgorzelec,[294] Wrocław and Frankfurt an der Oder[295] will now all belong to Poland. But even though the Neisse—in Polish Nysa—is now in Poland, no one will call it by the Polish name, since the sound of "Neisse" is "nicer", whereas "Nysa" sounds too much like a whimper.[296] While we get some German territory, the British and Americans get the German gold[297] and German patents, while the USSR gets reparations from East Germany plus a proportion of the industrial product of West Germany for a certain time. The treaty stipulates also that the Germans must quit Polish territory, but that this expulsion is to be carried out "in a humane and orderly fashion." Although the treaty makes

---

[292]Leopold Infeld (1898–1968), Polish physicist. He was a docent at the University of Lwów (1930–1933), and then worked at Cambridge University (1933–1934), Princeton University (where he collaborated with Einstein), and the University of Toronto, Canada (1938–1950), returning to Poland in 1950, where he became a professor at Warsaw University. Member of the Polish Academy of Sciences.

[293]Julian Stawiński (1904–1973), Polish lawyer, translator from English and Russian, and editor of anthologies of science fiction and fantasy. After a brief period overseas after World War II, he returned to Poland in 1947.

[294]A town in Lower Silesia on the Nysa Łużycka.

[295]Actually just that part of this city on the eastern side of the Oder, now called Słubice.

[296]The Polish "y" is pronounced like an English "i" as in "gill", but "harder" or "darker", that is, further back in the mouth, the way New Zealanders pronounce it. Steinhaus's original play on the words "Neisse" and "Nysa" was obscure.

[297]US forces found a vast treasure hidden in the Kaiseroda salt mines near Merkers, Germany in April 1945. However, the fate of Nazi gold transferred to various overseas institutions during the war remains cloaked in mystery.

no mention of the Soviet Union's usurpation of Polish lands in the east, the Soviets proclaim directly that these lands are merely being returned to their rightful owners, and moreover at the behest of the Polish government.

The following have returned to Poland: Kot, Romer, Kwiatkowski, Strasburger, Cardinal Hlond, and Estreicher.[298]

Kossowski got hold of a car somehow and left with his family.

Jan Dobrzański came by on Wednesday (August 8) and told us that the average productivity of industry in Silesia is at present a mere one percent of its prewar value. They have instituted a six-week course whose aim is to turn workers into factory managers.

After several postponements owing to a lack of trains, Stefa left yesterday for Kraków to inspect the apartment awaiting us there, and to see if she can turn it into the semblance of a home. This represents the first time in six years that we are apart.

Yesterday a man wandered in dressed in the uniform of an American soldier. It was Marceli Stark, former assistant at Lwów University! He had only managed to return to Poland on July 17. His had been a terrible ordeal. He had been captured by the Germans in Warsaw, and did the rounds of the camps: Majdanek, Płaszów,[299] Ravensbrück, Oranienburg,[300] and Sachsenhausen. As the Red Army approached he and his fellow prisoners were force-marched westwards as far as Schwerin. He had heard that Auerbach had been killed, that Sternbach had poisoned himself to death with Veronal, that Jacob had perished in the Pawiak in Warsaw, and that Pepis, Schreier,[301] Michał Kerner,[302] and Adolf Lindenbaum had also perished.

---

[298]Stanisław Kot (1885–1975), Polish historian and politician. A member of the Polish Government-in-Exile during World War II. Left Poland again in 1947. Tadeusz Romer (1894–1978), Polish diplomat and politician. Worked in the Polish Ministry of Foreign Affairs between the wars, becoming Minister in the Polish Government-in-Exile (1943–1944). After the war he settled in Canada (so did not, after all, return to Poland). Eugeniusz Kwiatkowski (1888–1974), Polish politician and economist. Interned in Romania 1939–1945, he returned to Poland after the war. Henryk Leon Strasburger (1887–1951), Polish politician. Member of the Polish Government-in-Exile during World War II. He was summoned back to Poland after the war, but decided to remain in London. August Hlond (1881–1948), Polish cardinal. For a time the primate of Poland. Became archbishop of Gniezno and Warsaw in 1946. Forced German officeholders in former German territories to resign their posts in favor of Poles in 1945. Voiced intolerant views of Jews in Poland. Karol Estreicher Jr. (1906–1984), Polish art historian. Director of the museum of Jagiellonian University 1951–1976. Involved in reclaiming Polish works of art looted by the Nazis during World War II.

[299]The Kraków-Płaszów concentration camp was a forced labor camp built by the SS in Płaszów, a southern suburb of Kraków. It was built over two Jewish cemeteries in the summer of 1942.

[300]Oranienburg concentration camp was one of the first detention facilities built by the Nazis after coming to power in 1933. It held mainly political opponents of the Nazis from the Berlin area. In July 1934, when the SA was suppressed, it was closed and replaced by the Sachsenhausen concentration camp.

[301]Józef Schreier (1908–1943), teacher in a Gymnasium in Drohobycz. Did serious mathematical research.

[302]Michał Kerner (1902–1943), teacher in a Gymnasium in Warsaw. Did serious mathematical research. Died during the liquidation of the Warsaw ghetto.

Stark himself had been on the brink of death several times. He survived only because at some stage the Germans discovered that he was a mathematician, and set him to solving differential equations for the *Luftwaffe*, counting the ephemeral spots on Jupiter in connection with a project at Studio Babelsberg,[303] and compiling a table of the sun's position minute-by-minute for use by pilots. He is bitter since he has lost his family, and most of his friends from before the war are either dead or gone abroad. He is completely averse to renewing acquaintance with former fellow socialists who now occupy positions of influence thanks in part to their possession of so-called "Aryan" papers, certifying their non-Jewishness. He is getting ready to travel illegally to Palestine through Italy, helped in this by the Left Poale Zion. I gave him a letter to pass on to Fraenkel.[304] He advised me to leave also. He described the apartment in Kraków where Knaster was living as "four bare rooms inhabited by Knaster and volumes of *Mathematical Monographs*;"[305] it has a bathroom and gas but no furniture.

A letter arrived from my sister Irena in Kraków. She asks me if I can help her find out how one rents a stall in the market place in Jasło. She thinks the authorities are prepared to offer dwellings in Silesia to replace those of former residents of Jasło destroyed by the Germans, and that this is certainly the case for those who have been displaced from their homes in Lwów. She also writes that Banach is unwell, but is to take up the mathematics chair at the Jagiellonian University of Kraków. Mazur and Żyliński are both members of the Union of Polish Patriots[306] and so active in repatriating exiled Poles. They have positions at the Ivan Franko University of L'viv,[307] as it's now called, as does Zarzycki.

The Americans have dropped an atom bomb on Hiroshima, weighing half a tonne with a blast equivalent to 20,000 tonnes of trinitrotoluene (TNT),[308] and costing some two billion dollars if one includes the expenses on research infrastructure, scientists, technical support, etc. Within a circle of radius 3 km everything was levelled with the ground. Truman made an announcement to Congress, and our newspapers called it "The greatest triumph of science: the atom bomb!"

---

[303] The world's oldest large-scale film studio, located in Potsdam-Babelsberg, and founded in 1912. Fritz Lang's *Metropolis* and Josef von Sternberg's *The Blue Angel* were shot there, as well as over a thousand films made between 1939 and 1945—for example, Leni Riefenstahl's propagandistic *Triumph of the Will*.

[304] Abraham Halevi Fraenkel (1891–1965), Israeli logician and set theorist, born and educated in Germany, moved to Palestine in 1929. An early Zionist. The first Dean of Mathematics at the Hebrew University of Jerusalem. The Zermelo–Fraenkel system of axioms for set theory is now standard.

[305] *Monografie Matematyczne*, a publishing venture begun in Lwów in 1932. The first volume was written by Banach.

[306] A political organization founded, with Stalin's connivance, in the USSR in 1943 by Polish communists. Between 1944 and 1946 it was involved in resettling Poles from the USSR in Poland.

[307] The Jan Kazimierz University of Lwów was renamed after the Ukrainian poet Ivan Yakovych Franko from January 1940 to June 1941, and then from 1945 on.

[308] Various sources put the blast equivalent at from 13,000 to 16,000 tonnes of TNT.

AUGUST 12, 1945. I have put the finishing touches to my paper on the quadratic price schedule—31 full-size pages—and shall send it to Stefa tomorrow via Mr. Pająk, with instructions to pass it on to Stark for him to take it abroad with him, in the hope that he will take the trouble to translate it into English and arrange for its publication in America or wherever.

On August 8 the USSR declared war on Japan. It may be worth mentioning that our newspapers recently reported that Japan had asked the USSR to present certain Japanese proposals in Potsdam, but the USSR did not oblige! Now they, the Soviets, want Truman to confirm to the Japanese that they were unaware of the existence of the atom bomb at the time of the Potsdam conference. They also stress the importance of their participation in the war against Japan, even though they have only been at war with Japan for three days! They are now ready to allow the Americans the use of bases on their territory. But such currying of favors doesn't fool anyone: Truman has been very explicit about restricting bomb know-how to America and Britain.

AUGUST 13, 1945. It now seems that Japan's surrender is imminent, with full sovereignty returned to the Mikado.[309]

From things that Stark related to me I have inferred that in Warsaw a telegram to North America costs 3 złotys a word, but in Łódź 58, and a telegram to Argentina and Brazil costs 4 złotys a word in Warsaw but in Łódź 162 złotys a word.

I have sent Stefa a bag with some of our things in it, together with a kilogram of veal, and a certificate of morality.[310]

At the university in Łódź mathematics is being taught by someone by the name of Jankowski, who is unknown to me. I have learned that Michałowicz has been appointed to the government's Scientific Committee. This is the same man who, between the wars, published in the proceedings of a conference on pediatrics a completely nonsensical critique of a paper on "patergometry", to which critique Mazur wrote a rebuttal, published in *Polish Mathesis*.[311]

The plane carrying the atomic bomb dropped on Hiroshima on August 6, 1945—may that date be etched in our memories!—was commanded by one Colonel Paul Tibbets. When, on August 9, another such bomb was dropped on Nagasaki, the newspapers used type of ordinary size to print the announcement that Nagasaki was no more.

Engineer Broch, formerly director of *Wspólnota*[312] (Community), tells me that thousands of non-Poles, mostly Russians, are trying to take advantage of the law of repatriation to come to Poland. It's not clear if this is planned infiltration or just a lot of poor souls seeking freedom from dialectical materialism.

---

[309] That is, Emperor Hirohito.

[310] Needed in connection with applications for certain kinds of employment.

[311] See Chapter 7.

[312] An organization formed to deal with the difficulties of Poles expelled from Poland or victims of World War II.

AUGUST 14, 1945. The Allies accepted the Japanese stipulation as to the sovereignty of the emperor on condition that the emperor be subject to the Allies. A high-level Japanese official was heard to declare that the capitulation was due to the dropping of the atomic bombs, so that "chivalrous contests are no longer possible", presumably referring to swordfights between samurai warriors. Was this gentleman aware of what Japan did in China in 1937—in particular, of its use of chemical weapons against the Chinese? Like the Germans, now that they are defeated they deplore the passing of knightly jousting in tournaments.

In Kraków there was a pogrom directed at Jews ostensibly because of "ritual murders"—probably organized by a *Volksdeutscher* and a *Reichsdeutscher*. The blood libel and other such arcane anti-Semitic nonsense must awaken love for the Poles' brothers-in-arms the Germans. And the Jews will go to Palestine—where they will encounter Adam Macieliński of Lwów, organizer of the ONR murders in Lwów in the "November days".[313] He has now become the leader of a squad of fighters sent by the Government-in-Exile to Palestine to buttress the British attempts to thwart the Jews' struggle to establish a state there.

Hans Frank, the Nazi *Generalgouverneur* of Poland, is to be tried in Lwów for his crimes.[314]

AUGUST 15, 1945. In Kraków the blood libel has been modernized: it is now being claimed that Jews abduct Christian children in order to use their blood for transfusions. For some reason this brings to mind the corpses one used to be able to see laid out in the Lwów prosectorium, and also Stark's mentioning the widespread rumor that every Jew returning from the camps has ten thousand dollars in his pockets.

AUGUST 16, 1945. A man by the name of Engel, from Łużna, has been released 24 hours after being arrested.

Zabierowski, the organizer of underground teaching in our administrative district during the war, and, immediately postwar, of a *Lycée*, has been dismissed from his post as director of district schools without a word of explanation or thanks.

Emperor Hirohito announced Japan's surrender on August 15.

In today's newspaper it is reported that yesterday, in the morning, a fifty-two-year-old Jewess, Róża Berger, not long returned from a camp, was killed by a bullet. On the other hand, there are absolutely no reports of abductions of Christian children.

Adam Ostrowski, Governor of the Kraków Voivodeship, was very clear in his statement condemning the anti-Semitic rampaging in Kraków—in contrast to the somewhat muddle-headed statement issued by the Rectors of tertiary institutions.

AUGUST 19, 1945. A couple of days ago I posed for a photograph with five of my pupils: Dudzik, Ochalik, and the Rzącas.

---

[313]Perhaps a reference to the activities of the ONR in the period 1936–1939 described in Chapter 7.

[314]In fact he was tried in Nuremberg, where he was found guilty of war crimes and crimes against humanity on October 1, 1946, and sentenced to death.

Yesterday evening I listened for the first time to foreign news on the radio. I learned that on August 17 a bilateral Polish-Soviet agreement was signed, ratifying the new eastern border between Poland and the USSR as running along the Curzon line, with a few concessions to Poland amounting to areas of some 30 km width in some places. Furthermore, Poland is to get 15 % of the reparations to be paid by Germany to the USSR, and is to supply the USSR with sixty million tonnes of coal at a predetermined price over a period of five years. This extraordinary agreement obtained by a pusillanimous government neither supported nor elected by the people of Poland, guided only by the weak wills of its members, illustrates "directed", as opposed to "direct" democracy.

I also heard a splendid jazz broadcast from London. The kind of freedom represented by that sort of music is dead forever here. There was also a report that the Americans are to build a bomber with range three times the maximum range of bombers hitherto.

In a letter I received yesterday, Stefa reports that Banach is still in Lwów and is very ill. She writes that the apartment in Kraków is nice although completely unfurnished. She says that Zdziebko, a former tenant of ours on land near the brickworks in Jasło, has built a makeshift workshop in the ruins of our house next to the market place, and has written to Mother asking to rent it from her. Finally, she says she has heard that Hemar[315] is writing songs for Metro-Goldwyn-Mayer, and has become a millionaire.

Mr. and Mrs. Stażniewicz, formerly of Polna, who some time ago went to Silesia in the hopes of bettering their lot, have just returned with stories of the incredible chaos reigning there: endemic lawlessness, roaming squads of fighting men, no livestock whatsoever, the favoring of local Germans, etc. A group of girls from Sosnowiec were encouraged to work on the harvest, but found themselves abandoned in the fields without food or shelter, exposed to rape by Soviet soldiers, and returned home in a great panic. Apparently the situation improves further west.

It seems that General MacArthur has returned to Manila, fulfilling an earlier promise, and that for the time being he is to rule over Japan as plenipotentiary.

I have conceived an interesting idea for a lecture, to be entitled "A few words about a few words."

The magazine *Tygodnik Powszechny* condemns the present anti-Semitic excesses and anti-Semitism generally, and to my mind strikes just the right note in doing so, unlike the analogous statement issued by the Rectors. The message of the *Tygodnik Powszechny* is all the more remarkable given the fact that it is a Catholic magazine, and therefore should reach much the same set of people as the Rectors' appeal was

---

[315] Marian Hemar (born Jan Maria Hescheles) (1901–1972), Polish-Jewish poet, journalist, and songwriter. Called "the bard of Lwów and the troubador of the London emigration." Participated in the defense of Lwów in the Polish-Ukrainian War and Polish-Soviet War (1918–1921). On the outbreak of World War II he fled Warsaw, eventually reaching the Middle East, where he signed up in the Polish Independent Carpathian Rifle Brigade. He settled in England, continuing to be popular in Poland with his weekly radio program broadcast by Radio Free Europe.

aimed at. In the meantime the ONR and the NSZ[316] make the same absurd criminal announcements as the Ukrainians.

This evening I am to be given a farewell reception at the Rzącas'.

AUGUST 23, 1945. I have spent the last two days in Jasło. I travelled by train for the first time in three years. The guard would not let me sit at the entrance to a railcar, but when I told him I was a Jaślanin, he let me sit on a bench in the guard's van. From afar you might be fooled into thinking Jasło still a real town, but on closer approach you see that it is really only a theatrical prop made up of broken *façades* without ceilings or roofs. The Germans spent two months mining and burning, so only a dozen or so buildings are left standing. However, there were plenty of people about, mostly come to town from the suburbs, many of which—Ułaszowice, for example—are largely intact. There were many wagons standing at the station. The trains were bringing people and goods from all over. There were stalls selling rolls, fruit, sausage, lemonade, and flowers, and among them young folk, tall and happy, milled about. The girls all looked very pretty to me, flitting about like butterflies amid somber ruins.

The first person to recognize me was Sowizdrzał. He and his family used to live next to the brickyard. They were then well known for thievery, but evidently the war had caused the general opinion of them to go up a notch. I went along the siding to the brickyard, where I found the vat in which the clay used to be mixed still in place, and from there went to look for the former neighbors of ours who were now holding a lease on the land around the brickyard. They were as surprised to see me as if I had fallen from the moon, but friendly enough. They had contracts they had made with the wartime German *Liegenschaftsverwaltung*[317] and receipts showing payments for the lease. They were naturally very concerned that I should not ask them to pay us again for the years for which they had already paid the German occupiers. From some time in 1944 there had been no one to take the payments, and this had troubled them. Thus I accepted payment for the first six months of 1944 in the form of the equivalent in złotys of 125 kilos of rye grain per morg, at the rate of 600 złotys per 100 kilos, and otherwise agreed to stick to the original agreement. Since then in Jasło 600 złotys per 125 kilos of rye grain was about half the going market rate they were satisfied in the extreme.

I went to Ułaszowice looking for Dr. Żelazny in the hope that he could put me up, but he was away in Nowy Sącz. So I buttonholed a passer-by and asked if he knew of a place for the night, and he readily offered me the use of his basement. This turned out to be Jan Kaznowski, a former court official who had known me by sight, and my Father personally. He received me very hospitably, gave me the use of a separate room—more like a partitioned-off area—a bed with clean bedding, supper, and breakfast. He agreed to watch over the parcel of land we owned on Ujejski

---

[316]See earlier footnotes. The point here is that these organizations were right-wing, nationalistic, and tending to anti-Semitism.

[317]"[Department of] Management of Lands"

Street in return for being permitted to take bricks from the ruins of the building that had stood there, and use the plot for a year free of charge.

Next day I went back to the market place, but failed to find anyone I knew; however, in what used to be our tobacco store I found Krystyna P., wife of the local photographer. She looked at me in disbelief and concern, but here, too, I quickly set things to rights—as with all those now dwelling in our former coach house, stables, and warehouses, still more or less standing. In our garden I found a gift from the Germans: a garage. The view from the market place is different now since the buildings whose walls bounded it on the eastern side are gone. I used to cultivate grapevines along these walls, and I could see green vestiges of the vine still persisting there.

The ruins of Jasło were in some places rather majestic, like small-scale portions of the ruins of ancient Rome: parts of walls, mountains of bricks, jumbled sheets of stucco work, girders at odd angles, and twisted railroad lines where once stood the offices of the District Council and the Polish Bank. It was as if the "destroyer of palaces" of Arabian folk tales had passed this way.

People who during the German occupation behaved ethically as regards the property of others will tend to come out better in the end than those who took undue advantage of the situation. I went to the office of the current *starosta*, which is now located in Gorajowice,[318] in a "palace" which had escaped destruction. An official, a Mr. Czulik, received me warmly, even though he knew me only by name. I had absolutely no difficulty re-establishing our property rights since Czulik's position on the legalities of restitution was completely fair. He will even reimburse us the rent collected from leaseholders from the time my Mother reported to him. In the land attached to the brickyard there had always been a certain laxity as to which plot was being leased by whom, and it appeared that the Germans had imposed order on the arrangements. Although their maps of the property, with precisely indicated sections, had been burned in the general conflagration they engineered before leaving, the boundaries they had established were fixed in the memories of the tenants, each of which kept jealous watch on his tenant neighbors lest encroachment occur—and furthermore the non-tenant neighbors had kept them all under close scrutiny. The areas of the sections are listed in contracts typed out in Polish on the back of pages printed in German.

On the train to Jasło I could identify myself at last not as Grzegorz Krochmalny but as Hugo Steinhaus, which felt like a rebirth into a more rational, less dangerous world, accompanied by a rather euphoric sense of relief. *Obiit Krochmalny, natus est Steinhaus!*[319]

We are now packing up in preparation for our departure three days hence.

---

[318] A village about 3 km north-east of Jasło.

[319] "Krochmalny has died, Steinhaus is born!"

The signing of the instrument of surrender by Japan to the Allies is to take place in Tokyo on August 31.[320] The earlier announcement by the Mikado and MacArthur's speech in Manila were unofficial preliminaries establishing basic conditions of the capitulation. MacArthur wants the official ceremony to be a demonstration of the military might of America, almost exactly six years to the day since the war began. For the sake of exactness, the ceremony should take place on September 1, the date in 1939 when the Germans attacked Poland, and not September 3, the date put about by the Germans as marking the beginning of the war. The American Secretary of State Byrnes,[321] backed by Ernest Bevin,[322] has come out with a statement to the effect that the Polish Provisional Government of National Unity does not have the support of a majority of Poles, and likewise the government of Romania. He also talks of the need to have western observers present at the forthcoming elections in Bulgaria, but the devious USSR counters with the argument that this would contravene Bulgaria's sovereignty.

Such is the untrustworthiness of the Soviets, who, after protecting Japan's rear from attack throughout the war, now do an about-face and rush in to occupy Manchuria![323] But there are limits to the reach of their duplicity for the US is to administer the Korean Peninsula south of the 38th parallel.

Lately all one gets by way of news from the BBC or Voice of America is tidbits in between endless discussions of the import of the atomic bomb.

Gołdówna came to see me after returning from Wrocław. She said she delivered my letter of acceptance of a founding role in the University of Wrocław, but that she had been waylaid by a gaggle of four Rectors who carried her off to a conference, and hadn't managed to see Kulczyński. She came away with the understanding that no one knows anything for certain, and moreover with the impression they don't even know they don't know. Nevertheless the press prints nonsensical articles about what the university portends for Wrocław and Poland.

AUGUST 28, 1945. On August 26, Dolek arrived in Berdechów with a three-ton, 75-horse-power Chevrolet truck, and loaded it up with his mother, me and my things, Captain Zygmunt Michałowski, Jan Igielski, Dr. Ślęczkowski, Stefan Dudzik, Marian Szafraniec's brother, and a half a dozen live piglets. We left Stróże

---

[320] The Japanese instrument of surrender was signed on the deck of the USS *Missouri* in Tokyo Bay on September 2, 1945.

[321] James Francis ("Jimmy") Byrnes (1882–1972), American statesman. US Secretary of State under Truman from July 3, 1945 to January 21, 1947, he was involved in several postwar peace conferences, taking a firm stand against the Soviet Union.

[322] Ernest Bevin (1881–1951), British trade union leader and Labour politician. Minister of Labour in the wartime coalition government, and Foreign Secretary in the postwar Labour Government.

[323] In 1931 Manchuria, part of northeastern China, was occupied by the Japanese, who set up the deposed Qing emperor of China as puppet emperor of "Manchukuo", which they used as a base for their invasion of the rest of China. After the war the Communist Chinese took over Manchuria from the Nationalists under Chiang Kai-shek (whose *Kuomintang* government had been an ally of the West throughout the war) with Soviet approval. The region then served as a communist staging ground for the Chinese Civil War, the communists under Mao emerging victorious in 1949.

at noon and by 5 pm we had reached the city limits of Kraków, travelling the final stretch on the excellent asphalted Myślenice highway. Kraków terrified me with its unaccustomed bustle, noise, and dirt. And in addition to that, when I entered the apartment—which we were taking over from Knaster—I sensed the atmosphere left behind him of neurasthenic fear of pogroms and the law of the jungle generally. But I manfully shaved my moustache, thus shedding the last vestiges of my Krochmalny identity.

We went over to Irena's place where I found Mother, Olga, Alina,[324] and Jagienka.[325] After dinner, Staszek Dawidowicz[326] dropped by. Later that evening Irena and Leonard[327] visited us in our apartment, as did Zosia Cieluchówna.[328] Yesterday we paid the Zagórskis a visit. I found Maria, Zagórski's wife, very attractive. Mrs. Gałczyńska[329] was there too.

I want to present my talk entitled "A few words about a few words", but I will wait until the Zagórskis return from Murnau,[330] in Bavaria, whither they depart tomorrow. In any case, consultations with electricians and doctors are more pressing. Lucek[331] has taken my paper on quadratic price scheduling to show to the relevant person in the Kraków Electrical Works.

---

[324] Alina Dawidowicz, daughter of Olga Chwistkowa.

[325] Daughter of Alina.

[326] Alina's husband.

[327] Leonard Zgliński, husband of Steinhaus's sister Irena.

[328] Daughter of the Cieluchs.

[329] Natalia Gałczyńska (1908–1976), writer and translator. Wife of the Polish poet Konstanty Ildefons Gałczyński (1905–1953).

[330] A Bavarian town on the edge of the Alps, about 70 km south of Munich.

[331] An acquaintance.

Index of Names

Abakanowicz (Abdank-Abakanowicz), Bruno, Lithuanian-born Polish mathematician and engineer, 157
Abraham, Max, German physicist, 60
Abrahamowicz, Dawid, Polish-Jewish Galician politician, 112
Abramowicz, Kazimierz, Polish mathematician, 392
Achender, Mrs., a friend of the Steinhauses, 372
Adamski, Jan, a friend of the author, brother of Stanisław, 120, 145, 179, 183, 266, 312, 325, 337, 375
Adamski, Stanisław, a schoolfriend of the author, 33, 34, 40, 47, 263, 415
Afanaseva, Tatyana Alekseevna, Russian mathematician, wife of Ehrenfest, 80
Ajdukiewicz, Kazimierz, Polish philosopher and logician, 135
Albano, an Italian lecturer in Göttingen, 72
Albert, Zygmunt, 292
Aleksandrov, Pavel Sergeevich, Soviet mathematician, 250
Aleksandrowicz, Halina, a Lwów acquaintance, 265
Aleksandrowicz, owner of a paper manufactory, 390, 423
Aleksandrowicz, Roman, a Lwów acquaintance, 288
Aletti, an Italian hotelier, 150
Altenberg, Alfred, a Lwów publisher, 141
Altenberg, Ewa, a Polish microbiologist, 218
Altenberg, Peter, Austrian writer and poet, 67
Alter, a clerk with the Philips Company, 280
Althoff, Friedrich, high-ranking Prussian official, 59
Ambronn, Leopold P., Göttingen professor of astronomy, 62
Ampère, André-Marie, French physicist and mathematician, 103
Anders, Władysław, Polish general, 412
Andrzejewski, Jerzy, Polish writer, 374
Andrzejewskis, the family of Jerzy, 378
Antecka, maiden name of Henryk Kołodziejski's wife, 64
Antoniewicz, a Judge in Lwów, 136
Apfel, a Jasło merchant, 27
Appel, Ewa, author of textbooks, 348
Aranda Arellano, Angela, a Mexican singer, wife of Adam Didur, 187
Archimedes of Syracuse, illustrious physicist and mathematician of ancient Greece, 345
Arciszewski, Tomasz, Polish socialist politician, 382, 391
Arco, Georg Wilhelm Graf von, German physicist, 82
Aristotle, Greek philosopher, 203
Artin, Emil, Austrian-American mathematician, 137
Askenase (Askenazy), Stefan, Polish-Belgian pianist, 180
Askenazy, Frau, mother of the pianist Stefan Askenazy, and of Olga (Ziemilska) and Tusia (Baderowa), 113
Ataulf, king of the Goths, 156
Attlee, Clement Richard, 1st Earl, British politician, 278, 440
Auber brothers, bandleaders in Iwonicz, 13

Auerbach, Herman, Polish mathematician, 130, 165, 253, 304, 310, 393, 445, 452, 454, 457
Auerbach, Marian, Polish philologist, 215
Auffenberg, Count Moritz von, Austrian general, 110
August II the Strong, a Polish-Lithuanian king, 310
August III, a Polish-Lithuanian king, 310
Axentowicz, Teodor, Polish painter, 307
Ayres, Frank J., Jr., American mathematician, 149

Bader, Emil, and wife Tusia, Lwów friends of the author, 143, 267
Bader, Mieczysław, brother of Emil, 288
Baedeker, Karl, 19th century German publisher of travel guides, 70
Baer, W., German lawyer, 57
Bainville, Jacques, French historian and journalist, 77
Balicki, a Jasło resident, 455
Balicki, Zygmunt, Polish politician, 47
Balk, Henryk, Polish-Jewish poet and literary critic, 309
Balzac, Honoré de, French novelist, 29
Balzer, Oswald Marian, lawyer, 42
Banach, Stefan, illustrious Polish mathematician, 52, 121, 129, 135, 142–144, 149, 153, 156, 164, 165, 181, 215, 238, 253, 303, 347, 425, 458, 461
Banachiewicz, Tadeusz, Polish astronomer and mathematician, 53, 63, 64, 76, 80, 107, 455
Banachowa, Łucja, wife of Stefan Banach, 164
Bandera, Stepan Andriyovich, Western Ukrainian nationalist leader, 287, 300
Baranowski, Dante, Venetian-born, Polish actor, 124
Baranowski, Tadeusz, Polish biochemist and medical professor, 178
Bardeleben, Count Kurt von, German chessmaster, 33
Barnaś, a Jasło *starosta*, 415
Bartel, Kazimierz Władysław, Polish mathematician and politician, 143, 149, 163, 165, 194, 197, 214, 229, 241, 291, 292, 392
Bartel, Mrs., wife of Kazimierz Bartel, 292
Baudelaire, Charles, French poet, 72
Bauer, a Venetian hotelier, 98

Baumann, Göttingen professor of philosophy, 61
Bazhan, Mykola, Ukrainian writer, poet, and politician, 453
Beck, Adolf, Polish physiologist, 316, 445, 454
Beck, Henryk, son of Adolf, 316
Beck, Józef, Polish politician and army officer, 114, 207, 220, 231
Beckowa, wife of Henryk Beck, 316
Behring, Emil Adolf von, German physiologist, 300
Belloc, Joseph Hilaire, Anglo-French writer, 221
Belyĭ, Andreĭ (actually Boris Nikolaevich Bugaev), Russian novelist and poet, 186
Bendt, F., 39
Benedikt, Moritz, an Austrian newspaperman, 37
Beneš, Edvard, Czechoslovak political leader, 441, 448
Benson, Rex ("Baby"), Canadian roommate of the author in Göttingen, 74
Berent, Wacław, Polish writer, 29
Berger, Róża, victim of a pogrom, 460
Bergman, Stefan, Polish-American mathematician, 394
Bergson, Henri-Louis, French philosopher, 326
Berling, Zygmunt Henryk, Polish general and politician, 377
Bermant, A. F., Soviet mathematician, 250
Bernays, Paul Isaac, Swiss mathematician, 237
Bernoulli, Jacob, Swiss mathematician, 218
Bernstein, Felix, German-Jewish mathematician, 61
Bernstein, Serge, a mathematician at the Bologna ICM, 153
Berson, a Lwów engineer, 221
Betuchówna, Irena, a student of the author in Berdechów, 367
Bevin, Ernest, British labour politician, 464
Białobrzeski, Czesław, Polish physicist and philosopher, 312
Bieberbach, Ludwig, German mathematician, 136
Bielski, a Polish count, 262
Biernacki, Mieczysław, Polish mathematician and chemist, 393
Biernats, friends of the Cieluchs, 352
Bierut, Bolesław, Polish communist political leader, 414, 438
Birkhoff, George David, American mathematician, 157

Birnbaum, Zygmunt, Polish-American mathematician, 394
Biskupski, a Ruthenian mineralogist, 234
Bismarck, Otto von, Prussian chancellor, 58, 66, 77
Bjørnson, Martinus, illustrious Norwegian writer, 29
Blaschke, Wilhelm, Austro-Hungarian mathematician, 137
Blassberg, Maksymilian, a Kraków medical specialist, 179
Blaton, Jan Antoni, Polish physicist, 143, 164
Bloch, Maurycy (Morcio, Moryś), a fellow student of the author's daughter, 211, 212, 217, 430
Blum, Dora, an acquaintance of the author, 375
Blumenfeld (assumed name Noga), Jan, son of Izydor, 347
Blumenfeld, Dr. Ignacy, a friend of the author, 146, 165, 189, 219, 371
Blumenfeld, Izydor, a plant manager, friend of the Steinhauses, 140, 250, 308, 350
Blumenfeld, Stanisława (Stasia), wife of Izydor, 296, 297, 308, 347
Blumenfelds, the family of Izydor, 297, 300, 308
Bobrowski, a landowner from the Jasło area, 26
Bobrzyński, Michał, Polish historian, governor of Galicia, 93, 94, 112
Bocheński, Aleksander, a Polish landowner, 311
Bochniewicz, a judge, 30
Böcklin, Arnold, Swiss symbolist painter, 160
Bohr, Harald, Danish mathematician, 80
Bohr, Niels, Danish physicist, 80
Bold, Jan Maciej, Polish legionnaire, 114, 119
Bolek, R., a gendarme in the Polish Legions, 111
Boltzmann, Ludwig, Austrian physicist and philosopher, 80
Bong, Alfred, a Polish army doctor, 214, 217
Boratyński, a priest in Wilczyska, 358
Borejsza, Jerzy (born Benjamin Goldberg), Polish-Jewish writer and communist activist, 261, 262, 415
Borek, Kazimierz, Polish mathematician, a student of the author, 218, 394
Borel, Émile, French mathematician, 48, 64, 100, 215, 312
Borkowa, mother of Kazimierz Borek, 303
Born, Max, German physicist, 80, 137

Borowicz (or Horowitz), Alfred (Fredek), brother of Olga Pamm, 372, 378, 442
Borsuk, Karol, Polish mathematician, 96, 185
Boruchowicz, Michał Maksymilian, Polish-Jewish writer, 261
Boy-Żeleński, Tadeusz, prominent Polish writer, translator, and poet, 25, 29, 247, 261, 266, 291
Brach, Miss, a Polish schoolteacher, 353
Brąglewicz, Tytus, Jasło merchant, 6
Brahms, Johannes, German composer, 58
Brandt, a Viennese landlord, 113
Branges, Louis de, French-American mathematician, 136
Brauer, Richard, German mathematician, 136
Braun, Stefania, Polish mathematician, 395
Brendel, Martin, German astronomer, 60, 79
Brentano, Franz, German philosopher and psychologist, 58
Broch, an engineer, 459
Broglie, Maurice de, French physicist, 150
Broniewski, Władysław, Polish poet and soldier, 240, 246, 247
Bronisława (Bronia) R., sister of Kac senior's wife, 197, 199
Brühl, Count Heinrich von, 18th century minister of the Elector of Saxony, 70
Bruno, Polish railroad official and German guard, 361
Bujakowa, Maria (*née* Łomnicka), a Polish artist, 266
Bujwidówna, a Kraków student, 97
Bukharin, Nikolaĭ Ivanovich, Russian Bolshevik revolutionary, 235
Bul, a refugee family in Stróże, 416
Bulanda, Edmund Jan, Polish archaeologist, 174, 306, 422
Bułat, an official in Zimna Woda, 328
Bura, an Austrian lieutenant, 114
Burdecki, Feliks, Polish science-fiction writer and German collaborator, 381, 382
Burg, Adam von, Austrian mathematician, 39
Bürger, Gottfried, German poet, 58
Burzyński, Włodzimierz, Polish engineering professor, 423
Byczenko, a rector of the Ivan Franko University, 300
Byrnes, James Francis, American politician, 464
Byron, George Gordon, 6th Baron, English poet, 273

Calvin, Jean, French theologian and pastor, 102, 208
Cantor, Georg, German-Jewish mathematician, 61, 63
Carathéodory, Constantin, Greek mathematician, 52, 60, 80
Catherine the Great, Empress of Russia, 454
Cauchy, Augustin-Louis, French mathematician, 62
Cavan, Lord, British field marshal, 188
Cejnar, an Austrian NCO, 114
Cellini, Benvenuto, Italian sculptor and goldsmith, 158
Chandler, an American operetta star, 136
Chardalias, a Göttingen student from Greece, 72
Charles I, last Austrian emperor, 154
Chebyshev, Pafnuty, Russian mathematician, 81
Chekhov, Anton, Russian writer, 30
Chełmoński, Józef Marian, Polish painter, 307
Chesterton, G. K., English novelist and poet, 12
Chiang Kai-shek, Chinese political and military leader, 211, 421, 464
Chmielowski, Piotr, Polish philosopher and critic, 33
Churchill, Sir Winston, British politician, 349, 357, 405, 406, 417, 420, 427, 440
Chwalibogowski, Artur, medical scientist, 170
Chwistek, Leon, Polish painter and mathematician, a brother-in-law of the author, 53, 121, 153, 165, 198, 200, 231, 253, 278, 302, 310, 313, 392, 416
Chyliński, Konstanty, Polish historian and politician, 185
Ciechanowski, son-in-law of the author's Lwów neighbor Tomanek, 312
Ciechanowski, Stanisław, Polish professor of medicine, 96, 291
Cięglewicz, Edmund Z., a Polish philologist, 113
Cieluch Senior, owner of a farm in Berdechów, 335, 338, 340, 344, 352, 353
Cieluch, Artur, a son of Cieluch Senior, 335, 338
Cieluch, Jan, a son of Cieluch Senior, 335, 338, 449
Cieluch, Tadeusz, a son of Cieluch Senior, 335, 352, 438
Cieluch, Zofia (Zosia), a daughter of Cieluch Senior, 335, 465

Cieluchowa, Stefania (Stefa), wife of Tadeusz, 335, 336
Cieluchowa, wife of Cieluch Senior, 453
Cieluchs, the family of Cieluch Senior, 335, 336, 340, 342, 352, 353, 365, 389, 397, 403, 404, 408, 421, 449, 451, 452
Cieszyński, a student in Munich, 84
Cieszyński, Antoni, Polish physician, surgeon, and stomatologist, 244, 291
Cleomenes, ancient Greek sculptor, 99
Conrad, Joseph, Polish-born English novelist, 374
Constantius II, Roman emperor, 101
Constantius III, Western Roman Emperor, 156
Courant, Richard, German-American mathematician, 80, 137, 152, 153, 206
Couturat, Louis, French mathematician, logician, philosopher, and linguist, 105
Cramér, Harald, Swedish mathematician, 206
Cripps, Sir Richard Stafford, British ambassador to the USSR, labour politician, 278
Curie-Skłodowska, Maria, Polish-French physicist and chemist, 393
Curzon, George, British politician, 406, 461
Custine, Marquis de, French writer, 273, 310
Ćwierczakiewiczowa, Lucyna, author of Polish cookbooks, 12
Cybis, Jan, a Kraków painter, 200
Cyganiewicz brothers, 186
Cyganiewicz, Stanisław Jan ("Zbyszko I"), wrestler, 32
Cyganiewicz, Władek ("Zbyszko II"), brother of Stanisław, also a wrestler, 32
Czacki, Tadeusz, Polish historian and numismatist, 200
Czartoryski, a noble Polish family, 103
Czekanowski, Jan, Polish anthropologist and statistician, 194
Czerkawski, a property owner near Jasło, 390
Czeżowski, Tadeusz, Polish philosopher and logician, 135
Czubiński, Polish mathematician, 395
Czulik, an official in Gorajowice, 463
Czyżewski, Julian, Polish geographer, 172

Dąbrowska, a Polish woman in Kowel, 116
Dąbrowska, Maria, Polish writer and human rights activist, 308

Index of Names

d'Alembert, Jean le Rond, French mathematician and physicist, 454
Dankl, Count Viktor, Austrian general, 110
D'Annunzio, Gabriele, Prince of Montenevoso, Italian writer and poet, 160
Dante Alighieri, illustrious Italian poet, 158
Darwin, Charles, English evolutionary biologist, 454
Daszyński, Ignacy, Polish politician, 113
Dawidowicz, Stanisław, husband of Alina, 416, 465
Dawidowiczowa, Alina, daughter of the author's sister Olga, 416, 417, 465
Dawidowiczówna, Agnieszka (Jagienka), daughter of Alina, 416, 465
Dawkins, Richard, British evolutionary biologist, 218
Decykiewicz, an Austrian bureaucrat, 124
Degas, Edgar, French impressionist painter, 102
Dehn, Max, German mathematician, 115
Delekta, owner of a sweet shop, 14
Demko, Katarzyna, a Lwów teacher of English, 291
Derechowski, a shoemaker near Berdechów, 345
Diamandówna, Wanda, a Polish photographer, 193
Dianni, Augusto, Italian tenor, 49
Dickens, Charles, English novelist, 29
Dickstein, Samuel, Polish-Jewish mathematician, 186, 251, 392
Diderot, Denis, French encyclopedist, 454
Didur, Adamo (Adam), Polish opera star, 145, 187
Didurówna (Załuska), Mary, a daughter of Adamo Didur, 187
Dietrich, a railroad machinist, 280
Dietzius, Aleksander, Polish engineer, 125, 146
Dimitrov, Georgi, Bulgarian communist politician, 180
Dirichlet, Johann P. G. Lejeune, German mathematician, 58
Dittersdorf, Leon, a brother-in-law of the author's wife, Helena's husband, 161, 220
Dmowski, Roman, Polish politician, 47
Döblin, Wolfgang (Vincent Doblin), German-French mathematician, 206
Doboszyński, Adam Władysław, Polish politician, 190

Dobromęski, Mieczysław, Polish attorney, 420, 447
Dobrzaniecki, Władysław, Lwów surgeon, 291
Dobrzańska, Agnieszka, a daughter of Stefan, 362
Dobrzańska, Maria (Maniuta), wife of Jan, 336, 351, 444
Dobrzańska, Zofia (Zosia), wife of Stefan, 341, 348, 355, 362, 364
Dobrzański, Jan, an acquaintance of the author, 298, 336, 362, 405, 407, 418, 425, 432, 444, 445, 457
Dobrzański, Jan, Polish historian and pedagogue, 279
Dobrzański, Stefan, a brother of Jan, 336, 341, 342, 348, 351, 355, 362, 364, 417
Dobrzańskis, the families of Jan and Stefan, 420
Dodd, Edward Lewis, American mathematician, 206, 210
Dodd, Mrs., wife of Edward, 210
Doerman, an actuary, 53
Domaszewicz, Aleksander, a Polish neurosurgeon, 221
Dönitz, Karl, German admiral, 425
Downarowicz brothers, 47
Dreyfus, Alfred, a captain in the French army, 36, 77
Druśko, Emil, Polish railroad guard and German collaborator, 360, 361, 383
Duda, Roman, Polish mathematician, 377, 425, 435
Dudzik, a neighbor in Berdechów, 363, 417, 418, 438, 441, 450
Dudzik, Stefan, a pupil of the author in Berdechów, 367, 460, 464
Dunaj, a restaurant owner in Jaśle, 25, 183, 200
Durand-Edwards, Louis, an American friend of the author in Göttingen, 72, 73
Dürer, Albrecht, German painter and engraver, 304
Durkheim, Émile, French sociologist, 46
Durlak, a Polish butcher and German informant, 344
Dywan, Stanisław, an inhabitant of Stróże, 341, 342, 353, 383
Dzerzhinskiĭ, Feliks Edmundovich, Belorussian director of OGPU, 235
Dziadura, a former foreman and Polish national activist, 319
Dzieduszycki, Wojciech, Polish politician, 112
Dziedzioch, a priest, 437

Dziewulska, a female acquaintance of the author, 84
Dziewulski brothers, 56
Dziewulski, Wacław, Polish physicist, 52, 57, 75
Dziewulski, Władysław, Polish astronomer and mathematician, 52, 60
Dziwiński, Placyd Zdzisław, Polish mathematician, 38, 129

Eddington, Sir Arthur, British physicist, 171
Eden, Robert Anthony, British politician, 357, 424, 428
Eggerth, Marta, Hungarian soprano, wife of Jan Kiepura, 187
Ehrenfest, Paul, Austrian-Dutch physicist, 80, 193
Ehrlich, Ludwik, Polish jurist, 267
Ehrlich, Paul, medical scientist, 82
Eidelheit, Meier (Max), Polish-Jewish mathematician, 253, 289
Eilenberg, Samuel, Polish-American mathematician, 394
Einstein, Albert, illustrious German-American physicist, 59, 60, 81, 85, 171, 190, 192
Eisenhower, Dwight D., American military and political leader, 420, 425, 427
Elgert, Gustaw, 203
Emil, a Grybów butcher and German informant, 344
Eminowicz, Ludwik, a teacher at the Gymnasium, translator and poet, 21, 93
Engel, a resident of Łużna, 460
Engels, Friedrich, German social scientist and political theorist, 236, 249, 416
Erasmus of Rotterdam, Desiderius, Dutch renaissance humanist, 396
Ernest Augustus, King of Hanover, 142
Esterhazy, Ferdinand, a major in the French army implicated in the Dreyfus affair, 37
Estreicher, Karol Jr., Polish art historian, 457
Euler, Leonhard, illustrious Swiss mathematician, 81, 454
Everett, Joseph D., English physicist, 58
Ewa, a friend of Helena Szmoszówna, 133

Fajans, Kazimierz, Polish-American physical chemist, 196
Fehr, Henri, Swiss mathematician, 206

Feller, William, Croatian-American probabilist, 140, 152, 153, 206
Felsztyn, Tadeusz, Polish legionnaire and mathematician, 123
Fermat, Pierre de, French mathematician, 62, 81
Ferraris, Galileo, Italian physicist and electrical engineer, 209
Fierlinger, Zdeněk, Czech politician, 448
Filiber, a German chief of police, 354
Fitzgerald, George Francis, Irish physicist, 85
Flaubert, Gustave, French novelist, 29
Fleck, Ludwik, Polish doctor and biologist, 218, 277
Florek, a house painter and mayor of Gorlice, 410
Ford, Henry, American industrialist, 159
Fourier, Jean-Baptiste, French mathematician, 59, 85
Fraenkel, Abraham Halevi, Israeli logician and set-theorist, 458
Fraenkel, Marcel, a cousin of the author, 220
Fraenkel, Mrs., wife of Marcel, 220
France, Anatole, French writer, 124
Franck, James, German physicist, 137
Frank, an NCO in the Polish Legions, 114
Frank, Hans, Nazi *Generalgouverneur* of Poland, 321, 323, 354, 369, 375, 387, 433, 460
Frank, Mrs., a Lwów acquaintance of the author, 214
Fränkel, Adolf, German-Israeli mathematician, 61
Frankl, F., a mathematician, 156
Franko, Ivan Yakovych, Ukrainian novelist, poet, and political activist, 251
Franz Ferdinand, Archduke of Austria-Este, 108, 109
Franz Joseph I, Austrian emperor, 24, 25, 93, 94, 109, 252
Fréchet, Maurice, French mathematician, 206
Friedjung, Heinrich, Austrian historian, 37
Fröschels, Dr., a Viennese Freudian, 111
Fryde, Halina, Polish poet, 261
Frydman, Ludmiła, an aunt of the author, 16
Frydman, Marceli, an uncle of the author, 16, 40
Frydman, Ryszard, a cousin of the author, 16, 40
Frydmanowa, a great aunt of the author, 16
Fryling, Jan, a student in Munich, 84
Fryze, Stanisław, Polish engineering professor, 423
Fuchs, Lazarus I., German mathematician, 46

Index of Names

Fulińska, wife of Benedykt, 313
Fuliński, a son of Benedykt, 312
Fuliński, Benedykt, Polish biologist, 244, 299, 301, 305, 310, 312, 313, 319, 340, 447
Fulińskis, the family of Benedykt, 299, 301, 302, 304, 306, 307, 310, 311, 313, 316
Fyda, a sublieutenant in the Polish Legions, 116

Gabryszewski, a Jasło *starosta*, 25
Gagarin, a Soviet official, 254
Gałczyńska, Natalia, wife of Konstanty, 465
Gałczyński, Konstanty Ildefons, Polish poet, 465
Galgótzy, Anton von, a general in the Imperial Austrian Army, 24
Gallé, Émile, French glasswork artist, 71
Galvani, Luigi, Italian physician and physicist, 155
Ganszyniec (Gansiniec), Ryszard, Polish philologist and cultural historian, 254
Garlicka, Lidia (Lidka), a daughter of the Garlickis, 338, 388, 451, 453
Garlicka, Wanda (*née* Cieluch), a daughter of Cieluch Senior, 338, 369, 387, 438, 439, 453
Garlicki, a son-in-law of Cieluch Senior, Wanda's husband, 338
Garlicki, Aleksander (Oleś, Olek), a son of the Garlickis, 338, 453
Garlicki, Maryś, a son of the Garlickis, 338, 453
Gaspary, a Kraków *starosta*, 121
Gaulle, Charles de, French general and political leader, 272, 424, 436, 445
Gauss, Carl Friedrich, illustrious German mathematician, 58, 62, 63
Gautsch, Paul, Austrian prime minister, 50
Gawlik, a Lwów university caretaker, 306
Gawroński, Andrzej, Polish orientalist, 191
Gębarowicz, Mieczysław, Polish historian and art historian, 304
Gebhardt, Karl, German physician at Ravensbrück and Auschwitz, 434
Geissler, Kurt, German mathematician, 82
Georg of Bavaria, Prince, 25
George II, King of GB and Elector of Hanover, 57
Gericke, Frau, Göttingen landlady, 71, 73, 83, 85, 87, 136

Gerstmann, Father Adam, Polish Catholic priest, professor at the University of Lwów, 184, 230
Geyter, Pierre de, Belgian socialist and composer, 276
Gierymski, Ignacy Aleksander, Polish painter, 307
Gil, Franciszek, Polish writer and journalist, 261, 266
Ginczanka, Zuzanna, Polish-Jewish poet, 261, 266, 374
Gładyński, a Lwów professor, 278
Glińskis, a family encountered in Tatarów, 224
Goebbels, Joseph, German politician, 202, 323, 348, 356, 368, 425
Goetel, Ferdynand, Polish novelist and playwright, 374
Goethe, Johann Wolfgang von, illustrious German poet, 30, 158, 160
Goetzendorf-Grabowski, Count, 97
Gogol', Nikolaĭ Vasilievich, Ukrainian-born Russian novelist and playwright, 251
Gołdówna, a Wrocław student, 453, 464
Gołyczs, neighbors of Vincenz in the Carpathians, 146
Gomółka, brother of Mrs. Zarańska, 450
Gomułka, Władysław, Polish politician, 438, 442
Gorayski, Dymitr, business partner of the author's father, 97
Göring, Hermann, German politician, 180, 188, 202
Gorky, Maxim, Russian/Soviet writer, 30
Górska, Halina, Polish writer and communist activist, 242, 261, 300
Gottwald, Klement, Czechoslovak politician, 441
Goursat, Édouard, French mathematician, 79
Grabowski, Witold, a radiology expert and Wrocław professor, 214
Grabski, Stanisław, Polish economist and politician, 46, 244, 260, 442
Grabski, Władysław, Polish politician, brother of Stanisław, 260, 442
Grączewski, Jan, Polish philosopher, 171
Graham, Jocelyn Henry Clive (Harry), English lyricist and versifier, 221
Grek, Jan, Lwów professor of medicine, 291
Grekowa, Maria, wife of Jan, 291
Grodkowski, a Lwów student, 173, 184
Gröer, Franciszek, Polish pediatrician and medical scientist, 113, 169, 170, 277, 291, 292

Gromskis, members of Twardowski's logic seminar, 135
Gromyko, Andreĭ Andreevich, Soviet politician, 424
Grouchy, Second Marquis de, French general, 184
Grubiński, Wacław, Polish writer, 332
Grünhut, Dr. Bernard, father of Józef, 268
Grünhut, Dr. Józef, a cousin of the author's wife, 264, 268, 292, 298
Grünhuts, the family of Józef, 268, 326, 347
Grużewski, Aleksander, Polish mathematician, 155
Grzędzielski, Jerzy, Polish medical professor, 289, 439
Grzesicki, a brigadier in the Polish Legions, 118
Grzymalski, Wiesław, Polish engineer and architect, 292
Gumiński, a standard bearer in the Polish Legions, 114
Guze, Joanna, Polish writer and translator, 278, 407, 415, 452

Haar, Alfréd, Hungarian mathematician, 153
Habsburgs, Austrian imperial family, 109
Hadała, a Gymnasium teacher, 454
Hadamard, Jacques, French mathematician, 100, 144, 153, 216, 439
Haeckel, Ernst, German biologist, 87, 437
Halaunbrenner, Jacob, mathematics student in Lwów, 142
Haldane, Lord, British War Minister, 58
Halina, tenant, with her husband, in the Steinhauses' Lwów home, 299, 309
Hals, Frans, Dutch Golden Age painter, 64
Hamann, Heinrich, German Border Police Commissioner, 338
Hamerski, Edward, head internist veterinarian in Lwów, 289
Handelsman, Marceli, Polish historian, 375
Harden, Maximilian, Berlin journalist, 67
Harris, Sir Arthur Travers, RAF commander-in-chief, 349
Hartman, Stanisław, Polish mathematician, 252
Hartmann, Eduard von, German philosopher, 58
Hartmann, Johannes Franz, German astrophysicist, 91
Harun al-Rashid, fifth Abbasid caliph, 171
Hasse, Helmut, German mathematician, 137
Hasselwanger, a German physicist, 197
Hata, Sahachiro, Japanese bacteriologist, 82

Hauptmann, Gerhart, German writer, 29
Hausdorff, Felix, German-Jewish mathematician, 152
Hearn, Lafcadio, American writer, 40
Hebda, a policeman, 456
Hecke, Erich, German mathematician, 137
Hedinger, a student in Munich, 84
Hegel, Georg Wilhelm Friedrich, German philosopher, 236
Heijermans, Herman, Dutch writer, 29
Heine, Heinrich, German poet, 58, 203, 226
Heisenberg, Werner, German physicist, 205, 211, 281, 454
Helon, a neighbor in Osiczyna, 317, 320, 324, 329
Helonowa, wife of Mr. Helon, 320, 321, 325
Hemar, Marian (born Jan Maria Hescheles), Polish-Jewish journalist, poet, and song-writer, 461
Herglotz, Gustav, Austrian mathematician, 52, 60
Hermite, Charles, French mathematician, 62
Heron of Alexandria, ancient Greek mathematician and engineer, 345
Hertz, Paul, German physicist, 60
Hertz, Paweł, Polish writer, 332
Herzberg, Jan, Polish logician, 198, 233, 249, 310, 375, 394
Hetper, Władysław, Polish mathematician, 197, 253, 393
Heurtier, Eugénie, the Steinhaus children's governess, 27
Hevesy, George Charles de, Hungarian radiochemist, 113
Heydrich, Reinhard, high-ranking German Nazi official, 298
Hilarowicz, Henryk, Lwów professor, surgeon, 289
Hilbert, David, illustrious German mathematician, 52, 61, 80, 81, 83, 85, 91, 92, 137, 153, 237
Himmler, Heinrich, Nazi leader of the SS, 321, 336, 353, 402, 425, 450
Hirohito, Emperor of Japan, 453, 459, 460, 464
Hirschler, Jan, Polish zoologist, 267, 452
Hitler, Adolf, 67, 80, 136, 139, 162, 179, 180, 187, 202, 204, 205, 212, 221, 224, 230, 235, 261, 277, 278, 289, 292, 298, 300, 309, 348–350, 357, 363, 368, 380, 417, 420, 424, 433
Hłasko, director of the oil firm *Małopolska*, 198
Hlond, August, Polish cardinal, 457
Hoborski, Antoni, Polish mathematician, 392

Index of Names

Hofman, Wlastimil, Polish painter, 307
Hofrichter, Oberleutnant Adolph, of the Imperial Austrian Army, 37
Hohenlohe, an Austrian prince, 41, 148
Hohenzollerns, dynastic Prussian family, 78
Hollender, Tadeusz, Polish humorist and poet, 266, 306, 374, 375
Holzer, Alfred, husband of Lila, 304, 305, 350
Holzer, Lila, a cousin of the author's wife, 304, 305, 350, 372, 413
Homme, a mathematician, 130
Hopf, Eberhard, Austrian-American mathematician and astronomer, 206
Hopkins, Harry Lloyd, American new deal politician, 439
Horowicz, Kazimierz, Göttingen student, actuary, 53, 57, 71, 84
Horowitzes, acquaintances of the author, 350
Hosiasson-Lindenbaum, Janina, Polish mathematician and philosopher, wife of Adolf Lindenbaum, 394
Huber, Maksymilian, Polish engineer, 195
Hubert, an acquaintance of the author, 222
Hupka, Józef, Polish politician, 112, 113
Huret, Jules, French journalist, 79
Hurewicz, Witold, Polish-American mathematician, 393
Huxley, Aldous, English novelist, 271
Huysmans, Charles-Marie-Georges, French novelist, 102

Ibsen, Henrik, Norwegian playwright, 29, 49
Igielska, Maria, wife of Jan Igielski, 336, 337, 429
Igielski, a neighboring farmer in Berdechów, 342, 357
Igielski, Adam, a Polish locksmith, a relative of the Igielskis, 345
Igielski, Jan (Janek), a son of the "elder Igielskis", 336, 337, 425, 464
Igielskis, Schoenemann's in-laws, the family of the "elder Igielskis", 336, 340, 342, 353, 455
Indruchowa, an aunt of Danuta Smalewska, 296
Indruchs, a Lwów family, friends of the Steinhauses, 296, 297, 311
Infeld, Leopold, Polish physicist, 456
Ingarden, Roman Kajetan, Polish hydraulic engineer, a counsellor in the Austrian empire, 120

Ingarden, Roman Witold, Polish philosopher and engineer, son of Roman Kajetan, 423
Ironside, William Edmund, British field marshal, 221
Irzykowski, Karol, Polish critic and writer, 21
Iwaszkiewicz, Jarosław, Polish writer and poet, 186

Jabayev, Jambyl, Kazakh folk poet, 276
Jabłonowski, Stanisław Jan, Polish nobleman and military commander, 263
Jabłonowski, Stanisław, a Polish hetman, 284
Jacob, Dr., an official of the Lwów Jewish community, 300
Jacob, Marian Mojżesz, a Polish actuary and professor, 253, 393, 457
Jacobsohn, Göttingen student, 56
Jagiełło, Władysław, King of Poland and Grand Duke of Lithuania, 410
Jałny, factotum of a Ukrainian police detachment, 366
Janaszek, Göttingen medical student, 53, 56
Janik, a tailor in Stróże, 418
Janiszewski, Zygmunt, Polish mathematician, 114, 125, 129, 130
Jankowski, Polish mathematician, 459
Jantzen, Kazimierz, Polish astronomer, 53, 84
Jaruzelski, Wojciech Witold, Polish military and political leader, 413
Jastrun, Mieczysław, Polish-Jewish essayist and poet, 261
Jastrzębiec, the assumed name of Mr. Tomczycki, 265
Jaworski, Władysław Leopold, Polish lawyer, 93, 111–113
Jaworski, Władysław, Polish poet, 245
Jaworskis, a Lwów family, 265
Jedlicz, Józef (actually Kapuścieński), teacher at the Gymnasium, poet and critic, 21
Jędrzejewicz, Janusz, Polish politician and educator, 168
Jełowicki, Stanisław, a major in the Polish Legions, 114
Jentyses, a Kraków family, 98
Jesus of Nazareth, 67, 69, 107
Joachim, Joseph, Hungarian violinist and composer, 58
Jodl, Alfred, German *Generaloberst*, 425, 427
Jolles, Stanisław, mathematician at Charlottenburg Polytechnic in Berlin, 51

Jordan, Camille, French mathematician, 79
Joseph II, Austrian emperor, 5
Joszt, Adolf, Polish chemist, 219
Joyce, James, Irish novelist, 212
Judt, Tony, British-American historian, 235
Julian the Apostate, Roman emperor, 101
Julius, Saint, a pope, 263
Juniewiczowa, daughter of the *starosta* Zoll, 184
Juryś, a carpenter, 40

Kac family, 199, 308
Kac, Marek (Mark), Polish-American mathematician, 192, 193, 197, 198, 200, 201, 207, 214, 308, 333, 394, 426
Kaczmarz, Stefan, Polish mathematician, 143, 147, 153, 156, 165, 182, 190, 246, 394
Kaczmarzowa, wife of Stefan Kaczmarz, 246
Kadyj, Henryk, brother of Józef, anatomy professor in Lwów, 41
Kadyj, Józef, a Jasło doctor, 10, 41
Kadyj, Stanisław, Polish eye specialist, son of Józef, 219
Kagan, Beniamin Fedorovich, Russian-Ukrainian mathematician, 105
Kamiński, a Polish forester, 311
Kamińskis, the family of Kamiński, 311
Kant, Immanuel, German philosopher, 45
Kapeliński, Tadeusz, Polish politician, 443
Karl Albrecht, Archduke, former owner of an estate at Żywiec, 409
Karolin, a man who had once lived with Mrs. Żywiecka, 322
Karpf, Józef, a Jasło doctor, 264
Karpiński, a Jasło acquaintance, 455
Karpiński, Franciszek, Polish poet, 156
Karpinski, Louis Charles, American historian of mathematics, 156
Kasterska, Maria, Polish writer, 144
Kawecki, Zygmunt, teacher at the Gymnasium, writer, 21
Kaznowski, Jan, a former Jasło court official, 462
Keitel, Wilhelm, German field marshal, 425, 426
Kellogg, Oliver Dimon, American mathematician, 153
Kempisty, Stefan, Polish mathematician, 393
Kępiński, Felicjan, Polish astronomer, 53, 62
Kępiński, Stanisław, Polish mathematician, 46

Kerner, Michał, Warsaw Gymnasium teacher, 457
Key, Ellen, Swedish writer, 29
Khinchin, Aleksandr Yakovlevich, Soviet probabilist, 140, 153
Khrushchov, Nikita Sergeevich, Soviet politician, 282
Kiedryn, a Ruthenian journalist, 220
Kiepura brothers, 186
Kiepura, Jan Wiktor, Polish tenor and actor, 145, 186, 187
Kiepura, Władysław, brother of Jan, also an opera singer, 145, 186
Kiernik, Władysław, Polish politician, 439, 442, 449, 452
Kiliński, Jan, a commander in the Kościuszko uprising, 263
Klawek, Aleksy, a priest and Lwów professor, 198, 420, 422, 437
Klein, Felix, German mathematician, 59–62, 79, 83
Kleiner, Juliusz, Polish literary historian, 49, 254
Klemensiewicz, Zenon, Polish linguist, 348
Klimt, Gustav, artist of the Viennese Secession, 166
Klobassa, a son of Stanisław, 400
Klobassa, Stanisław Karol, oil entrepreneur, 19, 31, 97, 400
Knaster, Bronisław, Polish mathematician, 96, 155, 163, 217, 241, 245, 253, 272, 300, 346, 347, 423, 425, 435, 445, 447, 449, 452, 455, 458, 465
Knasters, the family of Bronisław, 266
Knastrowa, Maria (pseudonym Maria Morska), wife of Bronisław, 306, 319, 435, 445, 447, 452
Knoff, Mrs., a acquaintance met in Krynica, 192
Koc, Adam Ignacy, Polish politician and journalist, 202, 203
Koch, Heinrich Hermann, German medical scientist, 169
Kochanowski, Julian, medical scientist, 170
Kocman—see Kolman, 237
Koebe, Paul, German mathematician, 60
Kolman, Ernst, Czech Marxist philosopher and mathematician, 237
Kolmogorov, Andreĭ Nikolaevich, Soviet mathematician, 192
Kołodziej, Antoni, Polish politician, 439
Kołodziejska (*née* Antecka), wife of Henryk, 64, 447

Index of Names

Kołodziejski, Henryk, Polish economist and historian, 64, 439, 442, 446
Komarnicki, Tytus, Polish diplomat and historian, 207
Kommehl, a doctor, 179
Komornicki, Father Dr. Władysław, a Lwów theologian, relative of the Ostrowskis, 289
Komorowski, Count, an oil man, 124
Kőnig, Dénes, Jewish-Hungarian mathematician, son of Julius, 63, 153
Kőnig, Julius (Gyula), Jewish-Hungarian mathematician, 63
Kontny, Piotr, a Polish archivist, 272, 284
Koppe, Wilhelm, German Nazi commander, 369
Korbulak, a corporal in the Austrian Imperial Army, 95
Korczak, Janusz (pen name Henryk Goldszmit), Polish children's writer, pediatrician, and social activist, 376
Korczak, Władysław's father, a *starostwo* commissioner, 34
Korczak, Władysław, a friend of the author, 34
Kornijczuk, Oleksandr, Ukrainian playwright and political activist, 231, 232
Korowicz, Henryk, Lwów economist, 291
Kościuszko, Andrzej Tadeusz Bonawentura, Polish engineer and military leader, 263, 276, 450
Kosko, Allan, Polish writer and poet, 261
Kossak, Karol, Polish painter, nephew of Wojciech, 223, 333
Kossak, Mrs., wife of Wojciech, 227
Kossak, Wojciech, Polish painter, owner of a guest house in Tatarów, 223, 226
Kossak-Szczucka, Zofia, Polish writer and resistance fighter, 374
Kossaks, the family of Wojciech, 224, 226
Kossowski, an engineer in Stróże, 367, 371, 440, 447, 449, 450, 453, 456, 457
Kossowskis, the family of Kossowski, 419, 450
Kostecki, Eugeniusz, a Lvovian, 291
Kostkiewicz, a Jasło engineer, 94
Kostrzewina, a woman from a village near Berdechów, 354
Koszyk, a property-owner in Berdechów, 352, 364
Kot, Stanisław, Polish historian and politician, 457
Kotarbiński, Tadeusz, Polish philosopher and logician, 374
Kott, Aniela, sister of Jan Kott, 378

Kott, Jan (Janek), Polish writer, professor, and the author's son-in-law, 199, 220–222, 240, 261, 262, 266–268, 273, 274, 292, 293, 297, 298, 316, 319, 325, 326, 340, 343, 347, 350, 355, 366, 367, 369, 370, 373, 377, 378, 385, 388, 389, 391, 402, 407, 414, 415, 428, 445
Kott, Kazimierz, a cousin of Jan Kott, 265
Kottowa, Kazimiera, mother of Jan Kott, 378
Kowalski, Władysław, Polish politician, 438
Kozar-Słobódzki, Mieczysław, Polish song-writer, 115
Kozłowski, Leon Tadeusz, Polish archaeologist and politician, 185, 194, 195, 241, 244, 409
Koźniewski, Andrzej, Polish actuarial specialist, 253
Krahelska, Halina, Polish writer and political activist, 375
Kramer, Henryk, a Jasło acquaintance, 264
Kramerowa, Greta, wife of Henryk, 264
Kramsztyk, Stanisław, physicist and popularizer of science, 24
Kraus, Karl, Austrian satirical writer, 29, 67–69, 78, 85, 113, 152, 206, 312, 346
Kretkowski, Władysław, Polish mathematician, 46
Kriegshammer, Baron, Austrian minister of war, 25
Krokowa, a neighbor of the Cieluchs, 352
Króo, Jan Norbert, Polish physicist and mathematician, 53, 121
Krukowski, Włodzimierz, Polish scientist and engineer, 291
Krygowski, Zdzisław, Polish mathematician, 129
Krzemickis, owners of a guest house in Kamień Dobosza, 220, 223, 376
Krzemieniewski, Seweryn, Polish botanist, 172, 184, 213, 231, 422, 423
Krzeptowski, Wacław, Polish highland leader, 433
Krzeszowski, a Polish policeman, 339, 345
Krzysikówna, Zofia, wife of Orlicz, 133
Krzysztoń, Ferdynand, a PPR activist, 421
Krzyżanowski, Adam, Polish economist and politician, 439, 440, 455
Kubala, a Polish policeman, 338, 339, 345
Kugelman, a Göttingen student, 71
Kuhn, Thomas, American physicist and philosopher of science, 218
Kukulski, a Jasło merchant, 225

Kulczyński, Stanisław, Polish botanist and politician, 172, 194, 230, 244, 303, 422, 423, 445, 446, 452, 464
Kulinowska, wife of Mr. Kulinowski, 351
Kulinowski, a Bobowa veterinarian, 351
Kuratowski, Kazimierz, Polish mathematician, 96, 143, 164, 196
Kurek, Jalu, Polish writer and poet, 261
Kuryłowicz, Jerzy, Polish linguist, 143, 313
Kurzweil, Max, artist of the Viennese Secession, 166
Kutrzeba, Stanisław, Polish jurist and politician, 439, 440, 444, 446, 455
Kwaśniewska, Iza, a tenant in the author's Lwów home, 229
Kwiatkowski, Eugeniusz, Polish economist and politician, 457

Łącek, Stanisław, 29
Lamb, Horace, English mathematician, 80
Lanca, a captain in the AK, 378
Lanckorońska, Karolina, Polish art historian, 267, 435
Landau, Edmund, German-Jewish number theorist, 62, 80, 82, 136, 153
Lanes, Salomon, a doctor, 179
Lang, Fritz, German film director, 458
Lange, a Göttingen student, 72
Lange, Friedrich A., German philosopher, 45
Langer—see Langner, 229
Langevin, Paul, French physicist, 150, 312
Langner, Władysław, Polish general, 229
Laska, Polish Police Commissioner for Bobowa, 338, 343, 352–354, 400, 401, 410, 414, 418, 447, 449, 455, 456
Laskowska, M., a sister of Halina Szpondrowska, 325
Lauer, Henryk, Polish mathematician, 100
Lauterbach, a Polish engineer, 159
Lavrentiev, Mikhail Alekseevich, Soviet mathematician, 250, 253
Lebesgue, Henri, French mathematician, 96, 100, 122, 144, 214–216, 220, 312
Łęczyński, a Jasło *starosta*, 26
Ledóchowska, a niece of Cardinal Rampolli, 19
Lefchenko, a Red Army captain, 244
Lehr-Spławiński, Tadeusz, Polish linguist, 143, 446
Leja, Franciszek, Polish mathematician, 149
Lejwa, a Polish engineer encountered in Krynica, 189, 190

Łempicki, Zygmunt, Polish literary theorist, Germanist, and philosopher, 203
Lenin, Vladimir Ilyich, 234, 235, 249, 255, 275, 285, 287, 416
Leonardo da Vinci, illustrious Italian painter, 133
Leopold Salvator, Archduke of Austria, 26
Lepecki, Mieczysław, Polish writer, an adjutant of Piłsudski, 204
Lepszy, a Kraków landlord, 96
Leray, Jean, French mathematician, 142, 216
Łeśko, a railroad worker in Osiczyna, 317
Leśniewski, Stanisław, Polish mathematician and philosopher, 171
Leśniowski, a victim of Emil Druśko, 383
Leszczyński, a Jasło *starosta*, 26
Lévy, Paul Pierre, French probabilist, 206
Lewandowski, a wagoner, 3
Lewin, Leopold, Polish poet, 332
Lewy, Hans, German-American mathematician, 152
Lichtenstein, Leon, Polish-German mathematician, 136, 150, 153, 181
Lie, Marius Sophus, Norwegian mathematician, 59
Limanowski, Mieczysław, Polish geographer, 172
Lindemann, Frederick Alexander, English physicist, 349
Lindenbaum, Adolf, Polish set-theorist, 394, 457
Lindenfeld, a Kraków friend of Holzer, 305
Lindpaintner, Otto, German pioneer aviator, 99
Linhardt, Mrs., sister of Tadeusz Rittner, 124
Lippmann, Walter, American journalist, 348
Lipschitz, the author's maternal grandfather, 36–38, 68, 85, 111
Lipski, a Polish clerk, 185
Listing, Johann, German mathematician, 59
Lobachevskiĭ, Nikolaĭ Ivanovich, Russian mathematician, 61, 105, 157
Loewenstein, business partner of the author's father, 97
Loewenstein, Mrs., a Lwów friend of the Steinhauses, 371
Łomnicka, Ewa, a daughter of Antoni, 222
Łomnicka, Irena (married name Wachlowska), daughter of Antoni, 52, 222, 265
Łomnicki, Antoni, Polish mathematician, 52, 64, 129, 140, 153, 164, 289, 292, 345, 392, 437
Łomnicki, N., a Polish lawyer, 264

Index of Names

Łomnicki, Zbigniew, Antoni's nephew, Polish mathematician, 153, 158, 165
Longchamps de Bérier, Bronisław, a son of Roman, 291
Longchamps de Bérier, Kazimierz, a son of Roman, 291
Longchamps de Bérier, Roman, Polish lawyer and professor, 173, 229, 291
Longchamps de Bérier, Zygmunt, a son of Roman, 291
Lorentz, Henrik, Dutch physicist, 85
Loria, Stanisław, Polish physicist, 307, 452
Łoś, Jan Stanisław, Polish landowner, diplomat, and academic, 409
Loth, Edward, Polish anatomist and anthropologist, 53
Lotheisen, Frau, Göttingen landlady, 136
Lotze, Göttingen factotum, 63
Lotze, Hermann, German philosopher, 58
Louis XV, French king, 310
Lubbe, Marinus van der, a Dutch communist, 180
Lubelski, Salomon, Polish mathematician, 395
Lubomirski, Prince Andrzej, owner of Przeworsk, 27
Lucek, a Kraków acquaintance, 465
Łukasiewicz, Jan, Polish logician and philosopher, 171
Łukawski, A., a former Austrian officer, 284
Lusia O. (assumed name Masłowska), an employee in Mr. Walter's office, 313
Luster, Dr., a Polish refugee, 345
Lutosławski, Wincenty, Polish philosopher, 28
Luzin, Nikolaï Nikolaevich, Russian mathematician, 123
Lysenko, Trofim Denisovich, Soviet biologist and agronomist, 250, 277
Lyusternik, Lazar Aronovich, Soviet mathematician, 282

MacArthur, Douglas, American general and field marshal, 461, 464
MacGarvey, William Henry, head of the oil firm *Karpaty*, 19, 31, 125, 397
Mach, Ernst, Austrian physicist and philosopher, 46, 150, 197, 234, 249
Macher, Belgian consul, 31
Machiavelli, Niccolò, Italian politician, historian, and writer, 112
Maciejowski, a lieutenant in the Polish Legions, 118
Maciejowski, Ignacy (pseudonym "Sewer"), Polish novelist, 19, 31

Macieliński, Adam, a member of the ONR, 460
MacIntosh, an oilman, 31
Mackensen, August von, German fieldmarshal, 120
Macudziński, business partner of the author's father, 97
Mączewski, Stanisław, Lwów gynaecologist and obstetrician, 289, 439
Mączka, Zofia, an inmate of Ravensbrück concentration camp, 434
Maeterlinck, Maurice, Belgian poet and playwright, 49
Majewska, a female acquaintance of the author, 84
Maksymiuk, Ołeksa, a neighbor of Vincenz in the Carpathians, 146
Małachowski, Roman, Polish chemist, 196, 302
Malaxa, N., a Bucharest engineer, 216
Malczewski, Antoni, Polish poet, 30
Malenkov, Georgiï Maksimilianovich, Soviet politician, 282, 283
Maliniak, a fellow boarder in Vienna, 113
Mandelbrojt, Szolem, Polish-Jewish mathematician, 144, 216, 393
Mao Tse-tung, Chinese communist leader, 464
Marchlewski, Polish lawyer, 53, 56
Marcinkiewicz, Józef, Polish mathematician, 192, 245, 393
Marczenko, a rector at the University of Lwów, 253
Marczewski, *see* Szpilrajn-Marczewski, 253
Maresch, Oskar, a cousin of the Dobrzańskis, 420
Maria Theresa, Austrian empress, 5, 41, 44
Maroszányi, a Jasło *starosta*, 26
Marshall, George C., American general and politician, 443
Marx, Karl, German-Jewish economist and philosopher, 232, 236, 249, 255, 416, 417
Marzecki, Józef, 171
Mastelko, a retiree living in Osiczyna, 317
Matuszewski, Ryszard, Polish writer, 340
Matuszewskis, the family of Ryszard, 385, 389
Maupassant, Guy de, French novelist, 29, 322
Mayakovskiï, Vladimir Vladimirovich, Russian/Soviet poet and playwright, 281
Mazur, Stanisław, Polish mathematician, 142, 143, 164, 171, 253, 300, 458, 459
Mazurkiewicz, a Polish legionnaire, 114, 117

Mazurkiewicz, Stefan, Polish mathematician, 96, 204, 446
Meisels, Emil, a Lwów radiologist, 197, 213, 214
Memling, Hans, German-Dutch painter, 64
Menasse, a merchant, 5
Mendelssohn, Felix, German composer, 58
Mengele, Josef, German physician in Auschwitz concentration camp, 359
Meredith, George, English writer, 29
Meyerson, Émile, Polish-French chemist and philosopher, 150
Meysztowicz, Jan, Polish diplomat and writer, 207
Michalakówna, Maria, a former administrator, 351, 361
Michałowicz, Mieczysław, Polish physician and politician, 170, 459
Michałowska, Antonina (Tosia), a resident with the Igielskis, 336, 405
Michałowski, Count Józef, son of the *starosta* Władysław, 184
Michałowski, Count Władysław, a Jasło *starosta*, 24–26, 34, 112, 145
Michałowski, Zygmunt, husband of Antonina, 405, 409, 451, 464
Michelson, Albert, American physicist, 80, 85–87
Mickiewicz, Adam, illustrious Polish poet, 30, 34, 49, 229, 263, 273, 276, 398, 455
Mickiewicz, Władysław, Adam's son, 103
Mięsowicz, Adam, grandson of Adam Sołowij, 291
Miętus, a Warsaw acquaintance of the author's daughter, 351
Mihály, Count Károlyi, Hungarian statesman, 154
Mikołajczyk, Stanisław, Polish politician, 368, 382, 391, 439, 442, 445, 446, 452
Mikuli, an oil man, 223
Miłosz, Czesław, Polish writer and poet, 413
Miłowska, Helena, actress, 49
Minc, Hilary, Polish communist politician, 441
Minkowski, Hermann, German mathematician, 52, 61, 62, 80–82
Mirski, J., a Paris acquaintance of the author, 144
Mirushov, a Soviet mathematician at the Bologna ICM, 153
Misiakiewicz, a pupil of the author in Berdechów, 411
Mix, Tom, American western movie star, 97
Młynarski, Feliks, Polish economist, 376, 402
Möbius, P. J., German neurologist, 68

Mocenigo, name of a prominent Venetian family, 99
Mogilski, owner of a hut near Stróże, 360, 383
Molotov, Vyacheslav Mikhailovich, Soviet politician, 225, 233, 250, 256, 275, 278, 289, 382, 422, 424
Monge, Gaspard, Comte de Péluse, French mathematician, 103
Montel, Paul, French mathematician, 144
Montgomery, Bernard L., British marshal, 420
Morgan, Justin, American horse breeder, 211
Morley, Edward Williams, American physicist, 85, 86
Mościcka, wife of Mościcki, 195
Mościcki, Ignacy, Polish chemist and politician, 115, 188, 189, 195, 231, 280, 412, 444
Moser, Koloman, artist of the Viennese Secession, 166
Mosler, S., Polish mathematician, 289, 394
Moszkowski, Alexander, Berlin humorist and journalist, 81
Motyka, Józef, Polish botanist, 436, 440, 453, 455
Mozołowski, Włodzimierz, Lwów professor of medicine, 298
Mravinesies, a friend of the author's father, 33
Mravinesies, Dziunia, a playmate of the author, 33
Mroczkiewicz, Mrs. (*née* Skulska), a refugee in Milanówek, 378
Müller, a *Volksdeutscher* informant in Bobowa, 344
Müller, Gustaw, a brother-in-law of the author, husband of Felicja, 179
Müllerówna, Muta, a niece of the author, Gustaw's daughter, 268, 293, 415, 417
Musse, Major-General Félix-Joseph, French military *attaché* to Poland, 215
Mussolini, Benito, 155
Muthesius, Hermann, German architect, 71
Mycielska, wife of Jan, also a painter, 200
Mycielski, Count Jan, painter and grandfather of the mathematician Jan Mycielski, 200

Nabokov, Vladimir, Russian-American novelist, 186
Nacht (actually Prutkowski), Józef, Polish actor and writer, 221, 261, 266, 279, 415, 452

Index of Names

Naglerowa, Herminia (pseudonym Jan Stycz), Polish writer, 247
Nałkowska, Zofia, Polish novelist, 412
Napoleon I (Bonaparte), 59, 92, 143, 177, 273, 331
Natanson, Władysław, Polish physicist and mathematician, 107
Neder, L., Lichtenstein's assistant in Leipzig, 139
Negrusz, Roman, Polish physicist, 167
Neugebauer, Otto, Austrian-American mathematician, 152
Neumann, Bernhard Hermann, German-British-Australian mathematician, 136
Neumann, John von, Hungarian-American mathematician, 48, 149
Neuwert-Nowaczyński, see Nowaczyński, Adolf, 31
Newton, Isaac, illustrious English physicist and mathematician, 45
Neyman, a Polish official at the League of Nations, 207
Nicholas I, Russian tsar, 273
Niemojewski, Andrzej, Polish writer, 47
Nietzsche, Friedrich, German philosopher, 28, 29, 45, 68, 72, 348
Nikliborc, Władysław, Polish mathematician, 143, 153, 165
Nikodým, Otto, Polish mathematician, 121, 452
Noether, Emmy, German-Jewish mathematician, 137, 152
Noga, Stanisława (Stajda), a servant of the Blumenfelds, 319
Nowaczyński, Adolf, Polish writer, 19, 31, 35
Nowakowski (born Tempka), Zygmunt, Polish writer and actor, 219
Nowicki, Jerzy, son of Witold, 289
Nowicki, Witold, Lwów professor of anatomy and pathology, 289

Oberländer, Ludwik (Lulu), a schoolfriend of the author, 40, 66, 120, 183, 265, 266, 273, 316, 350, 375
Oberländer, Zbigniew, son of Ludwik, 266, 350
Oberländers, the family of Ludwik, 266
Ochalik, Janusz (Januszek), a son of Dela, 336, 342, 353, 367, 460
Ochalik, Polish officer, husband of Dela, 429
Ochalikowa, Dela (née Igielska), a daughter of the "elder Igielskis", 336, 337
Ochmann, a Krosno businessman, 31
Ochorowicz, Julian, Polish philosopher, 46
Offenbach, Jacques, German-born French composer, 78
Ohlenhusen, Götz von, Hanoverian politician, 58
Okulicki, Leopold, Polish general, 444
Olczak, an acquaintance of the author, 280
Olszewicz, Bolesław, Polish geographer, 454
Olszewska, Mrs., a friend of Halina Szpondrowska, 357
Oppenheimer, Julius Robert, American physicist, 137
Oranowska, Greta, Polish actress, 246
Orlicz, Władysław Roman, Polish mathematician, 53, 130, 142, 143, 165, 253, 377
Ortwin, Ostap (actually Oskar Katzenellenbogen), Polish critic, 46
Osóbka-Morawski, Edward, Polish politician, 391, 420, 433, 438, 442, 444
Ossoliński, Józef Maksymilian, Polish nobleman, politician, and writer, founder of the Ossolineum, 33, 262
Ostaszewski family, owners of a villa in Iwonicz, 13
Ostern, Janina, wife of Paweł, 371
Ostern, Paweł, Polish biochemist, 178, 371
Ostrowercha, Bohdan, a Ruthenian poet, 220, 223
Ostrowska, Jadwiga, wife of Tadeusz, 291, 292
Ostrowski, Adam, Polish lawyer and diplomat, 407, 460
Ostrowski, Alexander Markowich, Ukrainian-born, German-Jewish mathematician, 137
Ostrowski, Stanisław K., Polish sculptor, 410
Ostrowski, Tadeusz, Lwów surgeon, 291
Ostrowskis, the family of Tadeusz, 289
Ottman, N., a Paris acquaintance of the author, 144
Otto, a brother of Witold, 315
Otto, Edward, Polish mathematician, brother of Franciszek, 193
Otto, father of Witold, 315, 326, 327
Otto, Franciszek, Polish mathematician, Bartel's assistant, 193, 197, 214, 221
Otto, Witold, a Lwów University clerk and owner of a house in Osiczyna, 306, 315, 316, 318, 326–328, 357, 445, 447
Ottos, the family of Witold, 313, 315, 316, 319, 322, 325, 327, 329

Ottowa, Antonina (Tosia), wife of Witold, 315, 316, 318, 319, 322, 324, 325, 327, 328, 357, 445
Ottówna, Elza, daughter of Witold, 315, 317, 328, 357, 445

Paderewski, Ignacy Jan, Polish pianist, composer, and politician, 186, 189
Paersch, Fritz, German banker, 376
Pająk, father of Zbigniew, 342, 422, 425, 442, 451, 459
Pająk, Zbigniew (Zbyszek), a pupil of the author in Berdechów, 342, 345, 352, 353, 422, 452
Pająkowa, mother of Zbigniew, 353
Pająkówna (Dziedziakowa), Hanna, a sister of Zbigniew, 353
Paliwodzianka, a friend of Elza Ottówna, 317
Pamm, Alfred, husband of Olga, 204, 208, 210
Pamm, Olga, a cousin of the author's wife, 204, 208, 361
Parandziejówna, a Lwów treasury official, 297
Parecki, Stanisław Franciszek, Polish cartoonist and poet, 261
Parnas, Dr. Józef, a Lwów lawyer, 198, 300
Parnas, Jakub, Polish-Jewish-Soviet biochemist, 140, 143, 178, 196, 274, 278, 298, 446
Parnicki, Teodor, Polish writer, 247, 332
Pasterczyk, Wacław, a schoolfriend of the author, 26
Pasternak, Leon, Polish satirist and poet, 261
Patton, George S., American general, 420
Paty du Clam, Armand, a French lieutenant-colonel, Dreyfus' main accuser, 37
Pawlikowska-Jasnorzewska, Maria, Polish poet, 186
Pawłowski, Stanisław, Polish geographer, 172, 347
Pazia, the Steinhauses' maid in Lwów, 229, 232, 267, 293
Peano, Giuseppe, Italian mathematician, 193
Pearson, Egon, English mathematician and statistician, 210
Peiper, Tadeusz, Polish writer and poet, 246, 332
Pepin, Frankish king, 101
Pepis, Józef, Polish logician, 395, 457
Perkins, an oilman, 31
Petelenz, director of a Lwów Gymnasium, 159
Peter the Great, Russian tsar, 276
Philips, Anton, a Dutch businessman, 280

Piast, a ruling Polish dynasty, 260, 408
Picard, Charles Émile, French mathematician, 79, 100, 150
Picquart, Major Georges, a supporter of Dreyfus, 37
Piekara, Arkadiusz Henryk, Polish physicist, 454
Pieniążek, a landlord in Lwów, 123
Pieńkoś, landowning family in Biała, 416
Piepes-Poratyńska, a Lwów student, 173
Pieracki, Bronisław Wilhelm, Polish army officer and politician, 169
Pilat, Stanisław, Polish chemist at the Lwów Polytechnic, 198, 291, 292
Pilatowa, Ewa, chemistry docent in Lwów, wife of Stanisław, 292
Piłsudski, Józef Klemens, Polish marshal and political leader, 109, 110, 112, 114, 168, 174, 184, 187–189, 204, 260, 412
Piotrowski, a Lwów stationer and administrator, 370, 390, 423
Piotrowskis, the family of Piotrowski, refugees in Stróże, 419
Pirquet, Clemens Peter Freiherr von, Austrian medical scientist, 169
Placidia, Aelia Galla, daughter of Theodosius I, 156
Planck, Max, German physicist, 60, 154, 171, 454
Plushch, a Ukrainian clinician, 274, 275
Poczobut, a Nazi victim, 388
Podgórski, a councillor in Nowy Sącz, 451
Poe, Edgar Allan, American writer, 73, 351
Poincaré, Henri, illustrious French mathematician, 46, 80, 114, 157
Pol, Wincenty, Polish poet, 12
Polański, Ukrainian naturalist, 295
Pólya, George, Hungarian-American mathematician, 206
Poniatowski, Juliusz, Polish politician, 200, 201
Pontrjagin, Lev Semenovich, Soviet mathematician, 156
Popov, Blagoi, 180
Poratyński, Jan, Lwów pharmacist and social activist, 309
Porter, an American professor in Göttingen, 87
Potocki, Andrzej, Polish magnate, husband of Zofia, 435
Potockis, Countess Zofia and her sister, inmates of Ravensbrück, 435
Pottier, Eugène, French socialist revolutionary and poet, 276

Index of Names

Prandtl, Ludwig, German physicist, 60
Praxiteles of Athens, 4th century Attic sculptor, 158
Praxmayer, an Austrian clerk, 5
Pringsheim, Alfred, German mathematician, 82, 83, 85
Pringsheim, father of Alfred, 85
Progulski, Andrzej, son of Stanisław, 291, 292
Progulski, Stanisław, a Lwów pediatrician, 291, 292
Proszek, Marian, Polish railroad official and German guard, 344, 346, 361, 401, 418, 423
Prus, Bolesław (born Aleksander Głowacki), Polish positivist and writer, 219
Przeborski, Antoni, Polish mathematician, 53, 195, 394
Przyboś, Julian, Polish essayist and poet, 261, 264, 266
Przybyszewski, Stanisław, Polish poet and writer, 28, 29
Pushkin, Aleksandr Sergeevich, illustrious Russian poet, 30, 186, 276
Putrament, Jerzy, Polish writer, poet, and politician, 261, 413
Puzyna, Prince Józef, Polish mathematician, 46, 96, 122, 123, 125, 129, 251
Pyrek, a teacher, 455

Raczkiewicz, Władysław, Polish politician, 391
Radbruch, Lydia, wife of Gustav Radbruch, German philosopher of law and Minister of Justice, 138
Radek, Karol, Polish-Jewish Marxist and political activist, 235
Rademacher, Hans, German-American mathematician, 137
Rado, Richard, German-British mathematician, 136
Rajchel, Mrs., a Lwów physician, 217
Rajchman, Aleksander, Polish mathematician, 100, 101, 199, 245, 253, 312, 393
Rajewski, Jan, Polish mathematician, 46
Raman, Chandrasekhara Venkata, Indian physicist, 210
Rawita-Gawroński, Franciszek, Polish historian and writer, father of Andrzej Gawroński, 191
Rawski, a Jasło *starosta*, 26
Raymond, a Vichy doctor, 152
Reccesvinthus, Visigothic king, 101

Redl, Alfred, head of counter-intelligence in the Austrian Empire, spy for Russia, 37
Reguła, a corporal in the Polish army, 214
Reinlender, an inspector in the Imperial Austrian Army, 25
Rejchan, son of Stanisław, 223, 227
Rejchan, Stanisław, Polish ceramic artist, 223
Rembrandt, Harmenszoon van Rijn, Dutch painter, 64, 74
Rencki, Roman, Polish professor of medicine, 178, 216, 244, 291
Renz, Herman, circus owner, 97
Rettinger, Mieczysław Karol, Polish literary critic and political activist, 375
Reymanowa, Maria, a Lwów nurse, 291
Reymont, Władysław Stanisław, Polish writer, 29
Řezníček, Ferdinand von, Austrian painter and caricaturist, 78
Ribbentrop, Joachim von, German foreign minister, 180, 220, 225, 233, 323, 382, 447
Richter, Friedrich A., German businessman, 32
Riefenstahl, Leni, German actress and film director, 458
Riemann, G. F. Bernhard, illustrious German mathematician, 59, 81, 100
Riesz, Frigyes, Hungarian mathematician, 160
Rittner, Tadeusz, Polish dramatist, 124
Robbins, Herbert Ellis, American mathematician and statistician, 80
Rodenbach, Georges, Belgian poet, 65
Rodziewiczówna, Maria, Polish writer, 29
Rogala, Wojciech, Polish geologist, 172, 303, 423
Rojek, a Lwów student, 213, 229
Rola-Żymierski, Michał, Polish communist military leader, 405, 448
Rolland, Romain, French novelist, 101, 118
Romaszkans, a family encountered in Tatarów, 224
Romer, Eugeniusz, Polish geographer, 92, 172
Romer, Tadeusz, Polish diplomat and politician, 457
Röntgen, Wilhelm, German physicist, 197
Roosevelt, Franklin Delano, American politician, 349, 356, 405, 421, 424
Rose, Maksymilian, Polish neuroanatomist, 187
Rosenblatt, Alfred, Polish mathematician, 392
Rosenbusch, a German chemist, 72
Rosenzweig, an engineer at the Lwów Polytechnic, 209

Rothfeld, Dr., an official of the Lwów Jewish community, 300
Rothfeld-Rostowski, Jakub, professor of neurology in Lwów, 225
Rotwand, T., a student in Munich, 84
Rouppert, General Stanisław, head of the Polish army health service, 217
Routh, Edward John, English mathematician, 80
Rouvroy, Freiherr von, an Austrian baron, 94
Rozenblum, P., Göttingen philosophy student, 53, 84
Rozwadowski, a counsellor in the Austrian empire, 123
Ruben, Monsieur, a Parisian landlord, 143, 144
Rubinowicz, Wojciech, Polish physicist, 164
Rudniański, Stefan, Polish philosopher and pedagogue, 254
Rudnicki, Adolf, Polish-Jewish writer, 261
Rudolf, husband of Stefania's sister-in-law Irena, 370
Rudzka-Cybis, Hanna, a painter, wife of Jan Cybis, 200
Rudzki, Maurycy Pius, Polish geophysicist, 80
Ruff, Adam, son of Stanisław, 291, 292
Ruff, Stanisław, Lwów surgeon, 291, 292
Ruffowa, Anna, wife of Stanisław, 291
Runge, Carl, German mathematician, 59, 81, 83, 91
Russell, Bertrand, English philosopher and logician, 135
Ruziewicz, Stanisław, Polish mathematician, 96, 129, 142, 153, 164, 165, 172, 182, 191, 220, 234, 291, 392, 437
Ryder, Göttingen medical student, 53
Rydz-Śmigły, Edward, Polish marshal, politician, painter, and poet, 188, 189, 215, 231, 284, 412, 448
Rylski, a farmer in Stróże, 395
Rylski, a major in the Polish Legions, 117, 118
Rząca, a farmer near Stróże, 347
Rząca, Adam, a son of Mr. Rząca, 367, 436
Rząca, Mrs., wife of Rząca, 364
Rząca, Stanisław, a son of Mr. Rząca, 367, 426, 427, 431, 436, 442, 455
Rząca, Tadeusz, a son of Mr. Rząca, 367
Rzącas, the family of Mr. Rząca, 431, 447, 460, 462
Rzewuska, Rena, a teacher of French, 267, 309

Sadlińska, Mrs., an acquaintance of the author in Kamień Dobosza, 222
Saint-Simon, Louis de Rouvroy de, French soldier, diplomat, and memoirist, 310
Saks, Stanisław, Polish mathematician, 96, 156, 203, 241, 248, 253, 275, 374, 375, 393
Sampero, Count, a Vatican priest, 99
Sandauer, Artur, Polish essayist and literary critic, 413
Sangnier, Marc, French politician, 115
Sanguszko, Prince, 38
Sapieha, Prince Adam Stefan, archbishop of Kraków, 188, 454
Sapieha, Prince, a Jasło *starosta*, 26
Sarnecki, a Polish legionnaire, 115
Sauckel, Fritz, German politician, 369
Sawicka, Mrs., a refugee in Milanówek, 378
Scaligeri, a noble Veronese family, 160
Schacht, Hjalmar, German banker and liberal politician, 205
Scharage, Chaim, a Polish surgeon, 214
Schauder, Juliusz Paweł, Polish-Jewish mathematician, 142, 153, 165, 216, 248, 253, 303, 393
Schenck, J., Polish mathematician, 130
Schiller, Friedrich, German poet, 30
Schochet, a Jasło acquaintance, 264
Schoenborn, Józef (Dyk), a cousin of the author, 31, 32, 39, 43, 47, 49, 66, 122, 184, 188, 258, 265, 277, 293, 302, 305, 350, 414
Schoenbornowa, Wanda, wife of Dyk, 415
Schoenbornówna (Żurkowa), Józefa (Dziutka), a daughter of Dyk, 350, 414–416
Schoenbornówna, Zofia, a daughter of Dyk, 350, 415
Schoenborns, the family of Dyk, 415
Schoenemann, Irena (Irka), a daughter of Witold, 336
Schoenemann, Kocia, a daughter of Witold, 336
Schoenemann, Witold, a friend of Stefan Dobrzański from Stróże, 336, 425, 426
Schoenemannowa, Stanisława (Stasia) (*née* Igielska), wife of Schoenemann, 336, 429, 431, 433, 435, 439, 445, 451
Schoenówna, a Lwów student, 133
Schopenhauer, Arthur, German philosopher, 45
Schorr, Moses, professor of Judaism, 191

Schreier, Józef, Gymnasium teacher in Drohobycz, 457
Schrödinger, Erwin, Austrian physicist, 281
Schulenburg, Friedrich-Werner Graf von der, German diplomat, 267
Schur, Issai, German-Jewish mathematician, 136
Schwarzschild, Karl, German physicist, 60, 80
Schwegler, Albert, German philosopher and theologian, 39
Seeliger, Hugo von, German astronomer, 83
Seifert, a German Gymnasium teacher, 74
Sępowicz, a priest in Polna, 420, 422
Serafin, a railroad porter in Stróże, 344
Sergescu, Petre, Romanian mathematician, 144
Sergiusz, Grand Prince, 18
Serret, Joseph, French mathematician, 79
Shakespeare, William, illustrious English poet and playwright, 94, 428
Shatunovsky, Samuil Osipovich, Russian-Ukrainian mathematician, 105
Shaw, Bernard, Anglo-Irish playwright, 351
Shnirelman, Lev Genrikhovich, Soviet mathematician, 282
Sholokhov, Mikhail Aleksandrovich, Soviet novelist, 282, 406
Sichulski, Kazimierz, Polish legionnaire and painter, 114
Sienkiewicz, Henryk, Polish writer, 16, 30, 91, 157, 189
Sienkiewicz, Klemens, director of the Gymnasium, 21, 33
Sienkiewicz, Teodor, Klemens' son, a lawyer, 30
Sieradzki, Włodzimierz, Lwów professor of legal medicine, 289
Sieroszewski, Wacław, Polish writer, 40
Sierpińska, Lesława (Lesia), daughter-in-law of Wacław, 273
Sierpiński, Mieczysław, son of Wacława, 264, 445
Sierpiński, Wacław, Polish mathematician, 63, 76, 80, 82, 96, 122, 123, 125, 142, 156, 165, 220, 435, 440, 445, 455
Sikorski, Władysław Eugeniusz, Polish military and political leader, 111, 356, 382, 412, 432
Silberman, Ludwik, engineer, 38
Simon, Hermann, German physicist, 60
Skarbek, Count Aleksander, Polish politician, 110
Skarżeński, a Polish mathematician, 283
Skiwski, Jan Emil, Polish writer and German collaborator, 381, 382

Składkowski, Felicjan, Polish physician, general, and politician, 207, 231
Sklepiński, a Lwów pharmacist, 141
Skowron, a Polish priest, 211
Skrzyński, Aleksander Józef, Polish politician, 409
Skulskis, acquaintances of the author, 350
Skuza, Wojciech, Polish poet, 332
Sławek, Walery Jan, Polish politician and army officer, 112
Ślebodziński, Władysław, Polish mathematician, 121
Ślęczkowski, a doctor in Berdechów, 464
Sleszyński, Jan, Polish mathematician, 105, 107
Słonimski, Antoni, Polish writer and poet, 157, 186
Słotwiński, Józef, director of the Gymnasium following Sienkiewicz, 33
Słowacki, Juliusz, illustrious Polish poet, 49, 200
Słowik, a *Volksdeutscher*, 420
Smalewska, Danuta, a friend of the author's daughter, 264
Smalewski, father of Lidka's friend Danuta, 225, 264
Smolińska, a Kosów acquaintance, wife of Smoliński, 177
Smoliński, a Kosów acquaintance, 177
Smolka, Franciszek, Polish papyrologist and librarian, 249
Smoluchowski, Marian, Polish physicist, 46, 80, 192
Sobieski, Jan III, king of the Polish-Lithuanian Commonwealth, 263
Soblik, leader of band of partisans, 395
Soddy, Frederick, English radiochemist, 196
Sokół-Sokołowski, Konstanty, Polish mathematician, 394
Sokołowski, a Göttingen student, 53
Sokólski, Stanisław (Staś), an acquaintance of Kama Speidlówna, 328, 329, 331
Sołoneńko, a tradesman, 280
Soloviev, Aleksandr D., Soviet mathematician and politician, 243, 250
Sołowij, Adam, Lwów gynaecologist and obstetrician, 289
Sołtysik, a Lwów surgeon, 220
Sommerfeld, Arnold, German physicist, 60
Sosnkowski, Kazimierz, Polish general and politician, 432
Sowizdrzał, a Jasło neighbor, 462
Speer, Albert, German architect and Nazi minister, 379

Speidlówna, Kamila (Kama), an acquaintance in Osiczyna, 327–329, 331, 357, 445
Spitz, Mme, a French inmate of Ravensbrück, 435
Spława-Neyman, Jerzy, Polish-American mathematician, 210, 392
Spohr, Louis, German composer, 58
Sroczyński, a Jasło oilman, 31
Sroka, an inhabitant of Stróże, 361
Srokowski, a Polish newspaper editor, 119
Staël-Holstein, Anne Louise Germaine de, Swiss-born French woman of letters, political activist, 78
Stakhanov, Alekseĭ Grigorievich, Soviet model worker, 256
Stalin, Joseph, 187, 231, 234, 236, 248, 249, 252, 259, 266, 275, 281, 282, 287, 356, 359, 368, 382, 388, 391, 400, 403, 405–407, 410, 414, 416, 425, 427, 428, 430, 433, 440, 442, 448, 458
Stańczyk, Jan, Polish politician, 439, 442
Stańczyk, Polish 16th century court jester, 93
Stark, Marceli, Polish mathematician, 249, 269, 310, 457, 459
Starzyński, Count Stanisław, Polish lawyer, 93
Staszek, a servant of the Fulińskis, 304
Stawiński, Julian, Polish lawyer and translator, 456
Staźniewicz, a couple from Polna, 461
Stechel, A., Polish mathematician, 395
Stefko, Kamil, Polish law professor, 184
Stein, a Jasło lawyer, 302, 303, 305
Stein, Heinrich Friedrich Karl vom und zum, Prussian statesman, 58
Steiner, Jakob, Swiss geometer, 59
Steinhaus, Bogusław, the author's father, a Jasło merchant, 3, 7, 10–14, 16, 18, 23, 26, 31, 34, 36, 38, 51, 94, 97, 109, 111, 120, 178, 179, 181–183, 202, 245, 401, 455, 462
Steinhaus, Ignacy, a brother of the author's father, 35, 36, 93, 94, 96, 108, 110, 113, 119, 120
Steinhaus, Józef, the author's grandfather, a grocer, 5, 14
Steinhaus, Władysław (Władek), Ignacy's son, a Polish legionnaire, 113, 114, 117, 119
Steinhauses, the family of, 38
Steinhausowa, Ewelina (née Lipschitz), the author's mother, 9, 10, 13, 14, 19, 27, 51, 64, 100, 118, 120, 178, 179,
181–183, 223, 225, 229, 299, 316, 319, 340, 343, 347, 417, 461, 465
Steinhausowa, Stefania (Stefa) (née Szmoszówna), the author's wife, 118, 122, 123, 125, 133, 136–138, 143, 153, 176, 177, 203, 220, 222–224, 228, 241, 267, 277, 293, 297–300, 309, 311, 316, 318, 343, 350, 353, 355, 364, 385, 396, 398, 411, 425, 429, 440, 449, 452, 455, 457, 459, 461
Steinhausowa, the author's great aunt, sister of Józef, 15, 23, 24, 182
Steinhausówna (Chwistkowa), Olga, the author's youngest sister, 100, 113, 299, 300, 303, 307, 309, 416, 425, 465
Steinhausówna (Kottowa), Lidia (Lidka), the author's daughter, 123, 133, 136, 146, 171, 176, 197, 198, 218–221, 223, 224, 228, 245, 264, 266, 267, 293, 309, 316, 319, 326, 340, 343, 347, 350, 351, 355, 363, 366, 367, 369, 370, 373, 376–378, 385, 389, 398, 407, 415, 421–423, 425, 435, 442, 445, 447, 449, 451, 452, 455
Steinhausówna (Müllerowa), Felicja, the author's eldest sister, 34, 179, 229, 309, 319, 340, 417
Steinhausówna (Zglińska), Irena, the author's middle sister, 33, 100, 113, 307, 319, 340, 363, 378, 425, 458, 465
Stempowski, Stanisław, Polish-Ukrainian writer and politician, 223
Stendhal (pseudonym of Marie-Henri Beyle), French novelist, 29, 102, 374
Stern, Abraham, Polish-Jewish scholar and inventor, 157
Stern, Anatol, Polish writer, poet, and critic, 246
Sternbach, Ludwik, Polish mathematician, 163, 310, 394, 457
Sternbachs, the family of Ludwik, 310, 445
Sternberg, Josef von, German film director, 458
Stettinius, Edward Reilly, American politician, 424
Stiller, Miss, a Polish schoolteacher, 353, 355
Stodola, Aurel Boleslav, Slovak technologist, 162
Stojałowski, Stanisław, a priest, 13
Stojowski, business partner of the author's father, 97
Stożek, Emanuel, a son of Włodzimierz, 291

Stożek, Eustachy, a son of Włodzimierz, 289
Stożek, Włodzimierz, Polish mathematician, 53, 121, 142, 153, 164, 289, 392, 437
Strasburger, Henryk Leon, Polish politician, 457
Strauch, von, a Göttingen student, 83
Strindberg, August, Swedish playwright, 29
Studencki, a Polish physician and legionnaire, 115
Studyński, Kiryło, a Ukrainian professor in Lwów, 252
Suchard, Phillipe, a Swiss chocolatier, 116
Suchowiak, Lech, a student, 84
Suchowiak, Wacław L., Polish professor of mechanical engineering, 216
Sudermann, Hermann, German dramatist and novelist, 29
Sutherland, an English aristocrat, 119
Suttner, Bertha von, Austrian pacifist, 37
Suvorov, Aleksandr Vasilyevich, Russian generalissimo, 276
Świętosławski, Wojciech, Polish physicist and politician, 194, 196
Świeżawskis, a family from near Przemyśl, 363
Sylvester I, pope, 141
Syzdek, a priest in Polna, 398
Szadurski, a Göttingen student, 53
Szafraniec, Marian, an assistant of Laska, 342, 343, 351, 449, 464
Szczepański, Stanisław, a Jasło artist, 200
Szczerbski, a Polish-German interpreter, 383
Szczęśniewicz, a teacher from Polna, husband of Mrs. Zarańska, 391, 398, 437, 438
Szczypcina, wife of Szczypta, 360, 401
Szczypta, Polish postal worker and German informant, 344, 346, 352, 360, 361
Szegő, Gábor, Hungarian-American mathematician, 152
Szemplińska, Elżbieta, Polish writer and poet, 261, 279
Szenwald, Lucjan, Polish poet and communist activist, 261, 377
Szmosz, Adolf (Dolek), a brother-in-law of the author, 307, 329, 336, 340, 342, 345, 351, 352, 357, 364, 367, 389, 399, 407, 408, 418, 421, 438, 440–442, 447, 451, 455, 464
Szmosz, the author's father-in-law, 176, 347
Szmoszowa, Krystyna (Krysia), wife of Dolek, 336

Szmoszowa, Paulina, the author's mother-in-law, 329, 335, 336, 361, 364, 369, 398, 399, 408, 455, 464
Szmoszówna (Dittersdorfowa), Helena (Hela), the author's sister-in-law, 133, 161
Szpilrajn-Marczewski, Edward, Polish-Jewish mathematician, 253
Szpondrowska, Halina, a fellow refugee in Osiczyna, 315, 316, 318–320, 322, 324, 325, 357, 445
Szpondrowski, Andrzej (Andrzejek), a son of Halina, 315, 318, 320, 325, 357
Szpondrowski, Stanisław (Staś), a son of Halina, 315, 318, 319, 325, 357
Szujski, Józef, Polish writer, historian, 38
Szumański, Wacław, a lawyer encountered in Krynica, 194, 376
Szydłowski, a Lwów neighbor, 292
Szymanowicz, a Kraków doctor encountered at Morszyn, 216
Szymkiewicz, Dezydery, Polish botanist, 317

Tacitus, Publius Cornelius, Roman senator and historian, 374
Tanev, Vassili, 180
Tapkowski, Dr. Tadeusz, a Lwów lawyer, 291
Tarasek, a pupil of the author in Berdechów, 411
Tarnawski, Apolinary, doctor at a Kosów sanitorium, 176
Tarnawski, Władysław, Polish philologist, 277
Tarnowski, Count Stanisław, Polish critic, historian, 35
Tarski, Alfred, Polish-American logician and mathematician, 147, 165, 215, 392
Teichmann, Ludwik Karol, Polish anatomist, 58
Tenner, Roman, lawyer, 48
Tetmajer, Kazimierz Przerwa, Polish poet, 28, 29
Thälmann, Ernst, German communist, 270
Theodoric the Great, king of the Ostrogoths, 156
Thilo, Heinz, German physician in Auschwitz concentration camp, 359
Thoma, Ludwig, German author and publisher, 78
Thomas, Fräulein, Göttingen landlady, 52, 57, 63, 71, 77
Thugutt, Mieczysław, Polish politician, 442
Tibbets, Paul, US Air Force pilot, 459
Tieck, Johann Ludwig, German poet, 58

Timoshenko, Semyon Konstantinovich, Soviet military leader, 250
Tirpitz, Alfred von, German admiral, 139
Tiso, Jozef, Slovak leader, 323
Tito, Josip Broz, Yugoslav revolutionary and politician, 430
Todt, Fritz, German engineer and senior Nazi, 379
Toeplitz, Otto, German-Jewish mathematician, 80, 138, 139, 152
Tokarski, Julian, Polish geologist, 172
Tokarzowa (*née* Cieluchówna), a daughter of Cieluch Senior, 335, 453
Tola, Mrs., a Lwów neighbor, 299
Tolstoĭ, Akekseĭ Nikolaevich, Russian and Soviet writer, 282
Tolstoy, Leo, illustrious Russian writer, 30, 107
Tomanek, a Lwów neighbor, 294
Tomaszewska, Mrs. (*née* Połaniecka), an acquaintance of the author, 220
Tomczycka (*née* Petold), a Lwów acquaintance, 264
Tomczycki (assumed name Jastrzębiec), husband of Mrs. Tomczycka, 265
Torgler, Ernst, 180
Tournay, Cécile de, a Hungarian lady, 99
Toussaint L'Ouverture, François-Dominique, Haitian revolutionary leader, 331
Towiański, Andrzej, Polish mystic, 28
Traugutt, Romuald, Polish military leader, 450
Trofimov, a Soviet political commissar, 272
Trotskiĭ, Leon Davidovich, Russian-Jewish Bolshevik revolutionary, 235
Truman, Harry S., American politician, 424, 427, 440, 450, 459
Trzeciak, Stanisław, Polish priest, 191
Trzecieski, an oilman, 31
Turno, Mr. and Mrs. Kinga, acquaintances of the author, 200
Tuwim, Julian, Polish poet, 186, 205
Twain, Mark, American writer, 24
Twardowski, Kazimierz, Polish philosopher, 44, 46, 135

Uchatius, Franz von, Austrian general, 115
Ukrainka, Lesya (born Larysa Kvitka-Kosach), Ukrainian poet and dramatist, 276
Ulam, Stanisław Marcin, Polish-American mathematician, 143, 164, 165, 245, 392
Usiejewiczowa, Helena, a Polish journalist, 266

Vallée-Poussin, Charles-Jean de la, Belgian mathematician, 144, 153
Vegt, Arnold van der, Dutch circus owner, 97
Verlaine, Paul, French symbolist poet, 71
Verne, Jules, French science fiction writer, 38
Verworrn, Max, Göttingen professor of physiology, 57
Vetulani, Kazimierz, Lwów physicist, 291
Victor Emmanuel III, king of Italy, 155
Vincenz, Stanisław, Polish novelist and essayist, 146, 177, 223, 234
Vlasov, Andreĭ Andreevich, Soviet general who defected to the Germans, 382, 430
Voigt, Waldemar, German physicist, 60
Volta, Alessandro, Italian physicist, 155
Voltaire (François-Marie Arouet), illustrious French writer and philosopher, 102
Volterra, Vito, Italian mathematician and physicist, 154
Voroshilov, Kliment Yefremovich, Soviet military and political leader, 250
Vulpius, Christian, German writer, 124

Wachlowski, Zenon, Polish lawyer, husband of Irena Łomnicka, 222, 246
Wacławski, a Lwów student, 173
Waldstätten, a general in the Imperial Austrian Army, 24, 26
Wałęsa, Lech, Polish trade-union organizer and political activist, 446
Walfisz, Arnold, Polish mathematician, 393, 395
Walpole, Sir Hugh Seymour, English writer, 351
Walter, a police commissioner, 313
Wanda, a sister of Maria Dobrzańska, 351
Wańkowicz, Melchior, Polish writer, 185
Warchałowski, T., lawyer, 39, 455
Ward, an English mathematician, 203
Ward, Morgan, American mathematician, 149
Ward, Olive, the English Ward's wife, 203
Waring, Edward, English mathematician, 81
Wartenberg, Mścisław, Polish philosopher, 45
Wasilewska, Wanda, Polish-Soviet novelist and communist activist, 247, 261, 266, 278, 403
Wasowski, Józef, Polish socialist politician, 402
Wasserberg, Dr. Ignacy, Polish physician and philosopher, an official at the League of Nations in Geneva, 207

Wasylewski, Stanisław, Polish writer and critic, 247, 261
Wat, Aleksander, Polish writer and futurist poet, 246, 332
Ważyk, Adam, Polish writer and poet, 261
Weber, Wilhelm Eduard, German physicist, 58
Węgrzyńska, a Jasło acquaintance, 266
Węgrzyński, Jan, a schoolfriend of the author, 40
Węgrzyński, Władysław, teacher at the Gymnasium, 21
Weierstrass, Karl, German mathematician, 46
Weigel, Józef, son of Kasper, 291
Weigel, Kasper, professor at the Lwów Polytechnic, 291
Weigl, Rudolf Stefan, Polish biologist, 300, 328
Weinlösówna, D., Polish mathematician, 395
Weiss, Ignacy, a schoolfriend of the author, 265, 302, 306
Weisses, the family of Ignacy, 302, 309
Weissowa (née Stein), wife of Ignacy, 302
Welles, Orson, American film director and actor, 157
Wellhausen, Göttingen biblical scholar, 57
Wells, Herbert George, English writer, 157
Wertenstein, Piotr, a cousin of Jan Kott, 265
Weyberg, Zygmunt, Polish mineralogist, 96
Weyl, Hermann, German mathematician and physicist, 153
Whitehead, Alfred North, English philosopher, 135
Whyburn, Gordon Thomas, American mathematician, 149
Wiatr, a Polish folklorist, 361, 418
Wiatr, a Wrocław student, 453
Wiatr, Narcyz, Polish lawyer and populist political activist, brother of the folklorist, 423, 424
Widajewicz, Julia, Benedykt Fuliński's mother, 319
Wiechert, Emil, German geophysicist, 60, 61, 80, 81, 85
Wied, Wilhelm von, a German prince, 114
Wieniawa-Długoszowski, Bolesław, Polish general, politician, physician, and poet, 188
Wiktorowa, business partner of the author's father, 97
Wilde, Irina, Ukrainian writer, 251, 272
Wilde, Oscar, Anglo-Irish writer, 29
Wilder, a fellow hiker, 148
Wilhelm II, last German Kaiser, 66, 78, 139, 252

Wilk, a reaper, 352
Wilkosz, Witold, Polish mathematician, 121, 392
Winawer, Bruno, Polish writer of social comedies and science fiction, 261, 377
Winkler, Johannes, a Swiss historian, 42
Wiśniewski, Count, religious teacher, 15
Wiśniewski, teacher at the Gymnasium, poet, 21
Witalis, name of some of the author's pupils in Berdechów, 411, 416, 449
Witkiewicz, Roman, professor at the Lwów Polytechnic, 291
Witkiewicz, Stanisław Ignacy, Polish painter, playwright, and poet, 309
Witkowickis, Lwów neighbors of the Steinhauses, 298
Witkowski, Bernie, Polish-American band leader, 187
Witkowskis, Lwów acquaintances of the author, 296, 298, 312
Witmans, acquaintances of the author, 425
Witos, Wincenty, Polish politician, 335, 439, 442, 449
Wittlin, Józef, Polish novelist and poet, 135, 186
Wöhler, Friedrich, German chemist, 58
Wojdysławski, H. Menachem, Warsaw topologist, 233, 253
Wojtaczek, a victim of Emil Druśko, 383
Wolfskehl, Paul, German industrialist, 81
Wolisch, a Lvovian, 291
Wołyński, a Viennese acquaintance, 113
Womela, Stanisław, teacher at the Gymnasium, poet, 21
Wong, Anna May, Chinese-American actress, 326
Woyrsch, Remus von, Prussian lieutenant-general, 111
Woźnica, Mrs., wife of Stanisław, 398
Woźnica, Stanisław, a neighbor of the Cieluchs, 336, 337, 352, 366, 388, 390, 419, 421
Woźnicas, the family of Stanisław, 416
Wundheiler, Aleksander, Polish-American mathematician, 394
Wundt, Wilhelm Maximilian, German physiologist, psychologist, and philosopher, 199
Wybicki, Józef, Polish jurist, poet, and political activist, 264
Wycech, Czesław, Polish politician and historian, 442

Wyspiański, Stanisław, Polish poet, playwright, and painter, 28, 87
Wyszomirska, a Kraków acquaintance, 378

Ząbek, a friend of the Igielskis from Jasło, 337, 353, 391, 455
Zabierowski, a representative of the wartime Polish underground teaching organization, 355, 401, 437, 460
Zaborski, a Polish merchant and bureaucrat, 370
Zagórska, Maria, wife of Jerzy, 465
Zagórski, Jerzy, Polish writer and poet, 374
Zagórskis, the family of Jerzy, 465
Żakowski, Józef, Polish politician, 439
Zakrzewski, son of the Zakrzewskis, 417
Zakrzewskis, a family in Rozprza, 114
Zalcwasser, Zygmunt, Polish mathematician, 395
Załuski, Count Ireneusz, 13, 187
Załuskis, the family of Count Załuski, dominant in Iwonicz, 13, 402
Zamorski, a Krynica hotelier, 182
Zamoyski, Count Władysław, owner of the lake *Morskie Oko*, 41
Zarańska, Mrs., mother of three of the author's pupils in Berdechów, 391, 398, 450
Zaremba, Jan, a teacher and owner of a property in Stróże Niżne, 355, 386, 390, 401, 442
Zaremba, Karol, a brother of Jan, 400, 401, 410, 414, 449, 455
Zaremba, Stanisław Krystyn, Polish mountaineer and mathematician, son of Stanisław Zaremba Sr., 245, 393
Zaremba, Stanisław, Polish mathematician, 92, 97, 105, 107, 149, 196, 392
Zarembas, the family of Jan, 367, 385, 390, 419, 426
Zarzycki (Zaricki), Miron, Ukrainian mathematician, 253, 458
Zawadzki, an acquaintance of the author, 129
Zawirski, Zygmunt, Polish philosopher and logician, 135
Zbierzchowski, Henryk (Nemo), Polish poet and playwright, 219
Zdziebko, a Jasło tenant, 461
Zeeman, Pieter, Dutch physicist, 85
Zegadłowicz, Emil, Polish writer, 28, 43
Żelazny, Józef, a Jasło physician, 455, 462
Zeltner, Hermann, German philsopher, 150

Zeno, Saint, a 4th century Veronese bishop, 160
Zermelo, Ernst, German mathematician, 61, 149
Żeromski, Stefan, Polish writer, 29
Zglińska (Mańczakowa), Barbara (Basia), daughter of the author's sister Irena, 340, 363
Zgliński, Leonard, Polish lawyer, a brother-in-law of the author, husband of Irena, 179, 363, 376, 465
Zgorzelska, Aleksandra, 409
Zhukov, Georgiĭ Konstantinovich, Soviet marshal and politician, 425, 427
Zięborak, Kazimierz, a Polish chemist and legionnaire, 115
Zieliński, Tadeusz Stefan, Polish philologist and historian, 202
Zielony (actually Grünwald), Leon, an uncle of the author's wife, 118, 229, 265, 267, 310, 316
Ziemilska, Olga (*née* Askenazy), wife of Dr. Ziemilski, 180, 309, 350, 371
Ziemilski, family doctor, 180, 214, 274, 303, 350, 371
Ziemilski, Jerzy, son of Dr. Ziemilski, 350, 371
Ziemilskis, the family of Dr. Ziemilski, 293
Ziereis, Franz, German *SS-Sturmbannführer*, 451
Zierhoffer, August, Polish geographer and geologist, 234
Zinoviev, Grigoriĭ Yevseevich, Russian-Jewish Bolshevik revolutionary, 235
Zippers, a Lwów family, 370
Żmigryder-Konopka, Zdzisław, Polish history professor and politician, 348
Zola, Émile, French novelist, 29, 37
Zoll, Antoni, a Jasło *starosta* and singer, 26, 145, 187
Zoll, Fryderyk, Jr., Polish lawyer, 98
Żongołłowicz, Bronisław, a Polish priest, 184, 185
Żórawski, Kazimierz, Polish mathematician, 149
Żukotyńska, Małgorzata Wanda, a sister-in-law of Dolek Szmosz, 309, 313, 423, 435, 436, 438, 446
Żuławski, Jerzy, teacher at the Gymnasium, writer, 21, 43, 49
Żuławski, Zygmunt, Polish politician, 439, 442
Zwisłocki, Tadeusz, Polish chemist and legionnaire, 115
Żychiewicz, Emil, fellow gymnasist, later bookseller, 29

Index of Names

Zygmund, Antoni, Polish-American mathematician, 96, 192, 203, 245, 393
Zygmunt August (Sigismund II Augustus, King of Poland), 56

Żyliński, Eustachy, Polish mathematical logician, 53, 129, 142, 153, 253, 458
Żywiecka, Mrs., a neighbor in Osiczyna, 317, 322, 329